Organic Chemicals from Biomass

BIOTECHNOLOGY

BIOTECHNOLOGY SERIES

1. R. Saliwanchik	*Legal Protection for Microbiological and Genetic Engineering Inventions*
2. L. Vining (editor)	*Biochemistry and Genetic Regulation of Commercially Important Antibiotics*
3. K. Herrmann and R. Somerville (editors)	*Amino Acids: Biosynthesis and Genetic Regulation*
4. D. Wise (editor)	*Organic Chemicals from Biomass*

Other volumes in preparation

A. Demain (editor)	*Biology of Industrial Microorganisms*
A. Laskin (editor)	*Applications of Isolated Enzymes and Immobilized Cells to Biotechnology*
S. Dagley (editor)	*Biodegradation*

Organic Chemicals from Biomass

Edited by

Donald L. Wise
Dynatech R/D Company
Cambridge, Massachusetts

1983
THE BENJAMIN/CUMMINGS PUBLISHING COMPANY, INC.
Advanced Book Program
Menlo Park, California
LONDON · AMSTERDAM · DON MILLS, ONTARIO · SYDNEY · TOKYO

Library of Congress Cataloging in Publication Data

Main entry under title:

Organic chemicals from biomass.

(Biotechnology series ; 4)
Bibliography: p.
Includes index.
1. Biomass energy. 2. Organic compounds.
3. Genetic engineering. I. Wise, D. L.
(Donald Lee), 1937- . II. Series:
Biotechnology ; 4.
TP360.072 1983 661′.8 83-2682
ISBN 0-8053-9040-5

MANUFACTURED IN THE UNITED STATES OF AMERICA

CONTRIBUTORS

Fredric R. Bloom
Bethesda Research Laboratories, Inc.
Gaithersburg, Maryland

Ronald P. Cannell
Stone & Webster Engineering Corporation
Boston, Massachusetts

Ellen L. Dally
Genetic Research Corporation
Columbia, Maryland

Stanton R. de Riel
Dynatech R/D Company
Cambridge, Massachusetts

Douglas E. Eveleigh
Cook College
New Jersey Agricultural Experimental Station
Rutgers University
New Brunswick, New Jersey

Reinaldo F. Gomez
Genentech, Inc.
S. San Francisco, California

Bruce K. Hamilton
Genex Corporation
Science and Technology Center
Gaithersburg, Maryland

Arthur E. Humphrey
Department of Chemical Engineering
Lehigh University
Bethlehem, Pennsylvania

Peter Kretschmer
Celanese Corporation
Summit, New Jersey

Peter F. Levy
Dynatech R/D Company
Cambridge, Massachusetts

David R. Light
Genentech, Inc.
S. San Francisco, California

Lars G. Ljungdahl
Center for Biological Resource Recovery
Department of Biochemistry
University of Georgia
Athens, Georgia

Anthony Moreira
Process Engineering
International Flavors & Fragrances
Union Beach, New Jersey

Janice A. Phillips
Department of Chemical Engineering
Lehigh University
Bethlehem, Pennsylvania

J. E. Sanderson
Dynatech R/D Company
Cambridge, Massachusetts

Anthony J. Sinskey
Department of Nutrition and Food Sciences
Massachusetts Institute of Technology
Cambridge, Massachusetts

Bradley R. Snedecor
Genentech, Inc.
S. San Francisco, California

H. W. Stokes
Department of Microbiology
University of Connecticut
Storrs, Connecticut

James R. Swartz
Genentech, Inc.
S. San Francisco, California

Godfred E. Tong
Stone & Webster Engineering Corporation
Boston, Massachusetts

J. Gregory Zeikus
Department of Bacteriology
University of Wisconsin—Madison
Madison, Wisconsin

CONTENTS

FOREWORD

The historical background, the present status of development, the basic fundamentals, and the potential of genetic engineering for organic chemical production from biomass are presented. The intention in the planning of this text was that contributed chapters depart from conventional methane and alcohol fermentations and deal with production of higher-valued organic chemicals. The text consists of discrete chapters by experts in a selected area of modern biotechnology, which provide the reader with insight into the rapidly emerging new field of industrial production of organic chemicals from biomass.

In essentially all chapters the historical perspective and present status of the specific biotechnology topic is reviewed. It will be interesting to the reader to bear in mind when reading chapters in this text that, prior to the development of the petrochemical industry, an array of organic chemicals were made via fermentation routes. Moreover, selected fermentable feedstocks, including industrial residues, were often used for bioconversion to organic chemicals. Some of these fermentable organic feedstocks are again being regarded in the light of modern biotechnology and are broadly termed "biomass." Thus biomass is used in the broadest sense in this text and is essentially considered to be any organic material fermentable to a more highly valued product.

The text is also intended to provide an excellent review of the fundamentals of organic chemical production from biomass by presenting the basic principles of thermodynamics involved and the biochemical pathways understood as basic to product formation. These fundamentals are integral to every chapter.

The objective in selecting the particular authors for the chapters in this text was to provide the reader with examples of specific organic chemical production, as well as the broad perspective on other organic chemicals. Thus contributed chapters from workers in the field provide a ready insight to those interested in moving into this field or in capitalizing on developmental results. Included are chapters discussing fermentations based on genetic engineering, chemical production by suppressing standard methane fermentation, and a number of other novel endeavors. Current progress and potential on more classical fermentations such as acetone–butanol, ethyl alcohol, and acetic acid are also included. In all cases these contributed chapters are intended to be tutorial in nature, with the objective of providing overall direction, rather than being alternatives to publication in standard archival journals.

This text is intended for the industrial executive, process engineer, or scientist questioning the direction and opportunities for organic chemical production from biomass.

CHAPTER

1

Organic Chemicals from Biomass: An Overview

Anthony J. Sinskey

OVERVIEW OF THE CHEMICAL INDUSTRY: INTRODUCTION

The chemical industry is one of the largest industries in the United States; its 1978 sales were over 3 billion dollars, and it has 1 million employees, representing nearly 5% of all manufacturing jobs in the United States (*Chemical & Engineering News,* 1979). Today's chemical industry traces its roots back to the German coal tar industry of the late 19th century. The gradual rationalization of chemical science from the 1890's through World War II, coupled with improvements in chemical technology and stimulated by the availability of cheap petroleum, caused the industry to expand dramatically from the 1940's through the 1960's. Some observers have predicted that the chemical industry is now reaching maturity, will cease rapid development, and will finally decline like such industries as steel. Others believe that new technical developments will stimulate demand for its final products and result in continued expansion in the decades ahead. Applied genetics may lead to such new technical developments.

The chemical industry transforms low-cost mineral raw materials into end-use products of greater value. The most important raw materials are petroleum, coal, minerals (phosphate, carbonate), and air (oxygen, nitrogen). Roughly two thirds of the chemical industry is devoted to producing inorganic chemicals such as minerals (lime, phosphate rock, potash, salt, sodium carbonate, and sulfur) and aluminum sulfate, ammonia, ammonium nitrate, ammonium sulfate, carbon dioxide, chlorine gas, bromine, iodine, hydrochloric acid, hydrogen, nitric acid, nitrogen gas, oxygen, phosphoric acid, phosphorus, sodium hydroxide, sodium phosphate, sodium silicate, sodium sulfate, surfuric acid, titanium dioxide, and urea.

The organic chemical industry transforms inexpensive and usually highly reduced raw materials—for example, petroleum, natural gas, and coal—into more valuable products that are in a higher oxidation state. The transformation is done in a series of physically separate steps, called unit operations, each of which normally increases the value of the starting material. Profit can be accrued at each step as the value of the product rises.

The major types of chemical operations performed in the chemical industry, are as follows.

- *Catalytic reforming:* A chemical can be processed at high temperature and pressure in the presence of a metal catalyst to make and break carbon bonds. This is the fundamental way by which a carbon skeleton may be rearranged. After rearrangement or reforming, the products are separated, usually by an equilibrium separation process such as extraction or distillation. A variety of products are produced in high-temperature reforming reactions with a nonspecific catalyst. The separations are frequently energy-intensive and may require elaborate and capital-intensive methods; as a consequence, these reactions are usually done by the oil refiners.
- *Reductions using hydrogen:* Because of the expense of hydrogen and of the ready availability of fully reduced raw materials, hydrogen is only used for specialized reactions. Over one third of the hydrogen produced in the United States is used for the reduction of nitrogen gas to ammonia. This is economical because nitrogen is not available in the reduced form from mineral sources, and ammonia is needed as a fertilizer and as a reactant for chemical synthesis.
- *Oxidations:* Hydrocarbons are oxidized with oxygen in air or oxygen purified in gaseous or liquid form. Sometimes a halogen, most notably chlorine, is used. In general, oxidation makes the products more polar and more water-soluble because carbon–oxygen bonds are more polar and thus more hydrophilic than carbon–hydrogen bonds. In addition, oxidation usually results in an increase in molecular weight.
- *Water elimination condensations:* Alcohols can react with organic acids to generate esters, and amines can react with organic acids to make amides. These reactions are particularly important building blocks for polymerization reactions.

- *Separations:* After any of the above reactions, the desired products must be separated from the reaction mixture containing by-products. Reaction conditions may be altered to optimize the product mix and generate more of a desired compound, or new technology may be developed to improve the product mix. In some cases, this is difficult or impossible, and new uses for by-products must be found. Either difficulty and expense in separation, or formation of large amounts of unusable by-products, can make economically attractive chemical reactions into uneconomical processes. Major changes in reaction processes are likely to cause major changes in separation process needs. Therefore, reaction and separation must be considered simultaneously.

The chemical industry is both a capital-intensive and a raw material-intensive industry. In addition, chemical processing uses large amounts of energy. Usually, the largest cost in chemicals production is the raw materials, for example, typically 50–80% of the manufacturing cost of commodity chemicals. The value-added ratio, which is the difference between the selling price and the materials cost divided by the materials cost, ranges from 0 to 1 for typical bulk chemical transformations. As a consequence of the raw material cost-intensiveness, a 20% reduction in production cost achieved by improved raw material conversion represents a major price breakthrough in the chemical industry. This is not true, for instance, in the electronics industry, where the cost of the silicon in the chips is a trivial portion of the manufacturing cost of a microprocessor.

Four main factors press for the continuing development of improved processes in the chemical industry; they are (1) the increasing raw material cost for raw material cost-intensive processes, (2) the capital-intensive nature of the business, (3) the additive value of serial chemical transformations, and (4) reducing pollution by developing more efficient conversion processes. By integrating more chemical reactions into a plant, a chemical company can reduce the cost of production and consequently achieve a higher return on investment.

The chemical industry is highly competitive, and new processes are introduced steadily. New processes develop in response to changing technology, changing raw material costs, and changing market demands. For instance, the demand for organic chemicals in the automobile industry has increased in response to the need to reduce automobile weight by replacing metals with plastics. The cost of petroleum has increased dramatically in recent years, and new technologies are constantly being developed to use this resource more efficiently and to utilize alternative resources such as coal and biomass. This chapter is concerned with the impact of applied genetics on the chemical industry. To assess this impact, it is useful to examine the chemical industry closely and explore how biotechnology may play a role in the industry.

OVERVIEW OF THE CHEMICAL INDUSTRY: BIOCHEMISTRY AND ENZYMES

The oxidation of organic chemicals yields energy, which, in an industrial process, is given off as heat. In living cells, energy from oxidation can be conserved by coupling oxidative reactions with other reactions such as synthesis of phosphate ester in the form of ATP. The ATP can then serve as an energy intermediary for use in the synthesis of new compounds, the transport of materials across membranes, cell motion, and other biological functions.

Green plants capture sunlight and use the energy to reduce carbon dioxide in the atmosphere to sugars. The plants then use some of those sugars to provide for their own energy needs, but the rest of the sugars are accumulated in what is now commonly called biomass. Therefore biomass includes starches, cellulose, lignins, and other materials. It is this biomass that is the foundation of all renewable resources produced by carbon dioxide fixation.

Enzymes promote certain chemical reactions and are catalysts in the same way that many of the metals used in industrial chemistry are catalysts. Enzymes are proteins and often are much less resistant to chemical and physical attack than metal catalysts. But more important, they are functionally different: (1) they can be extremely efficient catalysts, and (2) they are extremely specific, often catalyzing only a single transformation. It is this efficiency and specificity that make enzymes attractive as catalysts for industrial processes.

Although enzymes are catalysts, cells are sometimes called "living catalysts." This means that the cells can take a compound, oxidize it, and usefully recover some of the energy released. Cellular metabolism is restrained by the same thermodynamic rules as in chemical catalysis. For an excellent discussion see Chapter 9. This chapter considers how cells can be used as catalysts in chemical transformations and how the enzymes themselves are used in chemical transformations. The former is sometimes termed fermentation; the latter is sometimes known as biotransformation.

FERMENTATIONS

Many chemical substances are part of the common metabolic pathways shared by almost all life, from bacterial to humans. These biochemicals are called primary metabolites; they include many sugars, fatty acids, amino acids, and nucleic acids. Generally, primary metabolites are the substances that are least expensive to make by fermentation. While they are produced and consumed within the organism constantly, the amount is tightly controlled such that they are not produced in excess of the amount required for growth. Strains of microorganisms that overproduce primary metabolites are made by mutation to deregulate the metabolic controls. Process improvement is made by mutation to improve the rate and amount of product and the conversion yield of raw materials.

Other products of cells are known as secondary metabolites. Cells use enzymes that are not part of primary metabolism to produce substances for special purposes. Examples include the alkaloids and antibiotics. The carbon used to make the secondary metabolites can always be traced back to primary metabolites. However, unlike primary metabolites, secondary metabolites are not essential for growth. Secondary metabolites are also products of fermentation; but the conversion yields are generally lower, and genetic manipulation is more difficult.

Potential products of biotechnology may be classed in three groups.

- *Primary metabolites:* These are already made in substantial quantities by the cell, and isolation of strains to overproduce them is possible.
- *Secondary metabolites:* These substances, while produced by cells, are produced only under very tight genetic control. Substantial improvement in yields may be developed through applied genetics techniques; however, the improvements are often considerably more difficult to achieve than in the case of primary metabolites.
- *Nonmetabolites:* Many compounds, not normally present in cells, can in principle be made by enzyme-catalyzed transformations of primary and secondary metabolites; these compounds are potential candidates for production by fermentation. Not only do the problems associated with improving yields in both primary and secondary metabolites have to be addressed, but finding the necessary enzymes—often, in some organisms, isolating the DNA that codes for them, cloning it into the vector, and achieving expression—must be accomplished. Since this approach holds promise, many investigators are exploring significant commercial scale processes.

In summary, any compound found in intermediary metabolism or as a secondary metabolite or producible by enzymatic transformation of a metabolite is a potential candidate for production by fermentation.

Bioconversion technology may have impact on certain aspects of the chemical industry other than fermentation. Enzymes that can be isolated and used as catalysts to perform unique conversions will become more important. This is already being done on an economically significant scale for certain hydrolases, such as the amylases that are used to cleave corn starch into fermentable sugars. A more exotic use of enzymes is the oxidation of steroids for the synthesis or modification of certain hormones. The difficulties surrounding bioconversions are related less to producing the required enzymes than to technical problems associated with the use and stabilization of the enzymes through techniques such as immobilization.

Enzyme-catalyzed reactions are of four types (Walsh, 1979):

1. group transfers;
2. oxidations and reductions;
3. eliminations, isomerations, and rearrangements; and
4. those that make and break carbon-carbon bonds.

An examination of each type of reaction provides insight into the potential for application of enzyme catalysis to the chemical industry.

ENZYME-CATALYZED GROUP TRANSFERS

The common feature of these reactions is that some electrophilic group of a substrate molecule is transferred to an acceptor that is plentiful in the cellular environment. The most common transfer is to water, which is a ubiquitous nucleophile. These enzymes include esterases, nucleases and peptidases. Hydrolytic cleavage enzymes are the most commercially important enzymes, and they include both peptidases and amylases.

A second class involves transfer to ammonia rather than to water. Ammonia is too reactive to be allowed to remain free in the living cells; normally, ammonia is donated by the amide group of glutamine.

A third example is the transfer of phosphate groups. This is the main reaction by which chemical energy available through oxidation reactions is transferred from one molecule to another, as was discussed earlier.

ENZYMATIC OXIDATIONS AND REDUCTIONS

When one molecule is oxidized by an enzyme catalyst, another molecule must be reduced; the electrons removed from the molecule undergoing oxidation must be transferred to some acceptor in biological systems, undergoing a two-electron reduction to H_2O_2 and also a four-electron reduction to H_2O. Enzyme-catalyzed reductions show an obligate requirement either for coenzymes to act as electron acceptors or for a transition metal to act as a conduit for passage of electrons from the substrate to some acceptor molecule. The most important of these coenzymes or cofactors are nicotinamide, flavin, and pterin compounds.

Enzymes that incorporate molecular oxygen directly into organic molecules are ubiquitous and of high metabolic significance. Since these oxidation reactions are commercially important, such enzymes may become important to the chemical industry. The basics of enzymatic epoxidation reactions have been clearly summarized from the standpoint of enzyme technology. The ability of these enzymes to incorporate molecular oxygen into organic substrates efficiently and selectively is very enticing, since these reactions are often poorly accomplished when conventional chemistry is used. The use of such enzymes to carry out biotransformations is not without significant technical and economic problems. Low specific activity, cofactor recycling, and enzyme stability are significant technical problems limiting both the economic and technical applications of such enzymes to the chemical industry. Only when the enzymes can be produced cheaply with high specific activity and stabilized, and the cofactors can be either supplied cheaply or regenerated via coupled reactions will enzymatic epoxidation reactions become economically important to the chemical industry.

ENZYMATIC-CATALYZED ELIMINATIONS, ISOMERIZATIONS, AND REARRANGEMENTS

These reactions are generally related to acid–base chemistry. The common reaction is the elimination of a good leaving group to create a double bond.

These reactions include the addition of water to carbonyl groups, the hydrolysis of carboxylic and phosphate esters, the elimination of water from double bonds, many types of rearrangements, and substitution reactions.

ENZYMATIC REACTIONS THAT MAKE AND BREAK CARBON-CARBON BONDS

These reactions typically involve carbanion chemistry. They include decarboxylation of beta-keto and alpha-keto acids, carboxylation reactions, and condensation reactions; for example, an acetyl group from acetyl-CoA is used to add two carbon units, isoprene is used to add five carbon units, and folic acid is used to transfer one carbon unit.

Thousands of enzymes are known, and more are discovered each year. However, fundamentally new reactions are discovered only rarely. It would appear that, in principle, any functionalized hydrocarbon can be selectively oxidized and reduced, cyclized, rearranged, isomerized, or functionalized with nitrogen, sulfur, or phosphorus groups. Within that framework, the potential of biocatalysis appears to many to be limited only by the imagination of the biotechnologist.

OVERVIEW OF THE CHEMICAL INDUSTRY: ECONOMICS OF CHEMICAL TRANSFORMATIONS

Chemical transformation, fermentation, and bioconversion add value to the starting raw material and allow the producer to make a profit. Table 1.1 summarizes many important examples of transformations that are, or may become, economical. The column headed "Value-Added Ratio" is the quotient of the price of the product minus the price of raw material over the price

TABLE 1.1 Value Adding in Production of Major Organic Chemicals

Product	*Formula*	*Hydrogen Deficiency*	*Chiral Centers*	*Price 3/80 (per kg)*	*Price 3/80 (per kg carbon)*	*Present source*	*Yield*	*Value Added Ratio*	*Carbon Value-added Ratio*	*Domestic Production 1978 (metric tons)*	*Potential for Application of Biotechnology (de novo synthesis or bioconversions/ organism & enzymes)*
Acetaldehyde	C_2H_4O	1	0	\$0.54	\$0.99	Oxidation of ethylene	95%	0.17	0.84		Bioconversion from ethanol
Acetic acid	$C_2H_4O_2$	1	0	\$0.51	\$1.28	Oxidation of acetaldehyde	90%	−0.15	0.83	1,250,000	De novo from carbohydrate or bioconversion from ethanol
						Fermentation	95%	1.20	1.21		
Acetoin	$C_4H_8O_2$	1	0		(1.83×)	Reduction of diacetyl					De novo: *Aspergillus, Penicillium, Mycoderma;* from 2, 3-butaneodiol: *Mycoderma aceti:* from diacetyl: yeast
Acetone	C_3H_6O	1	0	\$0.57	\$0.92	Oxidation of cumene (80%)		−0.02	0.44	1,130,000	De novo: *Clostridium acetobutyliticum*
						Oxidation of Isopropanol (20%)	85%	0.07	0.11		
						Fermentation	95%	1.15	0.59		
Acetylene	C_2H_2	2	0	\$1.52	\$1.65	Natural gas	30%			110,000	Bioconversion
						$CaC + H_2O$	95%				
Acrolein	C_3H_4O	2	0	\$0.68	\$1.06	Oxidation of allyl alcohol		0.48	0.47		
Acrylic acid	$C_3H_4O_2$	2	0	\$0.94	\$1.88	Oxidation of propylene	95%	1.41	1.56	146,000	De novo: *Clostridium* or bioxidation of propylene
						CO + acetylene	80%	−0.50	−0.09		
Adipic acid	$C_6H_{10}O_4$	2	0	\$1.10	\$2.23	Oxidation of cyclohexane	90%	0.71	1.95	730,000	Bioxidation of fatty acids
						Oxidation of cyclohexanone		0.01	0.51		

Butadiene	C_4C_6	2	0	$0.66	$0.74	Petroleum	80%			1,580,000	Low
Butanol	$C_4H_{10}O$	0	0	$0.64	$0.77	Reduction of butanol	95%	0.13	−0.10	340,000	De novo: *Clostridium acetobutyliticum*
						Fermentation (6%)		1.19	0.40		
Butyl acrylate	$C_7H_{12}O_2$	2	0		(1.52)	Butanol, acrylic acid	95%			125,000	Yes
Butyl acetate	$C_6H_{12}O_2$	1	0	$0.81	$1.17	Butanol + acetic acid	95%	0.31	0.08	55,000	Yes
Butyraldehyde	C_4H_8O	1	0	$0.54	$0.81	Oxonation of propylene		0.46	0.88	352,000	Bioxidation of butanol
Cyclohexane	C_6H_{12}	1	0	$0.58	$0.68	Butanol + H_2 (80) Hydrocarbon	99%	0.18	0.25		Low
Cyclohexanol	$C_6H_{12}O$	1	0	$1.08	$1.50	Cyclohexanone + O_2	90%	0.67	0.99		Bioconversions
Cyclohexanone	$C_6H_{10}O$	2	0	$1.09	$1.48	Cyclohexanone + O_2	90%	0.69	0.96	552,000	
Dihydroxy-acetone	$C_3H_6O_3$	1	0	$17.00	$42.50	Fermentation		77	0.77		Fermentation glycerol: *Acetobacter*
Ethanol	C_2H_6O	0	0	$0.56	$1.07	Hydration of ethylene	95%	0.21	1.00	570,000 (syn.)	De novo: *Saccharomyces, Clostridium*
						Fermentation (40%)	95%	−0.48	0.85	380,000 (ferm.)	*Zymomonas*
Ethyl acetate	$C_4H_8O_2$	1	0	$0.68	$1.20	Ethanol + acetic acid	99%	0.26	0.01		Yes
Ethyl acrylate	$C_5H_8O_2$	2	0	$0.90	$1.65	Ethanol + acetic acid	99%	0.17	0.39	135,000	Yes
Ethylene	C_2H_4	1	0	$0.44	$0.51	Petroleum				11,700,000	De novo
Ethylene glycol	$C_2H_6O_2$	0	0	$0.68	$1.76	Hydration of ethylene oxide	85%	−0.27	0.03	1,760,000	Yes
Ethylene oxide	C_2H_4O	1	0	$0.79	$1.45	Ethylene + O_2	65%	0.17	0.85	2,250,000	Yes

(Continued)

TABLE 1.1 (Continued)

Product	*Formula*	*Hydrogen Deficiency*	*Chiral Centers*	*Price 3/80 (per kg)*	*Price 3/80 (per kg carbon)*	*Present Source*	*Yield*	*Value Added Ratio*	*Carbon Value-added Ratio*	*Domestic Production 1978 (metric tons)*	*Potential for Application of Biotechnology (de novo synthesis or bioconversions/ organism & enzymes)*
Formaldehyde	CH_2O	1	0	\$0.16	\$0.43	Oxidation of methanol	90%	−0.32	0.39	2,890,000	Yes
Glycerol	$C_3H_8O_3$	0	0	\$1.19	\$3.04	Oxidation of propylene	50%	0.61	2.54	59,000	De novo: "Protol" fermentation
						Cl_2, OH	80%	1.58	4.66		
						Propylene fermentation		4.41	4.53		
Isoprene	C_5H_8	2	0	\$0.34	\$0.39					83,000	Yes
Isopropanol	C_3H_8O	0	0	\$0.42	\$0.70	Hydration of propylene	70%	−0.20	0.14	780,000	Yes
Methanol	CH_4O	0	0	\$0.21	\$0.28	$CO + H_2$	60%			2,900,000	$CH_4 \rightarrow CH_3OH$
Methyl ethyl ketone	C_4H_8O	1	0	\$0.62	\$1.11	Oxidation of NGL	85%			300,000	
						Oxidation of butenes					
Methyl methacrylate	$C_5H_8O_2$	2	0	\$1.01	\$1.81	Acetone + HCN	85%				Yes
Propylene	C_3H_6	1	0	\$0.37	\$0.43					5,900,000	
Propylene glycol	$C_3H_8O_2$	0	1	\$0.80	\$1.69	Hydration of propylene oxide		0.21	0.59	250,000	Yes
Propylene oxide	C_3H_6O	1	1	\$0.66	\$1.06	Oxidation of propylene	93%	0.61	1.30	920,000	Yes
Styrene	C_8H_8	5	0	\$0.84	\$0.91	Benzene + ethylene	90%			3,200,000	Low
Vinyl acetate	$C_4H_6O_2$	2	0	\$0.68	\$1.22	Ethylene + acetic acid	90%	0.29	0.22	760,000	

of the raw material. Because the basic chemical industry is raw material intensive, one might expect that, in order to make a profit, all these ratios would have to be greater than zero. However, it is clear that this is not the case. A definition of value more subtle than price per unit mass must be developed.

The increase in value shown in Table 1.1 can be divided into two components: increase in degree of oxidation and increase in complexity. The column headed "Hydrogen Deficiency" is a quantitative measure of the degree of oxidation of the compound. It indicates the number of hydrogen molecules that must be added per molecule of compound if all double bonds and all rings are to be fully saturated. It is a better measure of the commercial value of a compound than is the cost of the carbon from which it was derived. The oxygen in these compounds is derived from the oxygen in air such that its cost includes only process costs. This is to be contrasted with the carbon, which was derived originally from a fossil fuel or biomass. Therefore, it is useful to compare the price on the basis of dollars per kilogram of carbon; this provides a better index for comparing value added in chemicals manufacture. The carbon value-added ratio was determined by analogy with the value-added ratio column, but prices are based on kilogram of carbon.

The source of value in chemical transformations is addition of complexity and/or functionality. A compound with a variety of functional groups (for instance, one that contains both a ketone and a hydroxyl group) is more difficult to make than one that contains only one type of functional group. Chemically speaking, the transformation necessary to make one functional group will often affect the other. This suggests why dihydroxyacetone has a much greater price than acetone, even though they are rather close chemically, that is, differing only by the fact that dihydroxyacetone has two additional hydroxyl groups. Also, note that dihydroxyacetone is normally made by fermentation, thus taking advantage of the ease with which living cells make multifunctional products. The resulting high price may also reflect the small volume of the market.

A second type of complexity is the presence of rings in compounds. A third type of complexity, and the most important to biology, is a chiral center; a carbon atom that has four different groups attached to it can exist in one of two forms: the left-handed or the right-handed form (see Fig. 1.1). These compounds are chemically identical and differ only in that one is a mirror image of the other. Compounds that have chiral centers and for which a demand exists for optically pure products are prime candidates for biological production.

Consider transformations (1)–(8) in Table 1.2. Ethylene, a basic petrochemical feedstock, is oxidized with air to acetaldehyde. The unsaturation between the carbons in ethylene is eliminated, but an unsaturation between carbon and the added oxygen is added so that the hydrogen deficiency remains the same. The molar increase in mass of the acetaldehyde as a

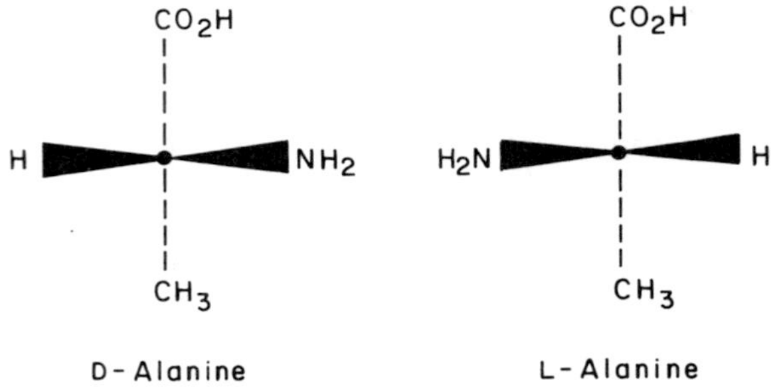

FIGURE 1.1. Left-handed and right-handed carbon atoms.

result of the oxygen addition explains why the carbon value-added ratio is significantly higher than the value-added ratio. Transformation (2) shows a further oxidation along the same pathway, yielding acetic acid. The hydrogen efficiency again remains the same, but another oxygen has been added so that the carbon value-added ratio is greater than the value-added ratio. This is a good example of why the carbon value-added ratio is a better guide to the real value. The cost per unit mass of acetic acid is actually lower than the cost per unit mass of acetaldehyde, so it might appear that the manufacturer is selling at a loss. Actually, the oxygen cost is low, and the carbon value-added ratio reflects the fact that a profit can be made at these prices for this transformation.

Transformations (1) and (2) in Table 1.2 are very similar in that they involve oxidation of compounds containing two carbon atoms. Why is the carbon value-added ratio so much higher for transformation (1)? The answer is that acetaldehyde is much more sensitive to further oxidation than acetic acid. The reaction conditions must be more tightly controlled, and the greater expense involved is reflected in the price of acetaldehyde.

Transformation (3) shows one of the few bioconversions currently being performed on a large scale. Sorbitol, made by catalytic hydrogenation of glucose, is oxidized by *Acetobacter suboxydans* to sorbose and then chemically converted to ascorbic acid. This reaction involves the oxidation of one specific hydroxide group out of four possible ones and the creation of a new chiral center. Therefore it is not surprising that a bioconversion process is most efficient.

Transformations (4), (5), and (6) show three commercially important fermentations. Hexoses (fermentable sugars such as glucose) are the raw material in each case. Acetic acid is a primary metabolite and is derived from a single reaction from the main pathway of anaerobic glycolysis. Glutamic acid is also a primary metabolite and is available in one step from the tricarboxylic acid cycle by a simple transamination. Penicillin is a secondary metabolite and is produced in large quantities by the pharmaceutical industry. It has two rings (including a strained beta-lactam ring) and four

TABLE 1.2 Adding Value in Selected Chemical Transformations, Fermentations, and Bioconversions

Transformation	*Substrate*	*Substrate Hydrogen Deficiency*	*Substrate Number of Chiral Centers*	*Product*	*Product Hydrogen Deficiency*	*Product Number of Chiral Centers*	*Value-added Ratio**	*Carbon Value-added Ratio**
1. Chemical	Ethylene	1	0	Acetaldehyde	1	0	0.17	0.84
2. Chemical	Acetaldehyde	1	0	Acetic acid	1	0	−0.15	0.83
3. Bioconversion	Sorbitol	0	4	Ascorbate	2	2	10.4	10.5
4. Fermentation	Hexoses	—	—	Acetic acid	1	0	1.32	1.33
5. Fermentation	Hexoses	—	—	Glutamic acid	2	1	17.0	16.6
6. Fermentation	Hexoses	—	—	Penicillin V	7	4	770.0	560.0
7. Chemical	Benzaldehyde	5	0	Phenylalanine	5	1	20.0	24.0
8. Fermentation	Hexoses	—	—	Phenylalanine	5	1	250.0	152.0

* Market price for chemicals, March 1980; fermentable sugars at \$0.22 per kilogram.

chiral centers. Note that, even for an antibiotic fermentation that has been constantly improved for 40 years, the value-added ratio is several hundred. This is significantly greater than the two other fermentation products.

Transformations (7) and (8) show a compound for which chemical and fermentation processes are currently competitive. Phenylalanine is one of the more expensive amino acids, currently costing about $55 per kilogram. The demand is not large now, but it is expected to increase in the future. The details of the two processes are discussed below, but suffice it now to say that improvements are possible in both chemical technology and biochemical technology for production of phenylalanine; it is not yet clear which will ultimately be more economical.

In summary, the chemical industry will be changed by biotechnology if strains of microorganisms, either as whole cells or as a source of enzyme catalysts, can carry out chemical transformations at competitive costs. Indeed, it is not merely a single chemical transformation that requires consideration, but an entire process. Like any other current fermentation process, a number of conditions must be satisfied.

1. The starting material, or fermentation substrate, must be relatively plentiful and inexpensive.
2. The product must be sufficiently valuable, for example, having a high carbon value-added ratio, to warrant production.
3. The conversion yield must be sufficiently high that the process is profitable.
4. Economic recovery of the product from starting materials and by-products must be possible.
5. Total income from production must be large enough to justify research and development costs.
6. Environmental and safety concerns must be satisfied.

Improvements in the manner in which biotechnology could have impact on the chemical industry may be brought about via genetic improvements in microorganisms and especially through the use of recombinant DNA technology.

THE PROSPECT FOR CHANGE: AN OVERVIEW OF NEW PROCESS INTRODUCTION IN THE CHEMICAL INDUSTRY

The development of biotechnology should be viewed not so much as the creation of a new industry, but as the revival and revitalization of an old one. Many basic organic chemicals were produced by fermentation before petroleum became relatively cheap. As recently as the 1950's, the majority of glycerol and a substantial quantity of industrial ethanol were made by fermentation techniques (see the section on ethanol). Now, of course, interest in industrial fermentation is greatly revived, particularly for ethanol (under the name "gasohol" when added to gasoline). Many specialty organic chemicals have always been made by biological methods; but now, with the rapidly increasing cost of petroleum, the potential to produce certain feedstock chemicals (methane, acetone, butanol, acetic acid, etc.) by biological processes is being reconsidered. It should be cautioned that the increase in petroleum cost alone is not sufficient cause to change processes, since the cost of sugars and other biologically derived raw materials will also increase with rising petroleum costs.

Public awareness of and interest in biotechnology have increased dramatically in the past few years. The surge of public concern over the safety of recombinant DNA that followed the Asilomar conference in 1975 has been replaced by great interest in the potential economic, environmental, and health benefits that biotechnology may offer. It was only in 1953 that the key concepts for understanding DNA structure were proposed and only in the 1960's that the basic mechanics of transcription and translation of DNA were elucidated. The discovery in the 1970's of restriction enzymes was quickly followed by techniques to manipulate and clone DNA. Many feel that the accelerating rate of scientific discovery will promote the development of an industry in the 1980's that could affect large segments of the chemical industry.

Less frequently discussed are the developments in technology necessary to exploit science. The mechanics of life are fascinating, but they do not necessarily lead to production of cheaper fuel or plastic. The technologies that are required to exploit microbes include development of more efficient and creatively engineered fermentation facilities and product separation processes. Microbes may produce molecules, but they will not isolate, purify, concentrate, mix, or package them for human use.

The use of enzymes for industrial processes has been discussed since their discovery. At present, enzymes have important but limited application in the pharmaceutical, chemical, and food industries. Perhaps the most spectacular success story in recent years has been the use of enzymes to produce high-fructose sweeteners from corn syrup.

Many problems in the application of biological processes remain to be solved. Recombinant DNA technology, as well as other genetic manipulation, allows one to transfer the gene coding for a particular enzyme to an organism that conceivably could produce the enzyme in a fermentation process more efficiently than the native organism. Additionally, genetic manipulation may even be used to change enzyme substrate specificities as well as the optimal pH and temperature of a particular enzyme. But genetic engineering does not completely circumvent the problems associated with building and operating a large fermentation or enzyme bioconversion facility.

The application of biotechnology in the modern setting of the chemical industry is unproven and underdeveloped. Research investment involves high risk and maintaining a long-term perspective, because even the most optimistic of the companies involved realized that significant profits from this research cannot be realized until, at best, the late 1980's.

The introduction of any new process in biotechnology or chemical technology should be evaluated carefully, including consideration of:

1. change in raw materials and change in their costs;
2. improvement in process and energy costs;
3. improvement in separation costs;
4. changes in market demand; and
5. long-term profitability relative to amortized research and development costs.

The chemical industry strives to find and fill new market demands. It strives to serve the established markets with higher quality and lower prices. It strives toward greater profitability. Biotechnology may help to some extent in achieving these goals.

FUTURE PRODUCTS AND PROCESSES IN THE CHEMICAL INDUSTRY

The strategy of chemical companies is to develop higher-valued products, specialty and performance chemicals with a high element of utility in their sales price, and to deemphasize commodity chemicals with their requirements for high capital intensity and lower profit margins. The drive toward specialty chemicals is not simply to improve financial performance. The market has an ever-increasing need for higher yields of existing products and an ever-present demand for new products.

The impact of applied genetics in the chemical industry will be primarily on high-value products.

ORGANIC CHEMICALS

Two modes of biotechnology will have an impact on chemical process development: fermentation and enzyme engineering. The first will primarily change resource entry into the chemical industry and affect the transition from nonrenewable (petroleum) to renewable (biomass) feedstocks (see Fig. 1.2). The second will have two functions: to allow the entry of fermentation-derived feedstocks into the chemical intermediate chains, and to compete directly with traditional chemical transformations.

Limitations of Fermentation

The source of feedstocks for almost all fermentation must be biomass. Biomass is composed primarily of polymers and 5- and 6-carbon sugars.

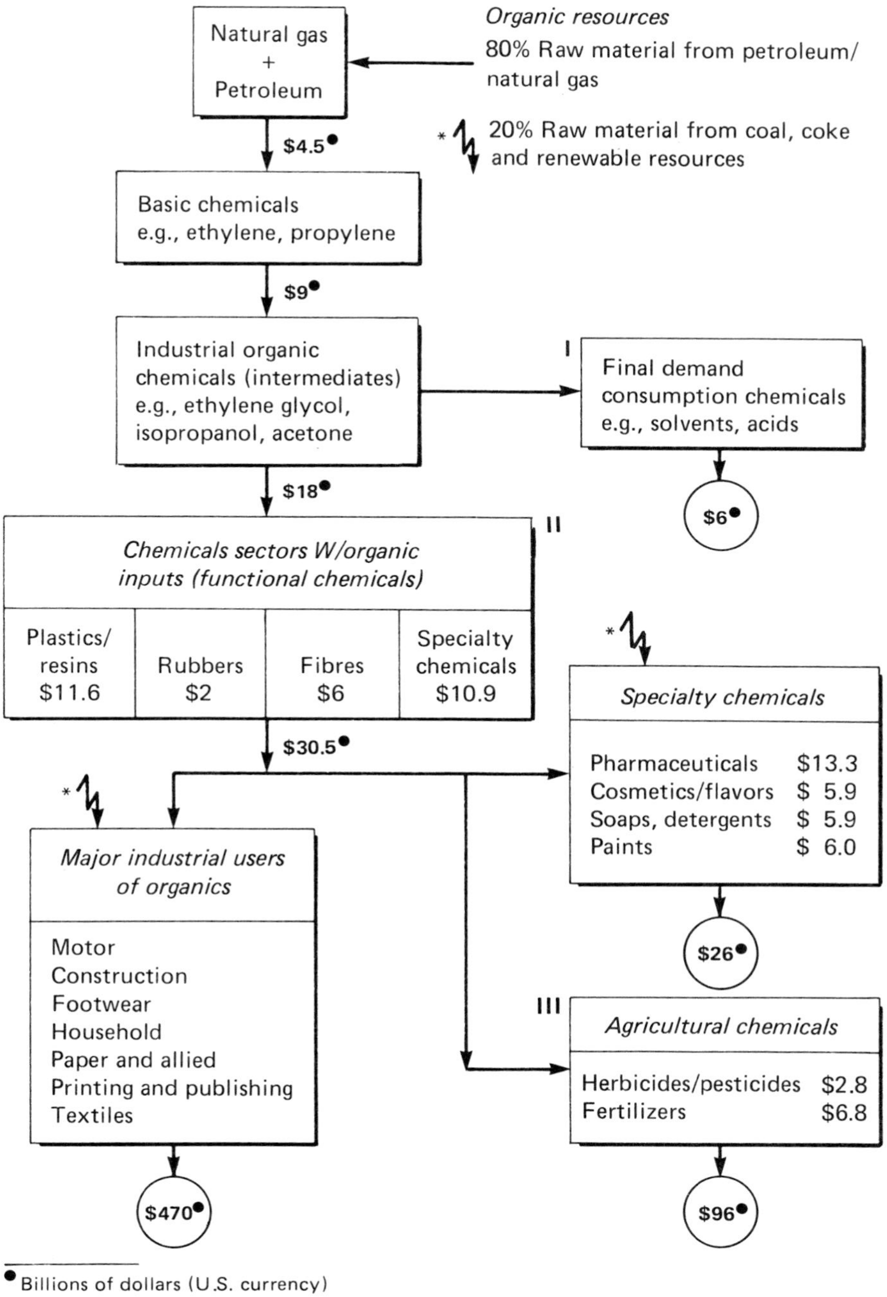

FIGURE 1.2. Flow of industrial organic chemicals from raw materials to consumption. Values of shipments are the estimated 1977 values. (Data sources: *US Industrial Outlook* United States Department of Commerce, 1978; *Kline Guide to Chemical Industry* Fairfield, N.J. Adapted from Tong (1979)).

These polymers must be either chemically or enzymatically hydrolized to the monomeric sugars before fermentation. Sugars can be fermented with no pretreatment. Starch and cellulose must be saccharified with acid or with enzymes before fermentation.

Problems with biomass utilization include:

1. competition with food needs for starch and sugar;
2. cyclic availability during this year;
3. biodegradability and associated storage problems;
4. high moisture content, resulting in high collection and storage costs;
5. requirement of mechanical or chemical processing for cellulosic biomass;
6. heterogenous composition of cellulosic biomass (for example, mixtures of cellulose, hemicellulose, and lignin); and
7. need for disposal of nonfermentable portions of biomass.

Two generalizations can be made about the problems associated with obtaining solutions of sugars from biomass that are suitable for fermentation. In the case of food-related biomass sources such as sugar, corn, and sorghum, few technological barriers exist for their conversion to fermentable sugars; however their cost requires subsidies to make fermentation to fuels and basic chemicals profitable. For cellulosic biomass sources such as agriculture residues, municipal wastes, and forest products, the economics still do not favor fermentation to fuels and basic chemicals because of technological barriers in collection, storage, pretreatment, fermentation, and residue disposal.

The nature of biochemical processes places further limits on fermentation. The starting material is partially oxidized carbon (for example, a carbohydrate, $C_nH_{2n}O_n$). Since the cell is living off the energy that is available from the oxidation of the sugar, the products are inherently oxygenated carbon compounds ranging from simple molecules such as ethanol and acetic acid to larger molecules such as lactic, fumaric, and citric acids (Fig. 1.3). Ethanol is in a lower oxidation state than sugar, but the fermentation produces a mole of CO_2 for each mole of ethanol, so that the net process is an oxidation. In general, the fermentation of a compound that is in a lower oxidation state than sugar requires a compensating production of CO_2. This generalization of thermodynamics is formalized for the theoretical yields for major fermentation, as shown in Table 1.3.

Expansion of Fermentation Chemical Industry

A schematic representation of fermentation processes in the chemical industry both at present and projected is shown in Fig. 1.4. The impact can be divided into three parts, according to how it has been and probably will be experienced.

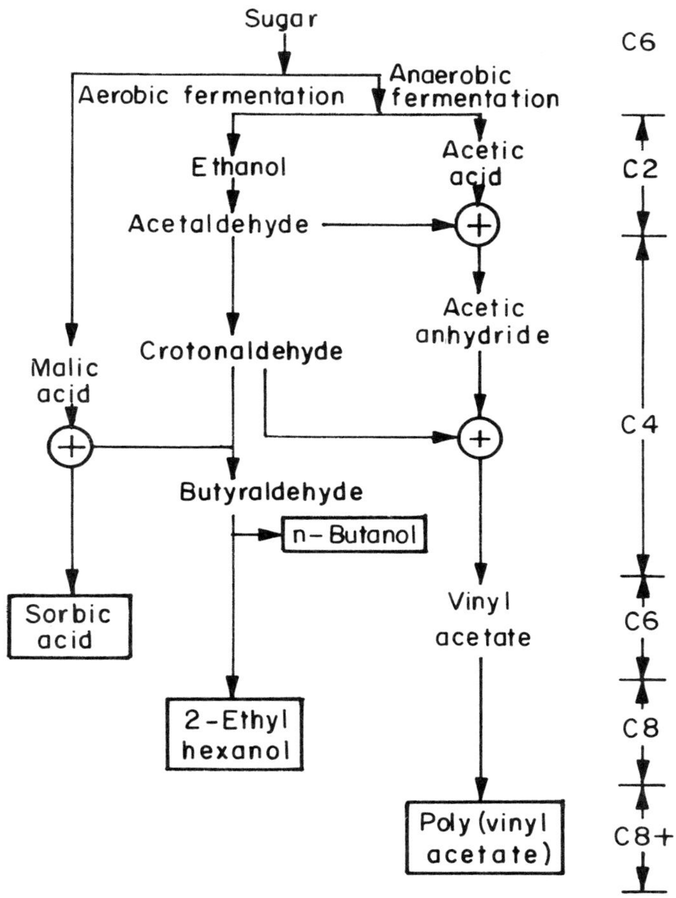

FIGURE 1.3. Oxygenated organic chemicals produced from fermentation chemical derivatives. From Tong (1979).

1. Aerobic fermentation to produce enzymes, vitamins, pesticides, growth regulators, amino acids, nucleic acids, and other specialty chemicals.
2. Anaerobic fermentation to produce organic acids and solvents.
3. Chemical modification of fermentation products of both aerobic and anaerobic fermentations.

Some examples of these are summarized in Table 1.4.

Fermentation is already well-established in the first category and should grow. Complex biochemicals, for example, chemicals containing multiple functional groups and chiral centers, are now made almost invariably by fermentation (that is, antibiotics and growth factors). The same is true for

TABLE 1.3 Theoretical Yields for Chemicals Produced from Glucose by Fermentations

Chemical	*Percent Overall Weight Yield*
Solvents	
n-Butanol, acetone, ethanol	41.2
n-Butanol, formaldehyde	67.0
n-Butanol, isopropanol	37.2
Glycerol, acetaldehyde	75.5
Ethanol	51.1
Acids	
Acetic, propionic	50, 74.4
Acetic	67, 100
Lactic	100
Citric	98
Methane	27.0
Amino Acids	
Glutamic acid	
Lysine	40–70
Aromatic	

enzymes. Amino acids and nucleotides, being somewhat less complicated molecules, are only in some cases fermentation products, but in some cases fermentation production may be expected to increase. Separate sections are devoted to these compounds.

If the first category is the realm of established fermentation, the second is the area of greatest current growth. Already 40% of the ethanol produced in the United States is from fermentation. The barriers, as discussed above, are related primarily to the need for a cheaper technology for converting cellulosics to fermentable sugars. New fermentations for acetic, propionic, and acrylic acids will help this area to expand.

The longest development time can be expected for compounds in the third category. A scheme developed by Tong (1979) for making some valuable products from fermentation chemicals is shown in Fig. 1.4. In addition, dehydration of fermentation ethanol provides access to the bulk of the chemical industry's products through the established chemical technology. The development of this area will be slow for the following reasons:

1. Under the present economic structure, petroleum is much cheaper as a source of chemicals than biomass processes. The relative price of oil over biomass would have to increase manyfold to make it competitive.
2. The limited availability of biomass and the huge demand for motor fuel will require that virtually all fermentation ethanol be used as a solvent or motor fuel.

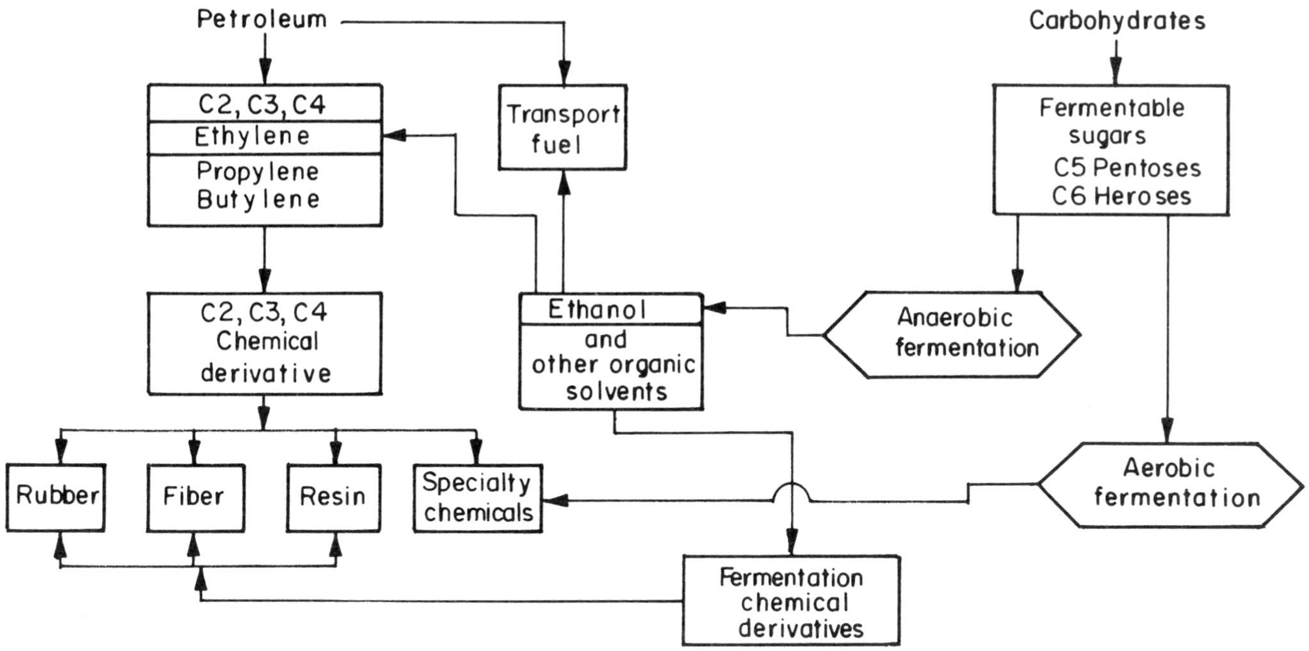

FIGURE 1.4. Diagram of alternative routes to organic chemicals. From Tong (1979).

TABLE 1.4 Expansion of Fermentation into the Chemical Industry

I. Aerobic Fermentation	*Examples*
Enzymes	Amylases, proteases
Vitamins	Riboflavin B_{12}
Pesticides	*Bacillus thuringiensis*
Growth regulators	Gibberellin
Amino acids	Glutamic, lysine
Nucleic acids	
Acids	Malic acid, citric acid
II. Anaerobic Fermentation	*Examples*
Solvents	Ethanol, acetone, *n*-butanol
Acids	Acetic, propionic, acrylic

FERMENTATION ROUTES TO CHEMICALS AND PREVIOUS IMPACT OF APPLIED GENETICS

Two types of fermentation processes have evolved. One type has been for the production of simple chemical products such as isopropanol, *n*-butanol, acetone, ethanol, 2,3-butylene glycol, glycerol, and the acids such as acetic, fumaric, and citric acid. The other type of process has been for the production of more complex chemicals such as antibiotics, enzymes, and hormones.

Traditional processes have evolved, to date, for the production of the small-molecular-weight chemicals. This section attempts to assess the role that genetic research has played in the development of such fermentation processes. As one will see, the role of genetic research has mainly been one of mutation and strain isolation. Very little genetic research has been employed to improve the industrial processes for small-molecular-weight chemicals because only a few are presently produced by fermentation. Therefore one can safely assume that little of the "new genetics" or genetic engineering has had an impact on the production of these small molecules.

ACETIC ACID

Acetic acid is a chemical that can be produced via biological means. For example, ethanol may be oxidized to acetic acid by *Acetobacter suboxydans;* this is the procedure by which vinegar or food-grade acetic acid is made. The theoretical yield is 0.67 g of acetate per gram of glucose. Annual production of vinegar was reported to be 125 million gallons in 1976; and according to *Chemical Week* (January 12, 1977), vinegar will be one of the fastest growing markets for ethanol (10% per year).

Other microbial processes for acetic acid are possible. One of the most attractive involves *Clostridium thermoaceticum,* wherein quantitative conversion of glucose to acetic acid is theoretically possible, that is,

$$1 \text{ glucose} \xrightarrow[\textit{thermoaceticum}]{\textit{Clostridium}} 3 \text{ acetic acid}$$

$$1 \text{ xylose} \longrightarrow 2.5 \text{ acetic acid}$$

The reason that the conversion is quantitative is that the organism is able to fix carbon dioxide. *Clostridium thermoaceticum* is one of the few bacteria that can use carbon dioxide as an electron acceptor and thus quantitatively convert glucose or xylose to acetate. In this case, the theoretical maximum yield is one gram of acetate per gram of glucose. For most chemical feedstock fermentations, the most important cost factors are the raw material cost, which is usually 50–70% of the manufacturing cost, and actual productivity (gram of product per unit volume per day). It is in the latter aspect that difficulties arise with the utilization of *C. thermoaceticum* for acetic acid production. The present productivities are extremely low because of two factors. One is that the cell mass concentration in the fermentation broth is low. The final concentration of acetate is also low, a maximum of 60 g/l, owing to the fact that growth and production of acetate are inhibited by acetate.

The question then becomes, Can biological processes compete with chemical processes? At present, acetic acid is produced by seven companies in the United States. Domestic capacity for production of synthetic acetic acid (1605 thousand metric tons, according to the SRI *Chemical Economics Handbook* (1977)) is split almost equally among three processes:

- *n*-butane oxydation (33%)
- methanol carbonylation (32%)
- acetaldehyde oxidation (32%)

From 1968 to 1978, production of acetic acid increased at an average rate of 2.7% per year, most growth occurring from 1968 to 1973 (6.9% average/year). The carbonylation process will increase in use and will probably become the major source of acetic acid in the United States by the mid-1980's.

The major cost components for two of the three major synthetic processes were summarized by SRI for 1978 as:

	Cents/lb. of Acetic Acid	
	Acetaldehyde Oxidation with Air*	*Low-Pressure* Carbonylation of Methanol*
Net variable costs	13.0	6.5
Labor	0.3	0.3
Other production costs	2.2	1.8
Total production costs	15.5	8.6

* Assumes 6.3/lb of methanol and 15.4/lb of acetaldehyde

It is obvious why the methanol carbonylation process will dominate the market for acetic acid. In view of the fact that methanol will be a major product of large-scale coal gasification processes, the carbonylation process begins to seem even more significant. Obviously, biological processes for production of acetate will have to compete with the above process technologies, and the raw materials for biological processes will have to be competitive with methanol as a feedstock.

There has been little significant genetic research of any kind with regards to the organism *C. thermoaceticum* for the production of acetic acid, and no commercial processes have ever been developed.

Acetobacter suboxydans can also be used for oxidation of other chemicals besides ethanol, as shown in Table 1.5 and described by Greenshields (1978).

Oxidation of mannitol to fructose was conducted on an industrial scale during the 1950's, but the process was discontinued because of the high cost of the raw material (mannitol) and other economic considerations.

A modern industrial fermentation utilizing *Acetobacter suboxydans* for preparing dihydroxacetone from glycerol has only recently been instituted on a commercial scale, despite the fact that this conversion was known in the late 19th century and that the usefulness of the product as a pharmaceutical became known in the early part of the 20th century. Although manufacture of dihydroxyacetone depends on efficient chemical purification with carbon treatment, resin treatment, and evaporation, the fermentation process has a high conversion efficiency (95–97% yield) and a pure product (0.2% ash).

The sorbose fermentation is of special interest, since L-sorbose is used in synthesis of L-ascorbic acid (vitamin C). The substrate (sorbitol) is a cheap raw material produced by electrolytic or catalytic hydrogenation of glucose. Again *A. suboxydans* is used, and 10–20% solutions of sorbitol plus accessory nutrients with vigorous aeration and agitation allow almost

TABLE 1.5 List of Substrates Oxidized and Products Formed by *Acetobacter suboxydans*

Substrate	*Product*
2,3-Butylene glycol	Acetylmethylcarbinol
Erythritol	L-Erythrulose
Ethylene glycol	Glycolic acid
D-Gluconic acid	D-2-Oxogluconic acid
D-Gluconic acid	D-5-Oxogluconic acid
D-Glucose	D-Gluconic acid
Glycerol	Dihydroxyacetone
D-Mannitol	D-Fructose
Perseitol	Perseulose
α-Propylene glycol	Acetol
D-Sorbitol	L-Sorbose

As described by Greenshields (1978).

quantitative conversion and a subsequent 70% yield after chemical purification to a crystalline product.

Perseulose, a rare ketoheptose, gluconic acid and 5-oxygluconic acid, acetylmethylcarbinol or acetoin, L-erythrulose, and D-tartaric acid have all been prepared industrially by *Acetobacter* species.

Except for some strain isolation procedures, no major applied genetic studies have been reported with respect to *A. suboxydans*. The organism is not listed in R. C. King's book entitled *Handbook of Genetics* (1974), and thus any strain improvement work may have been in industrial laboratories. Conner and Allgeier (1976) reviewed the microbiology of vinegar production. The bacterium is described as having an abnormally high mutability by some investigators, but others have disagreed. One investigator found that, with 39 strains of *Acetobacter,* there were no obvious changes in properties, and a controversy still exists. It appears that one of the reasons that genetic research has not had any impact with *Acetobacter suboxydans* is that it is still somewhat difficult to enumerate and handle the organism in the laboratory.

CITRIC ACID

Citric acid is the most important food acidulant; historically, it has held over 55–65% of the total food acidulant market. The two other important acidulants are phosphoric acid (used almost exclusively in cola-flavored beverages), with 20–25% of the total acidulant market, and malic acid (used primarily in newly marketed food and beverage products), with 5% of the market (SRI, 1977, 636,5021). Table 1.6 shows the amounts and uses of citric acid consumed in 1973 and 1976, along with projections for 1986.

TABLE 1.6 U.S. Citric Acid Consumption[1]

	1973		*1976*		*Projected 1986*	
	Millions of lbs.	*% of total*	*Millions of lbs.*	*% of total*	*Millions of lbs.*	*% of total*
Food and beverage uses	120	71	150	70	190	66
Pharmaceutical uses	30	18	24	11	30	10
Industrial and miscellaneous applications	20	11	40	19	70	24
Totals	170	100	214	100	290	100

From SRI, *Chemical Economics Handbook* (1977).

1. Total citric acid consumption is expected to continue growing at a rate of approximately 6–7% per year.

As of June 1977, only two companies were producing commercial quantities of citric acid in the United States. They are Miles Laboratories, Inc., and Pfizer, Inc. SRI estimates annual production capacity in 1977 as follows:

Company and Plant Location	*Annual Capacity June 1977 (Millions of Pounds)*
Miles Laboratories, Inc.	25
Dayton, Ohio	
Elkhart, Indiana	75
Pfizer, Inc.	165
Chemicals Division	
Brooklyn, New York	
Groton, Connecticut	
Southport, North Carolina	50
	315

For 20 years prior to 1970, the list price of citric acid remained fairly stable at 26–30 per pound. From July 1970 to March 1977, the list price increased from 32 to 59 per pound. The current price of citric acid is about 65/lb.

Citric acid is produced commercially by *Aspergillus niger* via one of three processes:

- the Kojo fermentation process (mainly used in Japan)
- liquid surface culture process
- submerged culture process

Miall (1978) has reviewed in detail the historical developments of these citric acid processes. He discusses the significant developments concerning trace metal requirements and elucidates the biosynthesis route of these industrially important fermentations.

The development of citric acid fermentation owes much to the work of Johnson and his colleagues at the University of Wisconsin, as does early development of penicillin fermentation (Miall, 1978). Relevant to this study is the fact that surprisingly little work has been published on improving citric acid producing strains of *A. niger* by mutation (Miall, 1978). In 1951 (British Patent No. 653,808), Miles Laboratories mentions the use of a mutant strain of *A. niger*. Weight yields of 100% have recently been reported of the use of *A. niger* mutants and solutions containing 160 g of sucrose per liter with added salts in surface culture (Miall, 1978). The mutants were obtained by irradiation of a strain for which a maximum yield of 29% was reported.

Commercial processes for production of citric acid by other organisms have recently been developed. The most important is the Takeda *n*-paraffin process. The most significant development is the formation of citric acid from hydrocarbons by *Candida lipolytica.* Although Kyowa Hakko Kogya patented the use of *Arthrobacter paraffineus* (British Patent 1,187,610) in 1970 and the Takeda Chemical Industries (British Patent 1,257,900) patented the use of strains of *Corynebacteria,* in 1971 most of the patents refer to the use of various strains of *Candida.* The earliest patents all claim production of citric acid by submerged fermentation from C_9 to C_{20} normal paraffin mixtures using *C. lipolytica* and other *Candida* strains. One patent (Takeda Chemical Industries, 1970, British Patent 1,211,246) mentions a 114% yield of anhydrous citric acid. From later patents, it is obvious that these organisms can also make substantial amounts of citric acid.

Takeda Chemical Industries in 1972 (British Patent 1,297,243) patented a procedure for isolating and testing mutant strains of microorganisms that will give improved yields of citric acid with less isocitric and other contaminating acids. In the patent, the mutants are primarily tested for growth on *n*-paraffins and on citrate, and those selected for further study grow on the former compound but not on the latter. An example is given of a mutant giving 133% yield of citric acid from C_{13} to C_{15} *n*-paraffin mixture with 17% of isocitric acid, whereas the parent strain gives a 61% yield of citric acid and 65% isocitric acid. Other examples using procedures to obtain mutants low in aconitate hydratase activity are also presented.

LACTIC ACID

Lactic acid is produced both chemically and by fermentation. Over the last 10–20 years, manufacture by fermentation has been under heavy competitive pressure from chemical processes. At one time there was the possibility that only the L-form of the acid would be allowed for food use, which would have been a boon to fermentation processes.

The organisms used for production of lactic acid are all species of *Lactobacillus,* most commonly *L. delbruckii.* The organism of choice depends on the substrate to be fermented and the type of acid to be formed. Starting materials may be glucose, sucrose, or lactose (whey), and the final product may be D,L- or L-lactic acid. The fermentation per se is very efficient and can result in yields of 90%, depending on the original carbohydrate. Since most of the problems in the manufacture of lactic acid lie in recovery and not in the fermentation step, few genetic studies have been employed for improvement of the industrial processes. Most of the genetic studies have been related to the fermentation of milk. Here one can find genetic work that has had important ramifications. Rates of acid production are critical and have resulted in improvements (Lawrence and Thomas, 1979).

Genetic approaches to control lysis of culture bacteria—probably the single most important factor in retarding milk fermentation—have had a significant economic impact on the dairy industry.

ADDITIONAL ORGANIC ACIDS

The industrial processes for itaconic acid are derived mainly from studies done at the Northern Regional Research Laboratory (NRRL) of the U.S. Department of Agriculture in Peoria, Illinois. A screening program resulted in the isolation of several species of *Aspergillus* that were able to form itaconic acid. A most promising species was *A. terreus,* and additional work at NRRL screened 30 cultures of *A. terreus* for production of itaconic acid in promising yields. Only one was found. Lockwood and Reeves (1945) screened 308 strains isolated from soil samples. Eleven of these strains made itaconic acid in yields greater than 45% of theoretical (which was accepted as one molecule of itaconic acid from one molecule of glucose consumed, or a 72% weight yield) (Miall, 1978).

Little additional work on strain improvement for the production of itaconic acid has been reported. The overall metabolism of itaconic acid has been elucidated by Cooper and Kornberg (1964) using a *Pseudomonas* sp. However, mutation experiments with *A. terreus* must have been conducted and evaluated, since Stodola et al. (1945) report that *A. terreus* mutants make not only itaconic acid but also itatartaric acid.

Pfizer is reported by Miall (1978) to be the only U.S. manufacturer of itaconic acid; 75% of itaconic acid is used in styrene butadiene copolymers and for lattices and emulsions in general.

FUMARIC ACID

Fumaric acid was at one time made by Pfizer on a commercial scale by fermentation with a strain of *Rhizopus,* but production was stopped when the economically more attractive synthesis by oxidation of benzene was developed (Miall, 1978; SRI, 1977).

In the 1940's, *R. nigricans* was patented as the production organism in submerged fermentation by Merck & Co. and Pfizer & Co. Again, NRRL evaluated strains for production and found a unique strain, *R. nigricans* 2582, which not only produced fumaric acid but also was able to synthesize invertase. Thus sucrose could be used as a fermentation substrate without prior hydrolysis.

As more substrates are evaluated for fermentation processes, one finds reports on soil isolates capable of forming fumaric acid from different substrates. For example, Yamada et al. (1970) isolated a species of *Candida* from soil called *C. hydrocarbofumarica,* which made fumaric acid from *n*-paraffins.

In 1960, about 23 $\times$ 10^6 pounds of fumaric acid was produced, and in 1975, 5.0 $\times$ 10^6 pounds (SRI, 1977). The price per pound in 1975 was 43¢/lb; in 1960, it was 23¢/lb. According to SRI, all fumaric acid is currently produced by isomerizing maleic acid, which is obtained directly from benzene oxidation or from maleic anhydride produced through oxidation of benzene. Tong (1978) reports that 15% of U.S. production of fumaric acid is by fermentation. Asymmetric addition of ammonia to fumaric acid catalyzed by aspartase-containing microorganisms, giving optically pure L-aspartic acid, is an industrially applied process (Leuenberger, 1978).

MALIC ACID

Fermentation processes are not currently economical for production of L-malic acid. L-malic acid can readily be made enzymatically from fumaric acid with whole cells or the isolated enzyme from *Brevibacterium ammoniagenes*.

Several species of microorganisms, mainly yeasts, have been shown to convert fumarate to L-malate, including *Candida utilis, Pechia membranaefaciens,* and *Pullularia pullulans,* as well as a species of *Brevibacterium*. In 1973, Tachibana and Murakami (1973) reported that *Schizophylum commune* can produce L-malate in high yield from ethanol.

Other tricarboxylic acid cycle acids could readily be made by fermentation if required, but no commercial use has yet been found (for example, α-oxoglutaric acid).

EPOXY SUCCINIC ACID

This acid has been isolated as a metabolic product of *Aspergillus fumigatus* (Miall, 1978). Apparently, epoxy succinic acid is formed by a direct oxidation of fumaric acid. The epoxy succinic acid can be further converted to mesotartaric acid by a variety of microorganisms, including *Pseudomonas putida*. The enzyme catalyzing the reaction has been isolated from *P. putida*.

KOJIC ACID

Kojic acid can be made in good yields by a number of aspergilli, particularly those of the *A. flavus-oxyzae* group. Kojic acid has a number of potential uses, including production of the flavoring agent maltol via kojic acid via comenic acid and 6-methyl comenic acid.

GLUCONIC ACID

A variety of microorganisms have been investigated for production of gluconic acid, including *Aspergillus niger, Acetobacter suboxydans,* various species of *Pseudomonas, Pullularia pullulans,* and a species of *Moroxella. Aspergillus niger* and *Acetobacter* sp. are the organisms with which some industrial processes have been pursued.

Strain development work again was undertaken at NRRL, and the research activity concentrated on *Pseudomonas,* 16 strains being examined. The researchers found that *P. ovalis* NRRL B-8 made only gluconic acid from glucose, and a 99% yield was eventually reached.

Recovery of gluconic acid is troublesome in this fermentation.

OXOGLUCONIC ACIDS

Further bacterial oxidation of gluconate leads to 2- or 5-oxogluconate with *Acetobacter melanogenium.* By employing *Pseudomonas albosesamae,* 2–5-dioxogluconate can be formed. The type of organism used primarily controls whether 2- or 5-oxogluconate is made. 5-Oxogluconate has never been made on a commercial scale, while 2-oxogluconic acid is made on a commercial scale as an intermediate in the production of erythorbic acid.

It is usually made by oxidation of glucose with strains of *Pseudomonas.* However, *Serriatia marcescens* NRRL B-486 has been used in pilot plant runs, where yields of 95–100% in 16 h are reported. Ten strains of *S. marcescens* all gave over 80% yields, and it is claimed that both yield and productivity are greater than when *Pseudomonas* sp. is used.

ERYTHORBIC ACID

Penicillium notatum can be used for the direct production of erythorbic acid from glucose. A strain was obtained following a mutation and screening program that made erythorbic acid in a medium containing 80 g of glucose per liter in about 40% yield in 5 days.

TARTARIC ACID

Besides conversion of 5-oxogluconate to tartaric acid by chemical catalysts, production of tartaric acid by fermentation with *A. suboxydans* or *P. fluorescens* has been reported.

Yamada et al. (1969) screened about 3600 strains of fungi and 4000 strains of bacteria and yeasts for tartaric acid production.

Six strains of bacteria, all isolated from Japanese persimmon, using a mannitol-based medium, were shown to produce L(+)-tartaric acid. Four strains were identified as *Gluconobacter suboxydans.* Kotera et al. (1972) reported on a mutation program with *G. suboxydans* and were able to obtain concentrations of 11 g/l tartaric acid, while the highest concentration reported by others was 6.0 g/l. Subsequent mutation program resulted in 14.68 g/l from 50 g of glucose, 35% of theoretical yield, in 6 days in shake flasks.

CONCLUSIONS ABOUT ORGANIC ACIDS

Only a few organic acids are made at present in significant quantities by fermentation. The main ones are acetic, citric, and lactic acids. These are used primarily in foods. The major reason for using fermentation is that food-use definitions require production by fermentation, which becomes competitive when the existence of multiple functionality or chiral centers makes chemical synthesis difficult (for example, L-lactic acid, citric acid). Genetics through mutation and selection has played an important role in the development of industrial processes for citric acid but less of a role for other acids. Most of the processes have been developed on an empirical basis. With some of the more specialized fermentation acids, classic microbial isolation and screening programs resulted in production of strains. Much of the recent published research on strain screening and mutation programs for the production of organic acids appears to be originating in Japan. The most notable accomplishment is the development of the citric acid process for production from *n*-paraffins by yeast.

ACETONE-BUTANOL

As the prices of petroleum-based raw materials increase and supplies diminish, the cost of producing acetone-butanol by the synthetic route will increase substantially. The question to be considered is whether production of acetone-butanol by fermentation will be the process of choice. A key issue appears to be the availability of a cheap raw material, since the previous biological process produced acetone-butanol at very near theoretical yields. The percent weight conversion yield attainable by fermentation of glucose is between 29% and 35%, the theoretical yield being 50%. Conversion efficiencies are then between 59% and 70% and depend on the specific strain used (Compere and Griffith, 1979).

The bacteria used for commercial production of acetone-butanol from carbohydrate raw materials are almost all of the genus *Clostridium* and are mainly strains of *C. acetobutylicum, butylicum,* and *pasteurianum*. Strain improvement research has had little impact on the acetone-butanol fermentation. The most recent biological studies were in the 1950's, when two problems were addressed. The first was that of lysis, and the second concerned the use of continuous culture processes for the production of acetone-butanol. Kutzenok and Aschner (1952) report on the degenerative processes in a strain of *Clostridium butylicum,* as do Finn and Nowrey (1959). Studies on continuous culture fermentations have been conducted mainly in the

Soviet Union and China. For example, Chiao et al. (1964) report on the continuous culture process used in China for production of acetone-butanol.

More recently, two specific problems have been addressed. The first is increasing solvent resistance of production strains by mutation and selection, but no significant findings have been reported to date. The second involves the recognition that phage infections can cause loss of active cultures.

In general, many basic genetic studies remain to be conducted with anaerobic bacteria such as *Clostridia*.

2,3-BUTANEDIOL

Mixed acid fermentations have been investigated intensely and can be divided into two types. The first uses those organisms resembling *Escherichia coli* in products produced, mainly volatile acids; a second group produces 2,3-butanediol as a major product. Several butanediol fermentations are possible. For example, glucose can be fermented to 2,3-butanediol, carbon dioxide, and hydrogen; or 2,3-butanediol, formate, and carbon dioxide; or 2,3-butanediol, glycerol, and carbon dioxide. Bacteria examined as sources of 2,3-butanediol are commonly strains of *Aerobacter, Bacillus,* and *Serratia* (Wood, 1961).

The pioneering development work on this fermentation was carried out largely at the National Research Council of Laboratories in Ottawa and in the U.S. Department of Agriculture laboratories at Peoria, Illinois. Hastings (1971) has summarized much of the early works done on this fermentation. Typically, weight conversion yields on glucose for 2,3-butylene glycol are 45%, with the theoretical value being 50%, which results in a 90% conversion efficiency (Tong, 1978). 2,3-Butylene glycol can be converted readily by chemical catalysis to 1,3-butadiene. Butadiene production in 1976 was approximated as 1.5×10^6 kg or 3.3×10^6 kg.

No significant genetic studies have been reported that can be related directly to the production of 2,3-butylene glycol except for standard strain isolation and improvement research.

GLYCEROL

Fermentation processes for manufacture of glycerol are also now being explored because of increased petroleum prices (Tong, 1978). Glycerol and fumaric acid are highly sensitive to crude oil prices owing to relatively low conversion yields and production volumes. Tong reports that, in 1976, 157×10^6 kg or 34.5×10^7 lb of glycerol was produced, and none of it was via fermentation processes.

Fermentation processes for the production of glycerol have been reviewed by Spencer and Spencer (1978). Two approaches can be used. One is the traditional use of stirring agents to control fermentation based upon the interruption of the normal yeast alcoholic fermentation. The second is to use microorganisms, especially yeast, that overproduce polyhydroxy alcohols in response to high osmotic pressures. The first is an anaerobic process, and the second is aerobic. No significant commercial ventures are reported that have employed the second method, even though Onishi reports via aerobic process a percent weight conversion yield of 43% glucose, with the theoretical yield being 76%. (The conversion efficiency is then 56%.)

Little applied genetics research has been reported on either process. The most recent work, which appears to be the standard situation, is that of Japanese investigators, who, when they were exploring hydrocarbon fermentation in great detail, investigated the use of various mutants for production of polyhydroxyl alcohols. For example, Tabuchi and Hara (1973) and Hattori and Suzuki (1974) report on studies employing *Candida lipolytica* and *Candida zeylanoides*. A glycerol-requiring auxotroph of strain KY6166 of *C. zeylanoides* was made via environmental manipulations to produce either erythritol or mannitol. The physiology of the osmophilic microorganisms has been a subject of intense review by Brown (1976). The cellular role of high polyol concentrations appears to be as a compatible solute that protects intracellular enzymes against inhibition or inactivation at biologically low levels of water activity. The genetic regulation of the enzymatic processes for production of such polyols largely remains open for investigation.

Furthermore, one could envisage novel fermentation processes based on the use of osmotolerant yeast, as well as halophilic bacteria or algae, since sterilization requirements and the use of a high concentration of sugar may reduce process costs.

CAROTENOIDS

The reported market for carotenoids is small at present, worldwide sales of B-carotene accounting for some 27 tons/year (Ninet and Renault, 1979). Fermentation-derived products have not, so far, been exploited commercially, mainly owing to limited markets and the relatively low prices of the competitive products.

MICROBIAL TRANSFORMATION OF STEROIDS AND STEROLS

In the last 30 years, literally thousands of steroid derivatives have been synthesized and tested for endocrine and other activities. Many of them were found to be active as anti-inflammatory and progestational agents and also as sedatives and anabolic and antitumor substances (Sebek and Perlman, 1979).

The value of hormonal and synthetic hormone pharmaceutical preparations in 1975 was reported to be 689 million dollars (see Table 1.7). Within that grouping the value of corticoids was 258 $\times 10^6$ dollars, estrogens 84.9 $\times 10^6$ dollars, and progestogens 159.6 $\times 10^6$ dollars.

To date, most of the research on microbial transformation has been directed to screening microorganisms for transformation of steroids and other types of reactions. The role that applied genetics has played in this research is difficult to assess. There is no doubt that some strain improvement experiments have been conducted by classical mutation and selection procedures, but actual hard data on economic impact have been somewhat difficult to obtain. An important consequence of the above type of screening program has been the determination of types of chemical reactions that can be carried out by microorganisms. Kieslich (1976) points out several reasons why the chemist would like to use a microbial system for organic preparations:

1. Specificity of microbial attack is possible; that is, microbial reactions can be used to attack positions in the molecule that are not affected by chemical methods.
2. Oxygen function of other substituents can be introduced stereospecifically or altered with the possible formation of optically active centers.
3. Reactions can be combined into a single fermentation step.
4. Mild conditions can be employed for compounds that are sensitive to heat, acids, and bases.

Two issues commonly arise. The first concerns the structure of the molecule of interest and its active centers; the second concerns the types of reactions to be conducted on the molecule. Simply put, the question is whether a microorganism can be identified to transform A → B. To date,

TABLE 1.7 Value of Pharmaceutical Preparations in 1975

Hormones and Synthetic Hormones	*Value (Millions of Dollars)*
ACTH	$ 3.5
Anabolic agents	5.8
Androgens	10.5
Corticoids	258.4
Estrogens	84.9
Insulin and antidiabetic agents	130.0
Progestogens	159.6
Sex hormone combinations	7.7
Thyroid and antithyroid preparations	18.4
Other	$ 10.4
Total	$689.2

From SRI, *Chemical Economics Handbook*, 566.1000A (1977).

there have been so many studies with so many substrates that classification schemes are required. Kieslich (1976) has classified possible microbial transformations with regards to structural features as follows:

1. Alicyclic compounds
2. Terpenoids
3. Aromatic compounds
4. O-Heterocyclic compounds
5. N-Heterocyclic compounds
6. Alkaloids
7. Di- and tri-nitrogen compounds
8. S,O,S-S-N and other heterocyclic compounds
9. Carbohydrates

What is the significance of this classification? In the first place, it allows one to evaluate and prepare a chemical transformation scheme; and tabulation clearly shows that most of the tools needed by organic chemists can be found in microbial systems. Second, many of the specific reaction products are the results of classical mutation studies of degradation pathways. The significant application of this knowledge is yet to come. Given the new genetic tools available, one has to assume that specific enzymes for a given organic transformation can be isolated and made available and that their activity in a variety of biological systems can be regulated and controlled.

In 1970, 430 different fungi, 700 bacteria, and 140 yeasts were used for conducting microbial transformations of cyclic nonsteroids (Kieslich, 1976). The economic significance of this type of research investigation is illustrated by the fact that the microbial transformation of the 11 α-hydroxylation of progesterone brought down the 1949 price of cortisone from \$200/g to \$6/g, which additional improvements reduced to less than \$1/g.

Microbial processes played a significant role in making specific steroids or precursors for further chemical conversion. The significant finding, according to Sebek and Perlman (1979), was that the critical C-11 could be oxidized by microbial preparations. For example, progesterone was shown to be oxidized to 11 α-hydroxyprogesterone initially by *Rhizopus arrhizus* and later by *Rhizopus nigricans* in almost quantitative yields (>85%). Other organisms were then reported to carry out other oxidations. For example, Perlman et al. (1952) reported that *Streptomyces argenteolus* oxidized progesterone to 16-α-hydroxyprogesterone. This reaction became important when 16-α-hydroxy-9-α-fluoroprednisolone was demonstrated to be an excellent anti-inflammatory agent.

Other screening programs were undertaken, and many additional disclosures of hydroxylations have resulted. In time, all available carbon atoms were found to be hydroxylated by microorganisms, a result confirming the versatility of the microbial world.

RIBONUCLEOTIDES

Disodium 5′-inosinate (IMP) and disodium 5′-guanylate (GMP) are flavor enhancers that can be used with foods. These ribonucleotides are produced commercially in Japan and South Korea, but neither is manufactured in the United States. Japanese production of these nucleotides is in the range of 3000–5000 metric tons per year (SRI, 1979, 583.6001Q).

IMP and GMP can be manufactured by enzymatic degradation of RNA. Yeast is the usual source of RNA, and the yeast is degraded enzymatically to 5′-phosphodiesterase. Alternatively, specific fermentation processes combined with chemical modification can be employed. Inosine can be produced via fermentation, for example, by employing mutant strains of *Bacillus subtilis*. The inosine is then phosphorylated to 5′-IMP. For production of 5′-GMP, the fermentation product 5-amino-4-imidazole-carboxamide riboside (AICAR) can be chemically converted to 5′-GMP. Other methods are possible, including chemical conversion of guanosine, enzymatic conversion from 5′-XMP and direct fermentation from glucose.

It is with regards to the fermentation processes that applied genetics has had a significant impact. The field really began in the 1950's, when three scientific and industrial developments were unfolding as follows:

1. Glutamic acid fermentation was successfully accomplished.
2. Knowledge of the biosynthesis of nucleic acids, regulation, and enzymatic degradation was accumulating rapidly.
3. Useful analytical methods for nucleosides, nucleotides, and nucleic acids were becoming established.

Thus in the 1950's, mutation research was initiated for the industrial production of primary metabolites with the following properties:

- high productivity of metabolites,
- lack of specific enzymes in the biosynthetic pathway,
- release or deregulation of metabolic pathway, and
- altered membrane permeability.

In 1960, the manufacturing of 5′-IMP and 5′-GMP by enzymatic hydrolysis of ribonucleic acid (RNA) was begun in Japan, and several years later a process for fermentation of 5′-IMP and 5′-GMP was also established in Japan.

The microbial production of nucleosides and nucleotides has been the subject of numerous reviews, most recently by Nakao (1979), Demain (1978), and Hirose et al. (1979). The role that applied genetics has had is interesting from a historical point of view. For example, the biosynthetic pathways of purine nucleotides were first established by scientists such as Buchanan in the 1950's (Buchanan and Hartman, 1969, 1970). Further, mutants of microorganisms were being isolated and characterized at about the same time. For example, Magasanik (1957) began classifying mutants into several classes according to their requirements for purine bases and intermediates accumulated by them. As summarized by Nakao (1979), Magasanik's classification was as follows:

Class 1:	Mutants requiring for their growth either adenine, hydroxanthine, xanthine, or guanine.
Class 1b:	Mutants accumulating aminoimidazole riboside (AIR).
Class 1c:	Mutants accumulating 5-amino-4-imidazole-carboxamide riboside (AICAR).
Class 2:	Mutants lacking 5′-IMP dehydrogenase and requiring xanthine or guanine for growth.
Class 3:	Mutants lacking 5′-XMP aminase, requiring guanine for growth and accumulating xanthosine in the culture medium.
Class 4:	Mutants showing specific requirements for adenine. These were divided into two groups, one for mutants lacking adenylosuccinate (SAMP) lyase and accumulating 5-amino-4-imidazole-N-succinocarboxamide riboside (SAICAR) and another for mutants lacking SAMP synthetase and accumulating inosine.

The industrially important strains used for the fermentative production of nucleic acid related substances have been isolated and improved on from various bacterial species on the basis of the above findings.

The key to the application of these mutants was the research that led to the understanding of the complex and fine controls of purine biosynthesis pathways. Mutational approaches were used to overcome these fine control mechanisms, which are described in detail by Gots (1971).

Bacillus subtilis is an organism with which overproduction of purine nucleotides has been pursued by elimination of biosynthetic control mechanisms. Two types of mutants are required for maximizing bypass of the control mechanisms. In the first, auxotrophic mutants are obtained and fed growth-limiting concentrations of the end product requirements, so that the intracellular levels of feedback inhibitors and repressors can be lowered. In the second type, constitutive mutants that are no longer subject to feedback repression or desensitized mutants are selected by resistance to toxic purine analogs. The reasons for these two types of mutants is indicated in the detailed studies reported by Ishio and Shiro (1973), namely, that two general control mechanisms are responsible for regulation of purine biosynthesis:

- feedback inhibition, and
- enzyme repression.

As summarized by Hirose et al. (1979), phosphoribosyl pyrophosphate amidotransferase, the first key enzyme in the purine nucleotide biosynthetic pathway, was competitively inhibited by AMP and ADP, while the inhibition by guanine derivatives was weak. Therefore adenine derivatives are regulators of phosphoribosyl pyrophosphate amidotransferase activity. With adenylosuccinate synthetase, the first enzyme of the AMP-specific pathway, AMP and ADP inhibited it strongly, and GMP inhibited it weakly. Reaction products adenylosuccinate and GDP were weak inhibitors. The last enzyme of the pathway, adenylosuccinate lyase, was also specifically and competitively inhibited by AMP and ATP, but by neither GMP nor IMP. Similarly, IMP dehydrogenase, the first enzyme of the GMP-specific pathway, was strongly inhibited by the end product, GMP, and by the reaction product, SMP, but was negligibly inhibited by AMP and ADP. GMP inhibition was found to be a competitive inhibitor with respect to the substrate IMP. The last enzyme of the pathway, GMP synthetase, was slightly inhibited by GMP and AMP.

In studying repression of enzyme formation by purine derivatives, it was established that three enzymes of the common pathway, phosphoribosyl pyrophosphate amidotransferase, adenylosuccinate lyase, and IMP transformylase, were repressed by the addition of the end-product derivatives such as adenosine and guanosine and only partially repressed by the intermediate derivatives such as inosine and xanthosine.

Adenylosuccinate synthetase of the AMP-specific pathway was found to be specifically repressed by adenosine, and IMP dehydrogenase of the GMP-specific pathway was found to be specifically repressed by guanosine. Xanthosine and inosine had no repressive effect on either enzyme.

The important question here is how genetics has had an impact on the industrial production of flavor enhancers. The impact has been greatest for the Japanese fermentation industry; and, for the most part, the application of genetic approaches had been actively pursued by both Japanese industrial and academic scientists. Detailed summaries of the findings have been presented and thoroughly reviewed by Demain (1978) and Hirose et al. (1979). Suffice it to say that the first genetic manipulations resulted in the development of strains capable of excreting between 9 and 12 g of inosine per liter of culture (Momose and Shiio, 1969). Subsequent manipulations increased the amount to 15 g of inosine per liter, and more recent studies report concentrations of 23 g of inosine per liter (Sasajima et al., 1970). Yields of up to 25% (g inosine/g glucose) have also been achieved (Hirose et al., 1979).

GUANOSINE

Several investigators have established that, in order to develop guanosine-producing microbial strains, the following regulatory mechanisms in purine biosynthesis would have to be addressed (Hirose et al., 1979):

1. adenylosuccinate synthetase deficiency;
2. GMP reductase deficiency;
3. lack of, or deficiency in, purine nucleoside hydrolyzing activity;
4. freedom of enzymes of GMP biosynthetic pathway, especially phosphorylribosyl pyrophosphate amidotransferase, IMP dehydrogenase, and GMP synthetase, from regulatory controls.

Guanosine-producing strains from *Bacillus subtilis, Bacillus pumilus, Bacillus licheniformis, Corynebacterium petrophilum, Corynebacterium guanofaciens,* and *Streptomyces griseus* have been reported that address one or more of the points listed above. Guanosine biosynthesis in *B. subtilis* has been reported on in detail by Hirose et al. (1979) and Demain (1978), and it has been established that appreciable accumulation of guanosine is dependent upon:

1. adenine auxotrophy to eliminate AMP feedback inhibition of phosphorylribosyl pyrophosphate amidotransferase;
2. removal of GMP reductase; and
3. bypassing of the inhibitory effect of GMP on the activity of IMP dehydrogenase.

Demain summarized the Japanese work in this area and observed that a mutant strain of *B. subtilis* was developed that produced 9 g of guanosine, 3 g of xanthosine, and 3 g of inosine per liter. More recent reports from Japanese scientists list guanosine concentrations of 16 g/l with a 20% yield from glucose with a newly developed strain of *B. subtilis* MG-1 (Hirose et al., 1979). This strain requires adenine and histidine, is GMP-reductase negative, and is resistant to methionine sulfoxide, psicofuranine, and decoyinine. This strain was derived from an inosine- and guanosine-producing strain of *B. subtilis,* which was a histine auxotroph and deficient in adenylosuccinate synthetase, GMP reductase, and nucleosidase.

Production of other nucleic acid-related products by fermentation procedures is actively being pursued by Japanese industrial and academic scientists. Products being pursued, as summarized by Hirose et al. (1979), are:

Product	*Accumulation*	*Microorganism*
ATP	1.5 g/l	*Brevibacterium ammoniagenes*
CDP-choline	35 μmoles/l	*Saccharomyces carlsbergensis*
FAD	1 g/l	*Sarcina lutea*
NAD	29 g/l	*Brevibacterium ammoniagenes*
Orotic acid	20 g/l	*Arthrobacter paraffineous*
CoA	2.9 g/l	*Brevibacterium ammoniagenes*
Cyclic AMP	4 μmoles/l	*Brevibacterium liquefaciens*

All of the above have limited use as drugs in Japan. The future markets for such products are not clear at this time.

In summary, the use of mutants designed to overlay the regulatory control that a microbial cell employs to balance synthesis of intermediates has had a major impact in the development of flavor enhancers for the Japanese fermentation/food industry.

INDUSTRIAL GASES

Carbon Dioxide

Most carbon dioxide is produced as a by-product of industrial manufacture of ammonia. The source of the hydrogen to reduce the nitrogen in air to ammonia is via steam reforming:

$$CH_4 + 2H_2O \rightarrow 4H_2 + CO_2$$

In addition, carbon dioxide is produced as a by-product in fermentations. About 1% of the U.S. industrial carbon dioxide was from fermentation in 1978. Production and removal of carbon dioxide can be significant technical problems in many types of fermentation processes.

Over one third of carbon dioxide used in the United States is used in refrigeration (solid CO_2, or dry ice). Other important uses are carbonating water for beverages, chemical synthesis, and maintenance of inert atmospheres. The major cost of carbon dioxide is in the postproduction steps: compression, purification, liquefaction, and, in the case of dry ice, snowing. The price of CO_2 is a function of locality. Because of the great abundance of carbon dioxide, it is unlikely to be more than a minor by-product of fermentation where local demand is sufficiently large. Table 1.8 summarizes production in 1978.

Ammonia

Ammonia is made industrially by the reaction of methane, air, and water. The methane and water react to form hydrogen (see above); then the hydrogen and nitrogen react:

$$N_4 + 3H_2 \rightarrow 2NH_3$$

About 15 billion metric tons of ammonia were produced this way in 1978.

Nitrogen can be "fixed" to ammonia by the nitrogenase enzymes using reducing potential (electrons) metabolically derived from the tricarboxylic acid cycle. It is estimated that 175 billion metric tons of ammonia are fixed by cells per year. One square meter of land planted with nodulated legumes (such as soybeans) can fix 1–30 g of nitrogen (Metzler, 1978).

Several prokaryotes are known to fix nitrogen; the most important of these are *Rhizobium* and blue-green algae. The infected roots of legumes

TABLE 1.8 Industrial Gases

Product	*Formula*	*Hydrogen Deficiency*	*Number of Chiral Centers*	*1978 Price (per kg)*	*Present Source*	*1978 Domestic Production (metric tons)*	*Potential for Applications of Biotechnology*
Ammonia	NH_3	0	0	\$0.13	Hydrogen + nitrogen	15,250	Enzyme reduction of N_2
Carbon dioxide	CO_2	2	0	—	CH_4 + H_2O	2,060,000	By-product of ethanol and butanol/acetone fermentation

develop nodules in which the *Rhizobium* bacteria degenerate into "bacteroids." Blue-green algae are also quantitatively significant nitrogen fixers; a square meter of rice paddy can contain algae that fix from 2 to 10 g of nitrogen per year.

The full 6-electron reduction is performed by a complex of enzymes and iron/molybdenum/sulfur groups that are inactivated by oxygen and require 12 ATP molecules to be hydrolyzed per N_2 reduced. The enzyme system is remarkable in the amount of energy it consumes and in the total lack of detectable partial reduction products. It is the subject of intense scientific inquiry (Walsh, 1979).

Microbial production of ammonia from nitrogen appears to be uncompetitive. Aside from the difficulties associated with the enzyme's sensitivity to oxygen and our near total lack of understanding of its mechanism, it consumes the equivalent of the energy of 4 kg of sugar to make 1 kg of ammonia. Since ammonia at present costs $0.13/kg and sugar costs $0.22/kg, the chemical Haber process does not appear to be threatened.

AMINO ACIDS

Amino acids are the building blocks of proteins. Twenty have the distinction of being incorporated into proteins manufactured on the ribosomes; others serve specialized structural roles (hydroxyproline), are important metabolic intermediates (homoserine), or are hormones and neurotransmitters (dihydroxyphenylalanine (DOPA)).

Many bacteria can synthesize all the amino acids they need; higher animals cannot. In general, the more complicated amino acids are required in the mammalian diet. In fact, if one of the essential amino acids is missing, protein synthesis halts, and significant quantities of the other 19 will be wasted. In humans, the essential amino acids are threonine, methionine, lysine, valine, isoleucine, leucine, arginine, histidine, phenylalanine, tyrosine, and tryptophan. The requirements vary in other animal species.

The alpha carbon of amino acids is chiral; it can exist in the D or L form (see Fig. 1.1). The vital form is L. In some cases, cells can convert the D to L, but it is the L that is used in protein synthesis.

All of the amino acids are used in research and in nutritional preparations. Most are used one way or another in the preparation of pharmaceuticals. Three are used in large quantities for two purposes: glutamate is a flavor enhancer (MSG); lysine and methionine are animal feed additives.

Amino acids are primary metabolites and are prime candidates for production via biotechnology. About half of the amino acids sold are now produced by fermentation. The leading country in acid fermentation processes is Japan. The fermentation industry in Japan now produces 14 amino acids at least partly by biotechnology processes.

Numerous basic studies have been conducted on the regulation of enzymes controlling the biosynthesis of amino acids in microorganisms. A number of these studies employ auxotrophic and regulatory mutants. Although many studies have been conducted with the bacterium *Escherichia coli,* Clarke (1979) has summarized studies that allow comparisons to be made with other microorganisms. In an excellent review, she discusses the theories of enzyme regulation, as well as gene expression. The basic concepts are then discussed with respect to the biosynthesis of aromatic amino acids, special attention being placed on tryptophan biosynthesis. Other, more practically oriented reports have also been summarized. For example, Hirose and Okada (1979) and Hirose et al. (1978) summarize the extensive studies done, mainly in Japan, on the production of amino acids by a variety of strains of bacteria. The successful use of mutant strains, that is, auxotrophic mutants and regulatory mutants, is described in great detail. Listed in Tables 1.9 and 1.10 are the mutations used for production of several amino acids. Although microbial production of all the amino acids has been studied by many investigators, glutamic acid and L-lysine are the ones produced in significant quantities by fermentation processes.

Table 1.11 gives production data for the amino acids. Production data

TABLE 1.9 Amino Acids Production by Auxotrophic Mutants

Amino Acid	*Requirement*	*Microorganism*
L-Citrulline	Arginine	*Bacillus subtilis* *Corynebacterium glutamicum*
L-Glutamic acid	Oleic acid Glycerol	*Brevibacterium thiogenitalis* *Corynebacterium alkanolyticum*
L-Homoserine	Threonine	*Corynebacterium glutamicum*
L-Leucine	Phenylalanine Histidine	*Corynebacterium glutamicum*
L-Lysine	Homoserine Threonine	*Corynebacterium glutamicum* *Brevibacterium flavum*
L-Ornithine	Citrulline	*Arthrobacter citreus*
L-Phenylalanine	Tyrosine	*Corynebacterium glutamicum*
L-Proline	Isoleucine Histidine Serine	*Brevibacterium flavum* *Brevibacterium* sp. *Kurthia catenoforma*
L-Threonine	Diamino-pimelic acid, methionine	*Escherichia coli*
	Methionine, valine	*Escherichia coli*
L-Tyrosine	Phenylalanine, purine	*Corynebacterium glutamicum*
L-Valine	Isoleucine or leucine	*Corynebacterium glutamicum*

From Hirose et al. (1978).

TABLE 1.10 Amino Acids Production by Regulatory Mutants

Amino Acid	*Selection*	*Microorganism*
L-Arginine	Canavanine	*Escherichia coli*
	Arginine-hydroxamate	*Bacillus subtilis*
	D-Serine, D-arginine, arginine-hydroxamate	*Corynebacterium glutamicum*
	2-Thiazolealanine	*Brevibacterium flavum*
	Substrate Specificity	*Serratia marcescens*
L-Histidine	2-Thiazolealanine, sulfaguanidine, α-amino-β-hydroxybutyric acid, ethionine, 2-amino benzothiazole	*Brevibacterium flavum*
	1-,2-,4-Triazolealanine 6-mercaptoguanine, 8-azaguanine, 4-thiouracil, 6-mercaptopurine, 5-methyltryptophan	*Corynebacterium glutamicum*
	Histidase−, 1-,2-,4-triazolealanine, 2-methylhistidine, 2-thiazolealanine	*Serratia marcescens*
	Revertant of histidine−	*Streptomyces*
L-Isoleucine	α-amino-β-hydroxyvaleric acid, O-methyl-threonine	*Brevibacterium flavum*
	2-aminoethyl-L-cysteine	*Brevibacterium flavum*
	4-azaleucine	*Bacillus subtilis*
L-Leucine	2-Thiazolealanine, methionine−	*Brevibacterium lactofermentum*
	isoleucine− α-Amino-butyric acid, revertant of isoleucine	*Serratia marcescens*
L-Lysine	2-Aminoethyl-L-cysteine	*Brevibacterium flavum*
	α-Chlorocaprolactam	*Brevibacterium flavum*
L-Phenylalanine	M-Fluorophenylalanine	*Brevibacterium flavum*
	Tyrosine−, *p*-fluorophenylalanine, *p*-amino-phenylalanine,	*Corynebacterium glutamicum*
L-Proline	Sulfaguanidine	*Brevibacterium flavum*
L-Serine	O-Metyl-serine, α-metyl-serine, *iso*-serine	*Corynebacterium glutamicum*
L-Threonine	α-Amino-β-hydroxy-valeric acid, methionine−	*Brevibacterium flavum*

(Continued)

TABLE 1.10 (Continued)

Amino Acid	*Selection*	*Microorganism*
L-Tryptophan	Phenylalanine-, 5-fluoro-tryptophan, 3-fluoro-tryptophan	*Brevibacterium flavum*
	5-Methyl-tryptophan, tyrosine–hydroxamate, 6-fluoro-tryptophan, 4-methyl-tryptophan, *p*-ethylphenylalanine, *p*-amino-phenylalanine tyrosine-hydroxamate, phenylalanine-hydroxamate	*Corynebacterium glutamicum*
L-Tyrosine	*m*-Fluoro-phenylalanine	*Brevibacterium flavum*
	3-Amino-tyrosine, *p*-amino-phenylalanine, *p*-ethyl-phenylalanine, tyrosine-hydroxamate, phenyl-alanine–	*Corynebacterium glutamicum*
L-Valine	2-Thiazolealanine	*Brevibacterium flavum*

From Hirose et al. (1978).

are given for Japan because in most cases U.S. domestic production is very small.

Glutamic Acid

MSG is used in large amounts as a flavor enhancer, particularly in oriental cooking. In 1978, the United States manufactured about 18,000 metric tons and imported about 11,000 metric tons. Of this amount, 97% was consumed by the food industry. The entire U.S. production is accounted for by the Stauffer fermentation plant in San Jose, California.

The microbes used in glutamic acid fermentation (*Corynebacterium glutamicum, C. lileum,* and *Brevibacterium flavum*) produce glutamic acid in 60% of theoretical yield. Thus there is some but not great potential for the use of applied genetics to improve the yield. In addition, many of the genetic approaches have been thoroughly investigated by industrial scientists.

Lysine

As was explained above, the lack of a single amino acid can retard protein synthesis, and therefore growth, in a mammal. The limiting amino acid is

TABLE 1.11 Amino Acids

Amino Acid	*Formula*	*Hydrogen Deficiency*	*Number of Chiral Centers*	*March 1980 Price (per kg) (pure L)*	*Present Source*	*Value-Added Ratio*	*1978 Domestic Production (metric tons)*	*Potential for Application of Biotechnology* (de novo *synthesis or bioconversion; organisms and enzymes)*
Alanine	$C_3H_7NO_2$	1	1	$80	Hydrolysis of protein chemical synthesis	—	10–50 (J)	
Arginine	$C_6H_{14}N_4O_2$	2	1	$28	Gelatin hydrolysis	—	200–300 (J)	Fermentation in Japan
Asparagine	$C_4H_8N_2O_3$	2	1	$50	Extraction		10–50 (J)	
Aspartic acid	$C_4H_7NO_4$	2	1	$12	Bioconversion of fumaric acid	—	500–1000 (J)	Bioconversion
Citrulline	$C_6H_{13}N_3O_3$	2	1	$250			10–90 (J)	Fermentation in Japan
Cysteine	$C_3H_7NO_2S$	1	1	$50	Extraction		100–200 (J)	
Cystine	$C_6H_{12}N_2O_4S_2$	—	—	$60	Extraction		100–200 (J)	
DOPA (dihydrophenylalanine)	$C_9H_{11}NO_4$	5	1	$750	Chemical		80–200 (J)	
Glutamic acid	$C_5H_9NO_4$	2	1	$4	Fermentation		18,000–100,000 (J)	De novo: *Micrococcus glutamicus*
Glutamine	$C_5H_{10}N_2O_3$	2	1	$55	Extraction		200–300 (J)	Fermentation in Japan
Histidine	$C_6H_9N_3O_2$	4	1	$160			100–200	Fermentation in Japan
Hydroxyproline	$C_5H_9NO_3$	2	2	$280	Extraction from collagen		10–50	
Isoleucine	$C_6H_{13}NO_2$	1	2	$350	Extraction		10–50 (J)	Fermentation in Japan
Leucine	$C_6H_{13}NO_2$	1	1	$55			50–100 (J)	Fermentation in Japan
Lysine	$C_6H_{14}N_2O_2$	1	1	$350	Fermentation (80%) Chemical (20%)		10,000 (J)	(80% by fermentation) de novo: *Corynebacterium glutamicum* and *Brevibacterium flavum*

TABLE 1.11 (Continued)

Amino Acid	Formula	Hydrogen Deficiency	Number of Chiral Centers	March 1980 Price (per kg) (pure L)	Present Source	Value-Added Ratio	1978 Domestic Production (metric tons)	Potential for Application of Biotechnology (de novo synthesis or bioconversion; organisms and enzymes)
Methionine	$C_5H_{11}NO_2S$	1	1	$265	Chemical acrulein	2.7	17,000 (D, L) 20,000 (D, L) (J)	
Ornithine	$C_5H_{12}N_2O_2$	1	1	$60			10–50 (J)	Fermentation in Japan
Phenylalanine	$C_9H_{11}NO_2$	5	1	$55	Chemical from benzaldehyde	20	50–100 (J)	Fermentation in Japan
Proline	$C_5H_9NO_2$	2	1	$125	Hydrolysis of gelatin		10–50 (J)	Fermentation in Japan
Serine	$C_3H_7NO_3$	1	1	$320			10–50 (J)	Bioconversion in Japan
Threonine	$C_4H_9NO_3$	1	2	$150			50–100 (J)	Fermentation in Japan
Tryptophan	$C_{11}H_{12}N_2O_2$	7	1	$110	Chemical from iadole	1.02	55 (J)	
Tryosine	$C_9H_{11}NO_3$	5	1	$13	Extraction		50–100 (J)	
Valine	$C_5H_{11}NO_2$	1	1	$60			50–100 (J)	Fermentation in Japan

a function of the animal and its feed. The major source of animal feed in the United States is soybean. The limiting amino acid for feeding swine is lysine; the limiting amino acid for feeding poultry is methionine.

L-Lysine is produced both by fermentation and by chemical synthesis. Fermentation has been gradually replacing chemical synthesis; of lysine produced worldwide is made by microbes. Currently, no lysine is produced in the United States by U.S. manufacturers. Imports were about seven thousand metric tons in 1979, mostly from Japan and South Korea.

Because of increased poultry demand, world demand for lysine is climbing. Eurolysine is spending 27 million dollars to double its production capacity in Amiens, France, to ten thousand metric tons. The Asian and Middle Eastern markets are estimated to increase to three thousand metric tons in 1985. Some bacteria produce lysine at over 90% of theoretical efficiency. Little genetic improvement is likely in conversion yield; however, significant improvement can be made in the rate and final concentration.

Recently, Shvinka et al. (1980) reported on the yield regulation of lysine biosynthesis in *Brevibacterium flavum*. Experimental data are presented showing how the increase in yield (gram of product per gram of substrate) can be controlled by enzyme induction and elimination of by-product synthesis. Consequently, they report close to theoretical yields for lysine production and concentrations of 78 g/l lysine-HCl with *B. flavum* in a molasses medium in pilot plant fermentors.

A summary of the cost differences for chemical and fermentation production is shown in Table 1.12.

Methionine

Methionine is the limiting amino acid for feeding soybean meal to poultry. However, there is an important difference from lysine: the fowl has an enzyme that can convert D-methionine to L-methionine, so a D,L mixture can be used. D,L-Methionine is produced by chemical synthesis. The process uses acrolein, methyl mercaptan, and hydrogen cyanide as raw materials.

Domestic consumption of D,L-methionine is 28,000 metric tons, of which about 17,000 metric tons is produced in the United States. The major producers are Deguss, DuPont, and Monsanto.

Microbial production of methionine appears to be unlikely for the following reasons:

1. Pure L-methionine is not required for poultry feed addition.
2. Chemical production is cheap.
3. Microbial production is difficult because fixation of sulfur is energy-intensive (thus requiring significant oxidation of carbohydrates) and the metabolic pathway to methionine is long and complicated.

Other Amino Acids

Figure 1.5 shows a simplified diagram of the central metabolic pathway with the branches leading to amino acids. The most likely use for recombinant DNA to increase yields of an amino acid would be to clone the gene(s) coding for the enzymes needed to synthesize the amino acid. Each pathway of Fig. 1.5 has a number of enzymes; thus difficulties proportional to the number of enzymes in a pathway are expected.

The aromatic amino acids represent a special problem. Figure 1.6 shows the feedback inhibition pathways for aromatic amino acid production. Because aromatization is metabolically expensive, the pathways are tightly controlled. Further controls exist at later steps. Genetic manipulation to improve yields of aromatic acids will require manipulation of these complicated control mechanisms.

Aspartic acid may be required in the coming decade in large amounts for the production of the sweetener aspartame (L-aspartyl-L-phenylalanine methyl ester). It is presently made by a very efficient bioconversion of fumaric acid.

Phenylalanine also will be required for the production of aspartame. The pure L-enantiomer will be required. Figure 1.7 shows the two ways in which L-phenylalanine can be made: by chemical production from benzaldehyde and acetylamino acetic acid or by fermentation from sucrose. The difficulty in making L-phenylalanine chemically is that an ordinary reduction catalyst for reaction produces D- and L-phenylalanine in equal amounts. The two optical isomers must then be separated (reaction 3). However, an optically active reduction catalyst can give a higher yield of one optical isomer than the other, in some cases over 90%. Compound III, because it is highly functionalized around its reduction center, is a good candidate to make the L-isomer exclusively. Monsanto reported in 1977 that the invention of a rhodium reduction catalyst for reaction yields L- and D-phenylalanine in a ratio of 97:3. Thus the presence of a phenyl group that is energetically expensive to produce microbially and the availability of an optically active catalyst to produce the L-enantiomer exclusively in a chemical process make it unlikely that fermentative production of phenylalanine will be competitive.

Proline is an amino acid for which improvements in production may be brought about by genetic engineering. Bethesda Research Laboratory announced in April 1980 that it had cloned one enzyme necessary to increase proline production.

Tryptophan is the third limiting amino acid for most feed uses (see Fig. 1.6). L-Tryptophan was the first amino acid that was shown to be essential for animal nutrition. However, the need of this amino acid is considerably less than the demand for the other essential amino acids. It is not used today as a feed additive; its price is prohibitive for this. However, should its price drop substantially, a sizable market could open up in this area. In 1974, the production of L-tryptophan in Japan amounted to between 10

TABLE 1.12 Summary of Recent Estimates of Primary U.S. Cost Factors in the Products of L-Lysine Monohydrochloride by Fermentation and Chemical Synthesis

	Cost Factors in Production of 98% L-Lysine Monohydrochloride					
	By Fermentation[1]			*By Chemical Synthesis*[2]		
	Requirement (units per unit product)	*Estimated 1976 Cost per Unit Product*		*Requirement (units per unit product)*	*Estimated 1976 Cost per Unit Product*	
		Cents per Pound	*Cents per Kilogram*		*Cents per Pound*	*Cents per Kilogram*
Total Labor[3]		8	18		9	20
Materials						
Molasses	44	7	16			
Soybean meal, hydrolyzed	0.462	4	9			
Cyclohexanol				0.595	17	37
Anhydrous ammonia				0.645	6	14
Other chemicals[4]		7	15		4	10
Nutrients and solvents					4	8
Packaging, operating, and maintenance materials		10	22		9	21
Total materials		28	62		4	90
Total utilities[5]		6	12		7	16
Total direct operating cost		42	92		56	126
Plant overhead, taxes, and insurance		10	21		10	21

TABLE 1.12 (Continued)

Cost Factors in Production of 98% L-Lysine Monohydrochloride

	By Fermentation[1]			*By Chemical Synthesis*[2]		
		Estimated 1976 Cost per Unit Product			*Estimated 1976 Cost per Unit Product*	
	Requirement (units per unit product)	*Cents per Pound*	*Cents per Kilogram*	*Requirement (units per unit product)*	*Cents per Pound*	*Cents per Kilogram*
Total cash cost		52	113		66	147
Depreciation[6]		16	35		13	28
Interest on working capital		1	3		1	3
Total cost		69	151		80	178

From SRI (1979, 583.3401).

1. By fermentation; assumes a 23% yield on molasses.
2. By chemical synthesis; assumes a 65% yield on cyclohexanol.
3. Total labor includes operating, maintenance, and control laboratory labor.
4. Other chemicals, for both the processes of fermentation and chemical synthesis, include assumed use of HCl (36%) and ammonia (29%). For fermentation, potassium diphosphate, urea, ammonium sulfate, calcium carbonate, and magnesium sulfate are also included. For chemical synthesis, nitrosyl chloride, sulfuric acid, and a credit for ammonium sulfate by-product are also included.
5. Total utilities for both processes include cooling water, steam, process water, and electricity. For chemical synthesis, natural gas is also included.
6. Depreciation is 10% per year of fixed capital cost for a new 20 million pound per year U.S. plant built in 1975–1976 at an assumed capital cost of 38.6×10^6 for fermentation and 32.5×10^6 for chemical synthesis exclusive of land costs.

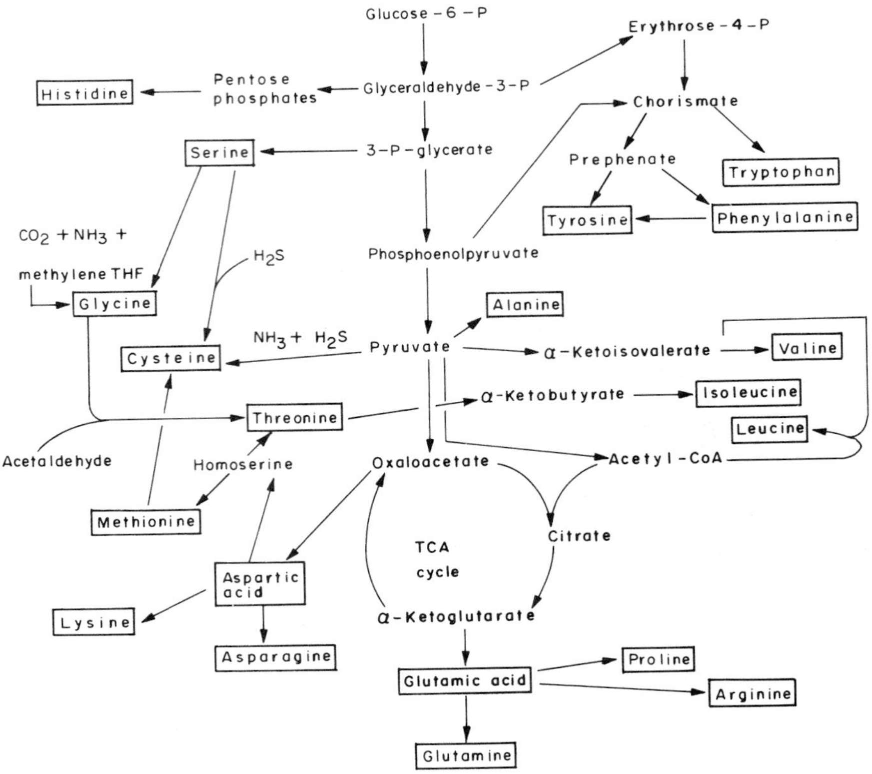

FIGURE 1.5. Biosynthesis of amino acids.

and 50 tons per year. Its major use was in the management of nitrogen metabolism in postoperative and burn patients by infusion with other amino acids and glucose.

Historically, L-tryptophan was manufactured by isolation from protein hydrolysates. Most proteins are about 1–2% L-tryptophan by weight. Obviously, the cost of this method today would be quite high; also, hydrolysis or protein often loses the optical activity in the individual amino acids. Today, however, the material is produced principally by chemical synthesis. The L-isomer can be separated by enzymatic methods. In one method, pancreatic enzymes are used to hydrolyze the methyl esters of D,L-tryptophan in aqueous solution, since they will act only on the L-form.

Unlike many of the other amino acids, L-tryptophan has not been produced in significant quantities by fermentation. Typically, one of two strategies has been used to produce the amino acid. One is the use of anthranilic acid as a precursor, and the other is the use of indole or its derivatives.

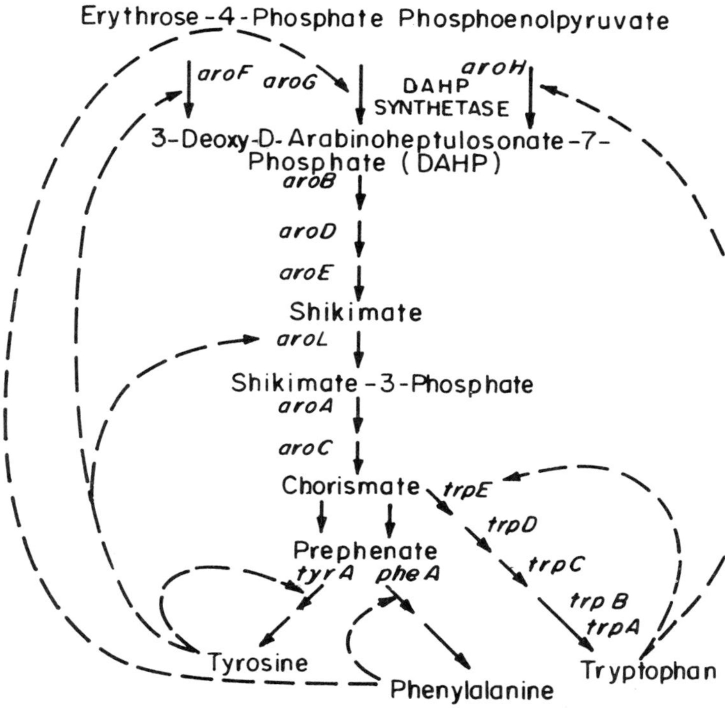

FIGURE 1.6. Outline of the pathway for the biosynthesis of the aromatic amino acids in *E. coli*. The dashed lines indicate the main sites of regulation by feedback inhibition and/or repression. The regulator gene *tyrR* is concerned with repression of DAHP synthetase (phe) and DAHP synthetase (tyr). The regulator gene *trpR* is concerned with the regulation of DAHP synthetase (trp) and the trp operon. *aroA*, EPSP (3-enolpyruvylshikimate-5-phosphate) synthetase; *aroB*, DHQ (dehydroquinate) synthetase; *aroC*, chorismate synthetase; *aroD*, DHQ dehydratase; *aroE*, dehydroshikimate reductase; *aroF*, DAHP (3-deoxy-D-arabinoheptulosonate-7-phosphate) synthetase (tyr); *aroG*, DAHP synthetase (phe); *aroH*, DAHP synthetase (trp), *pheA*, chorismate mutase-prephenate dehydratase; *tryA*, chorismate mutase-prephenate dehydrogenase (from Clarke, 1979).

The reason for utilizing these pathways is that no one had been able to obtain an organism that could accumulate tryptophan in significant amounts. Recently, however, this obstacle has been overturned; Nakayama (1976) reports the development of a strain of *C. glutamicum* that can accumulate up to 12 g/l of tryptophan using only cane sugar molasses as a carbon source.

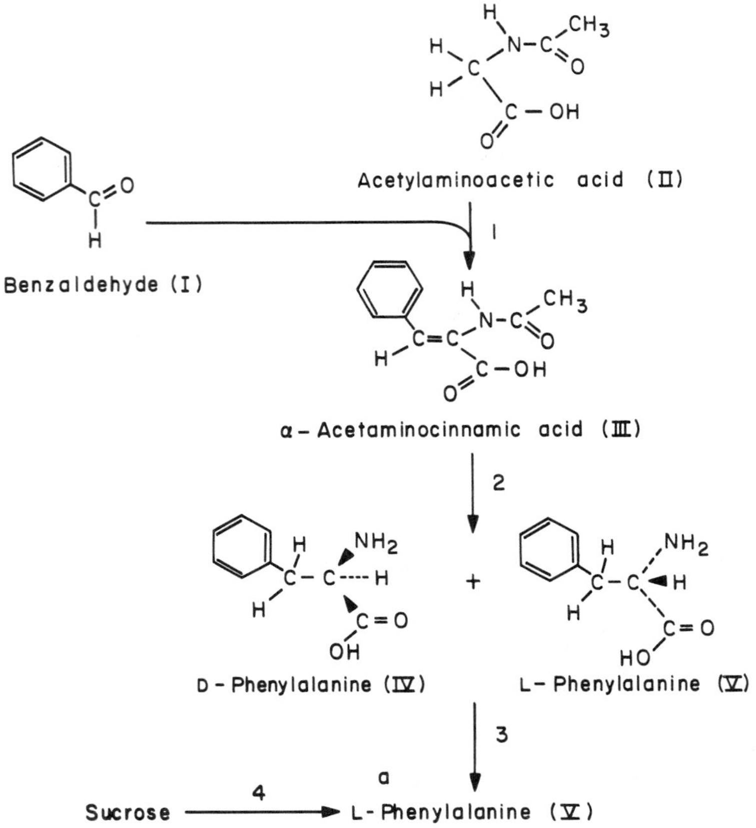

FIGURE 1.7. Chemical and fermentation production of phenylalanine.

BIOSYNTHESIS

Tryptophan, phenylalanine, and tyrosine share a common pathway originating from shikimic acid.

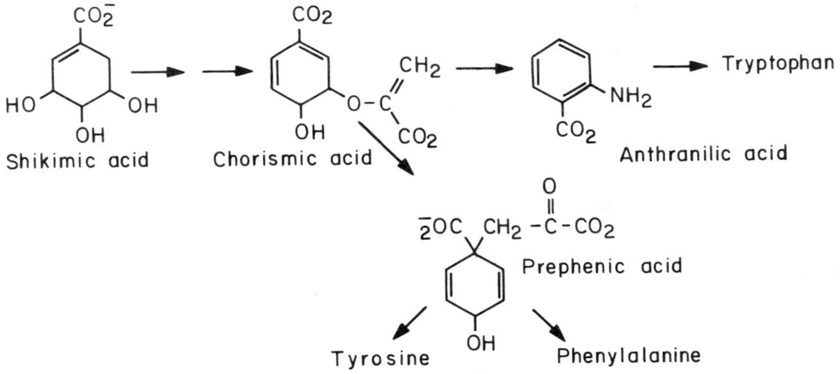

The pathway for the conversion of anthranilic acid to tryptophan is shown in Fig. 1.8. See Fig. 1.9 for further details as to the relationship between genes and enzymes of the pathway.

Tryptophan has been studied in many microorganisms and all are very similar. The genetics, however, appear to be somewhat diverse in *E. coli*, which is perhaps the most extensively studied. The operon consists of five contiguous cistrons with an operator and probably a promoter site. (See Fig. 1.10.)

The five polypeptides specified by the operon interact in the manner shown to produce three enzymes which catalyze the five steps intervening in the biosynthesis of L-tryptophan from chorismic acid. Repression of the operon requires the simultaneous presence of tryptophan—the product of an unlinked regulatory gene (*trpR*).

In looking for organisms to produce tryptophan, the following characteristics are desirable:

1. They should be unable to decompose anthranilate.
2. Synthesis of enzymes should be absent from repression by tryptophan.
3. Enzymes should not be feedback-inhibited by tryptophan.
4. They should be unable to decompose tryptophan.

For many years, *H. anomala* appeared to fulfill these requirements best. In the absence of indole or anthranilic acid, very little tryptophan is produced. With stepwise feeding of anthranilate and either glucose or ethanol as a carbon source, up to 6 g/l of tryptophan can be accumulated. However, by stepwise feedings of indole, ethanol, and anthranilate, up to ≈12 g/l of tryptophan can be accumulated. Incidentally, this is the saturation concentration of tryptophan in water.

Recently, however, it has been reported that a strain of *C. glutamicum* is unable to produce up to 12 g/l of tryptophan using glucose (from cane sugar molasses) as the sole carbon source. The strain was developed by producing a strain resistant to *phe, try,* and *tyr* analogs.

The strain 4MT-11, which showed a production of *try* inhibited by *phe* and *tyr* mutants resistant to analogs of *phe* and *tyr*, was selected and production was increased to 12 g/l. The genealogy is summarized as follows:

		try production (g/l)
KY 9456	$Phe^- Tyr^-$	0.15
↓	$5MT^r$, $TryHr^r$, $6FT^r$, $4MT^r$	
4MT-11	PFP^r	4.9
↓		
PFP-2-32	PAP^r	5.7
↓		
PAP-126-50	$TyrHr^r$	7.1
↓		
TX-49	$PheHr^r$	10.0
↓		
PX-115-97		12.0

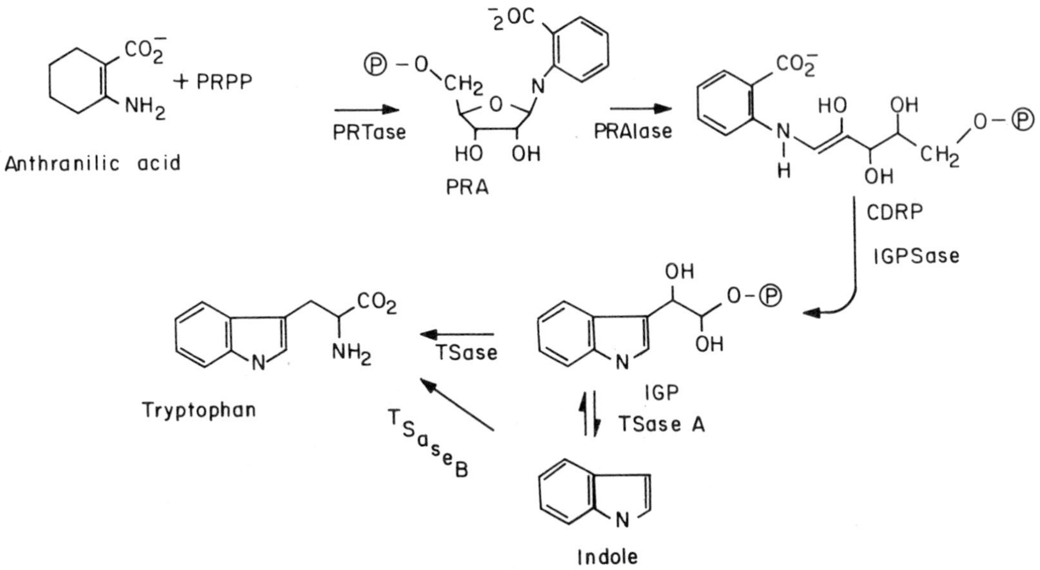

FIGURE 1.8. Pathway for the conversion of anthranilic acid to tryptophan.

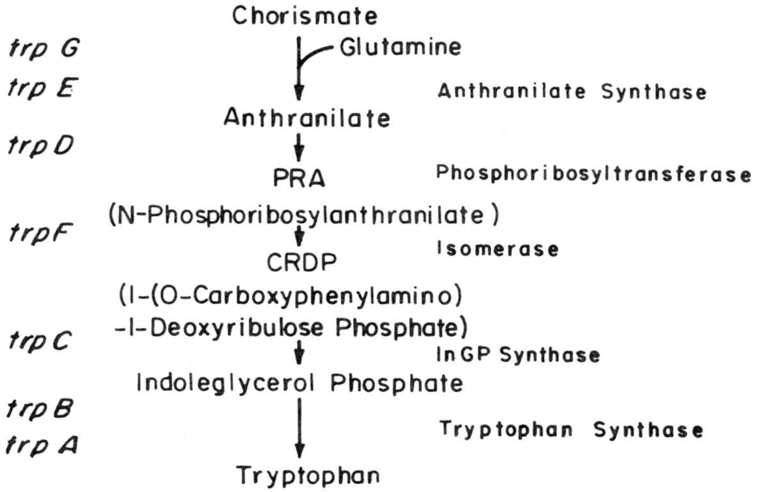

FIGURE 1.9. **Genes and enzymes of the pathway for the biosynthesis of tryptophan. In *E. coli,* the protein determined by *trpD* carries phosphoribosyltransferase activity and also the activity of the smaller subunit of anthranilate synthase determined by *trpG* in *B. subtilis* and *P. aeruginosa.* The *trpC* protein of *E. coli* carries PRA isomerase and InGP synthase activities, which are determined by *trpC* and *trpF* in *B. subtilis* and *P. aeruginosa,* respectively (from Clarke, 1979).**

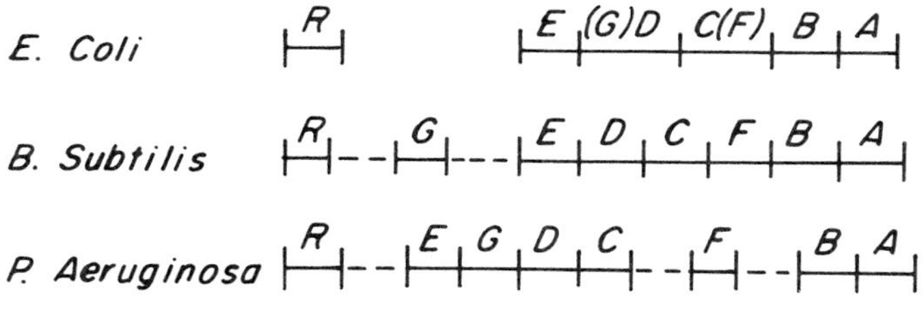

FIGURE 1.10. **Arrangements of genes for enzymes of tryptophan biosynthesis (see Figs. 1.8 and 1.9). The *trpR* regulator gene is involved in repression by tryptophan of all the enzymes in *E. coli* and *B. subtilis.* In *P. aeruginosa,* only the *trpEGDC* group of enzymes is repressed; *trpBA,* tryptophan synthase, is induced by indoleglycerol phosphate, and the *trpF* enzyme is constitutive (from Clarke, 1979).**

MONOMERS AND POLYMERS

A very large segment of the chemical industry is devoted to the production of plastics, fibers, and rubber; 4.3 million metric tons of fibers, 12 million metric tons of plastics, and 1.1 million metric tons of synthetic rubber were produced in the United States in 1978. With the exception of cellulose fibers, this entire mass was derived from petroleum. A breakdown of sales is shown in Table 1.13.

Microbial production of monomers was mentioned in the section on organic chemicals. Because the bulk of monomers are chemically simple and are presently available in high yield from petroleum, microbial production of most monomers is unlikely. The most likely exceptions are polyamides (chemically related to proteins), acrylics, and isoprene-type rubber.

The fact that present technology appears not to be replaceable by biotechnology should not lead one to the conclusion that the eventual impact on polymer production will be slight. Biopolymers represent a new way of thinking. Most of the important constituents of cells are polymers: proteins (polypeptides, from amino acid monomers), polysaccharides (from sugar monomers), and polynucleotides (from nucleotide monomers).

Cells normally assemble polymers with extreme specificity: the monomers occur in a well-defined order (for example, the AGTC sequences of DNA), and in addition, each monomer is added to the chain in a chemically specific way (for example, the head-to-tail structure of natural rubber from isoprene units). In contrast, synthetic polymers are randomly ordered (for example, the styrene and butadiene monomers of styrenebutadiene rubber are randomly ordered, the numbers reflecting only the proportions in the monomer mix), and each monomer is added to the chain in a haphazard way (for example, the nonspecificity of synthetic isoprenoid rubber). The higher quality of biological polymers is achieved by the specific order of monomers.

It appears that the ideal situation would be to imitate the biological production of polymers in all possible respects: a single biomachine to convert a raw material (for instance, sugar) into the monomer, polymerize it, then form the final product. Commercial dextrans are in fact produced by converting sucrose in polymeric glucose. Much experimental work must be done on the methods cells used to make structural polymers.

A more likely application is the development of new monomers for specialized applications. Polymer chemistry has largely been the study of modification of properties through the regulation of copolymerization reactions, addition of plasticizers, and regulation of temperature and pressure to yield a polymer with the desired properties. Conceivably, polymer chemists could modify functionalization and stereochemistry of the monomers. As with any new approach to refining a new technology, it is difficult to guess the impact now; but perhaps a valid analogy is the transition in the 1930's from exclusively cellulose-based polymers to chemical synthetic polymers, that is, at least 30 years.

PESTICIDES

Under primitive conditions, crop plants such as cereals and cabbages grow interspaced with other species as part of a complex community of plants

TABLE 1.13 Monomers and Polymers

Product	Domestic Production 1978 (thousand metric tons)	Potential for Biotechnology
	I. Plastics	
Thermosetting resins		
Epoxy	135	Fair
Polyester	544	Fair
Urea	504	Fair
Melamine	90	
Phenolic	727	Low
Thermoplastic resins		
Polyethylene		
Low density	3200	Low
High density	1890	Low
Polypropylene	1380	Low
Polystyrene	2680	Low
Polyamide, nylon type	124	Good
Polyvinyl alcohol	57	Fair
Polyvinyl chloride	2575	Low
Other vinyl resins	88	Low
	II. Fibers	
Cellulosic fibers		
Acetate	139	Fair
Rayon	269	Fair
Noncellulosic fibers		
Acrylic	327	Good
Nylon	1148	Good
Olefins	311	Low
Polyester	1710	Fair
Textile glass	418	Low
Other	7	
	III. Rubber	
Styrene-butadiene	628	Low
Polybutadiene	170	Low
Butyl	69	Low
Nitrile	33	Low
Polychlorophene	72	Low
Ethylene-propylene	78	Low
Polyisoprene	62	Good

and animals. Because of the dispersed distributions of wild plants, pests show adaptation that enable them to survive long enough or in sufficient numbers to get from host to host. They often produce large numbers of progeny or exhibit parthenogenesis (in which a female reproduces without fertilization), so that they stand a chance of survival in spite of the difficulties of finding a host on which to feed. As agriculture developed, this propensity showed that it could result in devastation. The pest that infected one plant in the field found that it had thousands of identical plants all around it; and so the ability to survive under difficult circumstances became the origin of widespread outbreaks of plant disease. Insecticides that worked by inhibiting insect respiration, such as DDT, were invented and used in large amounts. Their use was responsible for great improvements in disease control and crop protection, but the undesirable secondary effects are well known. The problem with DDT and related substances is due principally to the fact that they are concentrated as one moves higher in the food chain: birds ingest large amounts of DDT that have been absorbed by the insects that they eat. DDT becomes concentrated in the birds' bodies and causes effects such as the weakening of eggshells. Another major problem with DDT-like pesticides is the ability of the insects to gain resistance rapidly through mutation and natural selection.

Microbes could provide a range of new pesticide preparations that are harmless to man, ecologically compatible, highly effective, and selective. Microbes have been used in the past to control pests, and it is surprising that microbial pesticides have not been more systematically explored. The potential market for pesticides is indicated in Table 1.14.

The feeling of workers in the field seems to be one of enthusiasm combined with pessimism. Norris (1978) has published an excellent review on the microbial control of pest insects. The most general complaints are the unattainability of patents and the strict standards of environmental agencies more familiar with toxic pesticides than with behavior-modifying chemicals and biodegradable pesticides. Both of these questions should be addressed.

The impact of applied genetics on pesticide development can be divided into four areas:

1. natural toxins;
2. viruses;

TABLE 1.14 Pesticides

Product	*1978 Domestic Market (millions of dollars)*	*1978 Domestic Production (metric tons)*
Fungicides	$ 220	66,000
Herbicides	1,760	295,000
Plant growth regulators	223	3,300
Insecticides and rodenticides	1,040	272,000

3. bacteria, fungi, and protozoa; and
4. pheromones

Natural Toxins

Bacillus thuringiensis toxin grew in the 15-year period from 1962 to 1977 to sales of $13 million. The crystalline toxic compounds present on the outer coating of the bacterial spore destroy the larvae of a number of strains of insects. The World Health Organization (WHO) has recently shown great interest in the development of the new *B. thuringiensis* strain (H.14). It is known to be extremely lethal for the larvae of certain diptera, including some mosquitoes. WHO sees a possibility of using it to control the black fly, *simuliumdanosum,* the vector of *onchocerciasis* (river blindness).

The high density of the crystalline toxic compounds is a problem because they tend to sink to the bottom in water. A formulation is needed that will keep the toxic compounds afloat long enough to reach the target larvae downstream from the point of application. It is therefore necessary to put the maximum amount of crystal toxin in the primary powder used as a basis for an actual pesticide and then to formulate the latter in such a way as to give the desired residual effect under normal field conditions. From the commercial point of view, the development of such a formulation could probably enable the manufacturer to patent the material, a matter of great importance in the highly competitive pesticide industry.

As summarized by Norris (1978), the present effectiveness and future potential of microbiological insecticides depend on a subtle interaction between cost-effectiveness in comparison with established chemical agents, safety in the sense of absence of toxic effects on man and other animal and plant species, and efficacy in the field against the target insect.

Viruses

Many viral diseases of insects are known. Over 300 insect viruses are known to exist, and many of these rapidly cause epizootics among insect populations in nature. An infestation of sawflies in Canada during the 1930's was fought with a polyhedrosis virus grown in the larvae of sawflies and dispersed in water to the young spruce trees. The effort was considered very successful (Norris, 1978).

The use of myxomatosis virus to control rabbits is an example of an ill-considered application of virus to pest control. The virus was discovered in Uruguay in 1896 and introduced in 1926 to Australia, where rabbits imported from Europe had become a major pest. The attempt was a failure. Several further attempts failed before one in 1950 finally succeeded, probably because of favorable weather conditions. More than 80% of the rabbits of southeast Australia were wiped out in the first wave of the epidemic; in some areas, millions of them were found piled high against rabbit-proof fencing. Two years later, the virus was introduced into France. Two years after that, it crossed into England. It is estimated that, by the end of 1955, 90% of the rabbits in Britain had been killed. Since that time, the disease has remained present in those rabbit populations in epidemic form. The effect of natural selection has been to produce rabbits that are more resistant to the disease, and animals blinded by it may still survive and feed normally. The population of rabbits now is increasingly immune to the disease and may be developing tolerance.

Viral agents of pest control should be more carefully selected in the future, and the ecology of their use should be thoroughly understood. Although most viruses are characteristically host-specific, others are not.

Bacteria, Fungi, and Protozoa

Bacteria, true microbes, pose different issues. There are fewer bacterial viruses that are known to cause diseases in insects and that might have the ability to act as insecticidal agents. Approximately 100 species of bacteria are known to infect insects.

Popillia japonica is a Japanese beetle that damages crops and lawn grasses and that produces widespread outbreaks. The natural pathogen of this beetle, *Bacillus popilliae,* has been used in a successful demonstration of microbial pest control. Spores of *Bacillus popilliae* are eaten by the larva along with its normal diet of vegetation. The spores can resist the normal digestion process of the larva and in due course germinate and release the rod-shaped bacteria that produce the disease.

In cases in which bacterial infection can be identified in a pest species, similar applications of biological control can be used.

Two fungi are of special interest in the search for microbial control agents. They are *Beauvaria bassina* and *Metarrhizium anisopliae,* both of which were used on a somewhat experimental basis at the beginning of the 20th century. Their spores germinate on the body wall of the host. The growing fungus colony penetrates the insect larva through a breathing pore or spear pore. There are other fungi that could be used to control pests by producing a specific infection.

Pheromones

Insect pheromones have long been believed to be sex attractants, produced by females to attract males. Twenty years of research with pheromones has made it clear that they are not nearly so easy to categorize: in some cases, males release pheromones to attract females, and in many cases the communication is not related to sex and, in fact, may be interspecial. The low toxicity, high specificity, and fast degradation time of pheromones have long made them attractive as pesticides.

The EPA recently approved the use of a synthetically produced pheromone to control the pink bollworm caterpillar. The pheromone is sprayed on an infected cotton field to prevent the male insects from mating. It has been shown that resistance to this chemical through evolution is almost nonexistent.

Three applications of pheromones have been found in pest control. The first is mass trapping. This is a large-scale capture in pest traps baited with pheromones. Israeli farmers have had to spray their cotton fields with pesticides less often since they began trapping the Egyptian cotton lead worm in 1975. Mass trapping is also the basis of the Norwegian campaign against the spruce bark beetle. Workers at the State University of New York have reported progress in trials using the aggregation pheromones with the European elm bark beetle, which carries the fungus that carries Dutch elm disease. Rhodesian workers have painted cattle with a mixture of tick pheromone and acaricide. Small areas of the mixture attract unfed nymphs and adults, killing them before they bite the cattle, thus reducing the chance that they would spread disease.

A second approach is to fill the atmosphere with the pheromone to confuse the insects and to prevent the males from finding the females for mating. Tests in the Philippines on the use of mating disruption by pheromone inhibitors of the striped rice stem bore have been very encouraging. The next steps will probably be to microencapsulate the inhibitors for slow continuous release and to set up trials in the open fields.

The third application is to use pheromones to attract pests for population monitoring.

Although the pheromones have been known for 20 years, the complexity and subtlety of their action is only now being appreciated. Pheromones are needed in such small amounts that they will probably be manufactured chemically rather than biologically. However, biological production may be practical if the insect cells that secrete the pheromones can be cultivated to produce the pheromones *in vitro*.

MICROBIAL HORMONES

Related to the above chemicals are microbial hormones. Little effort has been put forth on possible applications of such compounds. Gooday (1978) has reviewed the recent scientific literature and defines microbial hormones as powerful metabolites that control fundamental processes in microorganisms. The mode of action of the few chemicals characterized thus far is concerned with bringing cells together and with diverting cells from activities concerned with growth to those concerned with reproduction. There are now a handful of microbial sex hormones that have been fully characterized as similar to the insect pheromones described above. One of the most studied microbial hormones is trisporic acid, which controls sexual differentiation in the *Mucorales*.

Although such chemicals have been the subject mainly of basic studies, it is possible that novel applications could be developed for control of microbial growth and reproduction when they are not wanted.

DIFFICULTIES

The inability of regulatory agencies to comprehend the difference between biological pest control and use of chemically toxic materials for pest control will have to be addressed. Many of the regulatory issues are discussed elsewhere. However, for emphasis, it should be realized that the safety issues are also related to the disruption of the natural ecological web caused by the sudden reduction in the number of pest animals, rather than solely by the toxicity caused by a compound that is applied in minute amounts, decays rapidly, and has high specific activity.

Another problem is the limited markets for these products. Since each product by its nature will be specific for a single pest or organism, the market will be small, and the development cost for each product will be very high.

Patent protection is the final problem. Formulations and methods of application will surely be patentable. Novel and reliable delivery systems are required for this technology. In the case of natural toxins and pheromones, this will probably provide a fair incentive on the part of industry to sponsor research. However, the patentability of viral and bacterial pest control agents is less clear. The protection of the commercial interests of the companies investing large amounts of money in developing biological pest control agents should be considered carefully.

DYES AND PIGMENTS

Domestic production of dyes and pigments is shown in Table 1.15.

The pigment and dye industries are an outgrowth of the coal tar chemistry of the late 19th century. Most of the compounds are referred to not by their chemical identity, but by a scheme related to their origin, application, pH, and color characteristics.

The principal desirable features of these compounds are their fastness and stability. Since biological pigments are usually sensitive to physical treatments and are biodegradable, it is unlikely that they will find significant markets among those currently dominated by coal tar dyes and pigments. However, many of the natural microbial pigments have never been tested and characterized for such applications.

TABLE 1.15 Dyes and Pigments

Products	*1978 Market (millions of dollars)*	*Present Source*	*1978 Domestic Production (metric tons)*	*Potential for Application of Biotechnology (de novo synthesis or bioconversion organisms and enzymes)*
Dyes, total	$734	Cyclic intermediates (aromatic)	18,000	De novo: very unlikely. Bioconversion: specialized applications possible.
Pigments, total	$319	Cyclic intermediates (aromatic)	34,500	De novo: very unlikely. Bioconversion: specialized applications possible.

REFERENCES

Brown, A. D. (1976) *Bacterial. Rev. 40,* 803–846.

Buchanan, J. M., and Hartman, S. C. (1969) *Adv. Enzymol. 21,* 199–261.

Buchanan, J. M., and Hartman, S. C. (1970) *Metabolic Pathways* 3rd ed., pp. 41–68.

Chemical and Engineering News (1979) Nov.

Chiao, J. S., Cheng, Y.-H., Shen, Y.-C., Wang, L.-W., and Lang, C.-P. (1964) *Acta Microbiol. Sinica 10,* 137–148 (in Chinese).

Clarke, P. H. (1979) in *Biological Regulation and Development* Vol. I, *Gene Expression* (Goldberger, R. F., ed.) pp. 109–170, Plenum Press, New York.

Compere, A. L., and Griffith, W. L. (1979) in *Developments in Industrial Microbiology* Vol. XX, *Evaluation of Substrates for Butanol Production* pp. 509–517.

Conner, H. A., and Allegeier, R. J. (1976) in *Advances in Applied Microbiology* Vol. XX, pp. 82–133, Academic Press, New York.

Cooper, R. A., and Kornberg, H. L. (1964) *Biochem. J. 91,* 82.

Demain, A. (1978) in *Economic Microbiology* Vol. II *Primary Products of Metabolism* (Rose, A. H., ed.) pp. 187–208, Academic Press, New York.

Finn, R. K., and Nowrey, J. E. (1959) *Appl. Microbiol. 7,* 29–32.

Gooday, G. W. (1978) in *Companion to Microbiology: Selected Topics for Further Study* (Bull, A. T., and Meadow, P. M., eds.) pp. 207–220, Longman, New York.

Gots, J. S. (1971) *Metabolic Pathways* Vol. V, 3rd ed., pp. 225–255.

Greenshields, R. N. (1978) in *Economic Microbiology* Vol. II, *Primary Products of Metabolism* (Rose, A. H., ed.) pp. 121–186, Academic Press, New York.

Hastings, J. J. H. (1971) in *Advances in Applied Microbiology* Vol. XIV (Perlman, D., ed.) p. 1, Academic Press, New York.

Hattori, K., and Suzuki, T. (1974) *Agr. Biol. Chem. 38,* 581.

Hirose, Y., and Okade, H. (1979) in *Microbial Technology* Vol. I, *Microbial Processes* (Peppler, H. J., and Perlman, D., eds.) pp. 211–240, Academic Press, New York.

Hirose, Y., Sano, K., and Shibai, H. (1978) in *Annual Reports on Fermentation Processes* Vol. II (Perlman, D., ed.) pp. 155–189, Academic Press, New York.

Hirose, Y., Enei, H., and Shibai, H. (1979) in *Annual Reports on Fermentation Processes* Vol. III (Perlman, D., ed.) pp. 253–274, Academic Press, New York.

Ishii, K., and Shiio, I. (1973) *Agr. Biol. Chem. 37,* 287.

Kieslich, K. (1976) *Microbial Transformations of Non-Steroid Cyclic Compounds,* John Wiley, New York.

King, R. C. (1974) *Handbook of Genetics* Vol. I, *Bacteria, Bacteriphages and Fungi,* Plenum Press, New York.

Kotera, Y., Kodama, T., Minoda, Y., and Yamada, K. (1972) *Agr. Biol. Chem. 36,* 1315.

Kutzenok, A., and Aschner, M. (1952) *J. Bacteriol. 64,* 829–836.

Lawrence and Thomas (1979) in *Microbial Technology: Current State, Future Prospects* (Bull, A. T., Ellwood, D. C., and Ratlidge, C., eds.), Cambridge University Press, Cambridge, England.

Leuenberger, H. G. W. (1978) in *Antibiotics and Other Secondary Metabolites: Biosynthesis and Production* (Hutter, R., Leisinger, R., Nuesch, J., and Wehrli, W., eds.) pp. 87–100, Academic Press, New York.

Lockwood, L. B., and Reeves, M. D. (1945) *Arch. Biochem. 6,* 455.
Magasanik, B. (1957) *Ann. Rev. Microbiol. 11,* 221–252.
Metzler, D. E. (1978) in *Biochemistry: The Chemical Reactions of Living Cells,* Academic Press, New York.
Miall, L. M. (1978) in *Economic Microbiology* Vol. II, *Primary Products of Metabolism* (Rose, A. H., ed.) pp. 48–111, Academic Press, New York.
Momose, H., and Shiio, I. (1969) *J. Gen. Appl. Microbiol. 15,* 399.
Nakao, Y. (1979) in *Microbial Technology* Vol. I, *Microbial Processes* (Peppler, H. J., and Perlman, D., eds.) pp. 312–354, Academic Press, New York.
Nakayama, K. (1976) *Proc. Biochem. 11,* (2),4.
Ninet, L., and Renault, J. (1979) in *Microbial Technology* Vol. I, *Microbial Processes* (Peppler, H. J., and Perlman, D., eds.) pp. 529–544, Academic Press, New York.
Norris, J. R. (1978) in *Companion to Microbiology: Selected Topics for Further Study* (Bull, A., and Meadows, P. M., eds.) pp. 459–480, Longman, New York.
Onishi, U.S. Patent No. 3,012,945.
Perlman, D., Titus, E., and Fried, J. (1952) *J. Amer. Chem. Soc. 74,* 2126.
Sasajima, K., Nogami, I., and Yoneda, M. (1970) *Agr. Biol. Chem. 34,* 381.
Sebek, O. K., and Perlman, D. (1979) in *Microbial Technology* Vol. I, *Microbial Processes* (Peppler, H. J., and Perlman, D., eds.) pp. 483–496, Academic Press, New York.
Shvinka, J., Viesturs, U., and Ruklisha, M. (1980) *Biotech. Bioeng. 22,* 897–912.
Spencer, J. F. T., and Spencer, O. M. (1978) in *Economic Microbiology* Vol. II, *Primary Products of Metabolism* (Rose, A. H., ed.) pp. 393–443, Academic Press, New York.
SRI (1977) *Chemical Economics Handbook,* SRI, Menlo Park, Calif.
SRI (1979) *Chemical Economics Handbook,* SRI, Menlo Park, Calif.
Stodola, F. H., Friedkin, M., Mayer, A. J., and Coghill, R. D. (1945) *J. Biol. Chem. 161,* 739.
Tabuchi, T., and Hara, S. (1973) *Nippon Nogeikogaku Kaishi 47,* 485.
Tachibana, S., and Murakami, T. (1973) *J. Ferment. Technol. 51,* 858.
Tong, G. E. (1978) *Fermentation Routes to C3 and C4 Chemicals,* CEP 70–74.
Tong, G. E. (1979) *Microbial. Technol. 1,* 173–179.
Walsh, C. (1979) in *Enzymatic Reaction Mechanisms,* W. H. Freeman, San Francisco.
Yamada, K., Minoda, Y., Komada, T., and Kotera, Y. (1969) in *Fermentation Advances* (Perlman, D., ed.) p. 541, Academic Press, New York.
Yamada, K., Furukawa, T., and Nakahara, T. (1970) *Agr. Biol. Chem. 34,* 670.

CHAPTER
2

Recombinant DNA Approaches for Enhancing the Ethanol Productivity of *Zymomonas mobilis*

Douglas E. Eveleigh
H. W. Stokes
Ellen L. Dally

I. INTRODUCTION

The recent worldwide energy crisis has sparked an interest in the development of alternative sources of energy, particularly for liquid transportation fuels. One alternative that has been seriously investigated and promises to be useful is the microbial fermentation of ethanol from renewable agricultural resources. The ethanol (fuel alcohol) can be used directly as a fuel and also as an octane booster (for general reviews, see Bente, 1980; Bungay, 1981; Chambers et al., 1979; Ferchak and Pye, 1981; Flickinger, 1980; Goldstein, 1980; Kosaric et al., 1980; Lipinsky, 1978, 1981; Office of Technology Assessment, 1980; Tsao, 1978; Vlitos, 1981; Weisz and Marshall, 1979). Today, fuel alcohol production from starch has come of age. Large-scale ethanol production in the United States is currently via fermentation rather than from ethylene derived from oil (R. Katzen, 1982, personal communication). However, if fuel alcohol production is to be increased from the current 0.5 billion gallons to the U.S. National Alcohol Fuels Commission's (1981) goal of 10–20 billion gallons by 1990, alternative and massive supplies of biomass are required. General biomass, rather than corn-derived starch, is the obvious choice, since it is relatively inexpensive; is ubiquitously available from agricultural crop residues, municipal, forestry, and food wastes, or from silviculture energy farms; and furthermore does not compete directly as a foodstuff. It has been projected that the United States can obtain 10% (8 quads) of its energy from biomass. Research on the utilization of biomass was initially focused on the cellulose component (Bungay, 1981; Douglas, 1982) but now is also directed to the hemicellulose component (e.g., xylan), which can comprise 20–40% of the feedstock (Detroy et al., 1981; Gong et al., 1981a, b; Jeffries, 1981; Schneider et al., 1981). Fuel alcohol (ethanol) has been focused on as one major product because of its convenience as a liquid transportation fuel.

Ethanolic fermentations are found in a wide variety of microorganisms, though yeasts have traditionally been used for alcohol production, especially in the wine and beer industries (Rose, 1978). The exclusive application of yeast is due, in part, to the homofermentative mode, which yields a single product, ethanol, often in relatively high yield; for example, *Saccharomyces carlsbergensis,* and *S. cerevisiae* produce 12% ethanol, while in a very slow fermentation, *S. sake* yields 20%. However, other microbes ferment glucose to ethanol as efficiently as yeast and in as high yield. They have recently attracted considerable attention for their potential application in fuel alcohol production. *Thermoanaerobacter ethanolicus,* a thermophilic bacterium, yields 6–10% ethanol (Wiegel and Ljundahl, 1981), while *Zymomonas mobilis* can yield 12% alcohol (Swings and DeLey, 1977). *Z. mobilis* has a range of other advantageous features that make it a promising organism for development in the industrial production of alcohol (Table 2.1); on this basis, its genetic potential has been explored.

TABLE 2.1 Industrial Attributes of *Zymomonas mobilis*

1. Efficient conversion of glucose to ethanol (95%)
2. Uncoupled growth and ethanol production
3. Fast anaerobic fermentation
4. High ethanol tolerance (12%)
5. Tolerance to low pH
6. High rate of ethanol production (160 g/l/h)
7. Prokaryote—easy to manipulate

II. *ZYMOMONAS MOBILIS:* TAXONOMY, PHYSIOLOGY, AND ETHANOL PRODUCTION

Zymomonas mobilis is an anaerobic, gram-negative bacterium that converts glucose to ethanol via the Entner-Doudoroff pathway. *Z. mobilis* strains have been isolated from many parts of the world. They are found in fermenting plant juices in the tropical regions of America, Africa, and Asia and are the initial fermentative agents in the production of native alcoholic beverages such as pulque and palm wines. *Z. mobilis* has also been isolated from spoiled beer and cider in Great Britain and most recently in Finland (Viikari et al., 1981). The biology of this bacterium has been recently reviewed (Dadds and Martin, 1973; Lawford et al., 1982; Montenecourt, 1983; Rogers et al., 1982; Stokes et al., 1981; Swings and DeLey, 1977).

A. Taxonomy

The taxonomy of *Z. mobilis* has had a checkered history. Since Lindner first described the bacterium, it has been placed into five genera and has had 20 different names at different times (Swings and DeLey, 1977). In 1936, the genus *Zymomonas* was created. In *Bergey's Manual of Determinative Bacteriology* (eighth edition), the genus *Zymomonas* was divided into two species: *Zymomonas mobilis* (Lindner) Kluyver and van Niel 1936, of the type isolated from *Agave* by Lindner, and *Zymomonas anaerobia* (Shimwell) Kluyver 1957, isolated from beer and cider. The latter species was divided into three varieties: *anaerobia, immobilis,* and *pomaceae*. The two species were differentiated by their ability to ferment sucrose; the varieties were differentiated by their mobility and growth in the presence of SO_2. An improved taxonomy has been proposed that is based on modern methods of analysis such as DNA base composition (DeLey and Swings, 1976), DNA homology, and comparison of protein electrophorograms (DeLey et al., 1976). Following a comparison of 138 phenotypic features of 38 *Z. mobilis* strains of diverse origins, it was proposed that all the strains be placed into one species, *Zymomonas mobilis,* with 35 strains in the subspecies *mobilis*. Three strains, isolated from cider, differed slightly from the others and were placed into the subspecies *pomaceae*. All strains, except the cider sickness strains, were claimed to be genetically and phenotypically identical in spite of their worldwide distribution. For this reason, Swings and DeLey (1977) have suggested that *Zymomonas mobilis* either is of recent evolutionary origin and has not yet had time to develop genetic diversity or is genetically a very stable genus. The currently available strains are listed in Table 2.2.

B. Physiology

Zymomonas mobilis ferments glucose and fructose via the Entner-Doudoroff pathway (Gibbs and DeMoss, 1951), followed by decarboxylation of pyruvate

TABLE 2.2 Origin and Availability of *Zymomonas mobilis* Strains

	Original Isolation	*A.T.C.C.*[1]	*N.R.R.L.*[2]	*N.C.I.B.*[3,7]	*Rogers*[4]	*Saddler*[4]
			Strains Used in This Study			
	Lindner, Fermenting *Agave* juice Mexico, 1924	10988	B-806	8938	ZM1	Z1
Ag11:	Gonçalves de Lima et al.[4] Fermenting *Agave* juice (Aguamiel), Mexico, 1951	—	B-4576	—	Ag11	—
CP3:	Gonçalves de Lima et al.[4] Fermenting sugarcane juice, (caldo-picado), Brazil, 1970	—	B-14022	—	—	—
CP4:	Gonçalves de Lima et al.[4]	—	B-14023	—	ZM4	—
			Other Available Strains			
Z6:	Fermenting *Elaeis* sap (palm wine) Kinshasa, Zaire, 1967[5]	29191[6]	B-4490	11199	ZM6	Z4
I:	Barker, Sick cider, Bristol, U.K. 1951 (deposit)[5]	29192[6]	B-4491	11200	—	Z3
	Bunker, Spoiled beer, London, U.K. 1951[5]	29501	B-1960 (via NCIB in 1957)	8227	—	Z
		29501	B-4492 (via NCIB in 1980)	8227	—	Z

1. American Type Culture Collection, Washington, D.C., U.S.A.
2. Northern Regional Research Laboratory, U.S.D.A., Peoria, Illinois, U.S.A.
3. National Collection of Industrial Bacteria, Aberdeen, Scotland.
4. See references.
5. See Swings and DeLey (1977).
6. ATCC 29191 (IFO 13756) and ATCC 29192 (IFO 13757) are also available from the Institute for Fermentation, Osaka, Japan, cited under the IFO designations.
7. Further Swings and DeLey (1977) strains are now deposited at N.C.I.B.

to ethanol. The discovery of this pathway in *Z. mobilis* was the first example of an anaerobic organism that uses a pathway found predominantly in aerobic bacteria (Kersters and DeLey, 1968). The conversion of glucose to ethanol varies with the strain, but the most efficient strains can achieve a 95% conversion, with nearly equimolar amounts of ethanol and CO_2 being produced. Some lactic acid is produced, with traces of acetaldehyde, acetylmethyl carbinol, acetic acid, and glycerol. The synthesis of higher alcohols by *Z. mobilis* has also been reported, although only trace quantities are produced compared to ethanol (Bevers and Verachtert, 1976).

A major disadvantage of *Z. mobilis* is its ability to ferment a very limited range of substrates. All strains reported to date can utilize only glucose, fructose, and sucrose as energy sources. Growth on sucrose is strain-dependent and, in some cases, appears to be inducible (Swings and DeLey, 1977). There have been occasional reports of strains capable of growth on raffinose (Dadds et al., 1973) and sorbitol (Millis, 1956), but *Z. mobilis* does not grow on any other carbohydrates tested: cellobiose, galactose, lactose, maltose, mannose, ribose, xylose, or starch. It may be possible to extend this narrow range of carbohydrate utilization by genetic manipulation to make *Z. mobilis* a more useful and versatile organism for ethanol production.

Z. mobilis has several attributes that make it industrially attractive (Table 2.1). *Zymomonas mobilis* grows rapidly under totally anaerobic conditions. In contrast, yeasts require the addition of controlled amounts of oxygen for sterol formation to maintain viability at high cell concentrations in continuous cell recycle systems (Cysewski and Wilke, 1977). Furthermore, all 40 strains tested by Swings and DeLey (1977) grew within the pH range 5–7, 90% of the strains growing at pH 3.9. Fermentation at acid pH values reduces the risk of growth of contaminating bacteria in the fermentation vessel, which would be advantageous during industrial production. The optimum temperature for growth of the 40 *Z. mobilis* strains tested by Swings and DeLey (1977) was 30°C, although 97% of the strains grew at 36°C, 74% at 38°C, and 5% at 40°C. In commercial production of ethanol, it would be advantageous to use thermotolerant strains, since at high temperatures, the fermentation would proceed more rapidly, and product recovery and cooling costs would be reduced.

Zymomonas mobilis strains are ethanol-tolerant at levels comparable to some strains of *S. cerevisiae*. Swings and DeLey (1977) report that 47% of their 40 *Z. mobilis* strains tested still developed in ethanol concentrations up to 10%. Rogers et al. (1980) state that ethanol concentrations of 70–80 g/l have been achieved in continuous culture and up to 130 g/l in batch culture with one strain of *Z. mobilis*. Recent studies have also shown that it is possible to improve tolerance of strains by mutation and selection (Skotnicki et al., 1981). The bacterium is also glucose-tolerant. Swings and DeLey (1977) reported that all 40 strains they tested grew at 20% glucose, and half grew at 40% concentrations after lag phases of 4–20 days. In although osmotolerant yeasts are well known, they are not efficient ..ᴜɪ producers (Haraldson and Bjorling, 1981).

Another advantage of *Zymomonas mobilis* is its "uncoupled growth." Under certain circumstances, glucose metabolism can continue even in the absence of cell growth. For example, when *Z. mobilis* cultures were starved for pantothenate, a vitamin required for growth by all strains, cell growth stopped, but the rate of fermentation of glucose remained constant (Belaich et al., 1972). Analogously, ethanol production can continue after growth rates decline following the addition of chloramphenicol to the growth culture (Lazdunski and Belaich, 1972). Uncoupled growth and glucose fermentation also occur if the growth temperature is raised above 33°C (Forrest, 1967). While uncoupled growth is clearly inefficient for the cell, it would be useful in the practical application of ethanol production, since fermentation still would occur efficiently in the absence of cell growth.

C. Ethanol Production

Several research groups have evaluated the potential of *Z. mobilis* for use in the production of industrial alcohol (for review, see Lawford et al., 1982; Rogers et al., 1982). Conversion of glucose to ethanol by *Z. mobilis* is efficient. Ethanol yields vary with each strain and with growth conditions, but all strains produce at least 1.5 moles of ethanol per mole of glucose. The more efficient ones produce up to 1.9 mol/mol, which is comparable to the best production by *Saccharomyces cerevisiae* (Swings and DeLey, 1977). However, *Z. mobilis* has been found to have higher ethanol productivities than yeasts (Lavers et al., 1981; Rogers et al., 1979, 1980). In kinetic studies of alcohol production of *Z. mobilis,* strain ATCC 10988 at high glucose concentrations (25%) in batch culture, specific glucose uptake rates (q_s) (5.45 g/g/h), specific ethanol productivities (q_p) (2.53 g/g/h), and ethanol yields (92.5% theoretical yield) were all significantly greater than for the yeast *S. carlsbergensis* (*uvarum*) (q_s = 2.08 g/g/h; q_p = 0.87 g/g/h; ethanol yield = 85.9%) (Rogers et al., 1979). The higher specific rates of glucose uptake and ethanol production by *Z. mobilis* are associated with lower levels of biomass production than for yeasts. This appears to be a consequence of the lower energy available to *Z. mobilis* for growth. The Entner-Doudoroff pathway produces only 1 mole of ATP per mole of glucose compared to the glycolytic pathway in yeasts, which produces 2 mol ATP/mol glucose. In a continuous culture system with cell recycle, maximum ethanol productivities of 120 g/l/h were achieved with *Z. mobilis* ATCC 10988 (Rogers et al., 1980). This is significantly higher than the greatest previously reported value (82 g/l/h) for *S. cerevisiae* in cell recycle with vacuum fermentation (Cysewski and Wilke, 1977).

From a series of studies in Australia using batch, continuous, and continuous culture with cell recycle of four strains (ATCC 10988, Ag11, ATCC 29191, and CP4), CP4 was judged the best overall for ethanol production

(Skotnicki et al., 1981). CP4 had the fastest growth rate and was most ethanol-tolerant (80 g/l). CP4 also had faster rates of ethanol production (200 g/l/h) in continuous cell recycle fermentation, which could sustain ethanol concentrations of 60–65 g/l (Lee et al., 1980). CP4 continued to produce ethanol at 42°C when grown on glucose, while other strains did not (Skotnicki et al., 1981).

Continuous culture systems using immobilized cells have been developed for *Z. mobilis*. Ethanol productivities of 120 and 132 g/l/h were obtained from 5% and 10% glucose concentrations from cells immobilized via flocculation with polystyrene beads (Arcuri et al., 1980). Using cells entrapped in borosilicate glass fiber, an ethanol productivity of 85 g/l/h was achieved from a 5% glucose concentration (Arcuri et al., 1980). Cells immobilized in a calcium alginate or κ-carrageenan gel produced 44 g ethanol/l/h or 53 g/l/h, respectively, from a 15% glucose concentration (Grote et al., 1980). The isolation of more ethanol-tolerant mutants and optimization of medium composition and environmental conditions should lead to improved productivities.

Two methods of converting cellulose to ethanol in one-step processes using *Z. mobilis* have been investigated. Growing *Z. mobilis* in sequential culture with the cellulolytic *Clostridium thermocellum* using 1% cellulose as substrate resulted in a 27% conversion to ethanol (Saddler et al., 1981). Eight strains of *Z. mobilis* were tested for their ability to produce ethanol in a simultaneous hydrolysis and fermentation process from a 5% cellulose substrate (Viikari et al., 1981). The cellulose was from culture filtrates of *Trichoderma reesei*. On the average, 80% of the theoretical ethanol yields were produced.

In review, *Zymomonas mobilis* has excellent potential for development in industrial ethanol production but can utilize only a restricted range of fermentable substrates. Extending this narrow range by genetic manipulation would increase this organism's usefulness and versatility for industrial use. This is possible by insertion of appropriate genes (amylase, cellulase) into *Z. mobilis* by using genetic engineering techniques. This approach also permits further enhancement of its ethanol productivity by incorporation of genes resulting in greater flocculation or in tolerance to high concentrations of ethanol and molasses. The general genetic approach has recently been reviewed by Ho (1980) and Stokes et al. (1983b). The initial focus in such a genetic approach has been to develop broad-host-range plasmid vectors that are stable in *Zymomonas*.

III. GENETIC STUDIES OF *ZYMOMONAS MOBILIS*

Studies of *Zymomonas* genetics have been initiated in order to develop the industrially attractive features of ethanol production outlined above. A comparison of strains has been made with regard to their resistance to antibiotics, their native plasmids, and their ability to accept and maintain foreign DNA. More recently, construction of cloning vectors for *Z. mobilis* has been attempted. These approaches are reviewed.

A. Resistance to Antimicrobial Agents

Resistance to antibiotics and antimicrobial agents is a useful marker for genetic studies. The generally available strains are remarkably resistant to a diverse variety of agents. Swings and DeLey (1977) revealed that 100% of the 38 strains tested were resistant to nine antibiotics tested. Resistance to seven other antibiotics depended on the strain, but no strain tested was capable of growth in the presence of tetracycline at 10 μg/ml. We have confirmed the general resistance pattern with four common strains (Table 2.3). Resistance was found to mercuric chloride, nalidixic acid, neomycin, streptomycin, and trimethoprim. Strains did show substantial variability in resistance to ampicillin, ranging from 20 μg/ml for Ag11 to 300 μg/ml for CP4. The general diverse resistance that is exhibited may be due to lack of uptake of the antibiotics by *Zymomonas*. It is known, for example, that resistance to aminoglycosides in many anaerobic bacteria is due to an inadequate electrical potential across the membrane to drive aminoglycoside uptake (Bryan and Kwan, 1981). This mechanism of resistance may be acting in *Z. mobilis*. The resistance characteristics are of limited use in genetic studies with *Zymomonas*. The sensitivity to tetracycline and chloramphenicol potentially allows for direct selection to plasmids containing resistance genes for these markers. Tetracycline has been a useful selective marker for R-plasmid transfer experiments, since the spontaneous resistance mutation frequency is less than 1×10^{-9} for all strains tested at Tc 20 μg/

TABLE 2.3 Minimal Inhibitory Concentration (MIC)[1] of Antimicrobial Agents to *Zymomonas mobilis* Strains

	CP3[2]	*CP4*	*Ag11*	*ATCC 10988*
Ampicillin	100	300	20	200
Chloramphenicol	10	25	5	10
Mercuric chloride	25	30	30	30
Tetracycline	5	5	5	5
Rifampicin	5	5	5	5
Nalidixic acid	100	750	250	750
Neomycin	4000	1000	500	500
Streptomycin	4000	2000	500	4000
Trimethoprim	>2000	>2000	>2000	>2000

1. MIC is defined as that level of antimicrobial agent (in μg/ml) that reduced survival to 50% compared to viable count on unsupplemented complete medium. Survival was determined by spotting 10-μl aliquots of decimal dilutions onto appropriately supplemented complete medium.
2. Strain designations, see Table 2.1.

ml. Rifampicin resistance is often associated with chromosomal genes and thus is not of immediate application.

B. Mutagenesis

In an effort to improve *Z. mobilis* genetically for ethanol production, several mutagenic agents were tested for their general effectiveness (Skotnicki et al., 1982). Nitrosoguanidine (NTG) was found to be a better mutagenic agent than exposure to UV light. After treatment with NTG, mutants of strain CP4 were selected for increased ethanol tolerance (15% wt./vol.), greater flocculation, and tolerance to higher temperature. Selection of mutants capable of "good growth" on media containing molasses was also made. Growth and ethanol production of *Z. mobilis* are normally poor on molasses, but one mutant was able to grow and produce ethanol almost twice as fast as the parent strain on molasses. Minimal media have recently been described (Lawford et al., 1982; Goodman et al., 1982) that now permit selection of auxotrophic mutants, and these latter will greatly facilitate genetic analysis and strain construction.

C. R-Plasmid Transfer to *Z. mobilis*

The ability to accept and stably maintain foreign DNA is a necessary attribute for *Zymomonas* to possess if it is to be used in the construction of strains by recombinant DNA technology. The conjugal transfer of four R-plasmids into four efficient ethanol-producing strains was first reported by Skotnicki et al. (1980). It was shown that *Z. mobilis* can act as a recipient in plasmid transfer from *P. aeruginosa* and *E. coli* and as a donor in plasmid transfer to other *Zymomonas mobilis* strains. Transfer frequencies, expressed as transconjugants per recipient cell, of 10^{-1}–10^{-4} were obtained with *Zymomonas* as a recipient. The apparent extremely high transfer efficiency should be interpreted cautiously, since plasmid transfer is regulated by the donor cell. Transfer frequencies are usually expressed as transconjugants per donor cell. Transfer of *Z. mobilis* chromosomal markers was also observed when mobilized by the plasmid R68.45 (Skotnicki et al., 1980).

We have similarly demonstrated the ability of the *Z. mobilis* strains to act as recipients for foreign DNA in conjugal transfer experiments utilizing the IncP-1 wide-host-range plasmids R68 and RP1 (Table 2.4) (Stokes et al., 1982, 1983a). Both plasmids transferred to all six strains, although the strains varied in their ability to act as recipients. The transfer frequency was similar for all of the *Z. mobilis* strains, with the exception of Ag11, in which transfer was up to 40-fold less (Table 2.4). The highest transfer occurred into CP3 and 10988. These results suggest that there are differences in the ability of *Zymomonas* strains to act as recipients of plasmid DNA. Tonomura et al. (1982) have recently reported conjugation frequencies in

TABLE 2.4 Transfer Frequencies of IncP-1 Plasmids R68 and RP1 to *Zymomonas mobilis*

Recipient: Z. mobilis	*Donor* *PA08 (R68)*	*PA08 (RP1)*
Ag11	1.5×10^{-7}	1.5×10^{-7}
CP3	2×10^{-6}	4×10^{-6}
CP4	5×10^{-7}	2×10^{-6}
ATCC 10988	4×10^{-6}	5×10^{-6}
ATCC 29191	1×10^{-6}	1×10^{-6}
ATCC 29192	1×10^{-6}	3×10^{-6}

The transfer frequency is expressed as transconjugants per donor cell.
The selection of R^+ *Z. mobilis* transconjugants was made on the basis of transfer of tetracycline resistance (20 μg/ml).

a similar range for RP4 from *E. coli* to *Zymomonas*. The ability of plasmid R68 to stably segregate to daughter cells in *Zymomonas mobilis* was also strain-dependent. The stability of R68 varied from 3% after 19 generations in ATCC 29191 to 92% after 16 generations in CP3 (Table 2.5).

On the basis of transfer frequency and plasmid maintenance strains CP3, CP4, and ATCC 10988 appear to be the best candidates for genetic development.

D. Native Plasmids

Most *Z. mobilis* strains contain plasmids (Table 2.6) (Rogers et al., 1982; Stokes et al., 1982, 1983a). The strains we have examined contain a large plasmid (46×10^6 daltons) (Table 2.6). Eleven other strains of worldwide origin also harbor large-plasmid DNA (40–50 megadaltons; unpublished observation). A recent report cites the larger plasmids to be in the 23 megadalton range (Tonomura et al., 1982). This discrepancy is under study. In addition, CP3 and CP4 harbor a plasmid of 21 megadaltons, while Ag11 and ATCC 10988 also harbor plasmids of 19 megadaltons and 16 megadaltons, respectively. Strains ATCC 10988 and Ag11 also contain unique smaller

TABLE 2.5 Stability of R68 in *Zymomonas mobilis*

Strain	*Stability (%)*	*Number of Generations*
Ag11 (R68)	74	15
CP3 (R68)	92	16
CP4 (R68)	82	15
ATCC 10988 (R68)	90	14
ATCC 29191 (R68)	3	19
ATCC 29192 (R68)	15	18

The selection of R^+ colonies was made on the basis of tetracycline resistance (20 μg/ml).

TABLE 2.6 Plasmids of *Zymomonas mobilis*

Strain	*Plasmids (molecular size, daltons × 10^6)*				
Ag11	pRUT111(46)	pRUT112(19)	pRUT113(6.6)	—	—
ATCC 10988	pRUT881(46)	pRUT882(16)	—	pRUT883(1.6)	pRUT884(1.1)
ATCC 29191	pRUT991(46)	pRUT912(21)	—	pRUT913(11)	—
ATCC 29192	pRUT921(46)	pRUT922(21)	—	pRUT913(11)	—
CP3	pRUT31(46)	pRUT32(21)	—	—	—
CP4	pRUT41(46)	pRUT42(21)	—	—	—

plasmids (pRUT884: 1.1 md; pRUT883: 1.65 md; pRUT113: 6.6 md). The molecular weight of these smaller plasmids has been confirmed by measurement of their contour lengths via electron microscopy (Stokes et al., 1983a). The two smallest plasmids from ATCC 10988 are particularly interesting, since they could be useful as cloning vectors. They have been tested by several restriction enzymes for single cutting sites. The 1.1-megadalton plasmid has a single cutting site for the restriction enzyme Sau 3A1. Sau 3A1 has a 4-base-pair recognition site, but it generates the same "sticky ends" as two other restriction enzymes, Bgl II and Bam HI (Fig. 2.1). By insertion of a replicon from *Zymomonas mobilis*, for example, the 1.1-megadalton plasmid from ATCC 10988, into an established cloning vector, it should be possible to increase the stability of that vector in *Z. mobilis*.

No phenotype has yet been assigned to the *Zymomonas* plasmids. Although all strains are resistant to a large number of antibiotics, none of these appears to be plasmid encoded.

A range of plasmids has been found in other *Zymomonas* strains. Interestingly, one strain, ZM1, that was derived from ATCC 10988 via a University of Queensland subculture 410, lacks plasmids (Skotnicki et al., 1982).

E. Relationship Between the Plasmids

On the basis of plasmid content, CP3 and CP4 are indistinguishable. To test more precisely the relationship between the *Z. mobilis* strains, plasmids from CP3, CP4, ATCC 10988, and Ag11 were compared following digestion by a number of restriction endonucleases. Gel electrophoresis banding patterns obtained by digestion of the high-molecular-weight plasmids from Ag11 and

RESTRICTION ENZYME	RECOGNITION SITE
Sau 3A1	5'.. ↓GATC .. 3' 3'.. CTAG↑.. 5'
Bgl II	5'.. A↓GATCT .. 3' 3'.. TCTAG↑A .. 5'
BamH I	5'.. G↓GATCC .. 3' 3'.. CCTAG↑G .. 5'

FIGURE 2.1 Cutting sites for construction of vector pBR322-88.

ATCC 10988 were unique. However, CP3 and CP4 yielded identical restriction digest banding patterns with three restriction endonucleases tested (Dally et al., 1982). Thus CP3 and CP4 appear as different isolates of the same strain. They were both isolated from fermenting sugarcane juice in a marketplace at roughly a 6-week interval by Gonçalves de Lima et al. (1970). This correlation is important, since Rogers et al. (1982) selected a strain ZM4 (derived from CP4) for further development on the basis of its ethanol productivity. We selected essentially the same strain (CP3) on the basis of the same criteria.

The homology of the large plasmids of strains Ag11, ATCC 10988, CP3, and CP4 has been assessed via hybridization (Stokes et al., 1983a). Blot hybridization of pRUT31 and pRUT32 to EcoR1 digests of plasmid DNA showed complete homology between strains CP3 and CP4. Ag11 and ATCC 10988 plasmid DNA also exhibit some homology with the individual CP3 plasmids pRUT31 and pRUT32. It has also been shown that the low-molecular-weight plasmid DNA in ATCC 10988 cross-hybridizes to plasmid DNA of the other three strains. In contrast, the pRUT113 plasmid appears unique, since it only hybridizes to DNA of Ag11.

Given the wide geographical distribution of these strains (CP3 and CP4 were isolated in Brazil, whereas Ag11 and ATCC 10988 were isolated in different parts of Mexico), the similarity in the molecular weights of the larger plasmids is interesting. One hypothesis for this similarity is that the strains tested here are descended from a common ancestor. This hypothesis is supported by the cross-hybridization of plasmids from different strains, though it could also result from the presence of similar insertion sequences. In addition to the large plasmids present in all strains, the 1.1- and 1.65-megadalton plasmids unique to ATCC 10988 also exhibit some interstrain homology. The interstrain relatedness of the larger *Z. mobilis* plasmids provides further support for the proposal by Swings and DeLey (1977) that *Z. mobilis* is of recent evolutionary origin or is genetically a highly stable genus.

F. Construction of Cloning Vectors for *Z. mobilis*

The features of a useful *Zymomonas* cloning vector are outlined in Table 2.7. Since most cloning experimentation has been done in *E. coli,* it is desirable first to select and characterize recombinant plasmids in *E. coli* prior to introducing them into *Z. mobilis*. Therefore it is necessary to construct a vector that contains both *E. coli* and *Z. mobilis* replicons. Second, it is necessary to introduce these vectors into *Zymomonas*. As *Zymomonas* transformation and transduction systems have yet to be described, conjugation must be utilized. In this approach, non-self-transmissible plasmids should be used as cloning vectors (NIH Recombinant DNA Guidelines), which can subsequently be mobilized into *Z. mobilis* through the use of a helper plasmid. Furthermore, since *Z. mobilis* is resistant to many

TABLE 2.7 Necessary Features of a Vector for Cloning in *Zymomonas*

1. The vector should contain *E. coli* and *Zymomonas* replicons
2. Transfer is via conjugation and subsequent mobilization
3. There should be selectable markers in *Zymomonas* (Cm, Tc)
4. The vector should contain convenient cloning sites

antimicrobial agents, the available markers are limited to chloramphenicol and tetracycline. Finally, the vectors must contain convenient cloning sites, that is, single cutting sites that will allow for selection of inserts on the basis of insertional inactivation of a resistance gene.

Initially, we proposed to introduce a *Zymomonas* replicon into the narrow-host-range *E. coli* vector pBR322, in order for pBR322 to be stably maintained in *Z. mobilis*. Using the protocol outlined in Fig. 2.2, the 1.1-megadalton plasmid (pRUT 884) from ATCC 10988 was inserted into the Bam HI site of pBR322. Clones that were ampicillin-resistant and tetracycline-sensitive were characterized by agarose gel electrophoresis. Plasmid pRUT-884 was labeled with ^{32}P by nick translation and blot hybridized to the purified recombinant plasmids to verify that the inserts were from pRUT-884.

Insertion of pRUT884 into the Bam HI site inactivated the tetracycline resistance gene in pBR322, the only marker on that plasmid that could be used for selection in *Z. mobilis*. Therefore it was necessary to introduce another marker into the pBR322-88 recombinant. The chloramphenicol resistance gene from pBR325 was inserted into the recombinant plasmid according to the method outlined in Fig. 2.3. Ampicillin-resistant, chloramphenicol-resistant, tetracycline-sensitive recombinants that contain both a chloramphenicol resistance gene and *Zymomonas* plasmid DNA were selected. The electrophoretic characterization of pBR322 and the two recombinant plasmids are illustrated in Fig. 2.4.

After construction of the chloramphenicol-resistant 322-88 recombinant plasmid, it was tested for transfer into *Zymomonas mobilis* (Table 2.8). Since neither pBR325 nor the recombinant plasmid is self-transmissible, a helper plasmid, pRK2013::Tn10, was used to mobilize the two vectors. Plasmid pRK2013 (Ditta et al., 1980) contains the origin of replication from the Col E1 plasmid and the transfer genes of the broad-host-range plasmid RK2. It will transfer to a number of gram-negative bacteria but will not replicate in them. The tetracycline resistance transposon Tn10 was added as an extra selective marker in *Z. mobilis*. As a control, the broad-host-range plasmid, R68, was transferred into *E. coli* and *Zymomonas*. Because of the greatly reduced transfer frequency of R68 into *Z. mobilis* compared

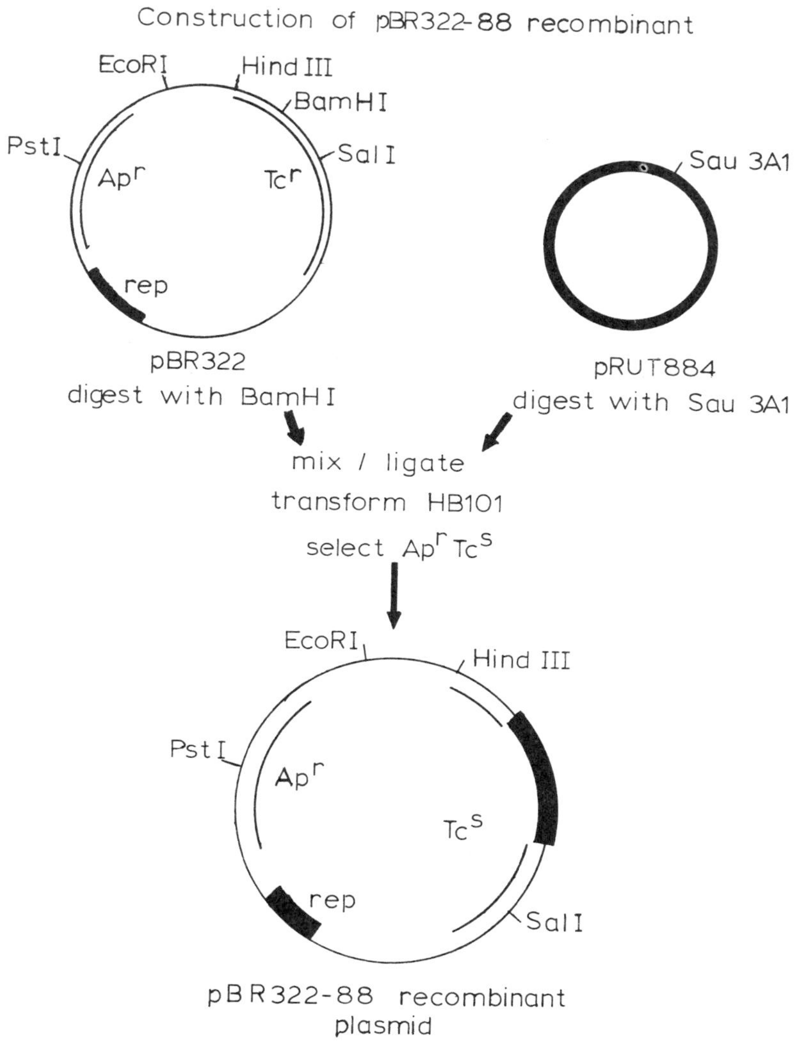

FIGURE 2.2 Protocol for the insertion of *Zymomonas* plasmid pRUT884 with pBR322.

with *E. coli,* it was not surprising that it was difficult to detect mobilization of the recombinant plasmid into *Z. mobilis* (Table 2.8).

Since cloning with narrow-host-range plasmids such as pBR322 and pBR325 was unsuccessful, broad-host-range plasmids were evaluated for cloning in *Z. mobilis*. The cloning vector, pKT 230, developed by Bagdasarian and Timmis (1982) was selected (Fig. 2.5). This plasmid is derived from the high-copy-number, broad-host-range *Pseudomonas* plasmid RSF1010.

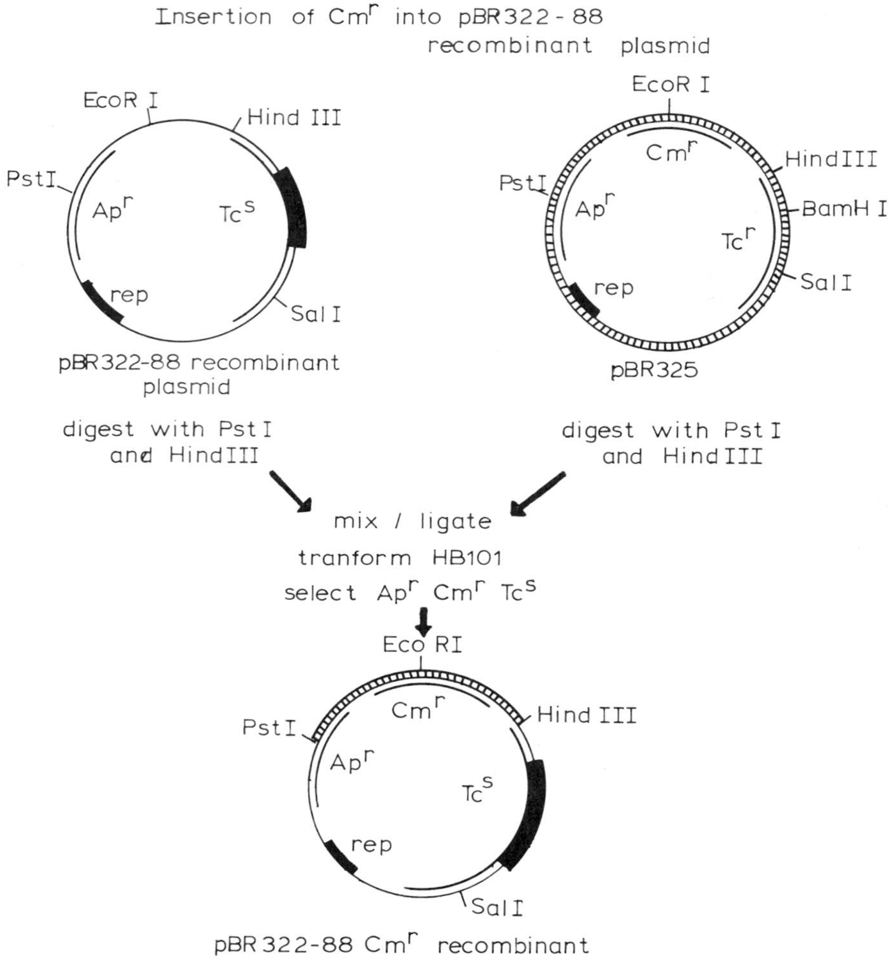

FIGURE 2.3 Protocol for cloning a gene for chloramphenicol resistance into pBR322-88.

It is non-self-transmissible but can be mobilized efficiently by several conjugative plasmids. It contains several single cutting sites that can be used for cloning. However, since neither of the resistance markers (kanamycin and streptomycin) on pKT 230 can be used for selection in *Zymomonas mobilis,* a first step was to clone for chloramphenicol resistance gene from pBR325 into this vector, following the protocol in Fig. 2.5. A partial digest of pKT 230 with Pst I and a total digest with Xho I would remove one of two fragments from this plasmid. A total digest of pBR325 with Pst I and Sal I would remove the chloramphenicol resistance gene. Xho I and Sal I do not have the same recognition sequences but do produce the equivalent

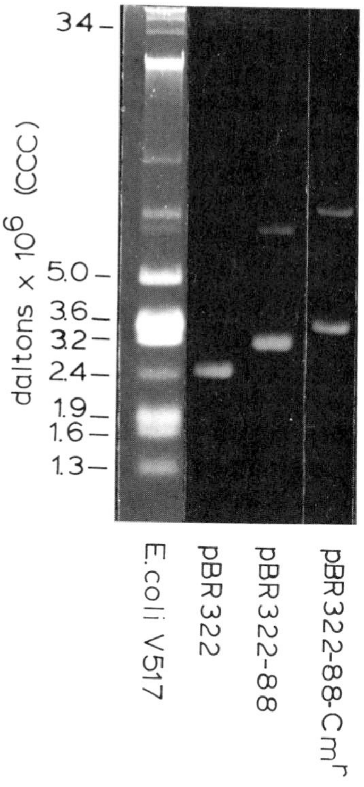

FIGURE 2.4 Relative mobility of vectors pBR322, pBR322-88, and pBR322-88 Cm^r. Plasmids from *E. coli* V517 were used for standard molecular weight markers (Macrina et al., 1978).

"sticky ends," so it is possible to ligate the fragments from the two plasmids generated by the digestions.

Two possible chloramphenicol-resistant, tetracycline-resistant plasmids can be formed, depending on which piece of pKT 230 is removed. The recombinant sequence shown in Fig. 2.5 was constructed and was confirmed by restriction analysis. If the smaller fragment from pKT 230 had been replaced, the recombinant would have two restriction sites for Hind III and Bam HI. If the larger fragment had been replaced, there would be only one site for each of those endonucleases. Fig. 2.6 confirms that the pKT 230-Cm^r recombinant has single cutting sites for Bam HI and Hind III. The pKT 230-Cm^r recombinant is mobilized into *Zymomonas mobilis* using pRK 2013:TN10 as the helper plasmid at high frequency and is greater than 95% stable unselected through 15 generations.

We now have at hand a vector that can be manipulated in *E. coli* and then transferred and maintained in *Z. mobilis*. Problems to be addressed include the questions of enzyme secretion and also of insertion and coordinate

TABLE 2.8 Attempts to Transfer Plasmid pBR322-88 to *Zymomonas* via Mobilization with pRK2013::Tn10[1]

			Comparative Controls		
Plasmid	*Recipient*	*Mobilization Frequency[2] (transconjugants/donor cell)*	*Plasmid*	*Recipient*	*Transfer Frequency (transconjugants/donor cell)*
pBR325	*E. coli*	1×10^{-4}	R68	*E. coli*	1×10^{-2}
pBR322-88 Cm^r	*Zymomonas*	$<10^{-9}$	R68	*Zymomonas*	1×10^{-7}

1. pRK2013::Tn10 was derived from pRK2013 in this laboratory.
2. Helper plasmid: pRK2013::Tn10.

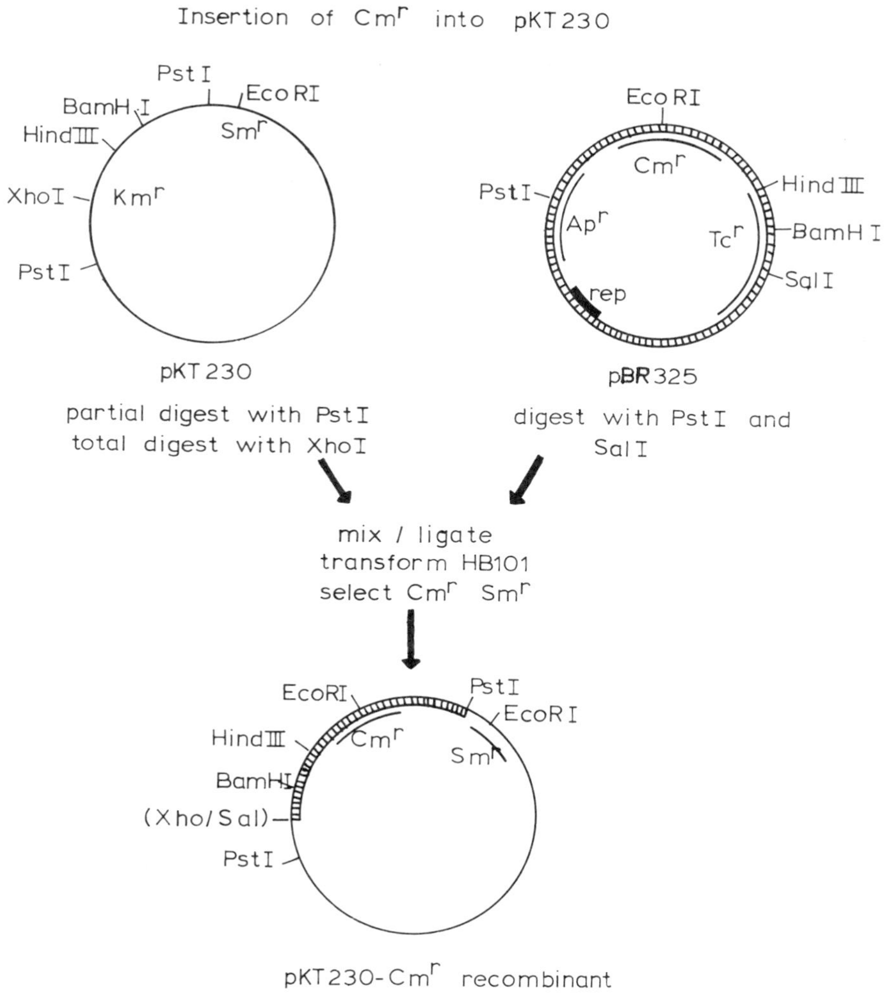

FIGURE 2.5 Protocol for insertion of a Cm^r gene into pKT 230.

activity of multienzyme (amylase, cellulase) systems. However, the way is now open for developing *Z. mobilis* through recombinant DNA techniques. Additional invertase genes can now be introduced. The potential of broadening the substrate range of *Z. mobilis* can now be entertained. For example, it may be possible to clone the genes for utilization of lactose, maltose, raffinose, starch, and cellulose in *Z. mobilis*. Further genetic approaches for increasing the ethanol productivity of *Z. mobilis* will include development of highly flocculent strains for use in cell recycle systems, removal of regulatory controls in the fermentation pathway, and enhancement of tolerance

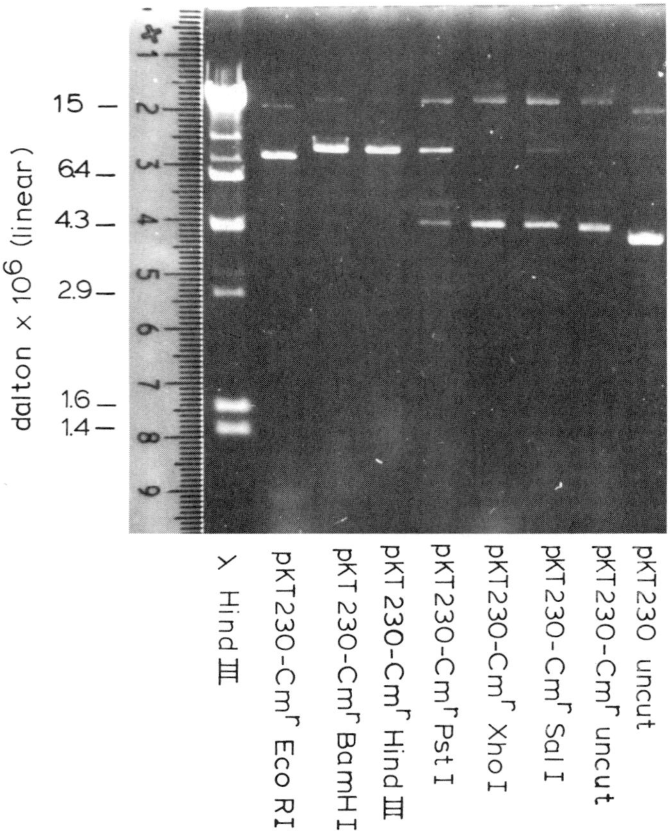

Enzyme	No. cutting sites
EcoRI	2
BamHI	1
HindIII	1
PstI	2
XhoI	0
SalI	0

RESTRICTION ANALYSIS OF pKT230-Cm^r RECOMBINANT PLASMID

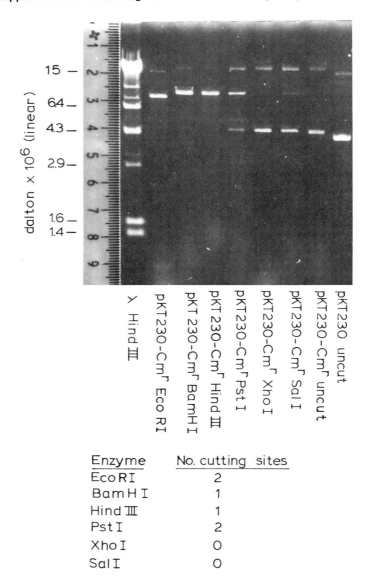

FIGURE 2.6 Restriction endonuclease digest analysis of pKT230-Cm^r recombinant plasmid.

to high sugar concentrations, to molasses, and even to ethanol (Stokes et al., 1983b).

ACKNOWLEDGMENTS

New Jersey Agricultural Experiment Station, Publication No. F-01111-2-82 supported by State funds and the U.S. Department of Energy, and Publication No. F-01902-1-82 supported by State funds and the U.S. Department of Agriculture.

REFERENCES

Arcuri, E. J., Worden, R. M., and Shumate, S. E., II (1980) *Biotechnol. Lett. 2,* 499–504.

Bagdasarian, M., and Timmis, K. N. (1982) *Curr. Top. Microbiol. Immunol. 96,* 47–68.

Belaich, J. P., Belaich, A., and Simonpietri, P. (1972) *J. Gen. Microbiol. 70,* 179–185.

Bente, P. F., Jr. (1980) *Solar Energy 25,* 297–400.

Bevers, J., and Verachtert, H. (1976) *J. Inst. Brew. 82,* 35–40.

Bryan, L. E., and Kwan, S. (1981) *J. Antimicrob. Chemother. 8* (Suppl. D), 1–8.

Bungay, H. R. (1981) *Energy, The Biomass Options,* John Wiley, New York.

Chambers, R. S., Herendeen, R. A., Joyce, J. J., and Penner, P. S. (1979) *Science 206,* 789–795.

Cysewski, G. H., and Wilke, C. H. (1977) *Biotechnol. Bioeng. 19,* 1125–1144.

Dadds, M. J. S., and Martin, P. A. (1973) *J. Inst. Brew. 79,* 386–391.

Dadds, M. J. S., Martin, P. A., and Carr, J. G. (1973) *J. Appl. Bacteriol. 36,* 531–539.

Dally, E. L., Stokes, H. W., and Eveleigh, D. E. (1982) *Biotechnol. Lett. 4,* 91–96.

Detroy, R. W., Cunningham, R. L., Bothast, R. J., and Bagby, M. O. (1981) Abstract. Third Symposium on Biotechnology in Energy Production and Conservation.

DeLey, J., and Swings, J. (1976) *Int. J. Syst. Bacteriol. 26,* 146–157.

Ditta, G., Stanfield, S., Corbin, D., and Helinski, D. R. (1980) *Proc. Nat. Acad. Sci. U.S.A. 77,* 7347–7351.

Douglas, L. (1982) in *Chemistry in Energy Production* (Wymer, R. G., and Keller, O. L., eds.), Oak Ridge National Laboratory, Oak Ridge, Tenn.

Ferchak, J. D., and Pye, E. K. (1981) *Solar Energy 26,* 9–16.

Flickinger, M. C. (1980) *Biotech. Bioeng. 22* (Suppl. 1), 27–48.

Forrest, W. W. (1967) *J. Bacteriol. 94,* 1459–1462.

Gibbs, M., and DeMoss, R. D. (1951) *Arch. Biochem. Biophys. 34,* 478–479.

Goldstein, I. S. (1980) "The Potential of Producing Petrochemical Feedstocks from Biomass," Special Report Paper, 43 pp., Solar Energy Research Institute, Golden, Col.

Gonçalves de Lima, O., De Araújo, J. M., Schumacher, I. E., and Cavalcanti Da Silva, E. (1970) *Rev. Inst. Antibiot. Univ. Recife 10,* 3–15.

Gong, C.-S., Chen, L. F., Flickinger, M. C., and Tsao, G. T. (1981a) *Adv. Biochem. Eng. 20,* 93–118.
Gong, C.-S., McCracken, L. D., and Tsao, G. T. (1981b) *Biotechnol. Lett. 3,* 245–250.
Goodman, E. E., Roger, P. L., and Skotnicki, M. L. (1982) *Appl. Environ. Microbiol. 44,* 496–498.
Grote, W., Lee, K. J., and Rogers, P. L. (1980) *Biotechnol. Lett. 2,* 481–486.
Haraldson, A., and Bjorling, T. (1981) *Eur. J. Microbiol. Biotechnol. 13,* 34–38.
Ho, N. W. Y. (1980) *Ann. Reports Fermn. Proc. 4,* 235–266.
Jeffries, T. W. (1981) *Biotechnol. Lett. 3,* 213–218.
Jeffries, T. W. (1983) *Adv. Biochem. Eng.* (in press).
Kersters, K., and DeLey, J. (1968) *Antonie van Leeuwenhoek J. Microbiol. Serol. 34,* 393–408.
Kosaric, N., Ng, D. C. M., Russell, I., and Stewart, G. S. (1980) *Adv. Appl. Microbiol. 26,* 147–227.
Lavers, B. H., Pang, P., MacKenzie, C. R., Lawford, G. R., Pik, J. R., and Lawford, H. G. (1981) in *Advances in Biotechnology* Vol. II. *Fuels, Chemicals, Foods, and Waste Treatment: Proceedings of the Sixth International Fermentation Symposium* (Moo-Young, M., and Robinson, C. W., eds.) pp. 195–200, Pergamon Press, New York.
Lawford, G. R., Lavers, B. H., Good, D., Charley, R., Fein, J., and Lawford, H. G. (1983) in *International Symposium on Ethanol from Biomass* (Duckworth, H. E., ed.) pp. 482–507, Royal Society of Canada, Ottawa.
Lazdunski, A., and Belaich, J. P. (1972) *J. Gen. Microbiol. 70,* 187–197.
Lee, K. J., Lefebvre, M., Tribe, D. E., and Rogers, P. L. (1980) *Biotechnol. Lett. 2,* 487–492.
Lipinsky, E. S. (1978) *Science 199,* 644–651.
Lipinsky, E. S. (1981) *Science 212,* 1465–1471.
Macrina, F. L., Kopecko, D. J., Jones, K. R., Ayers, D. J., and McCowen, S. M. (1978) *Plasmid 1,* 417–420.
Millis, N. F. (1956) *J. Gen. Microbiol. 15,* 521–528.
Montenecourt, B. S. (1983) in *Biology of Industrial Microorganisms* (Demain, A. L., ed.), Addison-Wesley, Reading, Mass.
Office of Technology Assessment (1980) ''Energy from Biological Processes'' (Bull, T. E., Project Director), Library of Congress, Catalog Card No. 80-600118.
Rogers, P. L., Lee, K. J., and Tribe, D. E. (1979) *Biotechnol. Lett. 1,* 165–175.
Rogers, P. L., Lee, K. J., and Tribe, D. E. (1980) *Proc. Biochem. 15* (August/September), 7–11.
Rogers, P. L., Lee, K. J., Skotnicki, M. L., and Tribe, D. E. (1982) *Adv. Chem. Biochem. Eng. 23,* 39–84.
Rose, A. H. (1978) *Economic Microbiology* Vol. I, Academic Press, New York.
Saddler, J. N., Chan, M. K.-H., and Louis-Seize, G. (1981) *Biotechnol. Lett. 3,* 321–326.
Schneider, H., Wang, P. Y., Chang, Y. K., and Maleska, R. (1981) *Biotechnol. Lett. 3,* 89–92.
Skotnicki, M. L., Lee, K. J., Tribe, D. E., and Rogers, P. L. (1981) *Appl. Environ. Microbiol. 41,* 889–893.
Skotnicki, M. L., Tribe, D. E., and Rogers, P. L. (1980) *Appl. Environ. Microbiol. 40,* 7–12.

Skotnicki, M. L., Lee, K. J., Tribe, D. E., and Rogers, P. L. (1982) in *Genetic Engineering of Microorganisms for Chemicals: Basic Life Science* (Hollaender, A., DeMoss, R. D., Kaplan, S., Konisky, J., Savage, D., and Wolfe, R. S., eds.) Vol. XIX, pp. 271–290, Plenum Press, New York.

Stokes, H. W., Dally, E. L., Spaniel, D. T., Williams, R. L., Montenecourt, B. S., and Eveleigh, D. E. (1981) *Biomass Digest 3,* 124–132.

Stokes, H. W., Dally, E. L., Williams, R. L., Montenecourt, B. S., and Eveleigh, D. E. (1982) in *Chemistry in Energy Production* (Keller, O. L., and Wymer, R. G., eds.) pp. 115–121, Oak Ridge National Laboratory, Oak Ridge, Tenn.

Stokes, H. W., Dally, E. L., Yablonsky, M. D., and Eveleigh, D. E. (1983a) *Plasmid 9,* 138–146.

Stokes, H. W., Picataggio, S., and Eveleigh, D. E. (1983b) *Adv. Solar Energy Res.* (in press).

Swings, J., and DeLey, J. (1975) *Int. J. Syst. Bacteriol. 25,* 324–328.

Swings, J., and DeLey, J. (1977) *Bacteriol. Rev. 41,* 1–46.

Swings, J., Kersters, K., and DeLey, J. (1976) *J. Gen. Microbiol. 93,* 266–271.

Tonomura, K., Kurose, N., Konishi, S., and Kawasaki, H. (1982) *Agric. Biol. Chem. 46,* 2851–2853.

Tsao, G. T. (1978) *Proc. Biochem. 13*(10), 12–14.

U.S. National Alcohol Fuels Commission (1981) "Fuel Alcohol: An Energy Alternative for the 1980s" (final report), 146 pp., ISBN 0-96-5762-0-7.

Viikari, L., Nybergh, P., and Linko, M. (1981) in *Advances in Biotechnology* Vol. II, *Fuels, Chemicals, Foods, and Waste Treatment: Proceedings of the Sixth International Fermentation Symposium* (Moo-Young, M., and Robinson, C. W., eds.) pp. 137–142, Pergamon Press, New York.

Vlitos, A. (1981) *Chemistry and Industry 9,* 303–310.

Weisz, J., and Marshall, J. F. (1979) *Science 206,* 24–29.

Wiegel, J., and Ljungdahl, L. G. (1981) *Arch. Microbiol. 128,* 343–348.

CHAPTER 3

Prospects for the Use of Genetic Engineering to Produce Industrial Chemicals with Strict Anaerobes

Bradley R. Snedecor
Reinaldo F. Gomez

I. INTRODUCTION

In this article, we will briefly review the salient features of anaerobes as producers of industrial chemicals. Then we will describe some of the current tools available to genetic engineers, as well as the state of genetic technology in strict anaerobes. Finally, we will discuss how genetic engineering could have an impact on the process of chemical production by anaerobes.

II. UTILITY OF ANAEROBES

The utility of anaerobes for the production of industrial chemicals has been reviewed by Zeikus (1980). The following paragraphs summarize the important points of that article as they apply to the topic at hand.

In recent years, the concern over petroleum shortages has caused government, as well as the industrial community, to search for alternative sources of fuels and chemical feedstocks. In addition, the increasing amounts of wastes produced by municipalities, agriculture, and industries have caused great concern.

The renewed interest in microbial fermentation processes, then, is related to our excessive dependence on petroleum and is due to three factors. First, there is the need to develop new technologies for the utilization of renewable resources and underutilized fossil reserves. Second, there is the need to improve energy conservation in waste disposal as well as to produce low-molecular-weight chemicals from sources other than petroleum. Finally, increasing prices of imported oil, as well as government-subsidized biotechnologies (such as the production of gasohol), are shifting the economics in favor of microbial fermentations.

The effective utilization of renewable resources such as biomass and wastes for chemicals and fuels presents a tremendous challenge to scientists and technologists. Within this community, microbiologists will occupy an important position in the development of new technologies. Biotechnology is thus a new and exciting discipline, and one of its future aims will be to develop novel and improved processes for the conversion of renewable resources and abundant nonrenewable resources to fuels and chemicals.

Why microbial fermentations? The term "fermentation," as it is defined by biochemists, refers to the anaerobic catabolism of organic matter, that is, the oxidation–reduction processes in which organic compounds, rather than oxygen, are used as terminal electron acceptors. This process results in the production of a variety of reduced compounds instead of CO_2 and H_2O.

Since some of the products of fermentations are chemicals with identifiable market potential, anaerobic metabolism forms the basis for the production of chemicals. Thus fermentations offer the potential for utilizing the catabolic features of anaerobes for large-scale production of chemicals and fuels.

The concept of using fermentation for the production of useful materials is not new. Alcoholic beverages and fermented foods have had a long history. In fact, fermentation was one of the first food preservation processes practiced by man. Ancient cultures utilized yeasts to make beer and wine and lactic acid bacteria for the preservation of milk.

Fuels have also been produced by fermentations. In 1939, in the United States, there were 36 industrial alcohol plants producing 200 million gallons of alcohol per year, of which 7%, or 14 million gallons, was derived from grain. During the Second World War, ethanol production by fermentation increased to a high of 637 million gallons per year in 1944. During this period, alcohol production from grain increased to 60% of the total.

Likewise, anaerobic digestion has been used for many years for waste treatment. Currently, anaerobic digestion is used throughout the world for the production of methane.

The acetone-butanol fermentation by *Clostridium acetobutylicum* is the only commercial process for production of industrial chemicals, by anaerobic bacteria, that uses a monoculture. The original observation that bacteria produce butanol was made by Pasteur in 1861. In 1914, Dr. Chaim Weizmann isolated the bacterium responsible for the fermentation and called it *C. acetobutylicum*. In 1915, he was issued British Patent No. 4845. Acetone was produced from corn during World War I for the manufacture of cordite. During the 1930's, plants were built in the United States and Japan. However, the process was discontinued in the 1960's in the United States owing to the unfavorable economics created by chemical synthesis of these products from petroleum feedstocks. Nevertheless, this process is currently used commercially in various countries, such as South Africa.

Some forms of biomass and other underutilized nonrenewable materials are resistant to anaerobic digestion and require some form of pretreatment. Pyrolysis of biomass and coal to syngas and its chemical conversion to methanol and acetic acid are economically feasible. Alternatively, biomass can be fermented directly into a variety of products or treated, prior to fermentation, to remove the nonfermentable materials, such as lignin, and also often achieve the decrystallization and depolymerization of the polysaccharides. Thus the products of biomass degradation and pyrolysis can serve as abundant substrates for fermentations. The products of such fermentations are diversified and could be utilized for a variety of purposes, such as chemical feedstocks and fuels.

The potential of anaerobes for industrial purposes lies in their diversified properties. In this regard, there are some points that are worth mentioning. First of all, little is known about the entire spectrum of anaerobic bacteria that may have importance in biofuels and biochemicals production. The point is that the search for these microorganisms in nature should continue, since it is highly unlikely that we have exhausted the pool of anaerobes that may have advantageous properties. For example, one could look for those which are able to decompose biopolymers in extreme environments.

The effects of air on these microorganisms, which are not able to use oxygen as a terminal electron acceptor, are quite varied. They range from lethality to no effect at all. However, the toxic effects of oxygen should not present a problem to large-scale industrial processes. The removal of oxygen from aqueous solutions is easy, and its solubility is low. In fact, anaerobiosis may have an advantage over strictly aerobic processes, since the latter are often limited by oxygen transfer problems. In addition, the fermentation of solid substrates, such as biomass, presents unique materials-handling problems that may make agitation and oxygen transfer an impossible task.

Thermophilic anaerobes have unique properties that make them particularly attractive for industrial purposes. These are: high metabolic rates, thermal stability of enzymes, increased end product to cell mass ratios, and a diversified metabolism. Examples of thermophiles that can produce acids, solvents, and methane are well known.

Anaerobes that are capable of growing at low pH and in the presence of high salt concentrations may also present some process advantages over those requiring less extreme conditions. For example, the ability to run fermentations under nonsterile conditions may be possible if the fermentation broth has either a low pH or a high soluble solids concentration. These conditions would limit the number of contaminants that may interfere with the fermentation.

The conversion of biomass by anaerobes can result in a variety of products. These include organic acids such as propionic, acetic, and lactic acid; alcohols and ketones such as ethanol, acetone, and butanol; gases such as methane; and other products of cell mass such as enzymes, vitamins, and possibly feeds and food materials that are high in protein. The point to remember is that large-scale fermentation processes will generate large amounts of by-products that must be effectively utilized in order to make them economically favorable.

III. MODERN GENETIC METHODOLOGY

The last two decades have seen the development of microbial genetics to an extent and in a direction that would have been difficult to predict. The remarkable progress made in this field has come about through a combination of advances in the areas of fundamental biochemistry, enzymology, and physiology. How these areas have come together to generate the discipline of "genetic engineering" is an interesting example of how basic sciences can evolve into the applied world.

A. Mutation and Selection

The ability to treat microorganisms with a variety of chemical agents and alter their genetic code has been recognized for a number of years. This had enabled microbiologists to "label" microorganisms with specific genetic traits, and this labeling allows the geneticist to recognize the desired microorganisms from a very large population. In addition, it is well known that the mutation of microorganisms in selected biochemical steps can result in strains that can perform in an improved fashion. The pharmaceutical industry has used this technique to improve many of their processes, such as the production of penicillin.

The principle of selection is defined as the application of cultural and environmental conditions to favor the growth of desired microorganisms in a mixed population. This is an important concept, since it allows microbiologists to select a given microorganism from among billions of other ones. Unfortunately, this principle cannot always be applied, and one is left with the alternative of having to screen for the desired microorganisms. Even here, technological advances have been made in recent years. Witness the development of various types of equipment that automate the process of screening.

The principles of mutation and selection have been the most widely used by the fermentation industry to improve microorganisms used in production processes. Although as we shall see in the next section, other, more sophisticated techniques are now available for genetic manipulation of microorganisms, it is expected that mutation and selection or screening will continue to be an important component in any strain improvement program. This is a logical expectation, given the simplicity and the proven record of the approach.

B. Traditional Gene Transfer Systems

The exchange of genetic material among closely related strains of microorganisms has been used by geneticists to study mechanisms of recombination, that is, the reassortments of a series of nucleotides along nucleic acid molecules. These reassortments can occur within one DNA molecule to produce deletions, inversions, transpositions, or duplications or between two separate molecules to produce molecules that are a combination of the two parental ones. The modes of gene transfer systems or recombinational events discussed in this section will be those occurring in in vivo recombination. In vitro recombinational techniques will be dealt with later. General recombination, as opposed to other types of recombination, requires a high degree of homology (similarity in nucleotide sequence) between the parental molecules. Recombinations can also occur between two molecules between which there is little or no homology. When the event is localized, in one of the two parental molecules, to a specific site, the recombination is said to be single site–specific. If it is localized in both parental molecules it is said to be double site–specific. In addition, there are recombinational events that not only do not require any homology, but also can occur at more or less random sites in the DNA molecule. These events will be discussed later.

The type of DNA involved in the recombinational event can be of various types: chromosomal, plasmid, and phage DNA. Plasmid DNA is an autonomous genetic element that replicates as a single unit, and its discovery has provided a major thrust to genetic engineering. Phage DNA is the genetic element that codes for the functions of bacterial viruses.

1. Transformation. Transformation is the process of uptake of purified DNA by bacterial cells. There are basically two types. The first type involves the development of competence, the cells' ability to interact with exogenous DNA, through specially evolved physiological states that can be obtained in the laboratory by growing the cultures under specific conditions. Examples of systems of this type include *Streptococcus pneumoniae, Bacillus subtilis,* and *Haemophilus influenza*. The second type of transformation system involves development of competence by artificial means, such as that induced in *Escherichia coli* by calcium treatment and the transformation of *Streptomyces* protoplasts in the presence of polyethylene glycol. Both types of transformation can be achieved with chromosomal DNA, plasmid DNA, or phage DNA (transfection).

2. Conjugation. Lederberg and Tatum (1946) first reported that when two auxotrophic strains are grown together, prototrophs can be recovered. This marked the discovery of unidirectional transfer of genetic material from one strain to another mediated by a sex factor. For the most part, the transfer of DNA through conjugation involves cell to cell interactions and DNA replication. This mode of genetic exchange achieves transfer of chromosomal DNA and plasmid DNA, such as R-factors (plasmids that carry antibiotic resistance determinants) and C-factors (plasmids that carry colicinogenic determinants).

3. Phage-Mediated Transfers

a. Phage Conversions. The presence of phages in culture often modifies the bacterial phenotype. If the phage is established as a stable prophage, that is, integrated into the bacterial chromosome, the term "lysogenic conversion" is applied. This process is arbitrarily differentiated from specialized transduction. Specialized transduction refers to the transfer of function to bacterial cells that normally are not found in phages in nature. The phage acquires the function through a rare recombinational event. When the phage provides a function that it normally carries in nature, or truly substitutes for a bacterial one, the process is considered to be phage conversion. An example of phage conversion is observed in the conversion of *Bacillus pumilis* asporogenous strains to sporulating ones (Bramucci et al., 1977).

b. Specialized Transduction. When phages that are integrated into the chromosome (prophages) are excised and "loop out," rare abnormal events result in the creation of hybrid phage–bacterial genomes. These hybrids can then be transferred to other bacterial cultures by subsequent infections followed by integration or plasmid formation in the recipient cells. This mode of transfer is said to be specialized because only those genes immediately adjacent to the integration site of the prophage are transferred.

c. Generalized Transduction. Unlike specialized transducing phages, generalized transducing phage preparations do not carry phage–bacterial DNA hybrids. The production of generalized transducing particles is thought to be the result of accidental packaging of bacterial DNA into phage particles during phage maturation. A wide variety of bacterial genes are subject to this mode of transfer. The generalized transducing particles can inject the bacterial DNA into recipient cells. These in turn must have the ability to recombine this DNA with chromosomal DNA in one way or another, and the new DNA must have the ability to establish itself as an autonomous plasmid.

C. "Modern" Recombinational Tools

The following descriptions deal with either new observations or modifications of standard procedures that have a direct effect on the expansion of flexibility in genetic manipulations of microorganisms.

1. Promiscuous Sex Factors. The P1R plasmids of *Pseudomonas* will transfer themselves by mating between virtually all gram-negative bacteria (Hopwood, 1979). These plasmids have the ability to mobilize chromosomal genes and therefore greatly extend host range. It has become possible to promote chromosomal recombination freely among three species of *Rhizobium* by means of plasmid R68.45, derived from *Pseudomonas aeruginosa* (Johnston and Beringer, 1977).

2. Transposable Genetic Elements. Translocating DNA elements are segments of DNA that can be transposed from one site to another within bacterial cells, between and within chromosomes, plasmids, or phage genomes, by illegitimate recombination (Hopwood, 1979). There are basically three classes of these elements (insertion sequences, transposons, and phage Mu). The simplest of these elements are insertion sequences. Transposons are transposable genetic elements that carry selectable markers. The most complicated of the elements is bacteriophage Mu-1, which in addition to being able to move from site to site can also occur extracellularly.

Perhaps the most useful of the translocatable elements are the transposons carrying drug resistance markers. The properties of drug resistance elements that are important for their usefulness in bacterial genetics are the following (Kleckner et al., 1977):

1. Translocatable drug resistance elements can be found inserted at a large number of sites on the bacterial chromosome.
2. Interrupted genes suffer complete loss of function.
3. The phenotype of the insertion mutation is completely linked to drug resistance in genetic crosses.
4. Insertion mutants can be recovered at high frequency after low-level "mutagenesis" by exposure to a translocatable element.

5. Insertion mutations can revert by precise excision with concomitant loss of drug resistance.
6. Insertions in operons are strongly polar.
7. Drug resistance insertions can generate deletions nearby.
8. Drug resistance elements can provide a portable region of homology.
9. Insertions behave as point mutations in fine structure mapping.
10. Insertions can be specifically obtained that are near but not within a gene of interest.

The transposons have a variety of uses in bringing about illegitimate recombination and thus the genetic engineering of new bacterial strains in vivo. These uses have been summarized by Kleckner et al. (1977) as follows:

I. Transfer of otherwise unselectable markers by virtue of the drug resistance associated with insertion mutations.

II. Isolation of mutants that lack an enzymatic function but have no readily scorable phenotype.

III. The obtaining and exploitation of drug resistance insertions near but not within regions of interest:
   - A. manipulation of markers with no selectable phenotype;
   - B. localized mutagenesis;
   - C. chromosomal mapping;
   - D. F′ mapping and complementation tests;
   - E. "cloning" of prokaryotic genes using in vitro recombination techniques.

IV. Use of drug resistance insertions as a portable region of homology:
   - A. construction of Hfr strains with origin and direction of transfer specified;
   - B. construction of new F′ episomes;
   - C. construction of specialized transducing phages;
   - D. deletions and duplications with predetermined endpoints specified by positions of insertions.

V. Generation of deletions with one endpoint at an insertion and the other endpoint unspecified.

VI. Selection of chromosomal duplications with translocatable drug resistance elements and their use in chromosomal mapping.

3. In vivo Recombination by Restriction Enzymes. Chang and Cohen (1977) reported that site-specific genetic recombination can be promoted in vivo by the EcoRI restriction endonuclease. The recombinant DNA molecules formed in vivo by ligation of DNA fragments that were cleaved intracellularly by the EcoRI enzyme were structurally and genetically indistinguishable from those constructed in vitro by the use of recombinant DNA methods. In addition, EcoRI-generated fragments of eukaryotic DNA unmodified by *E. coli* methylase have been shown to undergo ligation in vivo to segments of bacterial plasmid DNA.

4. In vitro Recombinant DNA Techniques. Several authors (e.g., Cohen, 1979) have summarized the techniques that are collectively known as recombinant DNA technology. These will not be reviewed again here. It is enough to say that the large array of techniques that can be combined to isolate, sequence, synthesize, modify, and join fragments of DNA makes it conceivable to design or obtain nearly any desired combination of DNA sequences, provided enough is known about the problem at hand to be able to specify what the desired DNA sequence is. Some of the ways this ability can be used to study or apply genes are reviewed by Wetzel (1980). In anaerobes, the ability to apply these techniques fully will depend on the development of means to introduce the products of in vitro DNA manipulation into these bacteria.

5. Protoplast Fusion. The fusion of mammalian cells, protoplasts of plant cells, and fungi has been a routine technique for a number of years. However, it was not until 1976 that protoplast fusion was described in bacteria (Fodor and Alfoldi, 1976; Schaeffer et al., 1976). Recombination through protoplast fusion in bacteria has, so far, only been reported in gram-positive bacteria, such as *Bacillus* (Fodor and Alfoldi, 1976; Schaeffer et al., 1976) and *Streptomyces* (Baltz, 1978; Hopwood et al., 1977).

The technique involves the generation of protoplasts, by removing the cell wall of bacteria, with combination treatments of antibiotics (for example, penicillin) and enzymes (for example, lysozyme) under isotonic or hypertonic conditions. The protoplasts are then induced to fuse with polyethylene glycol, followed by incubation under conditions that allow reversion of the protoplasts to the bacillary form (protoplast regeneration).

Protoplast fusion can be applied to establish recombination systems in strains with no efficient natural means of conjugation. One of the advantages of protoplast fusion is the high frequency at which recombinants are produced under nonselective conditions and in the absence of known sex factors. In *Streptomyces coelicolor,* frequencies as high as 20% have been observed (Hopwood, 1979). Another feature of protoplast fusion is that more than two strains can be combined in one fusion. Up to four strains have been fused to yield recombinants inheriting genes from all four parents. This approach should be extremely useful in accelerating strain improvement programs. Protoplast fusion has another advantage over conjugation. Since recombinants are so numerous, it is feasible to screen a random sample of the progeny for improved strains. In contrast, when natural conjugation is used, counterselectable markers must be introduced into each parent in order to select the rare recombinants prior to screening. This is a much more laborious procedure.

IV. GENETIC TECHNOLOGY IN ANAEROBES

Published reports relating to the genetics of anaerobes are very few.

In the *Clostridia,* conjugation has been observed in one species, *C. perfringens*. A growing number of isolates of *C. perfringens* have proved to be resistant to several antibiotics (Rood et al., 1978a). This raised the possibility that the rapid spread of varied antibiotic resistances is promoted in *Clostridium,* as it is in many other genera, by the conjugative exchange of antibiotic resistance plasmids. The presence of plasmids was demonstrated in *C. perfringens* (Ionesco et al., 1976), and in 1977, the exchange of antibiotic resistance plasmids between strains of *C. perfringens* by conjugation was demonstrated (Brefort et al., 1977). This observation was confirmed by Rood et al. (1978b). Transfer of antibiotic resistance has also been observed between strains of *Clostridium difficile* (Ionesco, 1980; Smith et al., 1981).

Phage are common in the *Clostridium* (e.g., Grant and Riemann, 1976; Mahoney and Katz, 1968), and an example of phage-mediated genetic transfer has been reported in *C. botulinum* (Eklund and Poysky, 1974). Protoplasts have been formed in several species, and transfection of protoplasts of *C. saccharoperbutylacetonicum* has been accomplished by Ogata et al. (1981).

The transformation of *C. butyricum* has been accomplished by H. Adler, C. Hadden, and A. Currasco at Oak Ridge National Laboratory (C. Hadden, personal communication). In *C. thermocellum,* mutagenesis has been studied (Gomez et al., 1981), and some auxotrophic mutations have been isolated (Mendez and Gomez, 1982). In addition, an insertion sequence has been isolated (see below).

In the non-spore-forming anaerobes, the situation is nearly the same as in *Clostridium*. Antibiotic resistance plasmids have been transferred between several species of *Bacteroides* by conjugation (*B. fragilis* to *B. fragilis* by Privetera et al. (1979), *B. fragilis* to *B. thetaiotaomicron* by Tally et al. (1979), *B. fragilis* to *B. uniformis* and vice versa by Welch et al. (1979)). Transfection has been accomplished in *B. thetaiotaomicron* (Burt and Woods, 1977), but transformation has not yet been reported, despite a contrary claim in one review (Macy and Probst, 1979).

In addition, however, antibiotic resistance plasmids have also been transferred by conjugation from *E. coli* to *B. fragilis* (Burt and Woods, 1976) and from *B. fragilis* to *E. coli* (Mancini and Behme, 1977), as well as from *B. ochraceus* to *E. coli* (Guiney and Davis, 1978).

In the last case, one of the antibiotic resistance markers involved (kanamycin resistance) was shown to behave as a transposon in *E. coli* (Amyes, 1980). This is the first example of a transposable DNA element isolated from an anaerobe. Preliminary evidence suggests the presence of another transposon in transmissible plasmid isolated from *B. fragilis* (Welch and Macrina, 1981).

While at the Massachusetts Institute of Technology, we pursued a line of research that culminated in the demonstration that *Clostridium thermocellum* contains an insertion sequence that functions in *E. coli* (B. S. Snedecor and R. F. Gomez, manuscript in preparation). *C. thermocellum* is a thermophilic anaerobe that grows on cellulose (Weimer and Zeikus, 1977). Each of these characteristics contributes to the interest in this microorganism.

As we stated earlier, increasing petroleum costs have led to increased interest in alternative raw materials, such as biomass, for producing chemicals and transportation fuels. This has led in turn to a resurgence of interest in anaerobic fermentation. Hence *C. thermocellum* is of interest as an example of an anaerobe.

Cellulose is the most abundant component of biomass. The fact that cellulose is the world's most abundant biological molecule provoked interest in its use as such a raw material long before the present widespread concern over petroleum shortages. Since it is a linear polymer of glucose, its use as a raw material for fermentation seemed conceptually simple, although practically difficult. *C. thermocellum* can be used as the agent for the breakdown of cellulose, so *C. thermocellum* is also of interest as a cellulolytic organism.

Finally, there is continuing interest in the application of thermophiles to fermentation processes. Consequently, *C. thermocellum* is also of interest as a thermophile.

Since mutation/selection is the only currently available method for improving this organism, these characteristics made *C. thermocellum* a suitable choice of an anaerobe for which genetic systems should be developed.

The first part of any genetic system for *C. thermocellum* has to be some method of introducing DNA into the organism; at this writing, that necessary part has not yet been developed. However, with the incorporation of a selectable marker, the insertion sequence that we isolated could by itself represent the other major part of a simple but versatile system for genetic manipulation of *C. thermocellum*.

As was stated, one of the first requirements for the development of a genetic system is the ability to move genetic information between strains. The following approach, suggested by the strategy used by Hinnen et al. (1978) to obtain the first transformation of yeast, was adopted.

The DNA of *C. thermocellum* was first cloned in *E. coli*. The strategy was to look for a selectable gene from the organism in *E. coli*—for instance, one that would complement an auxotrophic mutant—and then to use that cloned DNA as the source of a highly enriched DNA probe for detecting transformation of a similar *C. thermocellum* mutant. This has the advantage that it should allow detection of transformation events more rare than would otherwise be detected.

Recombinant plasmids were constructed that contained *C. thermocellum* DNA. These were cloned in *E. coli,* and the collection of clones was replica-plated to a medium that lacked a required amino acid, in order to screen for plasmids that would complement one of the amino acid auxotrophies (for proline and for leucine) of the host that was used.

One of the clones gave rise to a colony on a medium that lacked proline. The recombinant plasmid from this clone, called pTC1, contained a 2.3-kilo–base pair piece of *C. thermocellum* DNA.

At that point, it was hoped that the plasmid contained *C. thermocellum* genes for proline synthesis, which could complement the proline mutation in the *E. coli* host. To confirm this, the plasmid was reisolated and re-transformed into the host. It was apparent that most host cells receiving this plasmid did not become proline prototrophs, but that populations of cells containing the plasmid gave rise to prototrophs at a higher rate than cells receiving control plasmids. The plasmid pTC1 definitely had an associated phenotype, but this was the ability to generate genetic variants rather than complementation of a specific defect.

An indication of the mechanism by which this *C. thermocellum* DNA could cause genetic changes came when an attempt was made to enrich for cells that had lost the plasmid. This was done by selecting against cells containing the tetracycline resistance gene carried in the plasmid by the cloning vector. Tetracycline-sensitive cells were obtained with the expectation that these would have lost the plasmid pTC1. Instead, they contained derivatives of pTC1 that were altered in size.

The structure of these altered plasmids suggested that an insertion sequence present in pTC1 had catalyzed the generation of the altered plasmids. One derivative had acquired a second copy of what was later shown to be the 1.4-kilo–base pair insertion sequence, inserted into the tetracycline resistance gene. The other derivative contained a deletion that extended into the tetracycline resistance gene from one end of the original insertion sequence. This type of deletion is frequently caused by insertion sequences.

Transposable DNA elements cause a characteristic short duplication of the DNA sequence at each site of insertion and frequently carry inverted repeats at their ends. DNA sequencing of the insertion sequence at one insertion site (B. Snedecor, E. Chen, D. Henner, M. Dennis, and R. Gomez, manuscript in preparation) showed that this insertion sequence meets those expectations. It carries 7-bp inverted repeats at its ends and causes a 3-bp direct repeat at the site of insertion.

To show that this sequence is indeed capable of transposing between DNA molecules, mutants were obtained in which it could be shown by hybridization that a copy of the insertion sequence (and not the entire plasmid) had transposed to the chromosome of *E. coli* from the plasmid.

Hybridization was also used to confirm that the source of the insertion sequence was *C. thermocellum* and not *E. coli*. However, the insertion sequence is capable of directing protein synthesis in *E. coli* as determined by the maxicell technique (Sancar et al., 1979) and is also capable of transposition in *E. coli*.

In summary, an insertion sequence was isolated from *C. thermocellum* that will function in *E. coli*. With the in vitro incorporation of a selected gene, this could become usable as a transposon in *C. thermocellum* and perhaps other clostridia.

V. IMPROVEMENTS OF FERMENTATIONS BY GENETIC ENGINEERING

In any process, it is of course important to think about improvements. Basically, improvements in biotechnology can come from two sources. First, there is the application of engineering principles designed to control environmental parameters that can shift metabolic routes toward the desired products. This, of course, is the domain of the biochemical engineer. Second, the genetic modification of microorganisms is also a valid approach. Witness the success of the pharmaceutical industry in improving the production of antibiotics and other products by the application of simple genetic principles. On a more sophisticated level, the role and potential of recombinant DNA technology have attracted the attention of many sectors of our society as a way of developing new or improved biotechnologies.

The question is, Can we apply modern genetic technology to anaerobes? This is a problem. And this problem arises from our poor understanding of fundamental principles in anaerobes, not only of their genetics but also of some of the more important biochemical properties that one would like to manipulate. For example, the resistance, or rather the lack of it, of a bacterium to its own fermentation products often limits its ability to accumulate large amounts of a desirable material. This is indeed the case with many anaerobic bacteria. Another example is the fact that anaerobes, as well as facultative anaerobes, often accumulate a variety of products rather than the desired one. This is of course due to the fact that metabolism in these microorganisms was not designed for our convenience, but rather for the optimization of their energy metabolism according to ecological conditions. One would like to understand the principles underlying these heterofermentative properties in order to design rational approaches to their manipulation and ultimately maximize the production of the desired product. In addition, other fermentative properties, such as intrinsic enzymic activities and range of utilizable substrates, are areas that require further research before they can be improved.

Notwithstanding the paucity of biochemical data and genetic methodology in anaerobes, the following areas could be targets for genetic manipulations resulting in improved strains.

In fermentations aimed at producing low-value chemicals (less than

$1.00/lb), the cost of the substrate will be an important determinant in the overall process economics. Therefore it is important to utilize microorganisms that give the highest possible yields. Microorganisms could be engineered to utilize pathways that maximize yields and to direct as much of the substrate as is theoretically possible to the desired end product and away from unwanted products and cells mass.

The tolerance that a microorganism has to its own fermentation end products often limits the maximum concentration that can be achieved in a fermentation. This feature has a large impact on recovery costs of the end product, as well as on the fermentor capital cost, through its effect on productivity. Even though our understanding of the genetic traits that control the susceptibility of a microorganism to its own fermentation products is in its infancy, it is conceivable that, in the future, we will learn enough about these characteristics to allow us to undertake genetic manipulations.

The rate at which a product is produced is also an important consideration. Genetic manipulations aimed at the deregulation of metabolism and at the amplification of rate-limiting steps should result in microorganisms capable of fermenting substrates at increased rates.

Another aspect to consider is the ability of a microorganism to utilize a variety of substrates, such as starch, cellulose, and pentoses. Expanding the range of substrates that the microorganism can use and expanding the rates at which it can use them are goals that should be achievable through genetic manipulations.

Finally, the improvement of the harvesting properties of a microorganism (for example, by using filamentous variants) and the effective utilization of the cellular biomass are other areas that can be addressed by genetic engineering.

VI. CONCLUSIONS

Anaerobic bacteria undoubtedly present a significant potential for the production of chemicals and fuels. However, more research on both fundamentals and applications of anaerobic metabolism is needed before their true utility in biofuel and biochemical production can be adequately assessed. First of all, it is unlikely that we have exhausted nature's pool of anaerobes with interesting properties, in particular those able to function in extreme environments, such as thermophiles, halophiles, and acidophiles. In addition, more work is needed in order to understand and thus control fermentation processes. These areas include engineering designs, genetic methodology, and product recovery. It is also interesting to think about possible manipulations of known species by chemical control of fermentation parameters or by genetic manipulations of microorganisms that may result in new fermentation products.

ACKNOWLEDGMENT

This is contribution no. 106 from Genentech, Inc.

REFERENCES

Amyes, S. (1980) *J. Antimicrob. Chemother. 6,* 303–305.
Baltz, R. H. (1978) *J. Gen. Microbiol. 107,* 93.
Bramucci, M. G., Koggins, K. M., and Lovett, P. S. (1977) *J. Virol. 22,* 194.
Brefort, G., Magot, M., Ionesco, H., and Sebald, M. (1977) *Plasmid 1,* 52–66.
Burt, S., and Woods, D. (1976) *J. Gen. Microbiol. 93,* 405.
Burt, S., and Woods, D. (1977) *J. Gen. Microbiol. 103,* 181.
Chang, S., and Cohen, S. N. (1977) *Proc. Nat. Acad. Sci. U.S.A. 74,* 4811.
Cohen, S. N. (1979) in *Recombinant DNA and Genetic Experimentation* (Morgan, J., and Whelan, W. J., eds.) pp. 49–53, Pergamon Press, New York.
Eklund, M., and Poysky, F. (1974) *Appl. Microbiol. 27,* 251.
Fodor, K., and Alfoldi, L. (1976) *Proc. Nat. Acad. Sci. U.S.A. 73,* 2147.
Gomez, R. F., Snedecor, B. S., and Mendez, B. (1981) *Dev. Ind. Microbiol. 22,* 87.
Grant, R., and Riemann, H. (1976) *Can. J. Microbiol. 22,* 603.
Guiney, D., and Davis, C. (1978) *Nature 274,* 181.
Hinnen, A., Hicks, J., and Fink, G. (1978) *Proc. Nat. Acad. Sci. U.S.A. 75,* 1929.
Hopwood, D. A. (1979) *Proceedings of the Third International Symposium on the Genetics of Industrial Microorganisms* (Sebek, O. K., and Laskin, A. I., eds.) pp. 1–9, American Society for Microbiology, Washington, D.C.
Hopwood, D. A., Wright, H. M., Bibb, M. J., and Cohen, S. N. (1977) *Nature 268,* 171.
Ionesco, H. (1980) *Ann. Microbiol. 131,* 171.
Ionesco, H., Bieth, G., Dauguet, C., and Bouanchaud, D. (1976) *Ann. Microbiol. Inst. Pasteur 127,* 283.
Johnston, A. W. B., and Beringer, J. E. (1977) *Nature 267,* 611.
Kleckner, N., Roth, J., and Botstein, D. (1977) *J. Molec. Biol. 116,* 125.
Lederberg, J., and Tatum, E. L. (1946) *Cold Spring Harbor Symp. Quant. Biol. 11,* 113.
Macy, J., and Probst, I. (1979) *Ann. Rev. Microbiol. 33,* 561.
Mahoney, D., and Katz, G. (1968) *Can. J. Microbiol. 14,* 1085.
Mancini, C., and Behme, R. (1977) *J. Infect. Dis. 136,* 597.
Mendez, B., and Gomez, R. F. (1982) *Appl. Environ. Microbiol. 43,* 495.
Ogata, S., Choi, K., Yoshino, S., and Hyashida, S. (1981) *J. Fac. Agr. Kyushu Univ. 25,* 201.
Privetera, G., Dublanchet, A., and Sebald, M. (1979) *J. Infect. Dis. 139,* 97.
Rood, J., Maher, E., Somers, E., Campos, E., and Duncan, C. (1978a) *Antimicrob. Agents Chemother. 13,* 871.
Rood, J., Scott, V., and Duncan, C. (1978b) *Plasmid 1,* 563.
Sancar, A., Hack, A., and Rupp, W. (1979) *J. Bacteriol. 137,* 692.
Schaeffer, P., Cami, B., and Hotchkiss, R. D. (1976) *Proc. Nat. Acad. Sci. U.S.A. 73,* 2151.

Smith, C. J., Markowitz, S. M., and Macrina, F. L. (1981) *Antimicrob. Agents Chemother. 19,* 997.

Tally, F., Snydman, D., Gorbach, S., and Malamy, M. (1979) *J. Infect. Dis. 139,* 83.

Weimer, P., and Zeikus, J. (1977) *Appl. Environ. Microbiol. 33,* 289.

Welch, R. A., and Macrina, F. L. (1981) *J. Bacteriol. 145,* 867.

Welch, R., Jones, K., and Macrina, F. (1979) *Plasmid 2,* 261.

Wetzel, R. (1980) *American Scientist 68,* 644.

Zeikus, J. G. (1980) *Ann. Rev. Microbiol. 34,* 423.

CHAPTER

4

Design Factors for Construction of Competitive Production Strains and Manufacturing Processes

Bruce K. Hamilton

I. INTRODUCTION

Rapid recent developments in recombinant DNA technology (RDT) and, more generally, molecular biology are continuously making available new tools that can be powerfully employed in biological process conception, process development, and, ultimately, manufacturing. Amid these new developments, the fundamental commercial objective is, of course, unchanged: to implement and operate a process that competitively produces a marketable product. But a new organizational challenge is clearly evident: to create and foster an environment in which engineers, applied scientists, and business people interact in a vigorous give-and-take mode with the molecular biologists who are working on the frontiers of their science. The desired result is to shape and move new molecular biology into profitable and sustained commercial reality.

In the area of biological process conception (the analog of "process synthesis" in traditional chemical engineering (see Rudd et al., 1973)), nearly everyone in contact with the mass media is aware that RDT has made feasible new processes that were literally impossible just 5 years ago; the most obvious example is production of a higher eukaryotic protein such as a human interferon (HuIFN) by a prokaryotic production strain such as *E. coli.* But HuIFNs are a somewhat singular example (see Section II.A.2) because their scarcity and potential therapeutic value may result in large markets for very expensive products. When considering lower-value products such as organic chemicals (for example, ethanol or acetic acid), amino acids (for example, lysine), or even most specialty proteins (for example, human serum albumin or calf rennin) produced with microbes constructed through application of RDT, certain specific "design factors" become much more important than they are in the case of HuIFNs. Engineers, applied scientists, and their business associates must make sure that these design factors are understood by their molecular biology colleagues, so that the molecular biologists can develop and apply their science in pursuit of the necessary process objectives.

The same point holds true not only for biological process conception, but also for biological process development and improvement; biochemical engineers, and applied biologists in particular, must define what specifications have to be met by a production strain constructed through RDT in order for that strain to possess the characteristics necessary for it to be installed usefully into a profitable manufacturing process.

The product range susceptible to RDT methodologies is very broad; it includes not only proteins such as HuIFNs, human serum albumin, and calf rennin as already mentioned, but also antibiotics and antitumor agents, steroids, vitamins, amino acids, organic acids, nucleotides, solvents and liquid fuels, organic chemicals, and so on (see Table 4.1; see also Glick (1982) and *Impacts of Applied Genetics,* published by the Office of Technology Assessment, U.S. Congress, 1981). Even though this product range covers a wide spectrum, the "design factors" mentioned above apply in common for RDT microbes engineered to make such products profitably. Actually, these "design factors" have importance not only for the implementation of profitable manufacturing processes alone, but also for the initial design of applied molecular biology research and development programs that are intended to be the foundation of new and expanded biological process industries rivaling, in at least some instances, the scope and the scale of the chemical process industries (Shreve and Brink, 1977).

In this article, important production strain and biological process design factors will be specifically identified; their role in biological process conception, development, and improvement will be illustrated; and some of their implications for design of applied molecular biology research and development programs will be explored. In order to underscore the concrete relevance of these design factors, highlights from existing industrial microbiology and fermentation technology practice will be emphasized throughout the discussion.

TABLE 4.1 Some Product Classes Potentially Susceptible to RDT

1. Proteins (examples*: human insulin, HuIFNs, HGH, AGHs, HSA, AHF, urokinase, interleukins, rennin, other proteases, amylases, pectinases, penicillin acylase, glucose isomerase)
2. Antibiotics and antitumor agents
3. Steroids
4. Vitamins (example: B_{12})
5. Amino acids (example: lysine)
6. Organic acids (example: citric acid)
7. Nucleotides and related products (examples: IMP, GMP)
8. Solvents, liquid fuels, and organic chemicals (examples: ethanol, butanol, acetic acid)
9. Cells (examples*: yeast, SCP)
10. Polysaccharides (example: xanthan)
11. Sugars (example: fructose)
12. Pesticides
13. Biosurfactants

* abbreviations: HuIFNs = human interferons; HGH = human growth hormone; AGHs = animal growth hormones; HSA = human serum albumin; AHF = antihemophilic factor; SCP = single cell protein.

II. THE CHARACTER OF INDUSTRIAL MICROORGANISMS

A. High Titer

1. Titers Industrial Microorganisms Achieve Now. The titer (concentration) at which an industrial microorganism produces a desired product is often impressive (Table 4.2). For example, citric acid is made at up to about 150 g/l by the mold *Aspergillus niger* and also by yeasts such as various strains

TABLE 4.2 Titers for Selected Industrial Fermentations

Product	*Microorganism*	*Titer (g/l)*
Low-Molecular-Weight Primary Metabolites		
Citric acid	*Aspergillus niger, Candida*	~150
MSG	Coryneforms	>100
Ethanol	Yeasts	~100
Lysine	Coryneforms	75
Inosine	*Brevibacterium ammoniagenes*	30
IMP	*Brevibacterium ammoniagenes*	19
Secondary Metabolites		
Penicillin	*Penicillium chrysogenum*	>30
Macromolecules		
Cellulase	*Trichoderma reesei*	>20
Alkaline protease	*Bacillus subtilis*	14

of *Candida* (Bartholomew and Reisman, 1979; Miall, 1978). MSG (monosodium glutamate) is routinely made at over 100 g/l in commercial fermentations employing as a production strain one of several members of the coryneform group (Demain, 1980). Ethanol is produced by yeasts (particularly *Saccharomyces cerevisiae*), often at titers in the neighborhood of 100 g/l, although many alcohol distilleries run to only about 6 vol %, which is equivalent to 47 g/l (Demain, 1980; Venkatasubramanian and Keim, 1981; Esser and Schmidt, 1982). Lysine is made at over 75 g/l by coryneforms (French patent application No. 77 04611, Ajinomoto Co., February 17, 1977). The coryneform *Brevibacterium ammoniagenes* produces inosine at over 30 g/l, and IMP at 19 g/l or more (Hirose et al., 1979).

The products just listed (citric acid, MSG, ethanol, lysine, inosine, IMP) are all primary metabolites (Demain, 1980, 1981). Industrial microorganisms produce high titers of secondary metabolites as well (Demain, 1980, 1981). Penicillin, for example, is made by the mold *Penicillium chrysogenum* at over 30 g/l (Queener and Swartz, 1979).

In addition to the low-molecular-weight products so far mentioned, macromolecular products can be also made at high titer. The fungus *Trichoderma reesei* makes cellulase at over 20 g/l (Ryu and Mandels, 1980; Shoemaker, 1981). *Bacillus subtilis* produces alkaline protease at more than 14 g/l (Ikeda et al., 1974).

High titers can have a very favorable impact on manufacturing economics. Therefore one prime target during development of an industrial process for biological manufacture of a product should be production at high titer.

2. For RDT Microorganisms, How High Is High Titer? The answer to this question depends on the manufacturing and marketing economics of the product being considered. The importance of a first-pass manufacturing economics analysis at an early stage of project conception and evaluation cannot be overemphasized (the article by Bartholomew and Reisman (1979) describes how to go about an analysis of biological manufacturing economics). This cost analysis should be coupled with a rough estimate of market potential. In some cases, conclusions having to do with manufacturing and marketing economics are relatively obvious; for example, for an established product (such as lysine) with market expansion possibilities, it is likely that the product titer will rapidly need to be at least comparable to current technology (75 g/l for lysine; see Table 4.2). Eventually, new technology may be able to improve upon titers currently achieved. For the case of a product that has never before been made by a microbe but that can now be made (at least in principle) by a production strain constructed via RDT, consider the examples given below, which center on manufacture of higher eukaryotic proteins by use of RDT microorganisms.

Higher eukaryotic proteins in particular, and prokaryotic and eukaryotic

TABLE 4.3 "Value Groups" for Selected Higher Eukaryotic Proteins

Value Group	*Protein*	*Ballpark Value*	*Source*
High value	HuIFNs (α, β, γ)*	Billions of dollars per pound	Tissue culture
High value	Urokinase	Tens of millions of dollars per pound	Tissue culture
Moderate value	HSA	~$1000 per pound	Blood
Moderate value	Calf rennin	~$1000 per pound	Calves
Low value	Proteases, amylases, etc.	Tens to hundreds of dollars per pound	Fermentation

* α, β, and γ are classes of HuIFN.

proteins in general, can be categorized into "value groups" (see Table 4.3). The "high value" group includes HuIFNs. The value of an HuIFN before the advent of RDT was literally billions of dollars per pound (Demain, 1981); this high cost resulted from expensive production techniques (for example, tissue culture; see White et al. (1980)) and product scarcity. Because the historical cost of HuIFN is so high, HuIFN produced by RDT will be able to sell at a relatively high purchase price (at least initially), assuming that demand for product materializes at the levels some have predicted (Lescaze, 1980). Therefore production fermentor HuIFN titers need not be very great in relation to the protein titers of 10–20 g/l cited in Table 4.2. Quantitatively, suppose that an RDT *E. coli* HuIFN fermentation is run with a harvest dry cell weight (DCW) of 10 g/l and an expression level of 10% of cellular protein as the desired HuIFN. Assuming that 50% of the *E. coli* DCW is protein (Aiba et al., 1973), the HuIFN titer will then be only 0.5 g/l. Taking the specific activity of the HuIFN to be 2×10^8 units/mg (Zoon et al., 1980), 0.5 g of HuIFN per liter is equivalent to 10^{11} units per liter; one liter of such an RDT *E. coli* broth would contain 100 billion units of HuIFN. Note that, in 1978, the American Cancer Society paid two million dollars to purchase just 40 billion units of human interferon (*Journal of the American Medical Association,* 1979). Thus an *E. coli* fermentation that makes only half a gram per liter of HuIFN seems quite good because HuIFN sells for such an enormous price (at least it does now).

On the other hand, the ballpark sales price of human serum albumin (HSA) or calf rennin is about $1000 per pound; HSA and calf rennin are "moderate value," rather than "high value" proteins (Table 4.3). A manufacturing economics analysis will show that it is unlikely that it will be possible to produce and sell HSA or calf rennin profitably at $1000 per pound by using an RDT microbe that makes product at only 0.5 g/l; substantially higher titers will be necessary. Attainment of such high titers may require development of new RDT strategies (see Sections III and VII).

3. What Limits Titer? The magnitude of a product harvest titer can be influenced by at least five characteristics:

1. Amount of synthesizing system (for example, enzyme) generated
2. Relative activity of synthesizing system present
3. Stability of synthesizing system
4. Supply of substrate
5. Stability of product

The influence of whether or not the product is extracellular is covered in Section III and is excluded from explicit consideration here.

Three of the above characteristics focus on the "synthesizing system," which is typically composed of enzymes but may also include other components such as plasmids, chromosomes, mRNA, ribosomes, cofactors, the electron transport chain, and so on, depending upon how "synthesis system" is defined. To simplify discussion, "synthesis system" here will be restricted to enzymes, except where explicitly stated otherwise.

The amounts of individual enzymes generated (characteristic 1) can be affected by gene repression, induction, and dosage. The amount of enzyme or, more generally, the amount of protein generated is often termed the "expression level," for example, 10% of cellular protein as HuIFN (see Section II.A.2). Some ways in which RDT might be able to increase the expression level of desired proteins include introduction of (a) feedback repression–resistant mutant genes or resistant genes from nonhost sources, (b) nonhost constitutive genes, (c) nonhost high-level promotor(s) for relevant genes, and (d) multicopy plasmids coding for relevant genes (Jackson, 1981).

The relative activities of the enzymes present (characteristic 2) are affected by intrinsic turnover numbers for the enzymes; the individual activities can also be subject to inhibition. RDT might be able to affect relative activities by allowing the introduction of genes that code for (a) feedback inhibition–resistant enzymes or (b) enzymes with higher turnover numbers.

The stability of an enzyme (characteristic 3) can be affected by enzyme degradation (for example, by proteases, in which case the use of a protease-deficient mutant host could be beneficial) or by in vivo inactivation (for example, by oxygen (Friebel and Demain, 1977) or high temperature). In the instance when a specific enzyme is particularly susceptible to thermal inactivation, RDT might allow the introduction of a gene from a thermophile into the host so that thermostability of the gene product might be improved (Jackson, 1981).

Even if a copious amount of enzyme is present, active, and stable, little product will be made if the substrate supply is inadequate (characteristic 4). One way the substrate supply can be limited is by transport (Aiba et al., 1973; Bailey and Ollis, 1977). For highly soluble substrates and single-cell microbes, transport limitation would be most likely to occur at the cell membrane, especially if a transport apparatus is involved. For oxygen, transport limitation can occur in the transfer from the gas to the liquid phase. If the microorganism grows in pellets or some other non-single-cell structure, cells in the interior of the structure may well be subjected to substrate starvation.

Degradation of product (characteristic 5) obviously can also lower harvest titers. If the product is susceptible to enzymatic degradation, use of a host lacking the degradative enzyme is advantageous. For example, significant degradation of lysine does not occur in *Corynebacterium glutamicum,* which is employed as a lysine production strain. In *E. coli,* on the other hand, lysine can be decarboxylated to yield cadaverine (Nakayama, 1972).

Characteristics 1–4 can all be adversely affected by product toxicity. Product toxicity limits titer in fermentations for ethanol (Bu'Lock, 1979; Flickinger, 1980; Ho, 1980; Rose and Beavan, 1981), butanol and acetone (Moreira et al., 1981) and acidogenic (for example, acetic acid) and aromatic chemicals (Schwartz and Keller, 1982; Zeikus, 1980). Genetic approaches to developing toxicity-resistant strains are discussed in general by Jackson (1981), for specific cases involving antibiotic production by numerous workers (e.g., Bu'Lock et al., 1980; Unowsky and Hoppe, 1978), and for the acetic acid fermentation by Schwartz and Keller (1982). H. Y. Wang et al. (1981) have reviewed methods for continuous removal of a toxic product as it is produced. Continuous removal methods include application of membranes (dialysis, ultrafiltration), ion exchange, adsorption, vacuum distillation, and solvent extraction.

Consider now not only enzymes as components of the "synthesizing system," but also plasmids coding for those enzymes. Then, obviously, the importance of "plasmid stability" comes into focus. Detailed discussion of the influence of plasmid stability is given elsewhere (Jackson, 1981; Imanaka and Aiba, 1981; Imanaka et al., 1980; Dwivedi et al., 1982; Jones et al., 1980; Helling et al., 1981).

B. High Yield

1. Use Low Cost Raw Material Efficiently. Particularly for the manufacture of low-cost products (for example, ethanol, citric acid, MSG, lysine), industrial microorganisms must have the ability to convert a high percentage of low-cost substrate (for example, glucose equivalents in molasses) into the desired product instead of biomass, CO_2, and by-products. High yield (conversion) is important because raw material costs for fermentation can account for a very significant proportion of total manufacturing cost. For ethanol from corn, raw material costs can constitute at least 40–70% (depending on how by-product credits are allocated) of the total manufacturing cost (King, 1979; Venkatasubramanian and Keim, 1981; Weiss and Mikulka, 1981; Esser and Schmidt, 1982); even for penicillin, raw material costs are significant (Swartz, 1979).

Table 4.4 lists yield information for selected industrial fermentations. Data on both actual yields and theoretical maximum yields (see Section II.B.2) are given. For example, in the case of ethanol, one mole of glucose theoretically can be converted into two moles of ethanol (and also two moles of CO_2); on a weight basis, the weight of the ethanol product would be 51.1% of the weight of the raw material sugar. This theoretical maximum yield cannot actually be attained owing to utilization of sugar for yeast growth, maintenance, and by-product formation. In practice, the actual ethanol conversion achieved is quite high, 90–95% of the theoretical maximum value (Kunkee and Amerine, 1970). For citric acid, the actual conversion is 80% of theoretical maximum (Demain, 1980). Kinoshita and Nakayama (1978) cite a theoretical maximum yield for MSG fermentation of 81.7%, and on the basis of their discussion, the actual yield is probably 50–65%; actual conversion is therefore 60–80% of theoretical maximum. The theoretical maximum yield for lysine fermentation is 69% (Shvinka et al., 1980), and Demain (1980) gives an actual yield of 25–30%. For the case of lysine, then, actual conversion is only 35–45% of theoretical maximum, so that there is an appreciable margin for improvement (see Section II.B.3). Cooney (1979) gives a theoretical maximum yield of 110% for penicillin fermentation and an actual yield of 9–12%. (The penicillin case illustrates how, on a weight basis, theoretical maximum yield can exceed 100%, since the yield is calculated only in relation to sugar, but portions of the product molecule come from nonsugar substrates—for example, ammonium sulfate.)

The products considered in Table 4.4 are listed from top to bottom in order of decreasing actual conversion. It is interesting to note that the order from top to bottom also corresponds with the order for increasing value; that is, the higher the actual conversion, the lower the unit cost.

TABLE 4.4 Yields for Selected Industrial Fermentations[1]

Product	*Theoretical Maximum Yield*[2]	*Actual Yield*[2]	*Actual Conversion*[3]
Ethanol	51.1%[4]	45–48%[4]	90–95%
Citric acid	107%[5]	~85%[5]	~80%
MSG	81.7%[6]	50–65%[6]	60–80%
Lysine	69%[7]	25–30%[7]	35–45%
Penicillin	110%[8]	9–12%[8]	8–11%

1. Yields are given in terms of glucose equivalents in substrate; actual substrate usually is a carbon source such as molasses or hydrolyzed starch.
2. Given as a weight conversion: grams of product per gram of glucose equivalents in substrate × 100%.
3. Actual yield as percent of theoretical maximum yield.
4. Grams of ethanol per gram of glucose × 100%.
5. Grams of anhydrous citric acid (free acid) per gram of glucose × 100%.
6. Grams of glutamic acid (free acid) per gram of glucose × 100%.
7. Grams of lysine (free base) per gram of glucose × 100%.
8. Grams of penicillin G (sodium salt) per gram of glucose × 100%.

Yields in Table 4.4 are calculated on the basis of glucose equivalents in the carbon substrate employed. The actual major carbon substrate used industrially is usually a low-cost material such as molasses or hydrolyzed corn starch (Zabriskie et al., 1980; Solomons, 1969). Currently, alternative industrial carbon substrates being investigated include lignocellulosics (Dunlap and Chiang, 1980; Fan et al., 1981; Cooney et al., 1978), methanol (Tani and Yamada, 1980), and whey (Bernstein et al., 1977).

During strain and process development, it is important to keep in mind that actual commercial fermentations must work on actual commercial substrates. Attractive results obtained with expensive reagent-grade chemicals in the laboratory may not be reproducible with low-cost industrial substrates (e.g., see van Vuuren and Meyer, 1982; Skotnicki et al., 1982).

2. Carbon Allocation and Theoretical Maximum Yield. In the course of synthesis of a product by fermentation, the carbon source is allocated to (1) biomass, (2) product, (3) by-product, and (4) maintenance energy (Pirt, 1975). The maximum theoretical yield can be calculated by considering the hypothetical situation in which all consumed carbon source passes through only the pathway(s) required for product synthesis. If the biochemical pathway for a particular product is known, it is then possible to calculate the theoretical maximum yield for the product, that is, the amount of product that results when all carbon source is channeled to product synthesis, and none is allocated to form biomass or by-product or to provide maintenance energy.

In calculating theoretical maximum yield, it is necessary to trace the biochemical pathway step by step through the enzymatic reactions involved. If, for example, a net amount of ATP is consumed, that ATP must be regenerated, perhaps by conversion of the carbon source via respiration to CO_2. If, in the course of biosynthesis, net amounts of ATP are consumed while an excess of NADH is generated, the ATP can be regenerated through oxidative phosphorylation, but an appropriate value for the P/O ratio (Pirt, 1975) must be assumed. Finally, ATP consumed for active transport (of glucose, for example) should be taken into account. Consideration of all these factors in the calculation of theoretical maximum yield is illustrated in the analysis of penicillin biosynthesis performed by Cooney and Acevedo (1977) and in the analysis of lysine biosynthesis performed by Shvinka et al. (1980).

3. Example Improvement via RDT: SCP from Methanol. ICI workers have reported a yield improvement in the production of SCP from methanol as a result of application of RDT (Windass et al., 1980). The biochemical basis for this yield improvement is shown in Fig. 4.1. The SCP microorganism is *Methylophilus methylotrophus*. This organism normally assimilates ammonia nitrogen into biomass by employing two enzymes, glutamine synthetase

Parent microorganism

1) $\text{Glutamate} + NH_4^+ + \text{ATP} \xrightarrow{\text{GS}} \text{Glutamine} + \text{ADP} + P_i$

2) $\text{Glutamine} + \alpha\text{KG} + \text{NAD(P)H} \xrightarrow{\text{GOGAT}} 2\ \text{Glutamate} + \text{NAD(P)}$

Overall:

$$NH_4^+ + \alpha\text{KG} + \text{ATP} + \text{NAP(P)H} \longrightarrow \text{Glutamate} + \text{ADP} + P_i + \text{NAD(P)}$$

RDT microorganism

$$NH_4^+ + \alpha\text{KG} + \text{NAD(P)H} \xrightarrow{\text{GDH}} \text{Glutamate} + \text{NAD(P)}$$

FIGURE 4.1 Biochemical basis for RDT yield improvement in production of SCP from methanol.

(GS) and glutamate synthase (GOGAT), to convert ammonia and α-keto-glutarate (αKG) into glutamate. The ICI workers deleted this ammonia assimilation route from their SCP microorganism and, by RDT, substituted the conversion of ammonia and αKG into glutamate via glutamate dehydrogenase (GDH). By this substitution, nitrogen assimilation became more energy efficient (utilization of one less mole of ATP per mole of ammonia assimilated; see Fig. 4.1), and consequently, the RDT SCP microorganisms gave 4–7% higher yield on carbon than its parent.

4. Example with RDT Potential: Lysine. As was mentioned in section II.B.1, for the lysine fermentation there is substantial room for improvement in yield; RDT approaches might be able to provide routes to that improvement. The general biochemical basis of one RDT yield improvement strategy could be as follows: particular pathway enzymes in the biosynthetic route leading to lysine (see Fig. 4.2) are probably rate-limiting in present industrial fermentations. RDT should provide means for specifically increasing cellular activities of such rate-limiting or bottleneck pathway enzymes. By increasing cellular activities of bottleneck enzymes, yield should increase as the result of the directing of more carbon toward the desired product (lysine) and away from side reactions, and also possibly as a consequence of improved CO_2 recapture (see below).

The biochemical detail of the general strategy just outlined for improving lysine yield might be fleshed out as follows (refer to Fig. 4.2). First, aspartokinase (AK) must be released from feedback inhibition. This has been accomplished before by the use of homoserine auxotrophs or analog-resistant mutants (Demain, 1971; Kinoshita and Nakayama, 1978; Hirose et al., 1978).

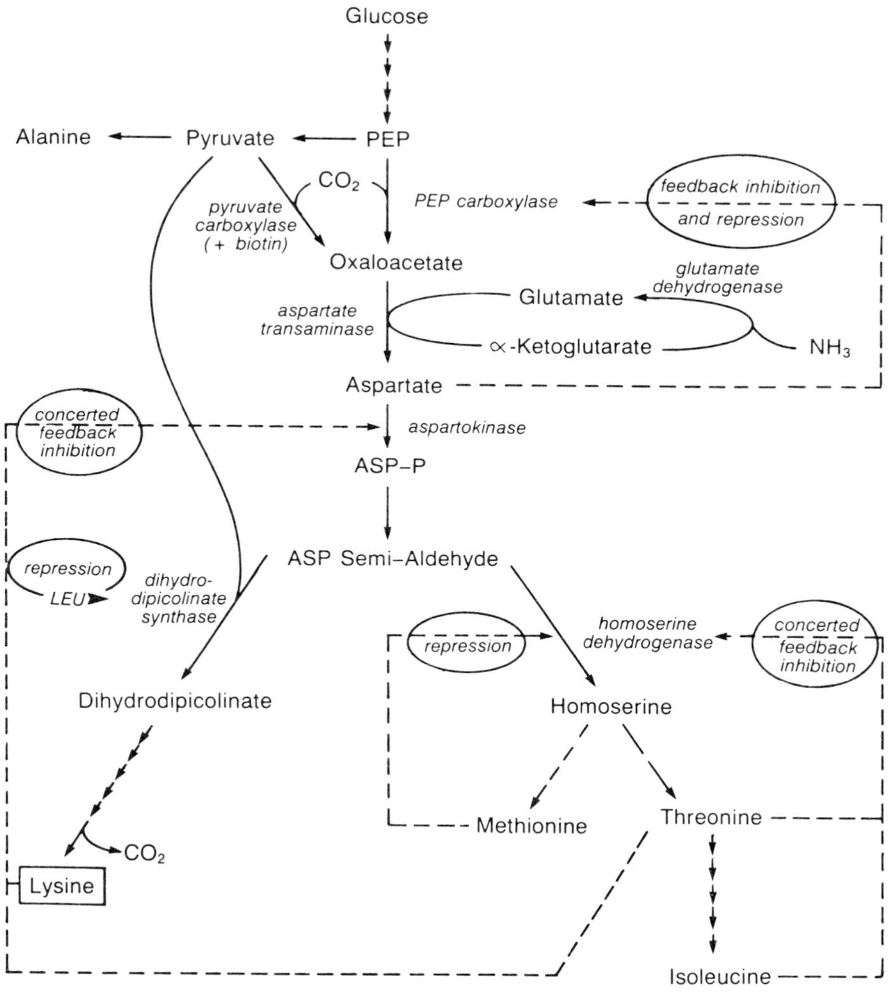

FIGURE 4.2 Pathway and regulatory controls for biosynthesis of lysine in coryneform bacteria. Based on combination of data obtained from *Corynebacterium* and *Brevibacterium* (see references cited in text).

Next, workers at Ajinomoto (Tosaka et al., 1979a, b) have reported that a supply of aspartate is rate-limiting in the biosynthesis of lysine from glucose in *Brevibacterium*. *Brevibacterium* synthesizes aspartate principally from oxaloacetate (and glutamate) via aspartate transaminase (Shiio and Ujigawa, 1978; Shvinka et al., 1980). Thus it should be possible to improve lysine production by increasing cellular levels of aspartate transaminase (AT) and pyruvate carboxylase (PC). These two enzymes lead directly to aspartate. PC is a more desirable enzyme to amplify than PEP carboxylase because

PEP carboxylase is subject to feedback inhibition by aspartate (Tosaka et al., 1979b). Amplification of PC may also increase the fixation of CO_2, thus enhancing CO_2 recapture. Enhanced CO_2 recapture will result in an improved yield, since less substrate carbon will be lost to the atmosphere as CO_2.

A third enzyme to amplify (in addition to AT and PC) is dihydrodipicolinate synthase (DDS). First, the gene for DDS might be cloned so that it is under the control of an operator that is insensitive to repression by leucine. Normally, leucine represses DDS synthesis (Tosaka et al., 1978a). Second, the cloning of DDS should be highly advantageous because DDS is the first enzyme in the lysine-specific pathway branch (Fig. 4.2). Finally, amplification of DDS will direct not only aspartate semialdehyde but also pyruvate toward dihydrodipicolinate. This channeling should result in the production of more lysine and less alanine from pyruvate (see Fig. 4.2 and Tosaka et al. (1978b)).

Thus an example of a genetic engineering strategy (that is, what genes to clone) for the lysine fermentation can be summarized as follows (Table 4.5). First, obtain an AK insensitive to feedback inhibition. Then clone the genes for PC, AT, and DDS such that the cellular levels of these lysine pathway enzymes are amplified. This strategy assumes that physiological studies would confirm that PC, AT, and DDS are rate-limiting in the lysine parent production strain.

C. High Productivity

1. Productivities for Batch and Fed-Batch Operation. Rapid production rates lower product cost. For the case of a batch fermentation, if turnaround time for medium batching, sterilization, and final product harvest is neglected, a strain that attains a given harvest titer in one day obviously will be able to supply twice as much product on an annual basis as will a strain that takes 2 days to reach the same harvest titer. The resulting impact on unit

TABLE 4.5 Lysine Example: Rationale for Genetic Engineering Strategy (what genes to clone)

Problem	*Strategy*
Feedback inhibition of AK	Obtain AK resistant to feedback inhibition
Limitation of aspartate supply	Clone genes for PC, AT
Loss of substrate carbon as CO_2	Clone gene for PC
Repression of DDS by leucine (also, key position of DDS in lysine pathway)	Clone gene for DDS under control of leucine-insensitive operator
Loss of pyruvate to alanine	Clone genes for DDS and PC

AK = aspartokinase, PC = pyruvate carboxylase, AT = aspartate transaminase, DDS = dihydrodipicolinate synthase. Pathway and regulatory controls are shown in Fig. 4.2.

production cost (for example, dollars per pound) can be substantial; in the limit, the unit cost for the case just mentioned would be cut in half, but considerations involving other factors (for example, raw materials, labor, and recovery) usually reduce the savings.

Fermentation production rates are often given in terms of amount of product produced per unit volume of fermentation broth per unit time—such as grams per liter per hour. For example, if a particular production strain attains a harvest titer of 50 g/l in a 50-hour fermentation, then the average volumetric productivity (with turnaround time neglected) is 1 g/l per hour. Given a set annual production target (for example, 1,000,000 kg per year), fermentation capacity (for example, liters of installed fermentor volume) is fixed by the average volumetric productivity of the fermentation. Thus to produce 10^6 kg per year of a certain product with a fermentation having an average productivity of 1 g/l per hour, the total plant fermentation capacity (neglecting turnaround and down times and inoculum tank volumes) will need to be about 120,000 l of working volume.

Productivities for selected industrial fermentations run using batch and fed-batch operation (Pirt, 1975) are listed in Table 4.6. For MSG, a titer of at least 100 g/l can be achieved in no more than 1.5 days (Bartholomew and Reisman, 1979), so the average productivity (neglecting turnaround time) is about 3 g/l per hour. Demain (1980) indicates that in the citric acid fermentation, 100 g/l of product can be obtained in 4–5 days, giving an average productivity of about 1 g/l per hour. For the lysine fermentation, 75 g/l of lysine can be obtained in 72 hours (French patent application No. 77 04611, Ajinomoto Co., February 17, 1977), corresponding to a productivity of about 1 g/l per hour. According to Queener and Swartz (1979), a 185-hour penicillin fermentation can reach a harvest titer of 25 g/l, so the productivity is about 0.14 g/l per hour.

2. Continuous and Fill-and-Draw Operation. Productivities may be substantially improved if a production strain can be employed in a continuous rather than batch culture (Gaden, 1981; Aiba et al., 1973). For example, for the ethanol fermentation, productivities in experimental systems (for example, continuous fermentation by *Zymomonas* with continuous cell recycle (Lee et al., 1980)) have reached as high as 200 grams of ethanol per liter

TABLE 4.6 Productivities for Selected Industrial Fermentations

Product	*Productivity (g/l per hour)*
MSG	~3
Citric acid	~1
Lysine	~1
Penicillin	~0.14

per hour. However, actual current commercial ethanol fermentation technology is based almost exclusively on batch culture, average productivities ranging only from about 1 to 4 g/l per hour (Venkatasubramanian and Keim, 1981; Esser and Schmidt, 1982). Few commercial products are manufactured today by the use of continuous culture for practical reasons explained elsewhere (Gaden, 1981; Aiba et al., 1973; Dawson, 1977; Srinivasan and Summers, 1981; Righelato and Elsworth, 1970), although continuous culture is applied on an industrial scale for production of single-cell protein and for waste treatment (Humphrey, 1981). Strain instability can be one major problem in application of continuous fermentation; and for the case of an RDT production strain, this problem can easily reduce to a plasmid instability situation (see Section II.A.3).

Most commercial fermentations are run as batch or fed-batch cultures, or a modification of continuous culture termed "fill-and-draw" can be used (Humphrey, 1981). Also, during the past decade, continuous immobilized biocatalyst reactors have been introduced and employed for the manufacture of a limited number of products. Immobilized biocatalyst technology is discussed in Section VIII.

3. Potential RDT Impact. Productivity of a production microorganism can be constrained by the same factors that limit titer (see Section II.A.4). Therefore RDT may provide means for improving productivity through introduction of multiple gene copies, repression-resistant genes coding for feedback-resistant and/or stable enzymes perhaps having higher intrinsic turnover numbers, and so on. As an example, the lysine fermentation RDT strategy given in Section II.B.4 might well improve not only yield, but also productivity.

III. THE IMPORTANCE OF BEING EXTRACELLULAR

Every one of the products listed in Table 4.2 is made as an extracellular product that is excreted into the broth. A primary importance of being extracellular is easily illustrated. If a product is intracellular, its titer obviously cannot exceed the concentration of the production microorganism. In the case of the MSG fermentation, typical cell concentrations are 10–20 g dry cell weight per liter. Therefore even if the cell were 50% glutamate, the concentration of MSG could not exceed 5–10 g/l if the product were constrained to be intracellular. Through excretion, a physical constraint on high titers is removed, so that for the MSG fermentation, the harvest titer can rise to over 100 g/l.

Additional benefits of excretion of product include reduction of feedback effects by the lowering of intracellular product concentration. Moreover, for an extracellular product, recovery and isolation are usually simplified.

In the design and construction of a production strain it is therefore a great advantage if the strain can be engineered to make the product as an extracellular one. This "extracellular" design factor plays a central role in the choice of the production microorganism and in providing incentive to develop RDT systems for gram-positive microorganisms (for example, *Bacillus* and coryneforms) that are already prodigious excretors in nature and in existing industrial fermentations (see Sections IV and VII).

IV. THE EXPORT PROBLEM

Although it is desirable to have the production strain produce the product as an extracellular one (Section III), export into the fermentation broth is not easily achieved with certain potential host strains that might otherwise appear attractive because their molecular biology is well studied. For example, while *E. coli* does secrete proteins into the periplasmic space and outer membrane (Kreil, 1981), it does not seem to export proteins into the culture fluid unless the cell is lysed or damaged (Sherwood and Atkinson, 1981). In general, it appears to be a reasonable assertion that gram-negative microorganisms do not export products as well as do gram-positive ones or certain fungi. As anecdotal evidence in support of this assertion, examine Table 4.2. Not one of the high-titer production microorganisms listed is a gram-negative bacterium, while gram-positive bacteria (coryneforms including *Brevibacterium*, and *Bacillus*) and fungi (*Aspergillus*, yeasts, *Penicillium*, *Trichoderma*) do appear in the table.

Product export has not been investigated in any detail for much of a range of either hosts or products. Even in *E. coli*, which can only secrete and not export proteins, the protein secretion mechanism is not well understood (Kreil, 1981). Yet for practical application, product export is mandatory in many cases.

V. THE DRY CELL WEIGHT FACTOR

It is important to realize that cell concentrations routinely achieved in molecular genetics laboratories are usually far from the maximum cell concentrations that can be employed in a biological manufacturing process. For example, baker's yeast can be grown in fermentors to cell concentrations in excess of 60 g/l dry cell weight (DCW) (H. Y. Wang et al., 1979). Even the work horse of the molecular genetics lab, *E. coli*, can be grown in fermentors to over 45 g/l DCW (Bauer and White, 1976).

If a product is made as an intracellular one, and if the percent of DCW that is the desired product is independent of DCW concentration, then the higher the DCW concentration, the better. When the product is extracellular, so that product titer is not constrained by DCW, it is advantageous to have low DCW so that substrate is not wasted in making biomass (see Section II.B.2 and the MSG example in Section III). Another advantage of low biomass making large amounts of exported product is that the biomass disposal problem is reduced.

VI. ADDITIONAL DESIGN FACTORS

A. Nonpathogenicity

Use of a nonpathogenic host is obviously highly advantageous not only from the viewpoint of manufacturing an uncontaminated product, but also to assure worker and environmental safety. Most commercial RDT work published up until now is in *E. coli,* an endotoxin-producing host, only because RDT is by far most advanced in *E. coli.* The fact that *Saccharomyces cerevisiae, Bacillus subtilis,* and *Streptomyces* do not produce endotoxins is one reason to switch to these alternative hosts. Use of a host classified by the FDA as GRAS (generally recognized as safe), including *S. cerevisiae,* can be especially attractive when the product is to be applied as a feed additive or in food processing.

If possible, all insertions and plasmids introduced into the host should originate from nonpathogenic and nontransformed cells. This policy will avoid complications that might otherwise be encountered with regulatory agencies.

B. Low Oxygen Demand and Heat of Fermentation

Transfer of oxygen and heat (cooling) for fermentor operations costs money (Finn, 1969; D. I. C. Wang et al., 1979). It is interesting to note that oxygen demand and heat generation by fermentation (excluding dissipation of agitation energy) are usually related by direct proportion (Cooney et al., 1968). Manufacturing cost will be lowered if volumetric oxygen demand and heat of fermentation for a production strain can be decreased while titer, yield, and productivity are maintained.

C. Operating Temperature

As a consequence of the heat of fermentation, large fermentors usually do not need to be heated to be maintained at the desired operating temperature; they must almost always be cooled. If a production strain permits the fermentor operating temperature to be high, then cooling costs are reduced, especially if refrigeration units can be eliminated. On the other hand, oxygen supply can become a problem at high operating temperatures (e.g., see Bauer and White, 1976).

D. Cell Morphology

Single-cell morphology, as opposed to pellets, mycelial structures, etc., favors low-cost transfer of oxygen and heat as well as mixing. It also should be recognized, however, that non-single-cell morphology can aid low-cost separation of biomass from culture broth during recovery.

E. Ease of Metabolic Control

In many cases, it can prove convenient, if not advantageous, to be able to separate a fermentation into "growth" and "production" phases. To facilitate this two-phase fermentation, use of a simple "trigger" such as a temperature shift or addition of a low-cost inducer is desirable.

VII. CHOICE OF PRODUCTION MICROORGANISM

The ideal host microorganism to be genetically engineered to construct a production strain for a particular desired product would be characterized by:

I. host potential for high product titer, yield, and productivity (see Section II); included here, among other items, are:
   A. resistance to product toxicity, lack of product-degrading enzymes, high potential for plasmid stability (Section II.A.3);
   B. the ability to export product efficiently (Sections III and IV);
   C. the ability to achieve appropriate cell density (Section V);

II. the ability to grow on, and make product using, any one of a range of industrial carbon substrates (for example, molasses, starch, lignocellulosics, whey, and methanol) under industrial conditions, preferably continuously (Sections II.B.1 and II.C.2);

III. lack of pathogenicity and/or endotoxins (Section VI.A);

IV. low oxygen demand and heat of fermentation (Section VI.B);

V. appropriate operating temperature (Section VI.C);

VI. desirable cell morphology (Section VI.D);

VII. ease of metabolic control (Section VI.E);

VIII. existence of RDT for the host; this last factor is particularly constraining (see below).

Unfortunately, no microorganism currently meets all the above specifications. Several important candidate microbial hosts are listed in Table 4.7; each has special applications, advantages, and disadvantages as follows.

TABLE 4.7 Some Candidate Microbial Hosts for RDT Manufacturing Systems

Microbial Host	*Special Applications*	*Advantages*	*Disadvantages*
Escherichia coli	Manufacture of high-cost human proteins	Best-known molecular biology and RDT	Makes endotoxins; does not excrete proteins; not widely employed industrially
Pseudomonas	Hydrocarbon transformation; pollutant degradation	Carries out wide range of biotransformations	Few published reports on RDT
Bacillus	Protein excretion at high titer	Can excrete proteins at high titer; industrial microorganisms; no endotoxins; of all gram-positives, *B. subtilis* has best-characterized molecular genetics	RDT lags far behind *E. coli*
Saccharomyces cerevisiae	Brewing, food yeast, power ethanol	Industrial microorganism; among the best genetically characterized eukaryotic cells; can produce glycosylated proteins; no endotoxins	In general, does not excrete proteins into the broth

A. *Escherichia coli*

By far, *E. coli* is the microorganism with the best-known molecular genetics and RDT. *E. coli* is therefore an extremely attractive host strain, especially to molecular geneticists. Many RDT accomplishments in *E. coli* have already been publicized, including cloning and expression of the foreign gene products listed in Table 4.8 (Sherwood and Atkinson, 1981; Dally et al., 1981; Doel et al., 1980; Beppu et al., 1982; Hung, 1982; Heyneker, 1982; Goeddel, 1982; Okada et al., 1982). Moreover, RDT has been used to reconstruct *E. coli*'s own homologous gene system for overproduction of products such as tryptophan (Aiba et al., 1982), proline (Kretschmer et al., 1982), and threonine; results for threonine overproduction have been particularly impressive, with titer reportedly reaching 55 g/l (Debabov, 1982). In addition to the advanced state of RDT for *E. coli,* another advantage is that the U.S. NIH RAC (Recombinant DNA Advisory Committee) guidelines facilitate its use.

Disadvantages of *E. coli* include its production of endotoxins (Section VI.A) and its inability to excrete certain molecules, in particular proteins (Section IV), so that DCW constrains titer (Section III). Also, *E. coli* has not in the past been widely employed as an industrial microorganism for large-scale fermentation (note its absence from the examples given in Table

TABLE 4.8 Some Foreign Gene Products Expressed in *E. coli*

α, β, and γ Human interferons
Human insulin
Human thymosin α-1
Human growth hormone
Human somatostatin
Urokinase
α-Neoendorphin
Bovine growth hormone
Calf prorennin
Hepatitis B core and surface antigens
Foot-and-mouth virus surface antigen
Fowl plague virus surface antigen
Aspartame precursor

4.2). Some of the few cases for which *E. coli* has been employed in commercial manufacturing include production of 6-APA via *E. coli* penicillin acylase (Queener and Swartz, 1979), aspartic acid from fumaric acid and ammonia via *E. coli* aspartase (Chibata, 1978), and *E. coli* L-asparaginase for cancer chemotherapy (Pratt and Ruddon, 1979).

B. *Pseudomonas*

Pseudomonads are capable of a wide variety of biotransformations of compounds of many types, particularly hydrocarbons; therefore many potential applications are apparent (Chakrabarty, 1976; Ribbons and Williams, 1982). However, few reports of cloning in pseudomonads have appeared (Ribbons and Williams, 1982; Shapiro et al., 1981).

C. *Bacillus*

An especially attractive asset of *Bacillus* is that members of this genus are able to excrete proteins into the broth (Dean and Kaelbling, 1981; Debabov, 1982), titers in at least some cases easily being higher than 10 g/l (see Table 4.2 and Section II.A.1). Also, bacilli are used as industrial microorganisms for the production of enzymes, peptide antibiotics, and insect toxins (Dean and Kaelbling, 1981; Debabov, 1982). Bacilli do not produce endotoxins; and in Japan, *B. subtilis* (*natto*) is used as a human food (Hardy et al., 1981). Of all the gram-positives, *Bacillus subtilis* has the best-characterized molecular genetics; in fact, it has been claimed that *B. subtilis* is the most widely studied prokaryote other than *E. coli* (Gryczan, 1982). However, the state of the art of RDT in *Bacillus* lags far behind accomplishments with *E. coli* (Gryczan, 1982; Dean and Kaelbling, 1981).

The genes coding for hepatitis B core antigen and the major antigen of foot-and-mouth disease virus have been cloned and expressed in *B. subtilis;* the hepatitis B core antigen expression level was <0.1% of cellular protein, while the major antigen of foot-and-mouth disease virus had an expression level of about 1% of cellular protein (Hardy et al., 1981). A gene for human leukocyte interferon has been cloned and expressed in *B. subtilis* (Hardy et al., 1981). A foreign gene coding for a β-lactamase (from *B. licheniformis*) has been cloned and expressed in *B. subtilis,* and the β-lactamase gene product was excreted (Gray and Chang, 1981; Imanaka et al., 1981). Young's lab has reported cloning (Yoneda et al., 1979) and expression (Young, 1980) of a foreign gene coding for α-amylase (from *B. amyloliquefaciens*H) in *B. subtilis,* and also cloning and expression of the *E. coli* gene coding for thymidylate synthetase in *B. subtilis* (Rubin et al., 1980). Cloning and expression of several *E. coli trp* genes in *B. subtilis* has been achieved (Lovett et al., 1982).

D. *Saccharomyces cerevisiae*

This yeast is a eukaryote with which there is a great deal of industrial experience. Among the eukaryotes, *S. cerevisiae* is one of the best genetically characterized microorganisms, and several foreign genes have already been cloned and expressed in this host (Scott, 1982; Goeddel, 1982). *S. cerevisiae* produces no endotoxins and may be of use for manufacture of glycosylated foreign proteins (Lodish, 1981). *S. cerevisiae* produces secreted but cell-associated proteins, while *S. diastaticus* produces glucoamylase as a true extracellular, broth enzyme (Stewart, 1981).

Other important candidate RDT hosts on which much RDT work will be done in the future include *Streptomyces* (Bibb et al., 1980; Chater et al., 1981; Hopwood et al., 1981), which are industrially applied for manufacture of antibiotics; coryneforms (see, for example, Ajinomoto's Japanese patent No. 56-148295), used commercially for amino acid production; clostridia (see Chapter 3) for solvents and liquid fuels; and methylotrophs (Windass et al., 1980).

VIII. IMMOBILIZED BIOCATALYSTS

A. Rationale for Immobilized Biocatalysts

The goals of using immobilized biocatalysts (enzymes, cells, or organelles) whenever possible are fourfold:

1. increased yield of product made from raw material carbon feedstock,
2. increased productivity,

3. improved purity of the bioreactor product stream, and
4. reduced waste streams.

The ways in which immobilized biocatalyst technology meets each of these objectives are as follows:

1. *Increased yield.* Yield can be expected to increase as a consequence of reducing the portion of raw material carbon feedstock utilized for cell growth (to make biocatalyst) rather than product formation. Essentially, the immobilized biocatalyst approach provides a means for total or substantial biocatalyst recycle so that a given mass of cells containing biocatalyst, or biocatalyst isolated from cells, can be used continuously.

2. *Improved productivity.* Productivity should increase over that of a traditional fermentation plant because immobilized biocatalyst bioreactors are run continuously as opposed to the batch operation of the traditional fermentation process. Impact of fermentor turnaround times is nearly eliminated, since the portion of the process involving fermentors is greatly reduced. In addition, with a catalytically stable immobilized biocatalyst system, the fraction of plant time used for nonproductive cell growth is very greatly reduced.

3. *Purer product stream.* A particular advantage of immobilized biocatalyst reactors is that the biocatalyst is largely or completely retained in a bioreactor so that the product stream issuing from the reactor for feed to recovery operations will contain little if any biomass. In addition, in the case of membrane bioreactors, where biocatalyst is retained in the reactor by membranes having low-molecular-weight cutoffs, macromolecular material will also be retained in the bioreactor. Cellular macromolecules and debris will therefore not be present in the feed stream to the purification unit.

4. *Reduced waste streams.* In immobilized biocatalyst processes, production of cell biomass per unit of product is very greatly reduced, and so the quantity of cell biomass for disposal is markedly decreased. Likewise, wash water requirements are cut back (for example, washing of cells for recovery of residual product is eliminated), and therefore there is less liquid volume for disposal.

As a consequence of improved productivity, a purer product stream, and reduced waste streams, an immobilized biocatalyst reactor system should contribute to a decrease in fixed capital investment required to achieve a given production capability. With immobilized biocatalysts that are catalytically stable for prolonged time periods, the plant requirement for large fermentor capacity, which is costly, should be substantially reduced (e.g., see Venkatasubramanian et al., 1979). Consequently, liberated fermentor capacity in an already installed plant could be reallocated to the production of other income-producing fermentation products. Moreover, the purer product feed stream to recovery operations should increase recovery capacity in an already installed plant or decrease capital investment for recovery expansions or new recovery installations.

B. Overview of Applications of Immobilized Biocatalysts

Applications of immobilized biocatalysts can be divided into three categories: (1) single-step transformations involving no expensive cofactors, (2) single-step transformations involving expensive cofactors (for example, ATP, NAD(P)H, or tetrahydrofolates), and (3) multistep transformations involving perhaps an entire biosynthetic pathway.

To date, actual industrial applications of immobilized biocatalyst technology all involve only simple transformations requiring no expensive cofactors. Examples include the manufacture of HFCS (high-fructose corn syrup), 6-APA, L-aspartic acid, L-malic acid, and the resolution of D,L amino acids. Much recent research work on immobilized biocatalysts has focused on lengthy multistep pathway systems (see Table 4.9). Details on both actual operating processes and recent research are given in several excellent reviews (Chibata, 1978; Abbott, 1978; Jack and Zajic, 1977; Cheetham, 1980; Bucke and Wiseman, 1981).

C. Design Criteria for Immobilized Biocatalysts

In order to achieve optimal continuous operation, an immobilized biocatalyst and bioreactor must be designed with the following goals in mind:

1. The biocatalyst must be compatible with the environmental parameters (including the medium components) used in product manufacture. If necessary for a particular case, the biocatalyst and bioreactor design must result in sufficient oxygen transfer or CO_2 evolution capability.
2. The recovery of activity following the immobilization process should be high.
3. The catalytic density (activity per unit volume of catalyst preparation) should be high so that volumetric productivity is high.
4. The biocatalyst preparation should retain biocatalyst, but at the same time, it must be very permeable to both substrates and products. If the biocatalyst incorporates viable cells, the increase in biomass within the reactor, during its lifetime, should be kept to a minimum.
5. The biocatalyst should have good mechanical strength and should resist abrasion.
6. The biocatalyst should be chemically and thermally stable, and the useful lifetime of the biocatalytic activity should be long.
7. The preparation of the biocatalyst should be simple and, if ascepticity

TABLE 4.9 Examples of Published Studies on Immobilized Cell Processes Involving Lengthy Multienzyme Pathways

Product	*Substrate*	*Microorganism*	*Immobilization Method*	*Reference*
L-Glutamic acid	Glucose, ammonium salts	*Corynebacterium glutamicum*	Polyacrylamide gel	Slowinski and Charm (1973)
L-Glutamic acid	Glucose, urea, oxygen	*Brevibacterium flavum*	Collagen	Constantinides et al. (1981)
L-Isoleucine	Glucose, D-threonine	*Serratia*	Carrageenan gel	Wada et al. (1980)
Citric acid	Glucose, oxygen	*Saccharomycopsis*	Wood chips	Briffand and Engasser (1979)
Penicillin G	Glucose, ammonium sulfate, phenylacetate, oxygen	*Penicillium chrysogenum*	Polyacrylamide gel	Morikawa et al. (1979)
Bacitracin	Peptone, oxygen	*Bacillus* sp.	Polyacrylamide gel	Morikawa et al. (1980)
Glutathione	L-Glutamate, L-cysteine, glycine, glucose	*Saccharomyces cerevisiae*	Polyacrylamide gel	Murata et al. (1978)
α-Amylase	Meat extract	*Bacillus subtilis*	Polyacrylamide gel	Kokubu et al. (1978)
Butanol/acetone/ethanol	Glucose	*Clostridium acetobutylicum*	Calcium alginate gel	Haggstrom and Molin (1980)
Butanol/isopropanol	Glucose	*Clostridium butylicum*	Calcium alginate gel	Krovwell et al. (1980)
Ethanol	Glucose	*Saccharomyces cerevisiae*	Calcium alginate gel	Kierstan and Bucke (1977)
Ethanol	Glucose	*Saccharomyces cerevisiae*	Gelatin + glutaraldehyde	Sitton and Gaddy (1980)
Ethanol	Molasses	*Saccharomyces cerevisiae*	Not divulged	Ghose and Bandyopadhyay (1980)
Ethanol	Glucose	*Saccharomyces cerevisiae*	Carregeenan	Wada et al. (1979)
Ethanol	Glucose	*Saccharomyces cerevisiae*	Calcium alginate gel	Larsson and Mosbach (1979)
$CO_2 + H_2O$	Phenol	*Candida tropicalis*	Aluminum alginate gel	Hackel et al. (1975)
H_2	Malate, light	*Rhodospirillium rubrum*	Noble agar	Bennett and Weetall (1976)

is necessary, should not involve complex manipulations that might endanger the maintenance of catalyst ascepticity.

8. The biocatalyst must be inexpensive and should not present any disposal problems after the catalyst is spent.
9. The bioreactor product stream should contain a high concentration of product. The product yield from the carbon source should be high.
10. Reactor setup and operation should be simple and reliable.

IX. PRODUCT RECOVERY

Developments in RDT have placed an initial emphasis on reaction (for example, fermentation) aspects of the overall integrated manufacturing process, but more and more attention is being focused on product recovery: the obvious point that underscores the importance of recovery is that without recovery, there is no product.

Table 4.10 lists four points of view that are useful to consider when approaching the conception and development of a recovery process. Each of these general approaches is discussed very briefly below; several reviews cover various concepts about recovery of biological products in more detail (Belter, 1979; Aiba et al., 1973; Hutt, 1967; Peck, 1967; Paul et al., 1981; D. I. C. Wang et al., 1979; Bailey and Ollis, 1977).

The product type point of view (Table 4.10) is based on the idea that a member of a particular product class (see Table 4.1) stands a reasonable chance of being efficiently recovered by application of a variation of a recovery method conventionally employed for another member of the same class. For example, in the case of solvents and liquid fuels (such as ethanol and butanol), distillation is conventional, although significant effort is being expended to develop low-energy alternatives (Hartline, 1979; Remirez, 1980).

The product value point of view (Table 4.10) can be illustrated by examining Table 4.3. If no alternative is available, one can afford to use on a production scale a relatively costly isolation procedure (for example, immunoaffinity chromatography) to recover a "high value" product such as a human interferon. For a low-value product (for example, a protease), cost considerations force the use of lower-cost methods that yield product of lesser purity, which is quite acceptable for many product applications (e.g., see Faith et al., 1971).

TABLE 4.10 General Approaches to Product Recovery

1. Product type point of view
2. Product value point of view
3. Unit operations point of view
4. Physicochemical point of view

The unit operations point of view (Table 4.10) is based on the different individual separation methods (Table 4.11) that are available for application in a particular isolation scheme. These "unit operations" are the building blocks to be drawn on when assembling a recovery process. Important new developments in several unit operations areas (for example, scale-up of immunoaffinity chromatography via use of RDT to make large quantities of antibody) can be expected.

The most fundamental approach, of course, involves the physicochemical point of view (Table 4.12), in which the intrinsic molecular properties of the system are used as the basis for developing a recovery process. For separations of biomolecules, biological specificity is an important molecular property (for example, for affinity chromatography).

In actual practice, in conceiving and developing a recovery process from the ground up, it is advantageous to consider all the points of view just discussed (Table 4.10).

X. ECONOMIES OF SCALE

Plant scale greatly influences the manufacturing cost per unit of product made (for example, dollars per pound). The unit cost of product made on

TABLE 4.11 Product Recovery: Unit Operations Point of View

1. Cell disruption (*examples*: sonication, osmotic shock, grinding, lytic enzymes)
2. Flotation, foam fractionation
3. Filtration, centrifugation, flocculation, settling
4. Extraction (*examples*: liquid/liquid, liquid/solid)
5. Ion exchange
6. Adsorption (*example*: carbon decolorization)
7. Evaporation
8. Distillation
9. Precipitation
10. Crystallization
11. Membrane separation (*examples*: UF, dialysis, electrodialysis)
12. Chromatography (*examples*: affinity, gel, adsorption, chromatofocusing)
13. Isoelectric focusing
14. Magnetic separations
15. Drying (*examples*: tray, vacuum, spray, lyophilization)

TABLE 4.12 Product Recovery: Physicochemical Point of View

1. State of mass (solid, liquid, or gas)
2. Mass density
3. Solubility
4. Hydrophobicity
5. Adsorptivity
6. Ionizability
7. Charge density
8. Molecular weight
9. Vapor pressure
10. Particle size
11. Biological specificity

the laboratory bench is almost always many times the unit cost in the factory. Moreover, the general rule is: The larger the plant, the lower the manufacturing cost per unit of product produced (Chilton, 1960; Wessel, 1960; Schuman, 1960); lowering of unit cost results from several economies of scale. First, labor cost per manufactured unit (worker hours per pound of product made) is usually reduced as factory size grows. Second, construction cost for the complete plant per unit of product manufactured per unit time falls as design scale of a factory increases; this behavior can be correlated by a decimal power law. Third, raw material cost per unit of product falls as plant scale increases, owing to bulk purchasing and shipping.

In like manner, a single multiproduct plant is able to make products at a lower unit cost than several individual plants, each making a single product at the same scale as the multiproduct plant. The multiproduct plant is more cost-effective, especially if run on a sequential product campaign schedule, owing to economies in labor, utilities, equipment, and general plant services.

XI. CAPITAL INVESTMENT

Low manufacturing cost per unit of product made is of clear importance. Likewise, minimization of capital required for plant acquisition and startup is highly desirable. Even though a planned manufacturing facility might be designed to produce low-cost product, if the capital needed for construction of the plant is high, a formidable venture problem may be encountered in trying to raise that capital. More broadly, the overall objective obviously must be to keep all venture costs as low as possible, including capital needed for process research and development, product testing, product registration and clearance (where required), plant investment and startup, product launch, and general and administrative costs.

XII. CORPORATE INTEGRATION

The actual commercialization of a product requires the integration of multiple corporate functions (Giragosian, 1978). Fig. 4.3 identifies some of the necessary functions in flowchart form. In actual practice, the various functions laid out in Fig. 4.3 cannot be carried out sequentially; various activities must overlap in time. A matrix representation is perhaps therefore more appropriate. Fig. 4.4 is one such matrix representation actually implemented at Genex Corp. The first vertical column in Fig. 4.4 lists a possible mix of specific corporate groups (for example, Business Development, New Products Committee, Process Development) charged with defined missions for commer-

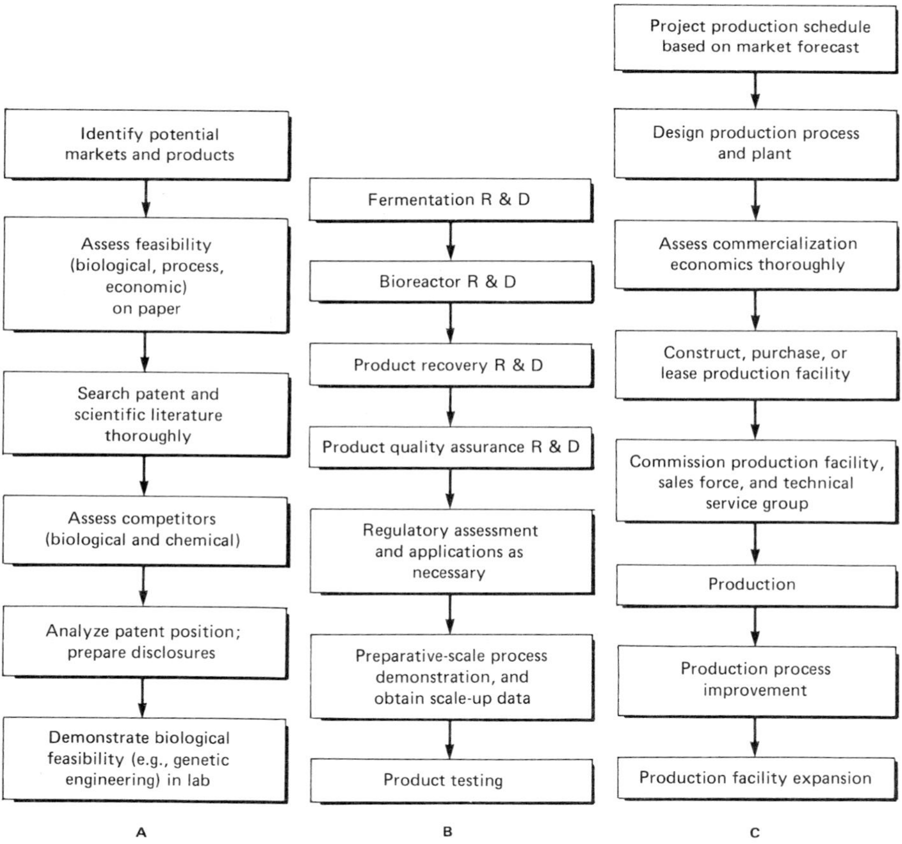

FIGURE 4.3 Flowchart for product commercialization. (a) Preliminary evaluation and demonstration of feasibility. (b) Process development and demonstration. (c) Final evaluation and commercialization.

Number of Ideas / *Stages* / *Functions*	*100* *Screening*	*10* *Concept Development*	*4* *Research*	*2* *Development*	*1* *Commercialization*
Business development	Catalyze	Make recommendation	Milestone reviews	Market tests, initial sales	Transfer to sales
New products committee	Solicit & screen new ideas	Refine & recycle ideas	Monitor		
Commercial lab	New ideas	Initial data	Product evaluation	Product	Application research
Market research	New ideas	Rough markets & projections	Full market study	Market update	Market update
Technology assessment	New ideas	Feasibility (technical & economic)	Milestone reviews	Monitor	
Genetic technology	New ideas	Consult	Conduct research	Strain improvement	Strain improvement
Process development	New ideas	Consult	Bench process	Scale-up	Process improvement
Manufacturing	New ideas	Consult	Monitor	Pilot plant plant design	Start-up production
Consultants	New ideas	Consult	Regulatory assessment	Regulatory approval	
Patent committee		Assess patent position, handle disclosures	Patent applications	Foreign filings	
Sales, distribution, and customer service	New ideas		Monitor	Plan launch	Launch, sell, deliver, service
Administrative services	Support	Support	Support	Support	Support
Technology planning committee		Decision	Decision	Decision	Monitor
Senior management committee			Monitor	Capital allocation	Decision

Source: Jacques Delente, Genex Corp.

FIGURE 4.4 Matrix representation of corporate integration necessary for successful commercialization.

cialization. For each product actually commercialized successfully (Fig. 4.4, column 6, uppermost right-hand corner), perhaps 100 new product ideas must be solicited and screened initially (top of second column). The source of these new ideas might be anywhere in the corporation, or it could be outside consultants, customer contacts, or publications. Perhaps ten out of 100 new product ideas have enough value to pass to the stage represented by the third column, where technical and economic feasibility and market potential of each idea must be roughly assessed, patent position analyzed, and disclosures prepared as appropriate. For the next stage, perhaps four out of the ten remaining new product ideas might prove good enough to enter a full-blown research stage (column 4), which includes not only in-depth laboratory work, but also patent applications, regulatory assessment, a full market study, and product evaluation. From the full-blown research stage, 50% of the projects might proceed into development (column 5), perhaps only one out of a hundred making it to successful commercialization (column 6).

XIII. SUMMARY

Design considerations for construction of competitive biological manufacturing processes can be grouped into four major categories:

1. Strain characteristics
2. Bioreactor performance parameters
3. Product recovery scheme
4. Plant scale and capital investment

Careful consideration of design factors within each of these four major categories promotes practical and efficient process conception, development, and implementation. Table 4.13 lists important design factors for the "strain characteristics" category; each of these factors is examined in this article, as are bioreactor performance parameters, product recovery, and plant scale and capital investment.

The advent of recombinant DNA technology (RDT) affords the opportunity to construct improved biocatalysts for better production of many existing products that are currently manufactured either biologically or chemically. RDT also permits the construction of formerly nonexistent biocatalysts, for example, microbes with the ability to produce mammalian proteins. The most effective RDT processes will result from early incorporation of practical technological perspectives into RDT research and development strategies.

TABLE 4.13 Summary of Design Factors for Construction of Competitive Production Strains

1. High titer, yield, and productivity:
   a. Amount of synthesizing system (e.g., enzyme) generated (e.g., multiple gene copies resistant to repression)
   b. Relative activity of synthesizing system present (e.g., enzymes resistant to product inhibition in a host resistant to product toxicity)
   c. Stability of synthesizing system (e.g., plasmid stability)
   d. Transport of substrate into cell
   e. Export of product from cell (if possible)
   f. Stability of product (e.g., lack of product-degrading enzymes)
   g. Ability to achieve appropriate cell density
2. Low substrate cost and abundant substrate availability
3. Lack of host pathogenicity and/or endotoxins
4. Low oxygen demand and heat of fermentation
5. Appropriate operating temperature
6. Desirable cell morphology
7. Ease of metabolic control
8. Ease of genetic manipulation, including existence of RDT for host

ACKNOWLEDGMENTS

I thank my colleagues for contributing many thoughts on the points I have attempted to present in this article. My gratitude is also extended to my associates who transformed my handwritten drafts into typed copy.

REFERENCES

Abbott, B. J. (1978) in *Annual Reports on Fermentation Processes* Vol. II (Tsao, G., and Perlman, D., eds.) pp. 91–123, Academic Press, New York.

Aiba, E., Humphrey, A. E., and Millis, N. F. (1973) *Biochemical Engineering* 2nd ed. p. 25 and Chaps. 5, 6, and 13, Academic Press, New York.

Aiba, S., Tsunekawa, H., and Imanaka, T. (1982) *Appl. Environ. Microbiol. 43,* 289–297.

Bailey, J. E., and Ollis, D. F. (1977) *Biochemical Engineering Fundamentals* pp. 267–272 and Chaps. 4 and 8, McGraw-Hill, New York.

Bartholomew, W., and Reisman, H. B. (1979) in *Microbial Technology* 2nd ed., Vol. II (Peppler, H. J., and Perlman, D., eds.) pp. 463–496, Academic Press, New York.

Bauer, S., and White, M. D. (1976) *Biotech. Bioeng. 18,* 839–846.

Belter, P. A. (1979) in *Microbial Technology* 2nd ed., Vol. II (Peppler, H. J., and Perlman, D., eds.) pp. 403–432, Academic Press, New York.

Bennett, M. A., and Weetall, H. W. (1976) *J. Solid-Phase Biochem. 1,* 132–142.

Beppu, T., Nishimori, K., Kawaguchu, Y., Hidaka, Y., and Vozumi, T. (1982) Abstract I-IX-1 (p. 27) of the Fourth International Symposium on Genetics of Industrial Microorganisms, Kyoto, Japan, June 6–11.

Bernstein, S., Tzeng, C. H., and Sisson, D. (1977) *Biotech. Bioeng. Symp. 7,* 1–9.
Bibb, M., Schottel, J. L., and Cohen, S. N. (1980) *Nature 284,* 526–631.
Briffand, J., and Engasser, M. (1979) *Biotech. Bioeng. 21,* 2093–2111.
Bucke, C., and Wiseman, A. (1981) *Chemistry and Industry* (April 4), 234–240.
Bu'Lock, J. D. (1979) in *Symposium 29 of the Society for General Microbiology* (Bull, A. T., ed.) pp. 309–325, Cambridge University Press, Cambridge, England.
Bu'Lock, J. D., Demnerova', K., Kilgour, W. J., Knguseder, F., and Steinbuchel, A. (1980) *Biotech. Lett. 2,* 285–290.
Chakrabarty, A. M. (1976) in *Microbiology—1976* Schlessinger, D., ed.) pp. 579–582, American Society for Microbiology, Washington, D.C.
Chater, K. F., Bruton, C. F., Suarez, J. E., and Springer, W. (1981) in *Microbiology—1981* (Schlessinger, D., ed.) pp. 380–383, American Society for Microbiology, Washington, D.C.
Cheetham, P. S. (1980) in *Topics in Enzyme and Fermentation Biotechnology* Vol. IV (Wiseman, A., ed.) pp. 189–238, John Wiley, New York.
Chibata, I. (1978) *Immobilized Enzymes,* John Wiley, New York.
Chilton, C. H. (1960) in *Cost Engineering in the Process Industries* (Chilton, C. H., et al., eds.) pp. 282–284, McGraw-Hill, New York.
Constantinides, A., Bhatia, D., and Vieth, W. R. (1981) *Biotech. Bioeng. 23,* 899–916.
Cooney, C. L. (1979) *Proc. Biochem.* (May), 31–33.
Cooney, C. L., and Acevedo, F. (1977) *Biotech. Bioeng. 19,* 1449–1461.
Cooney, C. L., Wang, D. I. C., and Mateles, R. I. (1968) *Biotech. Bioeng. 11,* 269–281.
Cooney, C. L., Wang, D. I. C., Wang, S. D., Gordon, J., and Jiminez, M. (1978) *Biotech. Bioeng. Symp. 8,* 103–114.
Dally, E. L., Eveleigh, D. E., Montenecourt, B. S., Stokes, H. W., and Williams, R. L. (1981) *Food Technology* (July), 26–33.
Dawson, P. S. S. (1977) in *Annual Reports in Fermentation Processes* Vol. I (Perlman, D., ed.) pp. 73–93, Academic Press, New York.
Dean, D. H., and Kaelbling, M. J. (1981) *Ann. N.Y. Acad. Sci. 369,* 23–32.
Debabov, V. (1982) Abstract I-11-4 (p. 9) of the Fourth International Symposium on Genetics of Industrial Microorganisms, Kyoto, Japan, June 6–11.
Demain, A. L. (1971) *Symp. Soc. General Microbiol. 21,* 77–101.
Demain, A. L. (1980) *Biotechnol. Lett. 2,* 113–118.
Demain, A. L. (1981) *Science 214,* 987–995.
Doel, M. T., Eaton, M., Cook, E. A., Lewis, H., Patel, T., and Carey, N. H. (1980) *Nucl. Acids Res. 8,* 4575–4592.
Dunlap, C. E., and Chiang, L. C. (1980) in *Utilization and Recycle of Agricultural Wastes and Residues* edited by (Shuler, M. L., ed.) pp. 19–65, CRC Press, Boca Raton, Fla.
Dwivedi, C. P., Imanaka, T., and Aiba, S. (1982) *Biotech. Bioeng. 24,* 1465–1468.
Esser, K., and Schmidt, U. (1982) *Proc. Biochem.* (May/June), 46–49.
Faith, W. T., Neubeck, C. E., and Reese, E. T. (1971) *Adv. Biochem. Eng. 1,* 77–111.
Fan, L. T., Gharpuray, M. M., and Lee, Y. H. (1981) *Biotech. Bioeng. Symp. 11,* 29–45.
Finn, R. K. (1969) *Proc. Biochem.* (June), 17, 22.
Flickinger, M. C. (1980) *Biotech. Bioeng. 22* (Suppl. 1), 27–48.

Friebel, T. E., and Demain, A. L. (1977) *FEMS Microbiol. Lett. 1,* 215–218.
Gaden, E. L., Jr. (1981) *Scientific American,* September, 180–196.
Ghose, T. K., and Bandyopadhyay, K. K. (1980) *Biotech. Bioeng. 22,* 1489–1496.
Giragosian, N. H. (1978) *Successful Product and Business Development,* Marcel Dekker, New York.
Glick, J. L. (1982) in *Biomedical Institutions, Biomedical Funding and Public Policy* (Fudenberg, H. H., ed.) Plenum Press, New York.
Goeddel, D. (1982) Abstract I-IX-7 (p. 30) of the Fourth International Symposium on Genetics of Industrial Microorganisms, Kyoto, Japan, June 6–11.
Gray, O., and Chang, S. (1981) *J. Bacteriol. 145,* 422–428.
Gryczan, T. J. (1982) in *The Molecular Biology of the Bacilli* (Dubnau, D. A., ed.) 307–329, Academic Press, New York.
Hackel, V., Klein, J., Megnet, R., and Wagner, F. (1975) *Eur. J. Appl. Microbiol. 1,* 291–293.
Haggstrom, L., and Molin, N. (1980) *Biotechnol. Lett. 2,* 241–246.
Hardy, K., Stahl, S., and Kupper, H. (1981) *Nature 293,* 481–483.
Hartline, F. F. (1979) *Science 206,* 41–42.
Helling, R. B., Kinney, T., and Adams, J. (1981) *J. Gen. Microbiol. 123,* 129–141.
Heyneker, H. L. (1982) Abstract I-IX-3 (p. 28) of the Fourth International Symposium on Genetics of Industrial Microorganisms, Kyoto, Japan, June 6–11.
Hirose, Y., Sano, K., and Shihai, H. (1978) in *Annual Reports on Fermentation Processes* Vol. II (Perlman, D., ed.) pp. 155–189, Academic Press, New York.
Hirose, Y., Enei, H., and Shibai, H. (1979) in *Annual Reports on Fermentation Processes* Vol. III (Perlman, D., ed.) pp. 253–274, Academic Press, New York.
Ho, N. W. Y. (1980) in *Annual Reports on Fermentation Processes* Vol. IV (Tsao, G. T., ed.) pp. 235–266, Academic Press, New York.
Hopwood, D. A., Thompson, C. J., Kieser, T., Ward, J. M., and Wright, H. M. (1981) in *Microbiology—1981* (Schlessinger, D., ed.) pp. 376–379, American Society for Microbiology, Washington, D.C.
Humphrey, A. E. (1981) in *Advances in Biotechnology* Vol. I (Moo-Young, M., Robinson, C. W., and Vezina, C., eds.) pp. 203–210, Pergamon Press, New York.
Hung, P. (1982) Abstract I-IX-2 (p. 27) of the Fourth International Symposium on Genetics of Industrial Microorganisms, Kyoto, Japan, June 6–11.
Hutt, R. (1967) in *Biochemical and Biological Engineering Sciences* Vol. I (Blakebrough, N., ed.) Academic Press, New York.
Ikeda, S., Tobe, S., Niwa, K., Ishizahu, A., and Hirose, Y. (1974) *Agr. Biol. Chem. 38* (12), 2317–2322.
Imanaka, T., and Aiba, S. (1981) *Ann. N.Y. Acad. Sci. 369,* 1–14.
Imanaka, T., Tsunekawa, H., and Aiba, S. (1980) *J. Gen. Microbiol. 118,* 253–261.
Imanaka, T., Tanaka, T., Tsunekawa, H., and Aiba, S. (1981) *J. Bacteriol. 147,* 776–786.
Jack, T. R., and Zajic, J. E. (1977) *Adv. Biochem. Eng. 5,* 125–145.
Jackson, D. A. (1981) in *Trends in the Biology of Fermentations for Fuels and Chemicals* (Hollaender, A., et al. eds.) pp. 187–200, Plenum, New York.
Jones, I. M., Primrose, S. B., Robinson, A., and Ellwood, D. C. (1980) *Molec. Gen. Genet. 180,* 579–584.
Journal of the American Medical Association (1979) *242,* 2829.
Kierstan, M., and Bucke, C. (1977) *Biotech. Bioeng. 19,* 387–397.

King, S. R. (1979) *Gasohol: Ethanol from Plant Matter as Motor Fuel,* report published by F. Eberstadt and Co., Inc., New York.
Kinoshita, S., and Nakayama, K. (1978) in *Economic Microbiology* Vol. II (Rose, A. H., ed.) pp. 209–261, Academic Press, New York.
Kokubu, T., Karube, I., and Suzuki, J. (1978) *Eur. J. Appl. Microbiol. Biotechnol. 5,* 233–240.
Kreil, G. (1981) *Ann. Rev. Biochem. 50,* 317–48.
Kretschmer, P. J., Bloom, F., Detuch, A., Jessee, J., Rushlow, K., Wareheim, F. D., Funk, C., Smith, C. J., Vanleeuwen, D., Finn, D., and Hartley, J. (1982) Abstract 0-11-5 (p. 41) of the Fourth International Symposium on Genetics of Industrial Microorganisms, Kyoto, Japan, June 6–11.
Krovwel, P. G., van der Laan, W. F. M., and Kossen, N. W. F. (1980) *Biotechnol. Lett. 2,* 253–258.
Kunkee, R. E., and Amerine, M. A. (1970) in *The Yeasts,* Vol. II, *Yeast Technology* (Rose, A. H., and Harrison, J. S., eds.) pp. 5–71, Academic Press, New York.
Larsson, P. O., and Mosbach, K. (1979) *Biotechnol. Lett. 1,* 501–506.
Lee, K. J., Lefebvre, M., Tribe, D. E., and Rogers, P. L. (1980) *Biotechnol. Lett. 2,* 487–492.
Lescaze, L. (1980) *The Washington Post* (June 15, 1980), A8.
Lodish, H. F. (1981) *Enzyme Microb. Technol. 3,* 178–188.
Lovett, P. S., Williams, D. M., and Duvall, E. J. (1982) in *Genetic Engineering of Microorganisms for Chemicals* (Hollaender, A., et al., eds.), pp. 51–57, Plenum Press, New York.
Miall, L. M. (1978) in *Economic Microbiology* Vol. II, *Primary Products of Metabolism* (Rose, A. H., ed.), pp. 47–119, Academic Press, New York.
Moreira, A. R., Ulmer, D. C., and Linden, J. C. (1981) *Biotech. Bioeng. Symp. 11,* 567–579.
Morikawa, Y., Karube, I., and Suzuki, S. (1979) *Biotech. Bioeng. 21,* 261–270.
Morikawa, Y., Karube, I., and Suzuki, S. (1980) *Biotech. Bioeng. 22,* 1015–1023.
Murata, K., Tani, K., Kato, J., and Chibata, I. (1978) *Eur. J. Appl. Microbiol. Biotechnol. 5,* 233–240.
Nakayama, K. (1972) in *The Microbial Production of Amino Acids* Yamada, K., Kinoshita, S., Tsunoda, T., and Aida, K. eds.) pp. 369–397, John Wiley, New York.
Okada, T., Ohsue, K., Tanaka, S., and Takano, I. (1982) Abstract P-1-27 (p. 96) of the Fourth International Symposium on Genetics of Industrial Microorganisms, Kyoto, Japan, June 6–11.
Paul, E. L., Kaufman, A., and Sklarz, W. A. (1981) *Ann. N.Y. Acad. Sci. 369,* 181–186.
Peck, W. C. (1967) in *Biochemical and Biological Engineering Sciences* Vol. I (Blakebrough, N., ed.) p. 465, Academic Press, New York.
Pirt, S. J. (1975) *Principles of Microbe and Cell Cultivation* pp. 66 and 74 and Chap. 21, John Wiley, New York.
Pratt, W. B., and Ruddon, R. W. (1979) *The Anticancer Drugs.* pp. 236–240, Oxford University Press, New York.
Queener, S., and Swartz, R. (1979) in *Economic Microbiology* Vol. III, *Secondary Products of Metabolism* (Rose, A. H., ed.) pp. 35–122, Academic Press, New York.
Remirez, R. (1980) *Chem. Eng.* (March 24), 57–59.

Ribbons, D. W., and Williams, P. A. (1982) in *Genetic Engineering of Microorganisms for Chemicals* (Hollaender, A., et al., eds.) pp. 211–232, Plenum Press, New York.

Righelato, R. C., and Elsworth, R. (1970) *Adv. Appl. Microbiol. 13,* 399–417.

Rose, A. H., and Beavan, M. J. (1981) in *Trends in the Biology of Fermentation for Fuels and Chemicals* (Hollaender, A., et al., eds.) pp. 513–531, Plenum Press, New York.

Rubin, E. M., Wilson, G. A., and Young, F. E. (1980) *Gene 10,* 227–235.

Rudd, D. F., Powers, G. J., and Sirola, J. J. (1973) *Process Synthesis,* Prentice-Hall, N.J.

Ryu, D. D. Y., and Mandels, M. (1980) *Enzyme Microb. Technol. 2,* 91–102.

Schuman, S. C. (1960) in *Cost Engineering in the Process Industries* (Chilton, C. H., et al., eds.) pp. 307–310, McGraw-Hill, New York.

Schwartz, R. D., and Keller, F. A., Jr. (1982) *Appl. Env. Microbiol. 43,* 117–123.

Scott, J. R. (1982) in *Genetic Engineering of Microorganisms for Chemicals* (Hollaender, A., et al., eds.) pp. 75–86, Plenum Press, New York.

Shapiro, J. A., Charbit, A., Benson, S., Caruso, M., Laux, R., Meyer, R., and Banuett, F. (1981) in *Trends in the Biology of Fermentations for Fuels and Chemicals* (Hollaender, A., et al., eds.) pp. 243–272, Plenum Press, New York.

Sherwood, R., and Atkinson, T. (1981) *Chemistry and Industry* (April 4), 241–247.

Shiio, I., and Ujigawa, K. (1978) *J. Biochem. 84,* 647–657.

Shoemaker, S. P. (1981) Abstract MICR 57, 182nd ACS National Meeting, New York.

Shreve, R. N., and Brink, J. A., Jr. (1977) *Chemical Process Industries* 4th ed. McGraw-Hill, New York.

Shvinka, J., Viesturs, V., and Ruklisha, M. (1980) *Biotech. Bioeng. 22,* 897–912.

Sitton, O. C., and Gaddy, J. L. (1980) *Biotech. Bioeng. 22,* 1735–1748.

Skotnicki, M. L., Lee, K. J., Tribe, D. E., and Rodgers, P. L. (1982) in *Genetic Engineering of Microorganisms for Chemicals* (Hollaender, A., et al., eds.) pp. 271–290, Plenum Press, New York.

Slowinoki, W., and Charm, S. E. (1973) *Biotech. Bioeng. 15,* 973–979.

Solomons, G. L. (1969) *Materials and Methods in Fermentation* pp. 116–123, Academic Press, New York.

Srinivasan, V. R., and Summers, R. J. (1981) in *Continuous Cultures of Cells* Vol. I (Calcott, P. H., ed.) pp. 97–112, CRC Press, Boca Raton, Fla.

Stewart, G. G. (1981) *Can. J. Microbiol. 27,* 973–990.

Swartz, R. W. (1979) in *Annual Reports on Fermentation Processes* Vol. III (Perlman, D., ed.) pp. 75–110, Academic Press, New York.

Tani, Y., and Yamada, H. (1980) *Biotech. Bioeng. 22* (Suppl. 1), 163–175.

Tosaka, O., Hirakawa, H., Takinami, K., and Hirose, Y. (1978a) *Agr. Biol. Chem. 42,* 1501–1506.

Tosaka, O., Hirakawa, H., Yoshihara, Y., Takinami, K., and Hirose, Y. (1978b) *Agr. Biol. Chem. 42,* 1773–1778.

Tosaka, O., Hirakawa, H., and Takinami, K. (1979a) *Agr. Biol. Chem. 43,* 491–495.

Tosaka, O., Morisak, H., and Takinami, K. (1979b) *Agr. Biol. Chem. 43,* 1513–1519.

Unowsky, J., and Hoppe, D. C. (1978) *J. Antibiot. 31,* 662–666.

van Vuuren, H. J. J., and Meyer, L. (1982) *Biotech. Lett. 4,* 253–256.

Venkatasubramanian, K., and Keim, C. R. (1981) *Ann. N.Y. Acad. Sci. 369,* 187–204.

Venkatasubramanian, K., Constantinides, A., and Vieth, W. R. (1979) in *Enzyme Engineering* Vol. III (Pye, E. K., and Weetall, H. B., eds.) pp. 29–41, Plenum Press, New York.

Wada, M., Kato, J., and Chibata, I. (1979) *Eur. J. Appl. Microbiol. 8,* 241–247.

Wada, M., Vehida, T., Kato, J., and Chibata, I. (1980) *Biotech. Bioeng. 22,* 1145–1180.

Wang, D. I. C., Cooney, C. L., Demain, A. L., Dunnill, P., Humphrey, A. E., and Lilly, M. D. (1979) *Fermentation and Enzyme Technology,* p. 210 and Chap. 12, John Wiley, New York.

Wang, H. Y., Cooney, C. L., and Wang, D. I. C. (1979) *Biotech. Bioeng. 21,* 975–995.

Wang, H. Y., Kominek, L. A., and Jost, J. L. (1981) in *Advances in Biotechnology* Vol. I (Moo-Young, M., ed.) pp. 601–607, Pergamon Press, New York.

Weiss, L. H., and Mikulka, C. J. (1981) *CEP* (June), 35–41.

Wessel, H. E. (1960) in *Cost Engineering in the Process Industries* (Chilton, C. H., et al., eds.) pp. 312–313, McGraw-Hill, New York.

White, R. J., Klein, F., Chan, J. A., and Stroshane, R. M. (1980) in *Annual Reports on Fermentation Processes* Vol. IV (Tsao, G. T., ed.) pp. 209–234, Academic Press, New York.

Windass, J. D., Worsey, M. J., Pioli, E. M., Pioli, D., Barth, P. T., Atherton, K. T., Dart, E. C., Byrom, D., Powell, K., and Senior, P. J. (1980) *Nature 287,* 396–401.

Yoneda, Y., Graham, S., and Young, F. E. (1979) *Biochem. Biophys. Res. Comm. 91,* 1556–1564.

Young, F. E. (1980) *J. Gen. Microbiol. 119,* 1–15.

Zabriskie, D. W., Armiger, W. B., Phillips, D. H., and Albano, P. A. (1980) *Traders' Guide to Fermentation Media Formulation* pp. 3–4, Traders Protein Division of Traders Oil Mill, Fort Worth, Texas.

Zeikus, J. G. (1980) *Ann. Rev. Microbiol. 34,* 423–464.

Zoon, K. C., Smith, M. E., Bridgen, P. J., Anfinsen, C. B., Hunkapiller, M. W., and Hood, L. E. (1980) *Science 207,* 527–528.

CHAPTER
5

Effects of Genetic Engineering of Microorganisms on the Future Production of Amino Acids from a Variety of Carbon Sources

Fredric R. Bloom
Peter J. Kretschmer

I. INTRODUCTION

Beginning with the isolation of the efficient L-glutamic acid–producing microorganism *Corynebacterium glutamicum* in 1957 (Kinoshita et al., 1957), many commercially important amino acids have been synthesized by using the gram-positive microorganisms *Corynebacterium* and *Brevibacterium*. In most of these production processes, the carbon source was sucrose, usually in the form of beet molasses, sugar cane molasses, or cornsteep liquor. However, the world demand for amino acids is growing despite the fact that many amino acids are relatively expensive. The predicted future decrease in cost of amino acids due to novel research involving recombinant DNA and molecular genetics will open the way to novel uses of, and thus create further demand for, amino acids as chemical feedstocks in processes hitherto considered economically unattractive. Such has been the case following the decreased cost of production of methionine and lysine (Office of Technology Assessment, 1981).

If the above projections prove to be true, alternate carbon sources to sucrose (from molasses) will become important for two reasons. The first is that the availability of cheap sources of beet molasses, sugar cane molasses, or cornsteep liquor may become limiting. The second reason is that, with the projected large decrease in cost of producing amino acids, even the cheapest molasses sources may become a substantial fraction of cost, and thus even cheaper carbon sources will become attractive.

This review will consider current and future feedstocks and the variety of microorganisms that potentially may convert these feedstocks not only to amino acids, but to other commercially important products as well.

Initially, we will discuss strategies currently employed for the production of amino acids from molasses, since these strategies will in large part be applied to any future microorganism of choice. The application of genetic engineering to these microorganisms, mainly of the *Corynebacterium* genus, will be discussed. We will then review methanol and ethanol as substrates for the production of amino acids. Since methanol and ethanol have been used as substrates for the production of single-cell protein, it is possible that the vast amount of knowledge gained in this area relating to the physiology and growth of the microorganisms could be applied to the production of cellular metabolites with appropriate genetic engineering. Finally, we will consider synthesis of amino acids using the cheapest of all carbon and energy sources, CO_2 and sunlight, and discuss the microorganisms that may be used in this area.

The use of alternative sources of substrate and microorganisms for the overproduction of cellular metabolites will depend largely on the use of recombinant DNA and molecular genetics, since, as we shall show, conventional methods of mutation and strain selection have not led to commercial processes for the production of amino acids. These tools will serve not only for the analysis of metabolic pathways in many diverse species, but also for the manipulation of these metabolic pathways.

We do not wish to imply that such research is necessarily short-term. In fact, practically no research involving recombinant DNA or genetics of amino acid pathways has been done for many of the microorganisms we will discuss. At this point it is therefore useful to consider briefly the current state of the art of recombinant DNA technology.

Initial recombinant DNA experiments were reported in the early 1970's. The first trans-species cloning experiments were reported in 1974 when the ampicillin gene of *Staphylococcus aureus* was cloned and expressed in *Escherichia coli* (Chang and Cohen, 1974). In the intervening 8 years, a tremendous explosion in molecular genetics has occurred. Scientists can now manipulate the DNA content of many organisms, including viruses, microorganisms such as bacteria and algae, and eukaryotes from yeast through to cells derived from humans.

The eventual curing of some genetic deficiencies in man is no longer considered impossible. However, until the early 1970's, the only means by which even microorganisms could be modified for improved product yield was by mutation and strain selection. Often, such yield improvement was a hit-or-miss program, with no explanation at the molecular level as to how the mutation in question resulted in increased yield. Thus although considerable effort was put into such programs, very little information was derived for application to future programs.

With the advent of recombinant DNA, two important goals could be realized. First, specific genes could be isolated and identified. The concentration of the gene product per cell could then be increased as a result of increasing the number of genes per cell. Second, basic questions as to how genes are controlled, how they are switched on and off, and how their products interact with metabolites within the cell could now be answered as a result of the ability to isolate specific genes. The answers to these questions, in contrast to the above mutation and strain selection programs, can be applied in many other programs of interest. As a result, genetic engineers can, in certain select microorganisms, increase the efficiency of product yield not just by increasing numbers of relevant genes per cell but also by specific overexpression of these genes to any desired level.

In other words, genetic engineers now know how to isolate genes, how to study their expression, and how to modify their expression within bacterial cells. The principles for these manipulations at the single-gene level have been established by using one microorganism, *E. coli*. In a sense, this has been the first phase of biotechnology, which occurred in the 1970's. The second phase of biotechnology will be to apply these initial lessons to the study of interactions and overexpression between a number of genes of a particular biosynthetic or degradative pathway as opposed to simple overproduction of any one protein. The second phase is more complex; first, one is often dealing with diversion of critical metabolites in the cell to only one pathway, and second, one must coordinate overexpression of a number of genes, and each gene must be appropriately regulated to achieve the maximal level of metabolite production without death of the cell.

The third and final phase of biotechnology, which will occur simultaneously with the second phase, is the application of lessons learned in the well-understood microorganisms to the more industrially important microorganisms. For example, manipulations of anaerobes and fungi, which are responsible for much of the world's important chemical activity in nature, will now be possible. It must be stressed, however, that the lessons learned in simple microorganisms will allow much more rapid experimentation with the more industrially important microorganisms. This is in direct contrast to the situation that existed with previous programs of simple mutation and strain selection. Thus there will be a cumulative effect to the research, even though it is done in different microorganisms, because in general the principles of gene regulation are the same from one microorganism to another.

Phases one and two are currently being pursued by many laboratories. The third phase has yet to begin in earnest, although a few laboratories are making preliminary efforts in this direction. These principles apply to the use of biomass and alternate carbon sources in the production of amino acids and other cellular metabolites. The timing and the extent of the effort put into such programs will determine the rapidity with which they will achieve practical application.

II. STRATEGIES FOR THE DEVELOPMENT OF AMINO ACID–PRODUCING MICROORGANISMS THAT UTILIZE MOLASSES

Cane or beet molasses is the preferred carbon source for the fermentative production of a number of amino acids such as L-glutamic acid and L-lysine. Molasses is derived as a by-product in the refining of cane or beet sugar. Little pretreatment is required for molasses to be used as a carbon source in the form of sucrose. The molasses is generally used directly upon addition of supplemental nitrogen and phosphate, dilution to a concentration of 4–20%, and sterilization (Forage and Righelato, 1979). However, the use of molasses in amino acid fermentations may have a number of disadvantages. The existence of potentially competitive future carbon sources such as ethanol and methanol in abundant quantities has prompted a search for amino acid excreters among ethanol- and methanol-utilizing microorganisms (see below). The disposal of the waste water from the large amino acid fermentation vessels adds to the expense of production. The availability of sufficient molasses supplies may also be in question, especially if the demand for amino acids continues to increase.

A. Genetic Manipulation

The commercial production of amino acids from molasses was reviewed recently by Kinoshita and Nakayama (1978). These authors have described in detail the derivation of a number of production strains as well as the theory behind strain development programs. Such strain improvements were the result of classic mutation and strain selection methods, including the isolation of analog-resistant and auxotrophic mutants. However, no system of genetic exchange exists in the majority of these organisms. Therefore the program was, of necessity, one of repeated mutagenesis and selection in the same strain background. These efforts achieved impressive results with a number of amino acids such as L-glutamic acid and L-lysine, although with the other amino acids, the yields were not as high (Kinoshita and

Nakayama, 1978). This has been attributed to a number of factors such as the complexity of the regulation of some biosynthetic pathways. For example, in some instances, mutations required to overproduce the amino acid could not be brought together in the same strain background by mutagenesis and selection (Kinoshita and Nakayama, 1978).

However, a number of general conclusions did emerge as a result of an analysis of this research. From these conclusions, a new strategy for the production of amino acid–excreting strains was proposed. This strategy centered on the use of transductional crosses to combine the desired mutations into one strain background (Kisumi et al., 1981). Of necessity, this strategy limits genetic engineering programs to those organisms for which a genetic system is available, such as the gram-negative bacteria *Salmonella, Escherichia, Klebsiella, Pseudomonas,* and *Serratia* and the gram-positive *Bacillus*.

To summarize research to date, mutations resulting in the following characteristics have been found to be very important for the overproduction of an amino acid:

1. Prevention of degradation of the amino acid by the production strain
2. Elimination of control of the biosynthetic pathway due to feedback inhibition by isolation of appropriate amino acid analog-resistant mutants
3. Restriction of carbon flow to alternate pathways
4. Increase in the availability of precursors
5. Increase in the synthesis of certain biosynthetic enzymes in the pathway

We will now illustrate these general rules by considering examples of strain improvement programs in *Corynebacterium, Brevibacterium,* and *Escherichia*.

1. Elimination of amino acid degradation. Many organisms are able to utilize certain amino acids as the sole source of carbon or nitrogen. These strains possess inducible enzymes such as tryptophanase, arginase, proline oxidase, and histidase, which are able to degrade the amino acid. Research on a number of production strains of *Corynebacterium glutamicum* indicates that these strains are defective in the degradation of the amino acid, although in most instances, the mutation was not deliberately isolated. Araki and Nakayama (1971) have postulated that the accumulation of histidine in a mutant strain of *C. glutamicum* is due in part to a lack of histidine degradation. More definitive results were obtained with strains of *Serratia marcescens* that possess a potent histidase activity (Kisumi et al., 1977). For example, when 10 g/l of L-histidine is added to a fermentation broth, the parental strain is capable of degrading 99% of the L-histidine within 48 h. A histidase-negative mutant was therefore deemed essential in constructing a histidine-excreting strain. Similarly, a mutant strain of *S. marcescens* that is defective for arginine degradation was isolated as the first step in the construction of an arginine excreter (Kisumi et al., 1978).

In summary, the general approach has been to isolate, as a first step, mutants that no longer degrade the amino acid.

2. Elimination of feedback inhibition. In general, wild-type strains of bacteria do not excrete amino acids into the medium. The biosynthetic pathways are so strictly regulated by feedback inhibition and repression that amino acids are not overproduced even within the bacterial cell. In order to reduce this strict regulation, amino acid analogs have been used to isolate strains in which these controls have been altered. This approach is feasible because certain enzymes in the biosynthetic pathway are subject to feedback inhibition by the amino acid that is the end product of the pathway. The analog acts as a false feedback inhibitor of the amino acid, subjecting the cell to a condition of amino acid starvation. Those mutants that are resistant to the effects of the analog often contain an altered enzyme no longer subject to the effects of feedback inhibition. As a result, the flow of carbon through the pathway is increased, and the amino acid is overproduced and excreted. Some amino acid analogs that have been used in this manner to select amino acid–excreting mutants are listed in Table 5.1

3. Restriction of carbon flow to alternate pathways. The simplest approach to blocking the flow of precursors to alternate pathways is to isolate auxotrophic mutants with biosynthetic blocks in those pathways. As an example,

TABLE 5.1 Amino Acid Analogs

Amino Acid	*Amino Acid Analog*	*Enzyme Affected*	*Organism*	*Reference*
L-Tryptophan	5-Methyltryptophan 6-Fluorotryptophan	Anthranilate synthetase	*Brevibacterium flavum* *B. flavum*	Shiio et al. (1975)
L-Phenylalanine	3-Fluorophenylalanine B-thienylalanine	DAHP synthetase Prephenate dehydratase	*B. flavum* *Bacillus subtilis*	Shiio et al. (1975) Pierson and Jensen (1974)
L-Histidine	1,2,4-Triazole 3-Alanine 2-Thiazolealanine	Phosphoribosyl ATP pyrophosphorylase	*Corynebacterium glutamicum* *C. glutamicum*	Araki and Nakayama (1974)
L-Proline	3,4-Dehydroproline	Glutamylkinase	*Escherichia coli*	Baich and Pierson (1965)
L-Lysine	S-(B-aminoethyl) L-cysteine	Aspartokinase	*B. flavum*	Shiio and Miyajima (1969)
L-Threonine	1-Amino B-hydroxyvaleric acid	Aspartokinase-homoserine dehydrogenase	*C. glutamicum*	Kase and Nakayama (1974)

C. glutamicum mutants that excrete lysine were isolated as strains requiring homoserine for growth (Shigeto, 1962). In these strains, the defective homoserine dehydrogenase prevents the drain of aspartate semialdehyde to methionine and threonine. The aspartate semialdehyde is therefore channeled toward lysine. Alternatively, in order to channel the precursors toward threonine, Kase and Nakayama (1972) isolated methionine auxotrophs. In the aromatic pathway of *C. glutamicum*, tryptophan excreters were isolated in a strain background auxotrophic for phenylalanine and tyrosine (Hagino and Nakayama, 1974), and tyrosine excreters were isolated as phenylalanine auxotrophs (Hagino and Nakayama, 1973).

A third example of altered carbon flow is provided by *C. glutamicum* and *Brevibacterium flavum*, which are able to convert 50% of carbohydrate in the fermentation medium to L-glutamic acid. As was discussed previously (Kinoshita and Nakayama, 1978), the basis for this overproduction is a diversion of carbon flow from the tricarboxylic acid cycle (TCA cycle) intermediate α-ketoglutarate. Because these strains do not synthesize significant quantities of α-ketoglutarate dehydrogenase, while levels of glutamate dehydrogenase are quite high, α-ketoglutarate is channeled to glutamate rather than to the next intermediate of the TCA cycle.

In summary, these examples provide examples of alteration of carbon flow by mutating genes for enzymes located in the biosynthetic pathway before formation of the amino acid.

4. Increasing availability of precursors. Apart from restricting carbon flow to alternate pathways, another approach to increasing the precursor concentration is simply to feed the precursor to the bacterium during fermentation, provided of course that the precursor can enter the cell. An example in which both approaches were used is the case of an *E. coli* strain doubly auxotrophic for methionine and valine (restricted carbon flow to alternate pathways), which excreted 8.4 g/l of L-threonine (Hirakawa et al., 1973). However, adding the inexpensive precursor, aspartate (5 mg/ml), at the start of the fermentation resulted in a 25% increase in yield to 10.6 g/l of threonine.

5. Increasing synthesis of key biosynthetic enzymes. Biosynthetic enzyme *activity* in a cell can be regulated by feedback inhibition of activity, as was discussed above (Section II.A.2). Alternatively, total activity in the cell can be regulated by the amount of enzyme *synthesized* by the cell. In particular, synthesis of many key amino acid biosynthetic enzymes is regulated at the level of transcription. Such transcriptional regulation is observed in general with synthesis of the first or branch point enzyme of the pathway.

One mechanism of transcriptional regulation is via a classic repressor protein, as in the case of the repression of synthesis of the tryptophan biosynthetic enzymes by the *trp*R repressor protein. Such repressors are now known for a number of amino acid biosynthetic pathways, including those of arginine, cysteine, and glutamine (Umbarger, 1978). In *E. coli,* the isolation of deletions that eliminate these repressor proteins is a relatively simple matter, whereas in the gram-positive microorganisms, such repressors have not been unambiguously identified, much less mapped on the chromosome.

Another mechanism of transcriptional regulation in *E. coli,* called attenuation, has been found to occur in a number of biosynthetic pathways, for example, tryptophan, phenylalanine, leucine, isoleucine–valine, histidine, and threonine (Yanofsky, 1981). Attenuation is actually the translational regulation of transcription termination (Yanofsky, 1981; Gardner, 1979). An excess of charged tRNA containing the particular amino acid (indicating a deficiency of cellular protein synthesis) reduces the extent of transcription of the gene by blocking the progress of the RNA polymerase. Direct evidence for attenuation has been provided by sequence analysis of the various promoter regions of *E. coli.* The known sequences have been summarized by Yanofsky (1981).

6. Summary—Past and Future. The preceding sections have dealt very briefly with the knowledge base gained from over 20 years of research on the genetics of amino acid pathways. Success in creating overproduction strains has until now relied on the accumulation, in one strain, of as many desirable mutations as possible, generally without the benefit of any type of genetic exchange system. Genetic engineering now promises the opportunity to endow any microbial system with that ability to precisely exchange desired genetic traits between strains. With such a powerful tool, the current knowledge base of amino acid pathways will be much more fully utilized and will result in greatly increased yields of amino acids.

B. Media Development

Many of the relevant concepts in this area have already been thoroughly discussed in the review by Kinoshita and Nakayama (1978). These media manipulations include:

1. the optimization of biotin concentration to improve the production of L-glutamic acid in *C. glutamicum,*
2. the addition of penicillin to the *C. glutamicum* fermentation broth to compensate for the presence of excess biotin in certain media such as molasses, and
3. the addition of surface active agents such as polyoxysorbitan trioleate or other fatty acid derivatives.

The above additions alter the fatty acid composition of the cell membranes and improve the permeability of *C. glutamicum* cells. This in turn allows leakage of glutamate from the cells into the medium. Alterations in cell membranes may serve, by some unknown mechanism, to improve the uptake of certain precursors that ordinarily do not enter the cell or that enter the cell at a rate that is limiting production.

A second area for research in media improvement involves the subject of carbon flow, already covered in Section II.A.3. Obviously, the choice and concentration of the carbon and nitrogen sources play a major role in the yield of a product. For example, Araki et al. (1974) were able to achieve higher yields of L-histidine by varying the ratio of molasses to total sugar (molasses + sucrose). However, another option that genetic engineering has made possible is to analyze the preferred medium (for example, molasses) and to overproduce the key enzymes responsible for utilization of the carbon source in question.

C. Application of Genetic Engineering to Current Amino Acid–Producing Strains

1. Gram-positive microorganisms. Screening programs in gram-positive bacteria are fast approaching their limits of usefulness in increasing the yields of amino acids through mutagenesis, selection, and media manipulation. It appears that the greatest difficulty encountered in this program has been the isolation of mutant strains in which the synthesis of branch point enzymes occurs at a derepressed rate. Further significant increases in productivity can be predicted to require the overproduction of these key enzymes in the biosynthetic pathways.

At present, the most viable approach to further improvement in production levels is the application of genetic engineering to these gram-positive production strains. No workable genetic exchange system exists in *C. glutamicum* or *B. flavum,* although *Bacillus subtilis,* another industrially important gram-positive microorganism, does possess an extensive genetic system including transformation, transduction, and a circular chromosomal map with 340 mapped genes (Dean and Dooley, 1981).

B. subtilis has been the focus of intense research both in academic laboratories and in private biotechnology companies. *Bacillus* species produce such antibiotic agents as gramicidin S, bacitracin, and polymyxin. In addition, such industrially important enzymes as alpha amylase, levansucrase, maltase, and penicillinase, as well as various proteases, are produced by *Bacillus* species.

A large number of cloning vectors now exist for the introduction of heterologous DNA into *B. subtilis*. The more frequently used plasmid vectors with their various restriction endonuclease sites have been reviewed recently

(Dean and Dooley, 1981). In addition, a number of "shuttle" vectors that can replicate in either *E. coli* or *B. subtilis* have been developed (Goldfarb et al., 1981). A number of bacteriophage vectors such as phi105, SP02, and Spp1 have also been utilized. Finally, libraries of *B. subtilis* DNA have been constructed in *E. coli* by using bacteriophage (Ferrari et al., 1981) or plasmid (C. Finn, personal communication) vectors.

B. subtilis is readily transformed by linear chromosomal DNA, in a process that converts the double-stranded DNA to single strands. Competent strains are not easily transformed with monomeric plasmid DNA. However, protoplasts will take up plasmid DNA. Chang and Cohen (1979) were able to transform double-stranded plasmid DNA into polyethylene-glycol-induced protoplasts of *B. subtilis*. Two significant problems were mentioned by Dean and Dooley (1981) that clearly require more concentrated research. The first is the instability of cloned *B. subtilis* DNA in *B. subtilis,* a problem that may disappear as more vectors are developed. There is no apparent stability problem in cloning DNA from other bacillus species into *B. subtilis*. The second problem is the general lack of expression of foreign DNA in *B. subtilis*. A solution to this problem will probably require recombining foreign DNA with a *B. subtilis* promoter system, as has proven successful in the expression of the *E. coli* chloramphenicol acetyl transferase gene in *B. subtilis* (Goldfarb et al., 1981).

In order to apply recombinant DNA technology to the commercially important amino acid–producing gram-positive microorganisms such as *Corynebacterium,* one can expect to take advantage of the already well-developed *E. coli* or *B. subtilis* systems. For example, one can imagine creating shuttle vectors capable of replicating in both *B. subtilis* and *Corynebacterium,* in order to use these vectors to clone relevant amino acid biosynthetic genes from *Corynebacterium* to *Bacillus* and vice versa. Such studies are underway in a number of laboratories and, as such, represent the third phase of genetic engineering discussed in Section I.

2. Gram-Negative Microorganisms. A number of recent studies have resulted in the successful genetic manipulation of gram-negative microorganisms for the overproduction of amino acids. These studies have utilized a number of the principles outlined in Section II.A and have taken advantage of the more advanced genetic systems of gram-negative microorganisms. Amino acid–excreting strains of *Serratia marcescens* have been constructed according to the strategy outlined by Kisumi et al. (1981). The appropriate mutations were combined in one strain background by utilizing transductional crosses with the bacteriophage PS20. The effectiveness of generalized transduction as a first step in the construction of useful industrial strains was shown by yields as high as 25 g/l for isoleucine, threonine, and arginine in different strains. Further improvements in yield will require the overproduction of a number of key enzymes in the biosynthetic pathways by cloning and manipulation of appropriate genes.

In 1979, Tribe and Pittard combined transductional crosses utilizing bacteriophage P1 with amplification of the tryptophan biosynthetic genes to construct a tryptophan-excreting strain of *E. coli* (Tribe and Pittard, 1979). Transductional crosses were used to combine mutations in which the key enzymes in the aromatic biosynthetic pathway, DAHP synthase and anthranilate synthase, were desensitized to end product inhibition. In addition, mutations were introduced into their strain that blocked the degradation of tryptophan and blocked the flow of the key intermediate chorismate to the other aromatic amino acids phenylalanine and tyrosine. Finally, the rates of synthesis of DAHP synthase and anthranilate synthase were increased by (1) inactivating the specific repressors that repress the synthesis of these key enzymes and (2) amplifying the *trp* operon on a low-copy-number plasmid. Tryptophan yields of 1 g/l were achieved, and the low yields were attributed in large part to the decay of DAHP synthase activity in the stationary phase of growth. The key aspect to this work was the attempt to amplify the synthesis of the *trp* biosynthetic enzymes by inactivating the *trp* repressor and by introducing a plasmid containing another copy of the *trp* operon into the production strain.

This strategy was carried one step further by Aiba et al. (1980). A *trp*R$^-$ (repressor defective) and tryptophanase$^-$ strain of *E. coli* was constructed and used as a recipient for the introduction of a multicopy plasmid containing the *trp* operon. The anthranilate synthase specified by this plasmid was insensitive to feedback inhibition by tryptophan. The plasmid also specified resistance to tetracycline, and provided tetracycline was included in the medium, the plasmid was stably maintained. After 27 h in a medium containing 5% glucose, 1% casamino acids, 0.3% ammonium chloride, 0.05% anthranilic acid, and 10 mg/ml of tetracycline, 5 g/l of tryptophan was produced. Anthranilic acid (a precursor) was fed to the culture at a rate of 50 mg/l per hour, and its addition was essential for a high rate of productivity.

In more recent studies, Debabov et al. (1981) cloned DNA fragments containing the genes A, B, and C of the *E. coli* threonine operon. The genes were cloned into the plasmid pBR322 from a strain resistant to feedback inhibition by threonine as a result of a mutation in the *thr*A gene. This plasmid, when transformed into a strain containing mutations blocking synthesis of threonine and partially blocking isoleucine–valine synthesis, resulted in yields of 20 gm/l of threonine after 48 h.

In studies by the authors and others at Bethesda Research Laboratories (unpublished results), similarly designed cloning experiments have resulted in excess proline production in *E. coli*. In these experiments, the three genes *pro*A, B, and C of the proline biosynthetic pathway have been cloned onto various multicopy plasmids. Strains containing these plasmids were selected for resistance to the proline analog, 3,4-dehydroproline, and it was found that the genetic information for this resistance phenotype was located on the plasmids. These plasmids, in conjunction with appropriate host cell mutations and media development studies, have resulted in proline yields in excess of 20 g/l.

In summary, it is clear that, in gram-negative bacteria, recombinant DNA technology offers a powerful tool in studies aimed at overexpression of amino acids. It is almost certain that such will be the case in gram-positive bacteria as well.

III. POTENTIAL METHODS AND MICROORGANISMS FOR THE FUTURE PRODUCTION OF AMINO ACIDS

A. Production of Amino Acids by Algae

The use of microorganisms to produce various chemicals from inexpensive carbon sources has been one of the major goals of industry. One intriguing possibility is the production of amino acids from carbon dioxide and sunlight by utilizing photosynthesizing algae. Certain species of eukaryotic green algae and cyanobacteria have been cultivated as a source of single-cell protein. This topic has recently been reviewed by Benemann et al. (1979).

Until recently, successful attempts to isolate mutants of algae resistant to amino acid analogs and to screen these mutants for excretion of amino acids were lacking. In 1975, Phares and Chapman (1975) were able to isolate mutants of *Anacystis nidulans* that were resistant to the tyrosine analogs fluorophenylalanine and 2-amino-3-phenylbutanoic acid. One class of mutants was shown to have a DAHP synthetase that was less sensitive to feedback inhibition by L-tyrosine. However, no data were presented on the ability of these mutants to excrete aromatic amino acids. More recently, Hall et al. (1980), in an extensive study, were able to isolate a number of mutants of cyanobacteria that excrete amino acids (see Table 5.2). However, the quantities excreted were only in the milligrams per liter range.

Riccardi et al. (1981) also isolated mutants of *Spirulina platensis* that were resistant to analogs of phenylalanine, methionine, and proline. A number of analog-resistant derivatives were resistant to more than one analog, possibly owing to a "metabolic interlock" (Jensen, 1969) involving several pathways. A number of resistant strains were found that excreted phenylalanine at levels up to 80 μg/l.

Thus the isolation of analog-resistant mutants that excrete amino acids, in addition to the isolation of a variety of auxotrophic mutants in cyanobacteria (see Table 5.3), suggests the possibility of genetically engineering strains of cyanobacteria for the overproduction of amino acids.

Key to the isolation of these mutants was the novel use of *B. subtilis* auxotrophs to screen the analog-resistant derivatives of cyanobacteria for excretion of amino acids. A standard technique for the isolation of amino acid–overproducing mutants is to screen analog-resistant mutant colonies

TABLE 5.2 Excretion of Amino Acids in Cyanobacteria

Strain	*Analog Resistance*[1]	*Amino Acid Excreted*	*Excretion Level*[2]
Anabaena (ATCC 29151)	—	Phenylalanine	19
	—	Tryptophan	0
	4-Fluorophenylalanine	Phenylalanine	64
	6-Fluorotryptophan	Tryptophan	122
Synechocystis sp. (ATCC 29108)	—	Phenylalanine	0
	4-Fluorophenylalanine	Phenylalanine	134
Synechocystis sp. 602	—	Phenylalanine	5
	2-Fluorophenylalanine	Phenylalanine	26
Synechococcus AN T×20	—	Phenylalanine	4
	3-Fluorophenylalanine	Phenylalanine	125
Anabaena sp.	Ethionine	Methionine	NG[3]
Synechocystis sp. (ATCC 29108)	Canavanine	Arginine	NG
Synechococcus sp. 602	Ethionine	Methionine	NG

From Hall et al. (1980)
1. In each case, the compound is an analog of the excreted amino acid. A dash indicates a wild-type strain.
2. Micromolar concentration per unit growth (O.D. 500 nm).
3. NG = not given.

for the ability to cross-feed a background lawn of organisms that are auxotrophic for that amino acid. Overproducing colonies are recognized by a surrounding halo of growth. However, this technique was not particularly successful with cyanobacteria until Hall et al. (1980) discovered that a *B. subtilis* auxotroph assimilated phenylalanine two orders of magnitude greater than a cyanobacterial auxotroph. By using the *B. subtilis* auxotroph as the

TABLE 5.3 Auxotrophic Mutants Isolated in Cyanobacteria

Strain	*Requirement*	*Reference*
Anacystis nidulans	Uracil	C. L. R. Stevens et al. (1975)
Anacystis nidulans	Phenylalanine, methionine, biotin, cysteine	Herdman and Carr (1972)
Synechococcus sp. 602	Phenylalanine	S. V. Shestakov, cited in Hall et al. (1980)
Agmonella quadruplicatum	Tryptophan (tryptophan synthetase$^-$)	Ingram et al. (1972)
Synechococcus cedrorum	Phenylalanine (prephenate dehydratase$^-$)	Kaney (1973)

indicator lawn, *Synechococcus* mutants capable of overexcreting phenylalanine were easily detected.

In order to establish a viable recombinant DNA system, one requires suitable cloning vectors. Numerous studies have found plasmids in various strains of cyanobacteria. Restaino and Frampton (1975) characterized plasmids of 30.5 and 5.3 Mdal in *Anacystis nidulans*. Roberts and Koths (1976) carried out a similar study in *Agmonella quadruplicatum*. More recent studies have revealed plasmids in a wide range of cyanobacteria (van den Hondel et al., 1979; Simon, 1978; Lau and Doolittle, 1979). Lau et al. (1980) investigated plasmids of the heavy metal–resistant *Synechococcus cedrorum, Synechococcus* sp. 6908 and 6707, and *A. quadruplicatum* PRG. Unfortunately, the presence or absence of the plasmids in these strains could not be correlated with the resistance or sensitivity of the strains to a variety of heavy metals or antibiotics.

A viable recombinant DNA system also requires a system for the transfer of genetic information. The possibility of plasmid transfer between species of cyanobacteria in nature was discussed by van den Hondel et al. (1979), who found that two quite distinct species (*Synechococcus* sp. 6707 and *A. nidulans*) contain plasmids that are identical in size and restriction pattern. Using standard DNA:DNA hybridization techniques, Lau et al. (1980) were able to demonstrate regions of homology between plasmids in different species.

A more organized approach to establishing a transformation system began with the observation that a number of strains of cyanobacteria could be transformed by chromosomal DNA (Astier and Espardellier, 1976; Devilly and Houghton, 1977; Grigorreva and Shestakov, 1976; Herdman, 1973). Recently, a transformation system in *A. quadruplicatum* was described by S. E. Stevens and Porter (1980), who were able to transform the gene encoding resistance to streptomycin at a frequency of approximately $10^5/\mu g$. Transformation of *A. nidulans* by plasmid DNA has been demonstrated by van den Hondel et al. (1980) using a procedure of Shestakov and Khyen (1970). The efficiency of transformation was 10^4–10^5 transformants per microgram of plasmid DNA.

The 5.3-Mdal plasmid, pUH24, found in several species of cyanobacteria, was selected as a potential plasmid-cloning vector. A transposable element coding for ampicillin resistance, Tn901, was introduced onto pUH24, resulting in an 8.3-Mdal plasmid. This plasmid can be transformed into *A. nidulans* with a frequency of 10^4–10^5 transformants per microgram of DNA. A smaller plasmid that removes part of the Tn901 transposon was derived by removing a BamH1 fragment. The new plasmid, pUC1, was 5.5 Mdal, conferred Ap resistance, and contained single BamH1 and Xho1 sites (van den Hondel et al., 1980).

Clearly, a great deal of genetic investigation and manipulation is required in order to establish a viable system for the industrial exploitation of the cyanobacteria. However, the above plasmid transformation experiments with expression of an antibiotic resistance gene together with a viable chromosomal transformation system indicate that the tools are available for such research to begin.

B. Production of Amino Acids from Ethanol

Ethanol is currently produced by chemical conversion of ethylene or by yeast fermentation of glucose from corn starch, processes that result in current costs of approximately $1.90 per gallon. However, recent research has indicated the possibility of cheaper production of ethanol in the future, namely production of ethanol and other chemicals from biomass (see below).

Ethanol has a number of advantages as a substrate such as purity, ease of storage, solubility, and high cell yield (Flickinger and Tsao, 1978). The present disadvantages of ethanol are its cost relative to alternative sources such as methanol and carbohydrates such as molasses and its toxicity at concentrations above 2–3%. Toxicity problems may be overcome by selection of mutant strains resistant to higher concentrations of ethanol. However, a significant reduction in the cost of ethanol will probably require the large-scale production of ethanol from biomass.

A number of recent reviews have described the use of microorganisms for the production of ethanol from renewable biomass sources (Zeikus, 1980; Eveleigh, 1981). However, it appears unlikely that the microorganisms currently being studied for the conversion of biomass to ethanol will significantly reduce the cost of ethanol in the near future. It is conceivable that recombinant DNA technology applied to organisms such as *Clostridium thermocellum* and *Trichoderma reesei* may eventually provide a commercial process for biomass conversion to lower-cost ethanol. Of more relevance to this review, however, may be the long-term possibility of using such cellulose-degrading microorganisms for the conversion of biomass directly to higher-value amino acids.

A significant number of screening programs have searched for amino acid–excreting microorganisms capable of assimilating ethanol. A summary of some of these studies is presented in Table 5.4. These and other studies on the production of amino acids from ethanol (with the possible exception of glutamate; see Table 5.4) suffer from a number of drawbacks. The yields of amino acids are quite low in comparison to identical fermentations using *n*-paraffins or sucrose as the sole carbon source. The fermentation times are also quite long (Table 5.4). In addition, concentrations of ethanol higher than 2–3% are toxic to the production microorganism, and thus ethanol must be continuously fed to the fermentation.

The possibilities for utilizing genetic engineering to improve the conversion of ethanol to amino acids will probably require the development of a recombinant DNA system in one or more of the bacteria listed in Table 5.4. A transformation system has been developed in *Acinetobacter calcoaceticus* (Juni, 1972). However, a more promising microorganism may be *Serratia*

TABLE 5.4 Production of Amino Acids from Ethanol

Amino Acid	Organism	Yield (g/l)	Fermentation Time (h)	Other Additions	Reference
Glutamic acid	*Brevibacterium* sp. B136	53.1 (60%)[1]	60–72	—	Oki et al. (1968)
Arginine	*Microbacterium ammoniaphilum* thiazolealanine^R	2.2 (2%)[1]	48	—	Kubota et al. (1974)
Arginine	*Brevibacterium flavum* guanine$^-$ thiazolealanine^R	18.4	48	Medium contained 1% glucose, 1% ethanol; 16% of ethanol consumed	Kubota et al. (1975)
Lysine	*Acinetobacter* threonine^R valine^R	4.1	96		Tanaka et al. (1975)
Lysine	*Acinetobacter calcoaceticum* 5-methyltryptophan^R ATCC 19606	6	168		Yakawa et al. (1981b)
Lysine, isoleucine	*Arthrobacter alkanicus* threonine^R ATCC 21657	6 2.1	168	Aspartate and polyoxysorbitan monooleate	Watanabe et al. (1973)
Lysine	*Nocardia lyena* homoserine$^-$ ATCC 21338	12	96		Tanaka et al. (1974)
Tryptophan	*Serratia marcescens* 5-methyltryptophan^R	0.012	168		Yukawa et al. (1981a)

1. Yield as a percentage of ethanol added.

marcescens, a gram-negative organism (a member of the enteric bacteria) with a relatively well-developed genetic system. This bacterium has already been used to select a number of well-characterized auxotrophic and analog-resistant mutants. A transduction system has already been developed and has been used to construct a number of amino acid–excreting strains using glucose as a carbon source (Kisumi et al., 1981; Kisumi, 1981). Plasmids from *E. coli* can be mobilized into *Serratia* (Bukhari et al., 1977), a finding suggesting that the extensive genetic engineering system of *E. coli* can be utilized to genetically engineer *Serratia*. Thus genetic engineering is already possible with at least one microorganism capable of converting ethanol to amino acids.

C. Production of Amino Acids from Methanol

The production of chemicals such as amino acids from microorganisms (methylotrophs) grown on methanol represents an attempt to utilize an inexpensive (11 cents/lb), readily available (38 billion lb/yr) carbon source. The biochemistry and enzymology of these microorganisms, which can also assimilate methane or formaldehyde to support cellular growth, have been reviewed recently (Colby et al., 1979; Tani et al., 1978b; Hamer, 1979).

Methanol is manufactured commercially by the oxidation of methane or by the reaction of hydrogen and carbon monoxide. Since it is a very cheap carbon source, methylotrophs have been utilized in a considerable number of studies on the production of single-cell protein (SCP) from methanol. Organisms that have been utilized in these studies include *Achromobacter, Methanomonas, Candida, Hansenula, Kloeckera, Pichia, Torulopsis,* and *Trichoderma* (Sahm, 1977; Wagner, 1977; Hamer, 1979). SCP processes have been described that involve the use of mixed cultures (Harrison et al., 1975; Cremieux et al., 1977). The advantage of mixed cultures is that the relatively low cell yields and long doubling times of single species fermentations are improved owing to the removal of toxic products by one of the microorganisms.

With the large amount of literature and knowledge on production of SCP from methanol, it is tempting to consider the application of genetic engineering for the construction of methylotrophs capable of producing fine chemicals, such as amino acids, from methanol. Such a possibility is further strengthened by recent experiments in which the methylotroph *Methylophilus methylotrophus* was successfully genetically engineered to improve the yield of SCP from methanol.

In this approach, Windass et al. (1980) cloned the glutamate dehydrogenase (*gdh*) gene of *E. coli* into a high-copy-number, broad-host-range plasmid, pTB70. The recombinant strain was mobilized (by conjugation) into a strain of *M. methylotrophus* that contained a temperature-sensitive mutation for the glutamate synthase gene. At the restrictive temperature, the *E. coli gdh* gene was able to complement the mutant, enabling growth on a methanol-minimal medium. More important, the use of the *gdh* gene rather than the native glutamate synthase gene for the process of ammonia assimilation saved one molecule of ATP per molecule of glutamate formed. Cell yield from methanol rose 4–7%. Such an example serves to illustrate the potential of genetic engineering within this group of microorganisms.

There have been a number of extensive studies on the production of amino acids by methylotrophs using methanol as a carbon source. Summaries of some of these studies are listed in Table 5.5, which illustrates that a large variety of amino acids are excreted by methylotrophs. Some amino acid–excreting strains such as *Methanomonas methylovora* were isolated following simple screening procedures (Kono et al., 1972), while others such as *Protaminobacter thiaminophagus* were found to excrete the relevant amino acid(s) following mutation and screening for resistance to amino acid analogs. In other words, successful strategies for the isolation of amino acid–excreting methylotrophs are identical to those used to isolate commercial strains of the *Corynebacteria*.

Similarly, the approaches to the genetic engineering of methylotrophs for the production of amino acids will be similar to those already developed

TABLE 5.5 Production of Amino Acids from Methanol

Amino Acid Produced	*Organism*	*Yield of Amino Acid (g/l)*	*Fermentation Time (h)*	*Other Additions*	*Reference*
Glutamic acid[1]	*Pseudomonas insueta* ATCC 21276	2.46	72		Shiio et al. (1971)
	Methanomonas methylovora ATCC 21369	10.5 (10%)[2]	48		Kono et al. (1972)
	Protaminobacter thiaminophagus	32.8	58		Nakayama et al. (1976)
	phe ATCC 21969	22 (13%)[2]	58		
Leucine	*Protaminobacter thiaminophagus* ATCC 21927	1.2[3]	96		Nakayama et al. (1975a)
Valine		1.0			
Lysine	*Protaminobacter thiaminophagus* L-threonine^R S-2 aminoethyl L-cysteine^R ATCC 21926	1.5	72		Nakayama et al. (1975b)
Serine	*Pseudomonas*		80–100	2% glycine	Wagner et al. (1977)
	DSM 672	5–6			
	DSM 673 gly$^-$	8–10			
	Arthrobacter globformis SK200 met$^-$	5.2	72	2% glucose 2% glycine	Tani et al. (1978a)
Tryptophan	*Pseudomonas* AM1	0.25	72	Indole—55% converted to tryptophan	Gatenbeck and Hedman (1976)
Tryptophan	*Methylomonas methanolica*	0.25	60	Indole—65% converted to tryptophan	
O-Methyl-homoserine	*Microcyclus eburneus* KY 7831 S-2 aminoethyl L-cysteine^R homoserine^R	3.8	72	Homoserine	Tanaka et al. (1980)

1. Other amino acids produced in trace quantities include alanine, isoleucine, leucine, threonine, valine, proline, glycine, lysine, and arginine.
2. Yield based on total amount of methanol added.
3. Other amino acids produced include aspartic acid, alanine, arginine, and lysine.

for gram-negative microorganisms (see Section II.C.2). The advantage of a wide-host-range plasmid as vector was amply illustrated by Windass et al. (1980) in the development of *M. methylotrophus* with increased SCP yield from methanol (see above). Such broad-host-range vectors allow cloning and manipulation within the versatile *E. coli,* followed by conjugation into a suitable methylotroph recipient. Such a procedure obviates the need for a plasmid DNA transformation system in the organism of choice, although plasmid DNA transformation is possible in some strains of *Pseudomonas* (Chakrabarty et al., 1975; Nagahari and Sakaguchi, 1978; Bagdasarian et al., 1979). A list of some broad-host-range vectors is given in Table 5.6

In summary, the major stumbling block to successful commercialization of methylotrophs as fine chemical producers may be the toxicity problems associated with methanol concentrations greater than 0.04 g/l. However, genetic engineering can be predicted to play a role in overcoming this drawback.

The future production of amino acids and other fine chemicals from methanol is an exciting possibility, given the facts that methanol is extremely cheap and abundant and that one industrially important methylotroph has already been successfully genetically engineered, with increased yield of product as a result.

D. Production of Amino Acids from Biomass

Large amounts of literature have recently reviewed attempts at the microbial conversion of biomass to chemicals, fuels (particularly ethanol), and proteins (Zeikus, 1980; Office of Technology Assessment, 1981; Eveleigh, 1981). Although there is obviously considerable variability in composition, biomass is considered to average 50–60% cellulose (a six-carbon sugar polysaccharide), 10–30% hemicellulose (a five-carbon sugar polysaccharide), and 10–30% lignin (a phenolic polymer).

For the microbial conversion of such biomass to chemicals and fuels to be economical requires efficient use of these components, either as a carbon source (cellulose and hemicellulose) or as a caloric fuel source (lignin). A number of microorganisms that produce cellulases and hemicellulases produce the five- and six-carbon sugar monomers, which in turn are metabolized to alcohols and organic acids. At present, it would appear that anaerobes offer the greatest hope that, with considerable research, amino acids or other cellular metabolites may be produced from biomass. Anaerobes are capable of growing efficiently on a wide variety of substrates, which in turn could be derived from biomass. In this sense, the anaerobic production of amino acids would be one step removed from biomass. Alternatively, anaerobes such as *Clostridium thermocellum* and *C. saccharolyticum* are capable of degrading cellulose and hemicellulose for energy; with appropriate manipulation, one can postulate the use of these micro-

TABLE 5.6 Broad-Host-Range Cloning Vectors

Plasmid	*Size*	*Copy Number*	*Drug Resistance*	*Restriction Endonuclease Sites*	*Reference*
R1162			Sm	HpaI, SstI	Bagdasarian
RSF1010	5.5 md	15–40 nonamplifiable	Su	EcoRI, BstEII PvuII	et al. (1979)
pMW79 (pBR322–RSF1010 hybrid)	8.4 md	multicopy nonamplifiable	Ap/Cb Sm/Tet	BamHI, SalI HindIII HpaI	Wood et al. (1981)
pFG7	8.4 md	multicopy amplified in *E. coli*	Sm/Su Ap Tet	HindIII, BamHI SalI, ClaI	Gautier and Bonewald (1980)
pRP301	54.7 Kb	low-copy conjugative	Derived from RP4		Windass et al. (1980)
pTB70	17.6 Kb	multicopy	Sm/Su Km	SalI	Windass et al. (1980)

organisms for the production of amino acids directly from biomass. It is for these reasons that we will look at what is known about amino acid production by anaerobes.

Very few studies have been reported on the production of amino acids by anaerobes. Allison (1969) and Sauer et al. (1975) have provided the most recent reviews of this subject in relation to ruminant anaerobes. From these studies, amino acid biosynthesis in ruminant microorganisms would appear to differ from aerobes such as *E. coli* not only in the regulation of synthesis, but also in the metabolic pathways of formation.

In a series of experiments designed to study precursors involved in anaerobic amino acid biosynthesis, Sauer et al. (1975) utilized labeled compounds to determine the role of reductive carboxylation in the formation of the two-oxo precursors of amino acids. These studies indicated that rumen anaerobes synthesized a large proportion of amino acids de novo, with reductive carboxylation reactions playing a major role in providing precursors of amino acids. In addition, this investigation supported the idea that one tricarboxylic acid cycle reaction, that of oxidation of 2-oxoglutarate (or α-ketoglutarate) to succinate, was effectively blocked in rumen bacteria. The conclusion was that pathways of amino acid biosynthesis described in aerobic bacteria may not necessarily be operative to any significant extent in rumen bacteria and that, in these organisms, the biosynthetic pathways of some amino acids should be reinvestigated.

The above studies, although concentrating on amino acid biosynthesis, did not address themselves to the question of extracellular production of amino acids. At least three groups have studied the excretion of amino acids by anaerobes. Hongo et al. (1974) and Uyeda et al. (1974) studied the production of alanine and *o*-butylhomoserine by clostridia, most notably *Clostridium saccharoperbutylacetonicum*. When glucose (5%) was used as the carbon source, this species generated 1 g/l of alanine after 4 days of incubation, in addition to the production of acetone, butanol, and ethanol (Hongo et al., 1974). Other species of clostridia isolated by these workers were able to excrete not only alanine but also *o*-butylhomoserine to levels of 300–400 mg/l for both compounds. The extracellular levels of these compounds were affected both positively and negatively by the addition of other amino acids to the medium. Thus with these anaerobes, it would appear that there may be amino acid transport systems capable of transporting some amino acids into the cells with resulting regulatory effects on biosynthetic pathways.

More recently, a group of Italian researchers have studied excretion of amino acids, in particular isoleucine, from bifidobacteria and clostridia. In earlier experiments (Matteuzzi et al., 1978; Crociani et al., 1977), this group studied isoleucine production in bifidobacteria and, using the antimetabolite valine analog, D,L-α-aminobutyric acid, were able to obtain levels of 9 g/l of isoleucine by the bacterium *Bifidobacterium thermophilum*. As a result of these initial studies, the capacity of a further 121 bifidobacteria and nine strains of saccharolytic clostridia were investigated for the ability to produce extracellular amino acids in a defined medium containing 2% glucose as the carbon source. From the clostridia species, alanine, valine, aspartic acid, and threonine were always produced to varying levels, one strain producing 228 mg/l of alanine. Similarly, with bifidobacteria strains, alanine, valine, and aspartic acid were generally formed, one strain excreting up to 410 mg/l of valine and 210 mg/l of alanine.

It must be remembered that the strains used in these studies were wild-type strains. The levels of excretion are very promising for future gene manipulation when one considers that wild-type aerobic bacteria allow very little excretion of amino acids into the medium (Umbarger, 1978). In other words, the mechanism of control of amino acid biosynthesis in anaerobes may be even more amenable to manipulation than that in aerobes.

In studies similar to those above, Stevenson (1978) studied freshly isolated rumen bacteria for the excretion of amino acids. The carbohydrate source was a mixture of 0.3% (w/v) of each of glucose, cellobiose, and soluble starch. All organisms studied (116 fresh isolates and 10 laboratory strains) produced some extracellular amino acids to a greater or lesser extent. In general, all strains accumulated one or more of alanine, valine, glutamic acid, aspartic acid, and glycine to levels of 50–170 mg/l, with other amino acids usually at 10 mg/l or less. These results are significant (as in the case of the Italian studies) because of the relatively large levels of excretion of some amino acids, but also, as was noted by Stevenson (1978), "the amino acids produced in the greatest abundance by fresh rumen isolates and known species in vitro (alanine, glutamic acid, valine, aspartic acid, and glycine) are the same as the free amino acids commonly found in rumen fluid in the greatest concentrations" (see also Wright and Hungate, 1967). A similar pattern of amino acid excretion has also been observed with another rumen microorganism, *Methanobacterium* spp. (Zehnder and Wuhrmann, 1977).

Microorganisms isolated in the first study of Stevenson did not accumulate large amounts of isoleucine. Therefore in the second study, Stevenson (1979) selected isolates of the first study that had been shown to excrete high levels of valine and studied their ability to excrete high levels of isoleucine in the presence of alpha aminobutyrate, an analog of valine. Two organisms were found that grew in the presence of 0.5% α-aminobutyrate. One was a known strain, *Eubacterium ruminatium*, and the second was a fresh isolate closely resembling this species. In the absence of α-aminobutyrate, these two organisms actively excreted valine during growth (90–195 mg/l) and accumulated small concentrations of isoleucine (3–7 mg/l). However, in the presence of 0.5–1.5% α-aminobutyrate, both microorganisms accumulated concentrations of isoleucine from 100–225 mg/l, while the normal accumulation of valine was unaffected. Lack of repression by the valine antimetabolite on valine synthesis could not be explained by lack of transport into the cell, since at the same time, it is obvious that the α-aminobutyrate did enter the cell and stimulate excretion of isoleucine. Thus as was discussed by Stevenson, lack of repression of valine synthesis indicates that the first enzyme of the valine biosynthetic pathway (which condenses α-ketobutyrate and acetaldehyde and which in aerobes is normally subject to repression by valine) is in these anaerobes not subject to regulation by α-aminobutyrate. With regard to the overproduction of isoleucine, at least part of the increased level of isoleucine would appear to be derived from the meta-metabolism of the α-aminobutyric acid through α-ketoglutarate to isoleucine. The amounts of isoleucine accumulated would represent approximately 25% of the observed degradation of α-aminobutyric acid. Thus the remaining 75% of isoleucine accumulated cannot be explained other than by alteration of the regulation of a biosynthetic pathway.

In summary, very little work has been done on the excretion of amino acids by anaerobic bacteria, both ruminant and nonruminant. However, studies to date have indicated the potential for anaerobes in the production of amino acids, particularly since the regulation of at least some amino acid production in these microorganisms does appear to be different from that of aerobic microorganisms. This fact, coupled with the ability of anaerobes to grow on a wide variety of substrates, to use a small amount of input energy for the creation of their own cellular mass, and to degrade cellulose and hemicellulose, makes the study of amino acid biosynthesis by anaerobes, particularly saccharolytic anaerobes, valuable for future consideration.

In the case of aerobic microorganisms, the gram-positive *Cellulomonas* species is also a possible candidate for genetically engineering a bacterium for amino acid excretion using biomass as a carbon source. *Cellulomonas* can be grown in an aqueous medium containing cellulose in the form of papermill white water (Carta, 1973), or sugar cane bagasse (Srinivasan and Callihan, 1971; Callihan and Clemmer, 1979). Srinivasan and Choi (1981) were able to isolate mutants of *Cellulomonas* sp. ATCC 21399 capable of growth in a minimal medium in the absence of yeast extract. The isolation of these mutants offered a considerable improvement in the cultivation of *Cellulomonas* for single-cell protein because the cost was reduced and because there was less opportunity for contamination. The ability to cultivate these mutant strains on minimal agar plates in the absence of yeast extract also allowed the authors to search for inhibition of *Cellulomonas* growth by amino acid analogs. For example, mutants resistant to the analog S-aminoethyl cysteine were isolated and analyzed. One mutant was capable of excreting 3 g/l of L-glutamic acid after 24 h in a medium containing 10 g/l of glucose and 0.5 g/l of ammonium sulfate. When Solka Floc (20 g/l) replaced glucose, the mutant excreted 7 g/l of L-glutamic acid after 36 h. A second mutant excreted 1.2 g/l of L-lysine after 36 h in a medium containing 20 g/l of Solka Floc. Thus these latter two strains were capable of synthesis and excretion of amino acids using cellulose in a defined medium.

In summary, past studies of cellulolytic microorganisms have concentrated on the natural production of solvents and organic acids. There are no recombinant DNA manipulation systems available, and as was summarized, only a few studies directed toward amino acid biosynthesis have been performed. These microorganisms, particularly those capable of degrading cellulose, should be studied further both for the development of recombinant DNA technology and for the genetics of amino acid biosynthesis.

IV. CONCLUSIONS

The application of genetic engineering to the production of amino acids will be one of the first commercial-scale trials for the application of this technology to a fundamental problem of future biotechnology—namely, how one manipulates, redirects, and overproduces metabolic pathways within an organism. Amino acid overproduction by microorganisms is an excellent model system for tackling such an intricate and complex problem for the following reasons. First, almost any microorganism can be induced to excrete amino acids, as we have described above. Thus one can build the knowledge base in this area by using microorganisms with well-established genetic systems and DNA technology, such as in *E. coli* and *B. subtilis*. If successful, these organisms can then be used for commercial production, or the technology base can be transferred to those useful commercial strains via the methods described above.

We have attempted to show in this review that, as genetic engineering technology is applied to other microorganisms, this knowledge base will likewise be expanded to the overproduction of metabolites other than amino acids, using carbon sources other than the traditional molasses.

REFERENCES

Aiba, S., Imanaka, T., and Tsunekawa, H. (1980) *Biotechnol. Lett. 2,* 525–530.

Allison, M. J. (1969) *J. Animal Sci. 29,* 797–807.

Araki, K., and Nakayama, K. (1971) *Agr. Biol. Chem. 35,* 2081–2088.

Araki, K., and Nakayama, K. (1974) *Agr. Biol. Chem. 38,* 2209.

Araki, K., Shimojo, S., and Nakayama, K. (1974) *Agr. Biol. Chem. 38,* 837–846.

Astier, C., and Espardellier, F. (1976) *C. R. Acad. Sci. 282,* 795–797.

Bagdasarian, M., Bagdasarian, M. M., Coleman, S., and Timmis, K. N. (1979) in *Plasmids of Medical Environmental and Commercial Importance* (Timmis, K. N., and Puhler, A., eds.), pp. 411–422, Elsevier, New York.

Baich, A., and Pierson, D. J. (1965) *Biochem. Biophys. Acta 104,* 397.

Benemann, J. R., Weissman, J. C., and Oswald, W. J. (1979) in *Economic Microbiology* Vol. IV, *Microbial Biomass* (Rose, A. H., ed.) pp. 190–206, Academic Press, New York.

Bukhari, A. I., Shapiro, J. A., and Adhya, S. L., eds. (1977) *DNA Insertion Elements, Plasmids and Episomes* Appendix B, Cold Spring Harbor Laboratory, New York.

Callihan, C. D., and Clemmer, J. E. (1979) in *Economic Microbiology* Vol. IV, *Microbial Biomass* (Rose, A. H., ed.) pp. 271–288, Academic Press, New York.

Carta, G. R. (1973) U.S. Patent No. 3,778,349.

Chakrabarty, A. M., Mylroie, J. R., Friello, D. A., and Vacca, J. G. (1975) *Proc. Nat. Acad. Sci. U.S.A. 72,* 3647–3651.

Chang, A. C. Y., and Cohen, S. N. (1974) *Proc. Nat. Acad. Sci. U.S.A. 71,* 1030–1034.

Chang, S., and Cohen, S. N. (1979) *Mol. Gen. Genet. 169,* 111.

Colby, J., Dalton, H., and Whittenbury, R. (1979) *Ann. Rev. Microbiol. 33,* 481–517.

Cremieux, A., Chevalier, J., Combet, M., Dumenil, G., Parlous, D., and Ballerini, D. (1977) *Eur. J. Appl. Microbiol. 4,* 1.

Crociani, F., Emaldi, O., and Matteuzzi, D. (1977) *Eur. J. Appl. Microbiol. 4,* 177–179.

Dean, D. H., and Dooley, M. M. (1981) *Recombinant DNA Technical Bull. 4*(4), 143–151.

Debabov, V. G., Kozlov, J. I., Zhdanova, N. I., Khurges, E. M., Yankovsky, N. K., Rozinov, M. N., Shakulov, R. S., Rebentish, B. A., Livshits, V. A., Gusyatiner, M. M., Mashko, S. V., Moshentseva, V. N., Kozyreva, L. F., and Arsatiants, R. A. (1981) U.S. Patent No. 4,278,765.

Devilly, C. I., and Houghton, J. A. (1977) *J. Gen. Microbiol. 98,* 277–280.

Eveleigh, D. E. (1981) *Scientific American 245* (3), 154–178.

Ferrari, E., Henner, D. J., and Hoch, J. A. (1981) *J. Bacteriol. 146,* 430.

Flickinger, M. C., and Tsao, G. T. (1978) *Annual Reports on Fermentation Processes* Vol. II, p. 23, Academic Press, New York.

Forage, A. J., and Righelato, R. C. (1979) in *Economic Microbiology* Vol. IV, *Microbial Biomass* (Rose, A. H., ed.) p. 301, Academic Press, New York.

Gardner, J. F. (1979) *Proc. Nat. Acad. Sci. U.S.A. 76,* 1706.

Gatenbeck, S. V., and Hedman, P. O. (1976) U.S. Patent No. 3,963,572.

Gautier, F., and Bonewald, R. (1980) *Mol. Gen. Genet. 178,* 375–380.

Goldfarb, D. S., Doi, R. H., and Rodriguez, R. L. (1981) *Nature 293,* 309–311.

Grigorreva, G. A., and Shestakov, S. V. (1976) in *Proceedings of the 2nd International Symposium on Photosynthetic Prokaryotes* (Codd, G. A., and Stewart, W. D. P., eds.) pp. 220–222.

Hagino, H., and Nakayama, K. (1973) *Agr. Biol. Chem. 37,* 2013–2023.

Hagino, H., and Nakayama, K. (1974) *Agr. Biol. Chem. 38,* 157.

Hall, G., Flick, M. B., and Jensen, R. A. (1980) *J. Bacteriol. 143,* 981–988.

Hamer, G. (1979) in *Economic Microbiology* Vol. IV, *Microbial Biomass* (Rose, A. H., ed.) pp. 315–360, Academic Press, New York.

Harrison, D. E. F., Wilkinson, T. G., Wren, S. J., and Harwood, J. H. (1975) in *Continuous Culture 6: Applications and New Fields* Dean, A. C. R., Ellwood, D. C., Evans, C. G. T., and Melling, J. (eds.) p. 122, Ellis Horwood Ltd., Chichester, England.

Herdman, M. (1973) in *Bacterial Transformation* (Archer, L. J., ed.) pp. 369–385, Academic Press, New York.

Herdman, M., and Carr, N. G. (1972) *J. Gen. Microbiol. 70,* 213–220.

Hirakawa, T., Tanaka, T., and Watanabe, K. (1973) *Agr. Biol. Chem. 37,* 123.

Hongo, M., Gan, B. H., and Uyeda, M. (1974) *Agr. Biol. Chem. 38,* 1805–1810.

Ingram, L. O., Pierson, D., Kane, J. F., Van Baalen, C., and Jensen, R. A. (1972) *J. Bacteriol. 111,* 112.
Jensen, R. A. (1969) *J. Biol. Chem. 244,* 2816–2823.
Juni, E. (1972) *J. Bacteriol. 112,* 917–931.
Kaney, A. R. (1973) *Arch. Microbiol. 92,* 139.
Kase, H., and Nakayama, K. (1972) *Agr. Biol. Chem. 36,* 1611.
Kase, H., and Nakayama, K. (1974) *Agr. Biol. Chem. 38,* 993.
Kinoshita, S., and Nakayama, K. (1978) in *Economic Microbiology* Vol. II, *Primary Products of Metabolism* (Rose, A. H., ed.) pp. 209–261, Academic Press, New York.
Kinoshita, S., Tanaka, K., Udaka, S., and Akita, S. (1957) *Proc. Internat. Symp. Enzyme Chem. 2,* 464.
Kisumi, M. (1981) Eighth Conference of the Association for Science Cooperation in Asia, Feb. 9–15, 1981.
Kisumi, M., Nakanishi, N., Takagi, T., and Chibata, I. (1977) *Appl. Environ. Microbiol. 34,* 465–472.
Kisumi, M., Takagi, T., and Chibata, I. (1978) *J. Biochem. 84,* 881–890.
Kisumi, M., Komatsubara, S., Sugiura, M., Nakanishi, N., Takagi, T., and Chibata, I. (1981) *J. Pharm. Dyn. 4,* S-46.
Kono, K., Toshikazu, O., Kitai, A., and Ozaki, A. (1972) U.S. Patent No. 3,663,370.
Kubota, K., Onoda, T., Kamijo, H., Yoshinaga, F., and Okumura, S. (1974) U.S. Patent No. 3,833,473.
Kubota, K., Onoda, T., Kamijo, H., and Okumura, S. (1975) U.S. Patent No. 3,878,044.
Lau, R. H., and Doolittle, W. F. (1979) *J. Bacteriol. 137,* 648–652.
Lau, R. H., Sapienza, C., and Doolittle, W. F. (1980) *Mol. Gen. Genet. 178,* 203–211.
Matteuzzi, D., Crociani, F., and Emaldi, O. (1978) *Ann. Microbiol. Inst. Pasteur 129B,* 175–181.
Nagahari, K., and Sakaguchi, K. (1978) *J. Bacteriol. 133,* 1527.
Nakayama, K., Kobata, M., Tanaka, Y., Nomura, T., and Katsumata, R. (1975a) U.S. Patent No. 3,907,637.
Nakayama, K., Kobata, M., Tanaka, Y., Nomura, T., and Katsumata, R. (1975b) U.S. Patent No. 3,907,641.
Nakayama, K., Kobata, M., Tanaka, Y., Nomura, T., and Katsumata, R. (1976) U.S. Patent No. 3,939,042.
Office of Technology Assessment (1981) Future Impact of Genetics on Industrial and Plant Genetics (report).
Oki, T., Sayama, Y., Nishimura, Y., and Ozaki, A. (1968) *Agr. Biol. Chem. 32,* 119–120.
Phares, W., and Chapman, L. F. (1975) *J. Bacteriol. 122,* 943–948.
Pierson, D. L., and Jensen, R. A. (1974) *J. Mol. Biol. 90,* 563.
Restaino, L., and Frampton, E. W. (1975) *J. Bacteriol. 124,* 155–160.
Riccardi, G., Sora, S., and Ciferri, O. (1981) *J. Bacteriol. 147,* 1002–1007.
Roberts, T. M., and Koths, K. E. (1976) *Cell 9,* 551–557.
Sahm, H. (1977) *Adv. Biochem. Eng. 6,* 77–103.
Sauer, F. D., Erfle, J. D., and Mahadevan, S. (1975) *Biochem. J. 150,* 357–372.
Shestakov, S. V., and Khyen, N. I. (1970) *Mol. Gen. Genet. 107,* 372–375.
Shigeto, M. (1962) *Nippon Nogei Kagaku Kaishi 36,* 809.

Shiio, I., and Miyajima, R. (1969) *J. Biochem. Tokyo 65,* 849.

Shiio, I., Otsuka, S., Kurasawa, S., and Uchlo, R. (1971) U.S. Patent No. 3,616,224.

Shiio, I., Sugimoto, S., and Nakagawa, M. (1975) *Agr. Biol. Chem. 39,* 627–635.

Simon, R. D. (1978) *J. Bacteriol. 136,* 414–418.

Srinivasan, V. R., and Callihan, C. D. (1971) U.S. Patent No. 3,627,095.

Srinivasan, V. R., and Choi, Y. C. (1981) U.S. Patent No. 4,278,766.

Stevenson, I. L. (1978) *Can. J. Microbiol. 24,* 1236–1241.

Stevenson, I. L. (1979) *Can. J. Microbiol. 25,* 1394–1400.

Stevens, C. L. R., Stevens, S. E., Jr., and Myers, J. (1975) *J. Bacteriol. 124,* 247.

Stevens, S. E., Jr., and Porter, R. D. (1980) *Proc. Nat. Acad. Sci. U.S.A. 77,* 6052–6056.

Tanaka, K., Ohshima, K., and Toloro, Y. (1974) U.S. Patent No. 3,839,151.

Tanaka, T., Hirakawa, T., and Takahara, K. (1975) U.S. Patent No. 3,920,520.

Tanaka, Y., Araki, K., and Nakayama, K. (1980) *Biotech. Lett. 2,* 67.

Tani, Y., Kanagawa, T., Hanpongkittikun, A., Ogata, K., and Yamada, H. (1978a) *Agr. Biol. Chem. 42,* 2275–2279.

Tani, Y., Kato, N., and Yamada, H. (1978b) *Adv. Appl. Microbiol. 24,* 165–186.

Tribe, D. E., and Pittard, J. (1979) *Appl. Environ. Microbiol. 38,* 181–190.

Umbarger, H. E. (1978) *Ann. Rev. Biochem. 47,* 533–606.

Uyeda, M., Gan, B. H., Takenobu, S., Ono, I., and Hongo, M. (1974) *Agr. Biol. Chem. 38,* 1811–1818.

van den Hondel, C. A. M. J. J., Keegstra, W., Borrias, W. E., and Van Arkel, G. A. (1979) *Plasmid 2,* 323–333.

van den Hondel, C. A. M. J. J., Verbeck, S., van der Ende, A., Weisbeck, P. J., Borrias, W. E., and Van Arkel, G. A. (1980) *Proc. Nat. Acad. Sci. U.S.A. 77,* 1570.

Wagner, F. (1977) *Experimentia 33,* 110–113.

Wagner, F., Sahm, H., and Keune, W. H. (1977) U.S. Patent No. 4,060,455.

Watanabe, K., Tanaka, T., Hirakawa, T., and Sasaki, M. (1973) U.S. Patent No. 3,759,789.

Windass, J. D., Worsey, M. J., Pioli, E. M., Pioli, D., Barth, P. T., Atherton, K., Dart, E. C., Byrom, D., Powell, K., and Senior, P. J. (1980) *Nature 287,* 396–401.

Wood, D. D., Hollinger, M. F., and Tindol, M. B. (1981) *J. Bacteriol. 144,* 1448–1451.

Wright, D. E., and Hungate, R. E. (1967) *Appl. Microbiol. 15,* 148–151.

Yanofsky, C. (1981) *Nature 289,* 751–758.

Yukawa, H., Osumi, K., Nara, T., and Takayama, Y. (1981a) U.S. Patent No. 4,271,267.

Yukawa, H., Osumi, K., Nara, T., and Takayama, Y. (1981b) U.S. Patent No. 4,276,380.

Zehnder, A. J. B., and Wuhrmann, K. (1977) *Arch. Microbiol. 111,* 199–205.

Zeikus, J. G. (1980) *Ann. Rev. Microbiol. 34,* 423–464.

CHAPTER
6

Development of a Biochemical Process for Production of Olefins from Peat, with Subsequent Conversion to Alcohols

Peter F. Levy
John E. Sanderson
Stanton R. de Riel

I. INTRODUCTION AND BACKGROUND

The sizable deposits of peat in the north central and coastal sections of the United States may provide raw materials for the production of significant quantities of fuels and chemicals.

It is a well-established fact that plant debris, which is primarily cellulosic in composition, will decompose into water, methane, and carbon dioxide when attacked by fungi and bacteria in the presence of oxygen. If, however, the debris falls into an oxygen-deficient environment such as a pond or saturated soil, the decomposition process is severely retarded, and peat may be formed. Soper and Osborn (1922) described the stoichiometry of peat formation:

$$C_{72}H_{120}O_{60} \rightarrow C_{62}H_{72}O_{24} + 8CO_2 + 2CH_4 + 20H_2O$$

Peat is classified on the basis of its botanical origin, fiber content, moisture content, and ash content. The U.S. Bureau of Mines recognizes three types of peat: moss peat, reed-sedge peat, and peat humus. Any peat derived from moss is classified as moss peat. Peat derived from reed-sedge, shrubs, and trees is known as reed-sedge peat. Any peat whose origin is largely unrecognizable is classified as peat humus.

Recent estimates indicate that peat covers approximately 152×10^6 hectares (375 million acres) of the earth's surface (Tibbetts, 1968). Sufficient quantities of peat reserves exist in the United States to make this a potentially valuable natural resource. A comparison with other energy reserves shows that peat reserves (1443 quads) are greater than the energy from uranium (1156 quads), shale oil reserves (1160 quads), and the combined reserves of petroleum and natural gas (1408 quads) (Rader, 1977). Only coal reserves (5000–10,000 quads) represent a greater energy reserve in the United States. The quantity of peat makes it a vast and largely untapped potential energy source, an alternative to imported petroleum.

The combustion of dry peat will release the greatest amount of energy (10,000 Btu/lb). Since peat is harvested from bogs, its moisture content is high (85–95%). Combustion requires predrying, a costly and energy-intensive process. Wet processing provides distinct advantages over any use that requires predrying. Recovery of a portion of the heat content of peat may be accomplished by wet oxidation, a process in which the peat is oxidized in its wet state (Othmer, 1978). Wet oxidation has been developed and utilized in the treatment of waste streams that are high in organic matter (Teletzke, 1964; Randall and Knopp, 1980). These reactions are typically run at elevated temperatures (250–300°C) at the highest feasible solids loading.

There are a variety of options for the wet processing of peat, as depicted in Fig. 6.1. Alkaline oxidation may be carried out exhaustively producing steam or, as proposed in this process, under conditions of limited oxygen to produce a dissolved substrate for anaerobic acetogenic bacteria. The lignin portion of peat is represented in Fig. 6.2 (Before). It consists of substituted aromatics linked by alkyl chains through ether bonds. The lignin polymers vary in size from 3000 to 10,000 molecular weight. Peat also contains a carbohydrate fraction varying between 5% and 40% of its dry weight (Waksman, 1930).

The pretreatment (alkaline oxidation) of peat is intended to render it more susceptible to microbial attack by disrupting its native structure. The products of the pretreatment are substituted single-ring aromatics, carboxylic acids, and methanol as shown in Fig. 6.2 (After). These water-soluble products are all potential substrates for anaerobic bacteria.

The pretreated material may then be fed to an anaerobic fermenter in which microbially mediated conversion to methane can proceed. Under controlled conditions, conversion to aliphatic organic acids may be en-

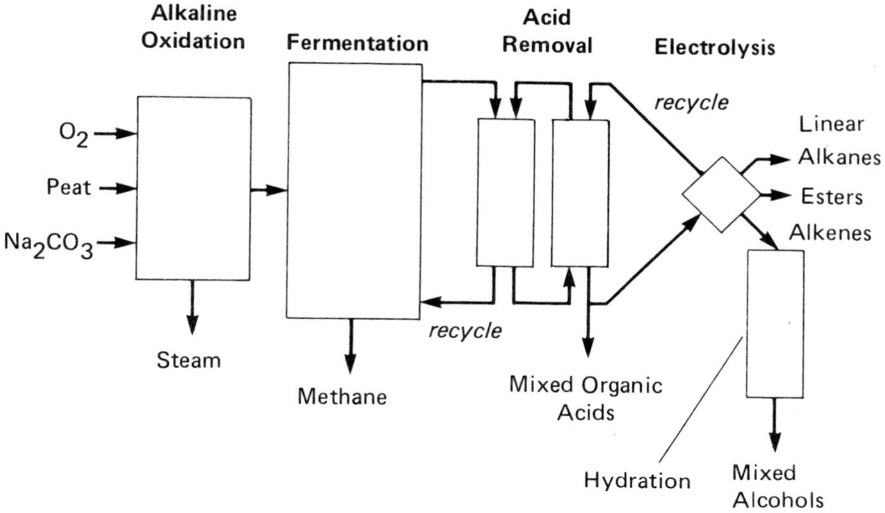

FIGURE 6.1 Potential options for peat wet processing.

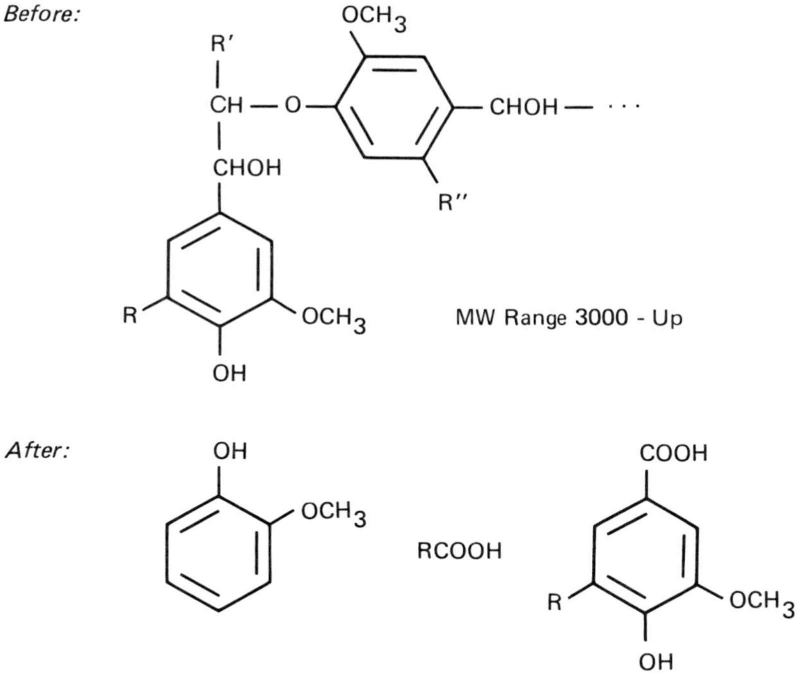

FIGURE 6.2 Solubilization and oxidation of lignin.

couraged. These organic acids can then be removed from the fermenter by liquid–liquid extraction and concentrated as the final product or electrolytically oxidized to form other products—linear alkanes, organic acid esters, or alkenes. The product spectrum of the electrolytic oxidation is determined by the conditions under which the reaction occurs. Further processing of the electrolysis product, such as hydration of olefins (alkenes) to alcohols, is another option.

The current program involves the production of a mixed alcohol fuel from peat. The process is shown schematically in Fig. 6.3. Following the limited alkaline oxidation, the pretreated peat is anaerobically fermented. The fermentation conditions are adjusted to encourage the formation of higher organic acids (butyric, valeric, caproic) instead of methane. This may require the addition of an electron acceptor such as sulfate or nitrate. The higher acids produced in the fermenter may be removed by liquid–liquid extraction into a hydrocarbon solvent and then concentrated by back-extraction into aqueous base. The acids are then electrolytically decarboxylated and oxidized to form mixed olefins. Butyric, valeric, and caproic acid mixtures will form a mixture of propylene, 1-butene, 2-butene (*cis-* and *trans-*), 1-pentene, and 2-pentene (*cis-* and *trans-*). The olefins may then be hydrated to form mixed alcohols. The fermentation, liquid–liquid extraction, and electrolytic oxidation steps of this process have already been demonstrated for polysaccharide substrates (Levy et al., 1981a, b). Work is currently being performed to adapt this technology to processing peat.

Subsequent hydration of these mixed olefins will produce an alcohol fuel comprised of propanol, butanol, and pentanol. Acid-catalyzed hydration

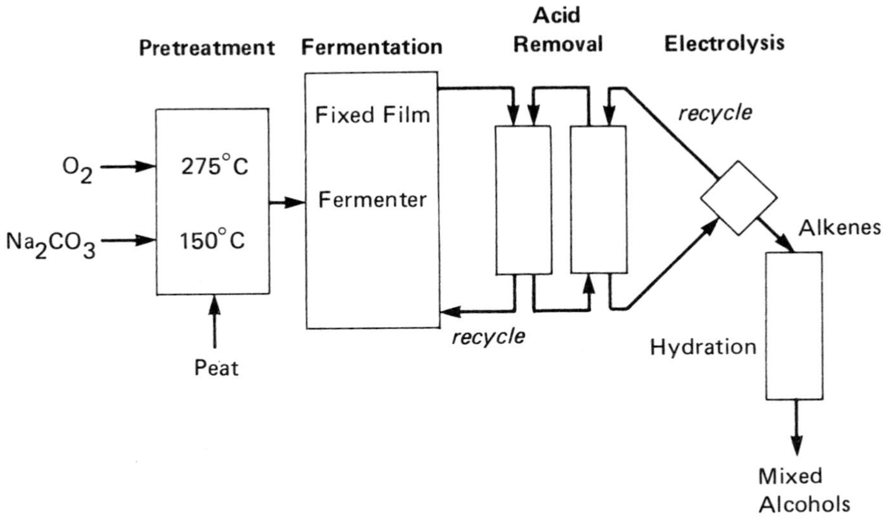

FIGURE 6.3 Process flowsheet.

TABLE 6.1 Energy Densities of Alcohol Fuels

Alcohol	*Heat of Combustion (Btu/lb)*
Methanol	9,600
Ethanol	12,820
Propanol	14,420
Butanol	15,530
Pentanol	16,350

of olefins is the current method of commercial production of most alcohols (Kirk-Othmer, 1980; Considine, 1974). Since this reaction is a well-known technology, it is not addressed further in this report.

Ethanol has been proved to be a suitable motor fuel in mixtures of up to 10% with gasoline. Because of ethanol's low energy density and high water solubility, use in higher percentages becomes more difficult. The higher energy densities (Table 6.1) and decreased water solubilities make the higher alcohols produced by this process better gasoline replacements than ethanol. The projected favorable energy balance for this process also represents an advantage over ethanol processes for liquid fuel production.

II. PRETREATMENT—LIMITED ALKALINE OXIDATION OF PEAT

A. Background

The oxidation of organic matter in the presence of water (wet oxidation) was developed for application to waste disposal in the paper industry (Flynn, 1979; Zimmerman and Diddams, 1960). Any substance capable of burning that remains dissolved in water can be oxidized at elevated temperatures. To obtain high conversions of organic matter to CO_2, temperatures in the vicinity of 250–300°C may be necessary. Oxidation of soluble sugar and paper pulp waste is shown as a function of reactor temperature in Fig. 6.4.

Suitable materials for wet oxidation processes are those which contain substantial amounts of water and cannot easily be concentrated to sustain combustion under conventional burning conditions. Although elevated temperatures are necessary, at sufficient solids concentrations, the wet oxidation reaction will be spontaneous above some threshold temperature. Wet oxidation technology may be readily applied to peat processing. This type of processing avoids the difficult and energy-intensive dewatering of peat necessary for direct combustion. Complete wet oxidation of peat and the vaporization of the associated water are self-sustaining at peat concentrations above 12%.

Wet oxidation of peat proceeds when the solubilized organic portion is exposed to oxygen at elevated temperatures. If the peat is not solubilized prior to oxygen addition, wet carbonization will occur, forming "peat coal."

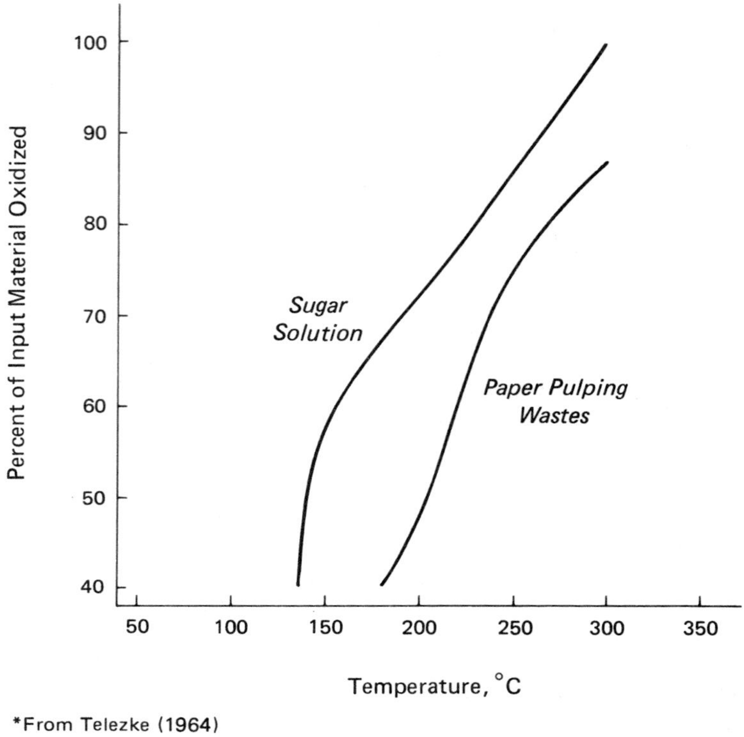

FIGURE 6.4 Wet oxidation of dissolved materials (from Teletzke, 1964).

Since the wet carbonized peat is easier to dewater mechanically and has a higher heat of combustion than raw peat, wet carbonization is being developed as a process to obtain fuel from peat (Mensinger, 1980).

Peat may be solubilized by the addition of alkali at moderate temperatures (75–150°C). The products of this alkaline hydrolysis of peat consist of many types of substituted phenols, such as condensed and uncondensed guaiacyl, as well as higher-molecular-weight entities. The substituent groups are carbon side chains containing 3–5 carbon atoms. Oxidation of the dissolved peat initially yields single-ring aromatics (aldehydes, alcohols, and carboxylic acids) with methoxy side chains. Low-molecular-weight alkyl carboxylic acids are also formed. Continued oxidation leads to the degradation of the aromatic nuclei (Oki et al., 1978). The methoxy content of the product also decreases with time, ostensibly because of the elimination of side chains from the aromatic nuclei. Complete oxidation will eventually result in the conversion of the dissolved peat products to carbon dioxide and heat evolved as steam. A summary of the oxidation pathways of peat is shown in Fig. 6.5. Upon addition of alkali and heat to the peat slurry, some organic material is dissolved, and some remains suspended. At elevated temperatures,

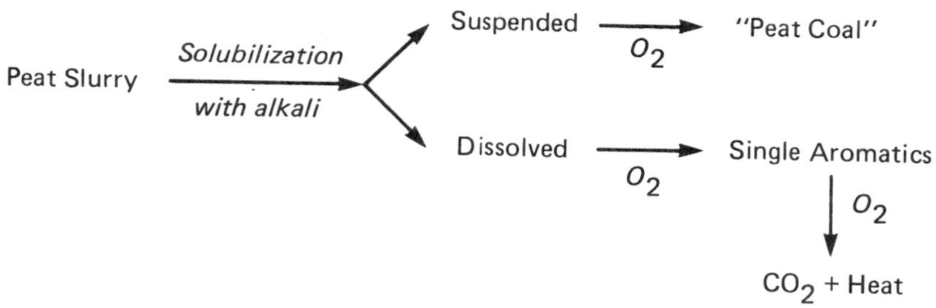

FIGURE 6.5 Peat oxidation pathways.

even in the presence of oxygen, the suspended material is "carbonized," that is, converted to "peat coal." The peat coal produced has a lower oxygen content and a higher heat of combustion than the raw peat feed (Mensinger, 1980).

When solubilized peat organics are subjected to a limited oxidation, single-ring aromatics and linear carboxylic acids are formed from the lignin portion of the peat. Methanol and ethanol may be formed from the carbohydrate portion as degradation products of the hemicellulose. In the presence of excess oxygen at temperatures between 275° and 300°C, the solubilized peat is completely oxidized to CO_2 and H_2O.

B. Determination of Peat Composition

Peat was donated by the Minnesota Gas Company for use in this experimental program. It is of the reed-sedge variety. An elemental analysis of the peat was performed by Galbraith Labs (Knoxville, Tennessee). Further characterization of the raw peat was performed as described by Levy et al. (1981a). The procedure is briefly described below.

The total ash content was determined by burning an oven-dried sample at 600°C in a muffle furnace and comparing weights before and after burning. A more detailed analysis was performed to determine the nature of the non-ash fraction. An oven-dried sample was extracted with 2:1::toluene:ethanol until constant weight was achieved (approximately 24 h). Then the nonextracted portion was hydrolyzed with 10 ml/g material of 72% H_2SO_4 at 30°C for 1 h, diluted with water to 6.5% H_2SO_4, and heated at 110°C for 3 h. The unhydrolyzed material was then filtered off, dried, and weighed. This material was identified as lignin and undissolved ash. The lignin portion was determined by burning in a 600°C muffle furnace. The acid-solubilized material was carbohydrate and soluble ash.

The heat of combustion of the raw peat was measured in a Parr Bomb Calorimeter on oven-dried samples. A summary of the composition of the peat used in our experiments is given in Fig. 6.6. The total ash content of the raw peat is approximately 10%. A portion of the ash (about 7%) is contained in the acid-solubilized material, the remainder of which (almost

Elemental Analysis (by weight)	
%C	48.47
%H	4.91
%N	2.80
%S	0.32
%O	31.75
Ethanol/Toluene Extractable	8.5%
Acid Solubilized Material	30.9%
Lignin	57.5%
Non Acid-Detergent Ash	3.1%
Heat of Combustion	8077 Btu/lb

FIGURE 6.6 Peat composition.

25% of the raw dried peat) is carbohydrate. This high carbohydrate content is reflected in the relatively low (8000 Btu/lb) heat of combustion.

C. Experimental Procedures for Solubilization and Oxidation Experiments

The solubilization and oxidation experiments were carried out in a 2-l PMD reactor purchased from Pressure Products Industries (Warminster, Pennsylvania). The reactor is capable of operating at up to 5000 psig at 340°C and is equipped with a gas booster pump, making the unit capable of operating with continuous gas flow. A magnetically coupled stirrer is supplied with the unit to provide mixing of the gas and liquid phases. The stainless steel reactor sits in an electric heating mantle, which is operated through a temperature controller. It is equipped with ports for liquid sampling.

Total solubilized solids are determined by drying a weighed sample, after neutralization and centrifugation at 12,000 × g for 15 min, in a 110°C oven until constant weight is obtained. The organic fraction is determined by burning the dried sample in a 600°C furnace and comparing weights before and after burning. Further characterization of the cooked peat consists of neutralization and successive ultrafiltrations through 300,000, 25,000, 10,000, and 500 nominal molecular weight cutoff membranes from Amicon Corp. (Lexington, Massachusetts). The total and organic solid content of the ultrafiltered samples are determined as described above. Volatile organic substances in the cooked liquor such as methanol, ethanol, acetic acid, and acetone are determined by gas chromatography. Filtered liquid samples are acidified and injected into a Perkin-Elmer Temperature Programmable GC equipped with a flame ionization detector and Chromasorb 101 packed glass columns. Product identification and quantification are made by comparison with standards.

Gas input in the peat oxidation experiments is determined from a calibrated rotameter on the gas feed line. All exhaust gases flow through a Wet Test Meter. Gas compositions (O_2, CO_2, N_2) are determined on a Fisher-Hamilton Model 29 gas partitioner using He carrier gas with thermal conductivity detectors and 30% di-2-ethyl hexyl sebacate and molecular sieve 13× columns in series.

The reactors are loaded with up to 1400 ml of a 4% peat/water slurry that has been blended in a Waring blender for about 3 min to produce a homogeneous suspension. Alkali (sodium carbonate, sodium hydroxide, or ammonium hydroxide) is added before sealing the reactor. Once sealed, the reactor is purged with N_2 and heated to operating temperature. Stirring is provided at 800 RPM for the solubilization experiments. Liquid samples are obtained from the sampling port at various times during the experiment.

Air or oxygen, used as the oxidizing agent, is not introduced into the reactor until solubilization is complete. For experiments with continuous gas flow, the gas input is determined from the calibrated rotameter. In other experiments, the reactor is pressurized with oxygen when it is at 150°C. The oxygen added is calculated from the reactor pressure. Stirring of the liquid phase does not commence until the desired oxidation temperature is reached. Slight temperature increases (10°C) typically are observed when stirring is initiated, due to the heat evolved when the oxidation reaction occurs.

Data collected in the solubilization and oxidation experiments include percent of input organic solids solubilized, size distribution of solubilized solids before and after oxidation, O_2 used, and CO_2 produced.

D. Results of Peat Oxidation Experiments

The initial step in the oxidation of peat is solubilization with alkali. High solubilization of peat organics (up to 90%) was achieved at moderate temperatures (up to 150°C). At 150°C, solubilization of peat was completed after 30 min at temperature. A series of experiments was performed to demonstrate the effect of base concentration on solubilization. All experiments were on 4% peat slurries at 150°C with sodium carbonate used as the base. The results (Fig. 6.7) indicate that the degree of solubilization is strongly dependent on base concentration up to 0.004 moles of base per gram of peat. Similar results were obtained when sodium hydroxide was used as the base. These results agree with earlier work performed at Dynatech with peat of a different composition (Dynatech, 1981).

Solubilization experiments using ammonia as alkali were also attempted. In all cases, low solubilization of peat was achieved with ammonia, probably because of its high volatility at elevated temperatures. A first-order rate constant of 0.33 h^{-1} was determined for solubilization of peat in excess

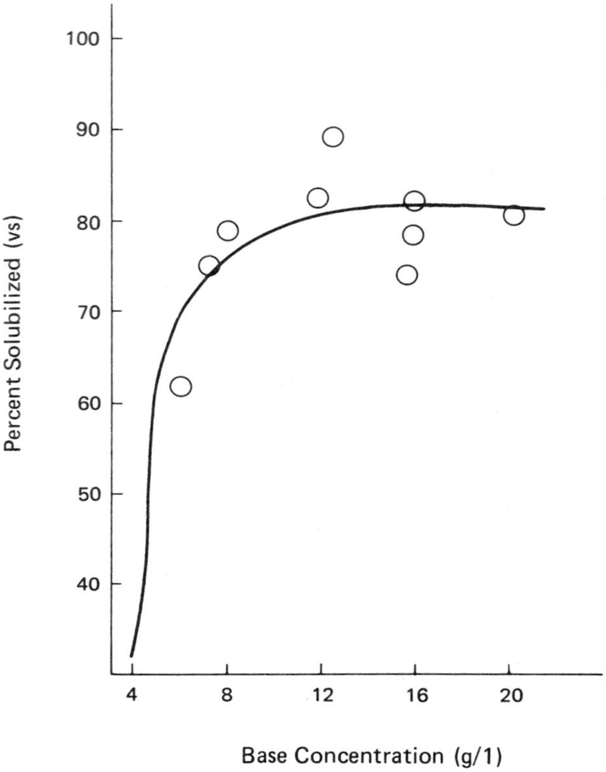

FIGURE 6.7 Peat solubilization with sodium carbonate.

sodium carbonate at 80°C (Fig. 6.8). At 80°C, 68% of the peat organics were solubilized. At higher temperatures, somewhat higher solubilization was achieved at faster rates.

Oxidation of solubilized peat was attempted in the presence of excess oxygen at temperatures between 120° and 300°C. The product organic material in three size ranges was measured:

1. High molecular weight: material that does not pass through 300,000 nominal molecular weight membrane.
2. Middle molecular weight: material between 300,000 and 500 nominal molecular weight membranes.
3. Low molecular weight: material that passes through 500 nominal molecular weight membrane.

The data from these experiments are plotted in Figs. 6.9–6.11 as yield of input volatile solids. At 120°C, there is very little change in the size distribution of the solubilized peat, though some O_2 is used and some CO_2 is produced.

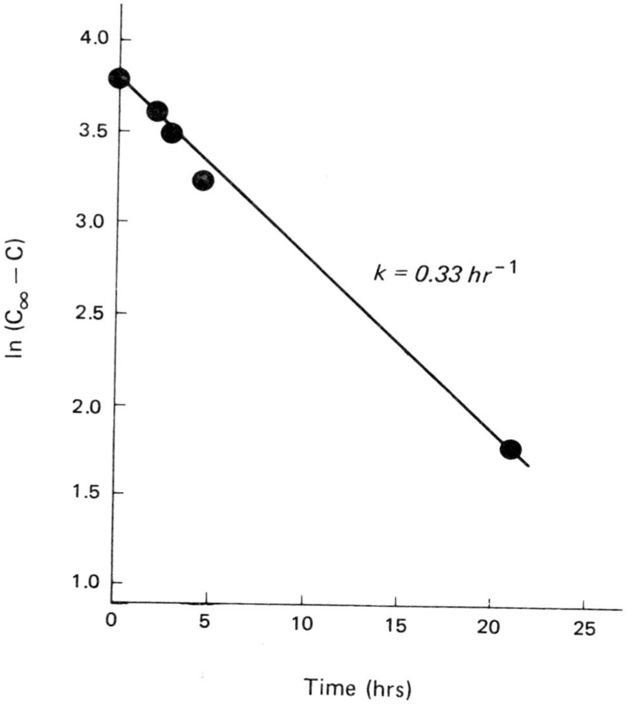

FIGURE 6.8 Rate constant of peat solubilization at 80°C with sodium carbonate (run #014183).

At 300°C, the highest oxygen consumption was observed, and over 90% of the input organic material was converted to CO_2 in 10 min.

In Fig. 6.11, a peak in the low-molecular-weight fraction appears at 275°C under limited oxygen addition. In this experiment, oxygen was added continuously at 0.25 l (STP)/min, maintaining an oxygen partial pressure of 3 psig. The oxidation was continued until 2.5 moles O_2/196 g of organic material was absorbed. The increase in low-molecular-weight material in this run was primarily due to an increase in production of liquid products (methanol, acetone, acetic acid). A small percentage of the starting material remained in the higher-weight fraction material, the rest being converted to CO_2.

E. Discussion of Results

Conditions for the partial and complete wet oxidation of Minnesota reed-sedge peat were determined. Solubilization of the peat organics with alkali is necessary before oxidation can proceed. Oxidation at low temperatures (120°C) produces very little change in the molecular weight distribution of

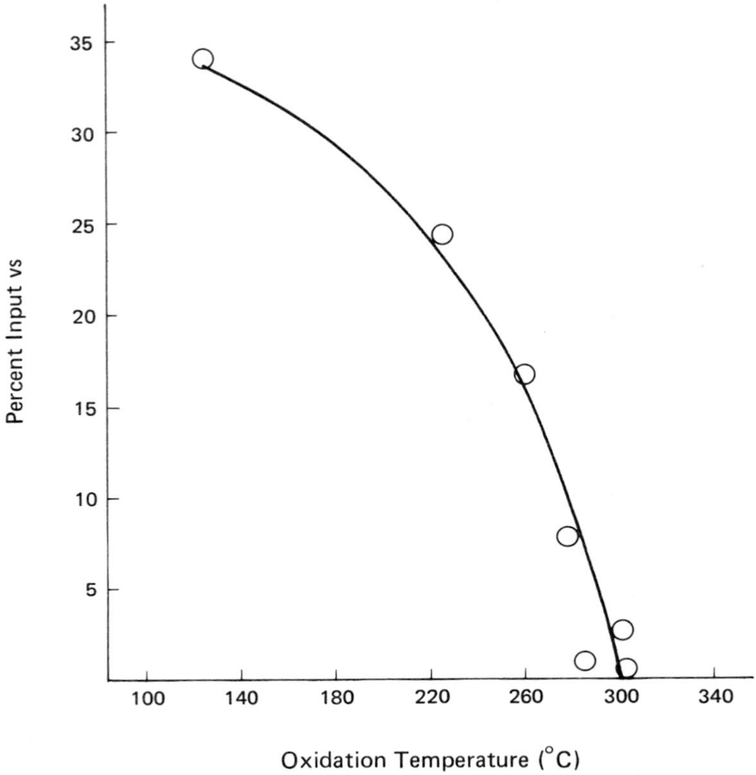

FIGURE 6.9 Yield versus oxidation temperature (high molecular weight).

the peat organics. At higher temperatures, the higher-weight fractions can be oxidized to a lower-weight dissolved product under conditions of limited oxygen availability. At 300°C, in the presence of excess oxygen, the solubilized material is completely oxidized to CO_2 with the evolution of heat.

Since the objective of our peat-processing program involves the conversion by anaerobic fermentation, the increase of a readily fermentable low-molecular-weight fraction of organic material is the goal of these pretreatment experiments. This fraction will contain truly dissolved material, including substituted single-ring aromatics, linear alcohols, and carboxylic acids, all potential substrates in anaerobic digestion. The experimental results to date indicate that this may be accomplished by operating the pretreatment reactor at temperatures in the range of 250°C with low O_2 partial pressures (up to 5 psig). Fairly short reaction times (less than 30 min) are anticipated.

Rates of solubilization were measured at 80°C. Results indicate that the reaction is first-order with respect to peat in the presence of excess base. At 150°C, the reaction is very rapid, making rate measurements difficult.

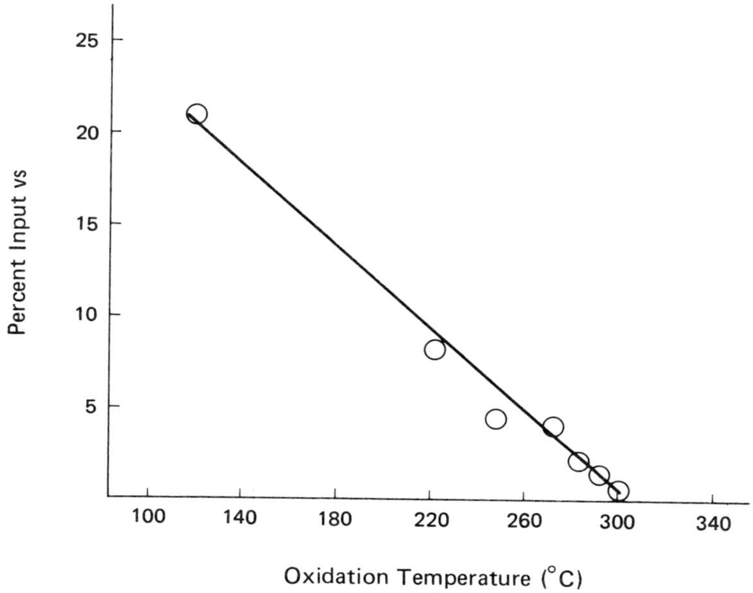

FIGURE 6.10 Yield versus oxidation temperature (middle molecular weight).

It is apparent that solubilization can be achieved with short residence times at higher temperatures (up to 275°C). No change in the solubilized material was observed at reactor temperatures up to 275°C without the addition of oxygen.

The reactor conditions that led to the complete oxidation of peat organics differ from the conditions that produce wet carbonized peat. The addition of alkali prior to oxidation produces a solubilized organic material that can be completely oxidized to CO_2. In the absence of alkali, the suspended organic material is only partially oxidized, producing a carbonized product.

III. FERMENTATION OF PRETREATED PEAT

A. Background

The anaerobic digestion of organic material to methane proceeds in two discrete biochemical stages. Initially, a class of microorganisms, called acetogens, metabolize the material, secreting acetic and other aliphatic organic acids as well as carbon dioxide and hydrogen. These products (acetic acid, CO_2, H_2) are metabolized by a second class of microorganisms, called methanogens, which produce methane and carbon dioxide as end products.

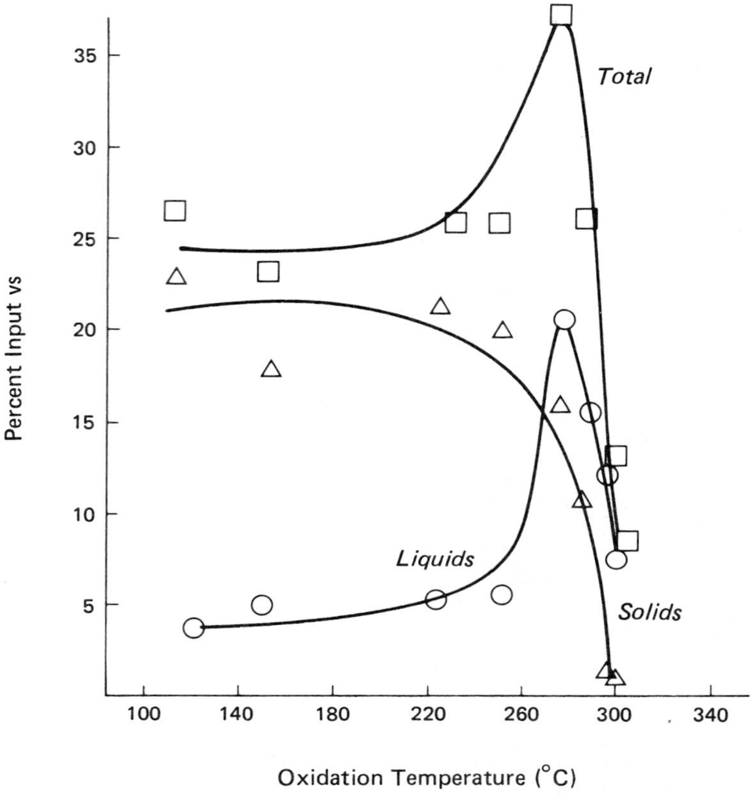

FIGURE 6.11 Yield versus oxidation temperature (low molecular weight).

The limited alkaline oxidation of peat results in an array of substituted single-ring aromatics, carboxylic acids, and methanol. These are all potential substrates for digestion by anaerobic microorganisms (Healy and Young, 1979; Healy et al., 1980; Ferry and Wolfe, 1976; Keith et al., 1978; Ruoff et al., 1980; Levy et al., 1981a). In model compound experiments, using 11 substituted aromatic compounds, Healy and Young (1979) demonstrated high conversions to methane as indicated in Table 6.2. Mixed microbial inoculum was used, and lag periods of a few days to a few weeks were experienced while the bacteria became acclimatized to the aromatic substrates. In some instances, cross-acclimatization (immediate acceptance of one substrate after adaptation to another) occurred, indicating the existence of similar metabolic pathways.

Cultures isolated from freshwater mud and sewage sludge samples and identified as *Acetobacterium woodii* have been found to be capable of anaerobic fermentation of methoxylated benzoic acid derivatives to acetic acid (Bache and Pfennig, 1981). However, this fermentation is believed to

TABLE 6.2 Anaerobic Degradation of Aromatic Compounds to Methane

Substrate	*Percent Converted*
Vanillin	72 ± 1.4
Vanillic acid	86 ± 2.8
Ferulic acid	86 ± 2.8
Cinnamic acid	87 ± 8.1
Benzoic acid	91 ± 7.8
Catechol	67 ± 1.6
Protocatechuic acid	63 ± 1.8
Phenol	70 ± 3.2
p-Hydroxybenzoic acid	80 ± 2.7
Syringic acid	80 ± 1.6
Syringaldehyde	102 ± 13.3

From Healy and Young (1979).

occur by demethoxylation rather than by ring cleavage. Substantial documentation of other anaerobic cultures that are capable of cleaving the aromatic ring of benzoic acid and its derivatives has been provided (Keith et al., 1978; Aftring and Taylor, 1981; Healy and Young, 1979; Healy et al., 1980; Ferry and Wolfe, 1976; Williams and Evans, 1975; Clark and Fina, 1952; Taylor et al., 1970; Evans, 1977). The biochemical pathways of the anaerobic catabolism of aromatic substrates have been postulated by a number of researchers.

In anaerobic respiration, bacteria use inorganic electron acceptors to metabolize substrates in the absence of O_2. Thus NO_3^- is reduced to N_2, SO_4^{2-}, to S^{2-}, and CO_2 to CH_4. Two types of cultures have been studied in some detail, those using nitrate as the electron acceptor and those using carbon dioxide. In both types of cultures, intermediates following ring opening include formic, acetic, propionic, and other aliphatic carboxylic acids (Taylor et al., 1970; Evans, 1977; Keith et al., 1978; Ferry and Wolfe, 1976). Reduction of nitrate continues to N_2, though nitrite does appear as an intermediate (Williams and Evans, 1975). Whether sulfate can act as a terminal electron acceptor for catabolism of aromatic substrates has yet to be demonstrated.

The pathway for degradation of the aromatic compound ferulic acid to methane has been described in detail (Healy et al., 1980) and is presented in Fig. 6.12. These workers added 10^{-4} molar 2-bromoethane sulfonic acid (BES) to the fermentation medium to specifically inhibit methanogenesis. In the presence of BES, the organic acid intermediates began to accumulate. The ability to produce organic acids by anaerobic fermentation when methanogenesis is suppressed has been demonstrated by using polysaccharide substrates (Levy et al., 1981a, b; Sanderson et al., 1979). As these fermentations proceed, organic acids build up by two-carbon additions onto

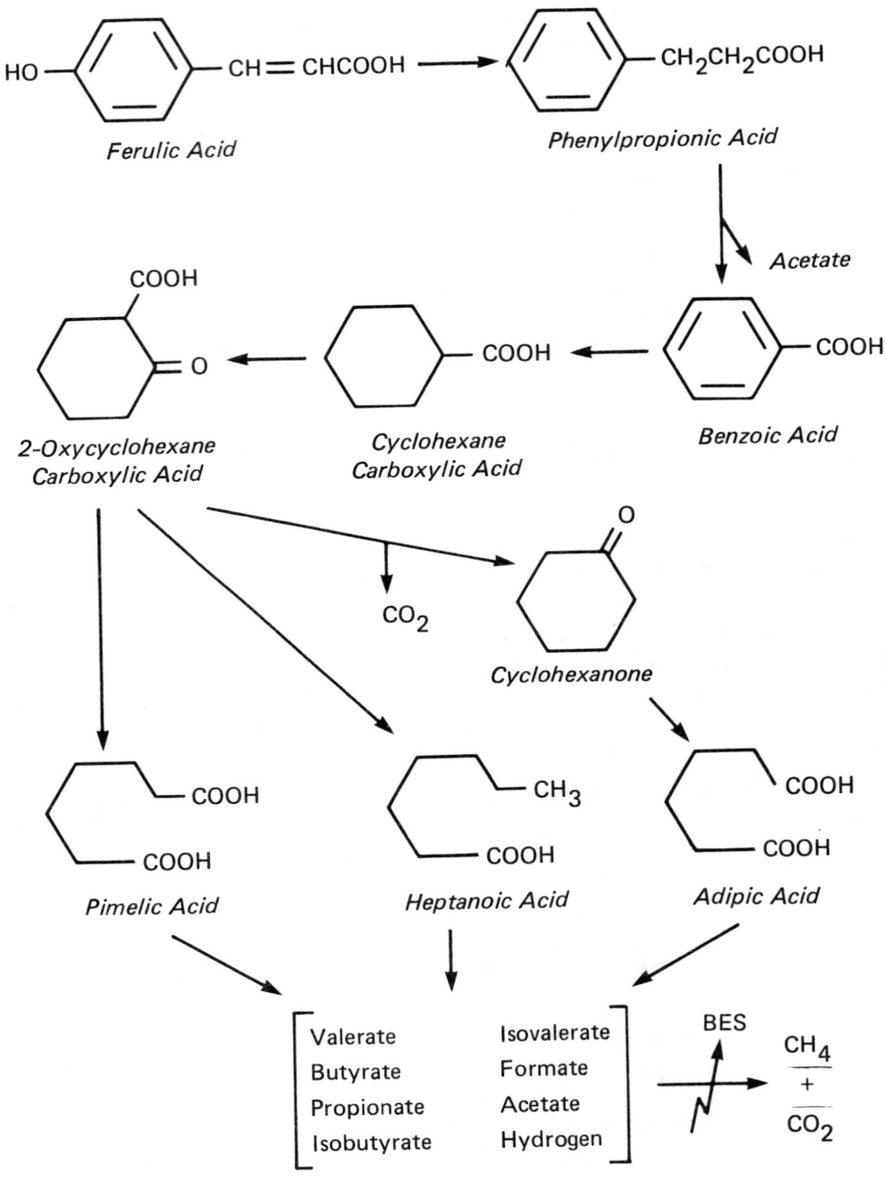

*Healy, Young, and Reinhard, *Appl. Environ. Micro.*, 39: 436-444 (1980).

FIGURE 6.12 Model for decomposition of ferulic acid to methane (from Healy et al., 1980).

acetic or propionic acid primers. Linear organic acids up to caproic (hexanoic) are formed. The effect of BES on organic acid accumulation is shown in Fig. 6.13. The substrate for these experiments was the freshwater plant Hydrilla. In the absence of BES (control), organic acids were degraded as methane was formed. When BES was added, no methane was formed, and organic acids accumulated in the fermentation liquor.

The fermentation of pretreated peat to methane has been demonstrated in the laboratory (Buivid et al., 1980; Ruoff et al., 1980; Dynatech, 1981). Procedures similar to those used by Healy and co-workers on model compounds were used. Sewage sludge was used as inoculum, and cultures were slowly acclimatized to the new substrate, which was introduced at low concentrations. Conversion of over 30% of the pretreated peat to methane and carbon dioxide was consistently observed in these experiments.

The objective of the current experimental program is to combine the anaerobic fermentation of peat with the technique of inhibiting methanogenesis to produce organic acids. This is to be achieved by developing mixed cultures that produce methane from pretreated peat and then altering the conditions of the fermentation so that only the acetogens can thrive. It is anticipated that besides the addition of BES, an electron acceptor such as nitrate or sulfate must be supplied. Model compound experiments, using benzoic acid as the substrate, are also being run to help gain a more basic understanding of the anaerobic digestion of the aromatic substrates.

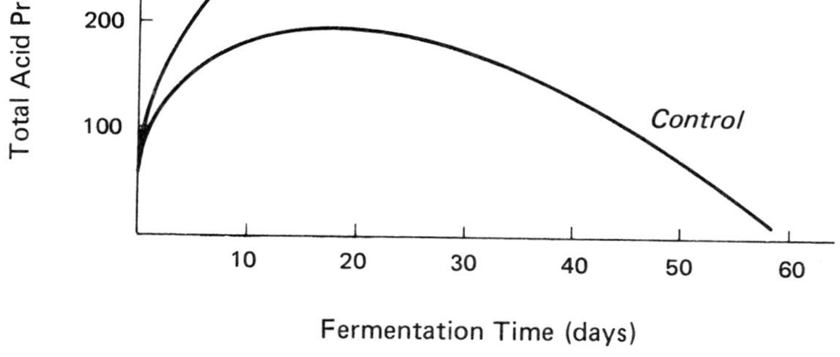

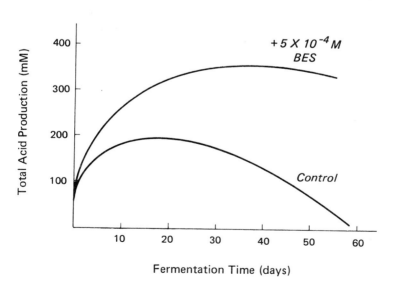

FIGURE 6.13 Effect of bromoethane sulfonic acid (BES) on organic acid production from Hydrilla.

B. Experimental Procedures

Two types of fermenters were used in the first year of this program. Chemostat fermenters (3-l liquid volume) with constant stirring were operated in a semicontinuous manner to develop cultures capable of metabolizing benzoic acid and pretreated peat. Serum vial fermenters (160-ml volume) were also employed to enable testing the effect of changing parameters of the fermentations.

The synthetic medium for model compound fermentations is prepared by diluting the following to 1.0 l with distilled water: 2.6 g of $NaHCO_3$, 1.0 ml of vitamin concentrate C-1, 1.5 ml of iron chloride concentrate C-2, 1.6 ml of phosphate concentrate C-3, and 15.2 ml of salt concentrate C-4. The contents of the concentrates are listed in Table 6.3. The model compound substrates are dissolved prior to addition of the concentrates. The medium is bubbled with N_2 to deoxygenate before the addition of inocula. In some experiments, a medium containing 2.0 g/l of KNO_3 as described by Williams and Evans (1975) is used. Peat that has been pretreated with alkali is diluted by 1:1 to 1:3 with water. Vitamin concentrate (C-1) and salt concentrate (C-4) are added at the same amounts as in the model compound fermentations. In some experiments, 10^{-3} M of 2-bromoethane

TABLE 6.3 Synthetic Medium Concentrates

Contents	*Amount*
C-1:	
Folic acid	2.0 mg/l
Pyridoxine·HCl	10 mg/l
Riboflavin	5 mg/l
Thiamine	5 mg/l
Nicotinic acid	5 mg/l
Pantothenic acid	5 mg/l
B-12	0.1 mg/l
p-aminobenzoic acid	5 mg/l
thioctic acid	5 mg/l
C-2:	
$FeCl_2 \cdot 2H_2O$	36.8 g/l
C-3:	
$(NH_4)_3PO_4$	26.7 g/l
C-4:	
NH_4Cl	13.1 g/l
$MgCl_2 \cdot 6H_2O$	1.20 g/l
KCl	40 g/l
$MnCl_2 \cdot 4H_2O$	1.33 g/l
$CoCl_2 \cdot 6H_2O$	2.0 g/l
H_3BO_3	0.38 g/l
$CaCl_2 \cdot 2H_2O$	0.18 g/l
$Na_2MoO_4 \cdot 2H_2O$	0.17 g/l
$ZnCl_2$	0.14 g/l

sulfonic acid (BES) is added to the medium as a methane suppressant. All fermentations are incubated at 37°C.

Either the substrate added to the fermentations is weighed (model compound) or a measured volume of known percent volatile solids is added (pretreated peat). Serum vial experiments to determine ultimate conversion are sequentially diluted by transferring 8 ml of an active culture to a second vial containing 72 ml of fresh medium. All transfers are performed anaerobically by syringe. This dilutes out any exogenous material contained in the inoculum. The screening procedure used for identifying active cultures is shown in Fig. 6.14.

Gaseous products (CO_2, CH_4) are determined by injecting 0.5 ml of headspace gas with a Pressure-Lok Syringe (Precision Sampling) into a Fisher Model 29 Gas Partitioner. Product peaks are identified and quantified by comparison with standard gas mixtures. The quantity of product in the fermenter headspace is determined from the percent of gas measured and the volume of headspace. Use of the Pressure-Lok Syringe enables gas above atmospheric pressure to be accounted for. Liquid products (that is,

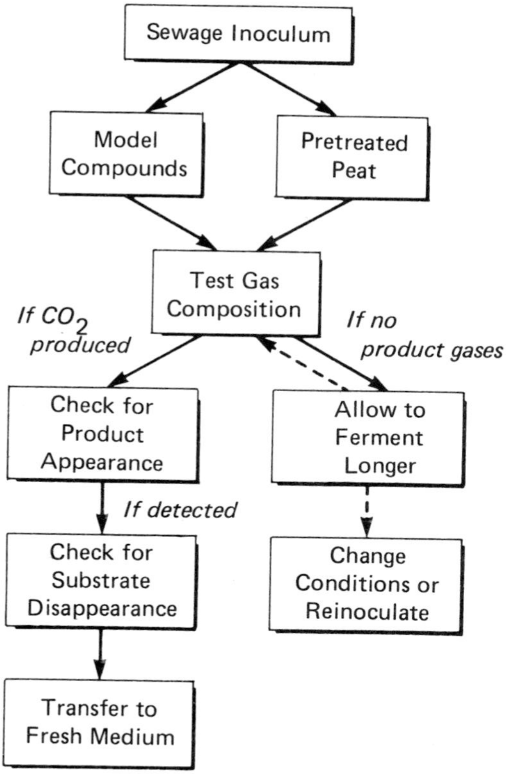

FIGURE 6.14 Screening procedure for active cultures.

organic acids) are measured by gas chromatography after extraction to remove interfering compounds. One milliliter of sample (or standard) is acidified by adding two drops of concentrated sulfuric acid and then extracted with 4.0 ml of diethyl ether. Three milliliters of the ether are transferred to a tube containing 1.0 ml of 1.0-N sodium hydroxide, and the organic acids are back-extracted into the base. The aqueous base is acidified before injection onto a GC column packed with Chromasorb 101. Organic acids are identified and quantified by comparison with standards that have been prepared in the same manner as the samples. Percent substrate conversion to products is determined by comparing the total weight of products detected with the weight of substrate added.

C. Results of Fermentation Experiments

Cultures were developed from sewage digester effluent that were capable of fermenting pretreated peat and benzoic acid to organic acids (primarily acetic and butyric) and methane. A representative list of conversions is given in Table 6.4. Cultures capable of quantitative conversion of benzoic acid (014022 and 014035) were developed. Peat, after pretreatment, could consistently be converted to organic acids in yields of 30% of the initial volatile solids loading. In these fermentations, the pH was always observed to increase as the experiment progressed. Values above pH 8.0 were not uncommon. Peat fermentations were successful in both media used. In the presence of excess nitrate, however, ultimate conversion of organic carbon to carbon dioxide was observed.

The use of peat that was pretreated under varying conditions (different temperature, alkali, oxygen consumption) did not give a clear-cut indication of any increase or decrease in fermentability. In Table 6.4, peat cooks 013843 and 014837 were at 150°C with sodium carbonate, 014043 was at

TABLE 6.4 Ultimate Conversion in Serum Vial Fermenters

Sample Number	*Substrate*	*Percent Conversion (VS)*
014021	5 mM benzoic acid	29.6
014022	5 mM benzoic acid	116
014027	Peat (013843)	31.8
014028	Peat (013843)	23.8
014035	5 mM benzoic acid	95.9
014037	Peat (014837)	34.4
014038	Peat (014837)	25.8
014043	Peat (014843)	27.8
014044	Peat (014843)	27.5
014506	Peat (014651-6B)	29.9
014507	Peat (014651-6B)	25.9
014513	Peat (014657-4)	32.7

150°C with sodium hydroxide, and 014651 and 014657 were at 250–275°C with sodium carbonate. All resulted in approximately 30% conversion to organic acids. It was observed, however, that the high-molecular-weight material (>25,000 nominal molecular weight) was not digested by the acclimatized bacteria, but the lower-molecular-weight fraction was digested. Higher conversions for the high-temperature treated peat were anticipated on the basis of these observations. Rate data have not yet been taken. This will be done after cultures have been further developed and the optimal pretreatment conditions have been determined.

D. Discussion of Results

The ability to ferment a model aromatic compound (benzoic acid) and pretreated peat to organic acids (particularly acetic, propionic, and butyric) has been demonstrated. This is an encouraging result in the development of this process, but many questions concerning these fermentations remain to be answered. Current experimental work is now underway to provide more information about this microbially mediated process.

Benzoic acid is being fermented in a chemostat-type fermenter with a 30-day retention time. Microbial conversion of benzoic acid to methane by adapted mixed culture anaerobes has been repeatedly demonstrated. The ability to recover at least a portion of the product as acetic acid has been demonstrated in this program. The ability to obtain quantitative yields of acetic and higher organic acids from benzoic acid is basic to the understanding of these fermentations. Since the thermodynamics of the proposed conversions are marginal, it is apparent that an electron acceptor must be supplied in order to maintain energy production for microbial growth. The use of excess nitrate as an electron acceptor has thus far resulted in the complete oxidation of benzoic acid to carbon dioxide and water. Whether limited quantities of nitrate or sulfate may be used to accomplish the stated goals will be investigated in the second year of this program.

Organic acids have been produced from peat with up to 30% of the substrate added being converted. Since data reported so far have been from serum vial fermenters, constant conditions were not maintained. Of particular interest in achieving higher conversions is maintenance of pH and nutrient availability. Fermentations of pretreated peat are now being carried out in a chemostat-type fermenter so that more constant conditions can be maintained. Nutrient requirements are being determined in separate serum vial experiments.

Once pretreatment and fermenter conditions have been more specifically determined, rate data will be obtained for operation in a CSTR-type fermenter. Since the substrate is dissolved before feeding to the fermenter, an anaerobic packed bed fermenter is appropriate for more efficient operation. This type of fermenter is packed with a solid support (such as small stones), which provides a surface for microbial attachment. Once attachment of bacteria is achieved (after approximately 6 months), much higher cell densities are attained than in traditional fermenters. Thus higher rates of product formation per reactor volume are observed. Bench-scale reactors of this type are now being started up for operation during the coming year on peat and benzoic acid substrate. The rate and efficiency of conversion to higher organic acids achieved will determine the economic viability of this process.

IV. ELECTROLYSIS

A. Introduction

1. Reaction Description. The electrochemical oxidation of aliphatic carboxylic acids is among the oldest of organic reactions. Kolbe (1849) described the decarboxylation and dimerization of organic acids to produce alkanes. Hofer and Moest (1902) described the decarboxylation of organic acids to produce alcohols. In addition, a variety of other products have been observed under both sets of reaction conditions, particularly when acids other than acetic are electrolyzed, including olefins, esters, nondimerized alkanes, and cyclopropane derivatives (Koehl, 1964; Utley, 1974).

The stoichiometry for the Kolbe and Hofer-Moest reactions are

$$RCOOH + R'COOH \rightarrow R\text{-}R' + 2CO_2 + H_2$$

$$RCOOH + H_2O \rightarrow R^{*}OH + CO_2 + H_2$$

respectively, where R* refers to R and isomers of R collectively.

The former (Kolbe) reaction is predominant at smooth platinum anodes with acetic acid and the longer chain fatty acids (C_6 and up) (Weedon, 1952). The second (Hofer-Moest) reaction is observed to predominate at carbon anodes (Eberson, 1973). In both cases, under most conditions, the main side product is olefin:

$$RCOOH \rightarrow R'HCCHR'' + CO_2 + H_2$$

where R′ and R″ are either hydrogen or organic radicals. With most other electrode materials and at low current densities, the primary side product (in aqueous solution to which this discussion is limited) is oxygen.

2. Reaction Mechanism. Speculation as to the mechanism of the Kolbe and Hofer-Moest reactions has been going on almost since the reactions were discovered. Brown and Walker (1891) presented a rudimentary mechanism involving the oxidation of organic acid anions, for example, acetate to radicals, as the primary anode process. Other theories, focusing on the fact that in the absence of organic acids water electrolysis will occur under the same conditions, assumed that oxidation of organic acid was not the primary electrode process. One of these, which was an elaboration of an early hypothesis presented by Schall (1896), suggested that active (singlet) oxygen is produced at the anode by water electrolysis, which subsequently oxidizes the organic acid to a peroxy acid. Observed products are formed from subsequent decomposition of the peroxy acid (Fichter, 1939). A similar mechanism proposed by Glasstone and Hickling (1934) suggests that the primary anode process is the oxidation of water to hydroxyl radicals, which then dimerize to form hydrogen peroxide or react directly with the organic acid to produce observed products. An argument supporting this theory is that anode materials that catalyze the decomposition of hydrogen peroxide produce oxygen, not Kolbe or Hofer-Moest products.

Subsequently, two series of papers were published that form the basis of the current view of the mechanism of the Kolbe reaction (Conway and Dzieciuch, 1963a, b, c; Dickinson and Wynne-Jones, 1962a, b, c). These papers present convincing evidence that oxygen evolution is prevented under Kolbe conditions by the formation of an adsorbed layer of organic acid or Kolbe intermediate, which prevents water from reaching active sites on the electrode surface. This insight coupled with the notion that the Hofer-Moest product and many of the other products observed are formed via a carbonium ion intermediate, which was suggested by Walling (1957) and confirmed elegantly by Corey et al. (1957), allowed Utley (1974) to construct a diagram similar to the one presented in Fig. 6.15.

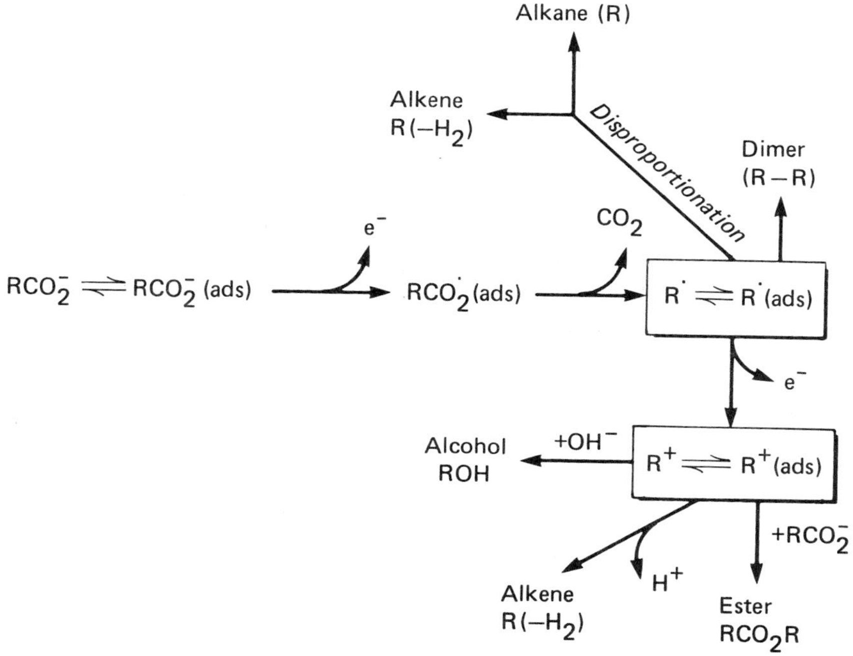

FIGURE 6.15 Anodic oxidation of carboxylate reaction mechanism.

Whereas this diagram fairly represents the current understanding of the mechanism of electrochemical oxidation of aliphatic organic acids, it does not explain many of the experimental observations made by many researchers and confirmed in this work. Some of these observations are as follows.

1. The ratio of alkane dimer to olefin in platinum increases with the length of the linear aliphatic acid (Petersen, 1900, 1906).
2. Alkane dimer is rarely observed using a carbon anode.
3. The Hofer-Moest reaction (alcohol formation) occurs at a lower potential than alkane dimer formation (Dickenson and Wynne-Jones, 1962a).
4. Olefin formation is favored at higher temperatures.
5. Oxygen formation is suppressed at lower current densities on carbon than on platinum.

An elaboration of this basic mechanism has been developed that is capable of explaining these observations. The basic hypothesis and supporting data are presented later.

B. Experimentation

1. Materials and Apparatus. The electrolysis reactions under investigation take place in a flow-through electrolytic cell manufactured by ECO, Inc. (Cambridge, Massachusetts). A schematic diagram of the cell is shown in Fig. 6.16. The cell has been modified and reinforced to allow it to be operated at pressures up to 200 psig. The electrodes are separated by expanded PTFE gaskets, giving an electrode spacing from 0.01 to 0.04 in. The exposed internal surface area of the electrodes is 36 cm^2. The inlet and outlet ports are both located on the cathode side of the cell. Reference ports on both the anode and cathode allow precise measurement of these respective overpotentials. A Simpson voltmeter is available to measure the several potential differences between the anode, cathode, and two reference ports. Potential differences in high-impedance circuits were measured at two input impedances and extrapolated to infinite input impedance.

The cell is integrated into a total system shown in Fig. 6.17. This system is capable of measuring and regulating liquid temperature and pH as well as current, potential difference, gas flow rate, and pressure. A Welker diaphragm pump draws the aqueous acid solution from a 1-l holding vessel and circulates it through the electrolytic cell. The liquid and generated gas return to the holding vessel, and the gaseous products exit through a back pressure regulator.

The liquid in the holding vessel is stirred continuously and monitored for temperature and pH. The system pressure is measured by a pressure gauge connected on-line between the electrolysis cell and the holding vessel,

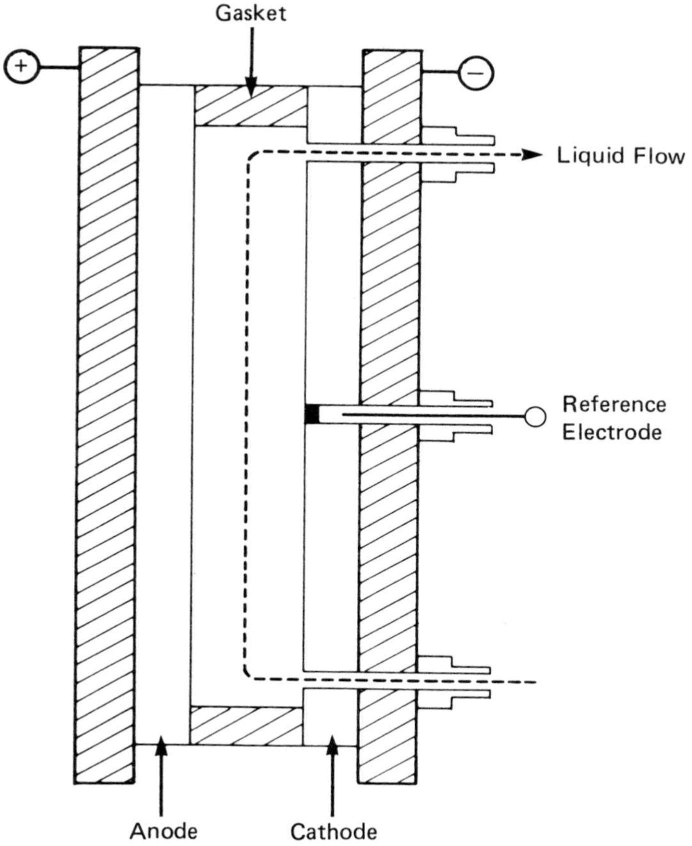

FIGURE 6.16 Schematic diagram of Electroprep Electrolysis cell.

as shown in Fig. 6.17. The electrolytic cell is connected to a Sorenson SRL10-100 DC power supply, capable of delivering up to 100 amp at between 0 and 10 V. The current is precisely monitored by a Sargent-Welsh ammeter connected in series between the anode and the positive terminal of the power supply.

There are two injection ports located on-line in this system. The first of these is used primarily for pH control, and the second is used for gas sampling. Water vapor was removed from the exit gas stream by using a "U" tube filled with indicating drierite. Once a satisfactory balance between decarboxylated products and carbon dioxide production was demonstrated, another "U" tube containing indicating Ascarite was added to the system to remove CO_2 from the exit gas to simplify the analysis of the gas composition. Gas production was measured either by water displacement or by a Wet Test Meter.

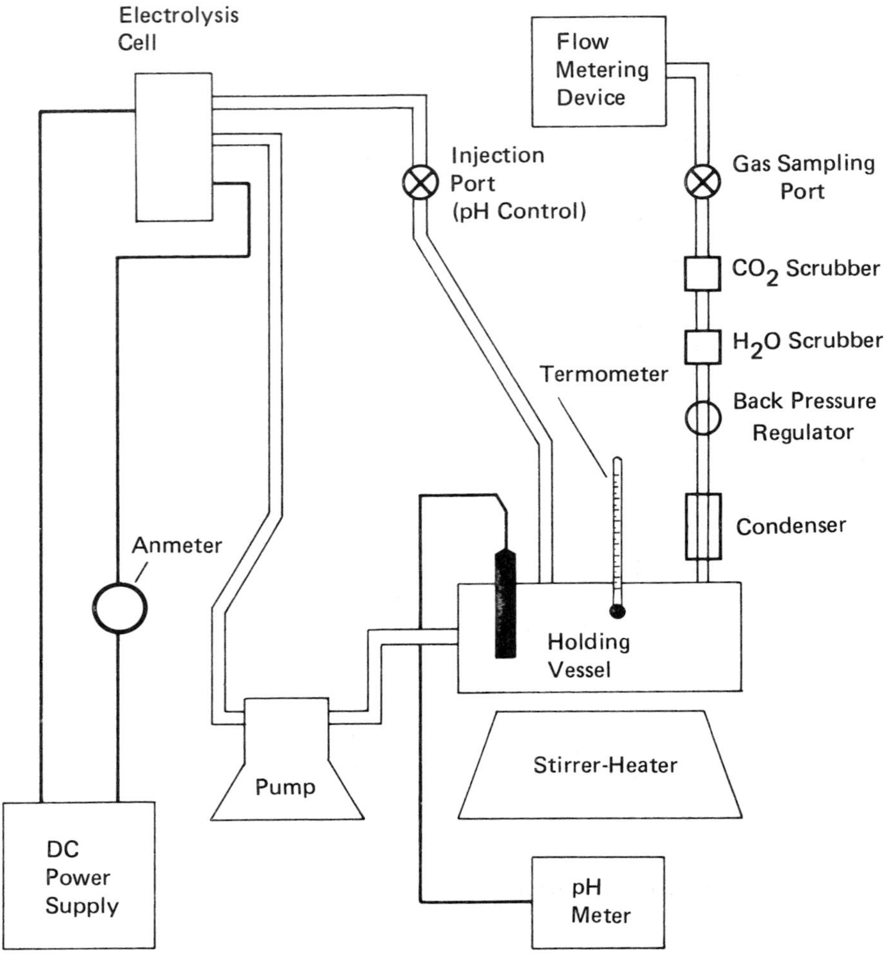

FIGURE 6.17 Experimental apparatus.

Experiments have been run up to 200 psi and up to 150°C. For experiments over 100°C, a condenser was added before the back pressure regulators to minimize water loss in the gas stream.

All reagents were reagent-grade materials obtained from major supply houses and were used without purification. In many experiments, sodium sulfate was used as a supporting electrolyte to reduce the electrical resistance of the electrolyte solution.

2. Product Analysis. The volume of gas produced was measured as described earlier. The composition of these gases was determined by injecting one sample into a Fisher Model 25V Gas Partitioner using argon carrier gas to determine the hydrogen concentration and another sample into a Fisher Model 29 Gas Partitioner using helium as the carrier gas to measure the concentrations of the other sample gases.

After the gas flow rate and composition are determined, the hydrogen flow rate is compared to the electron flow rate to insure against gas leaks. The electron flow rate may be calculated as follows:

$$e = \frac{60 \cdot I}{F}$$

where: e = electron flow rate (electrons/min)
I = current (amp)
F = Faraday's constant (96,500 coulombs/mole)

Since two electrons must be exchanged to produce a molecule of hydrogen at the cathode, the hydrogen flow rate should be half of the electron flow rate.

Other gaseous products (olefins, etc.) are separated and quantified by using a Gow-Mac 750 gas chromatograph equipped with a flame ionization detector. These products are separated on a ⅛″ × 20′ stainless steel column packed with *n*-octane on Durapak using nitrogen carrier gas at 48°C. The water-immiscible products are analyzed on a programmable Perkin-Elmer F30 gas chromatograph equipped with a flame ionization detector. Glass columns ¼″ in diameter × 6′ packed with Chromasorb 101 (no liquid phase) are used with nitrogen carrier gas to separate these products at 130–230°C. Water-soluble components are quantified by using the same instrument by injection of the acidified aqueous solution using the same temperature program.

Concentrations of organic acids before and after electrolysis are determined by potentiometric titration. A sample of the solution is acidified with hydrochloric acid and back-titrated with standard sodium hydroxide. The pH as a function of base added is plotted, and the organic acid concentration is determined from the difference between the inflection points above and below pH 4.75. At the same time, the amount of dissolved carbon dioxide may be estimated by measuring the difference in moles of base required to bring the sample and blank back to the starting pH.

C. Experimental Results and Proposed Mechanisms

The electrochemical oxidation of organic aliphatic acids in aqueous solution involves these competing reactions:

1. One-electron oxidation to produce radical products (alkane dimer): $E = +2.2$ V.
2. Two-electron oxidation to produce carbonium ion products (olefin, alcohol, ester): $E = +1.9$ V.
3. Oxidation of water to produce oxygen: $E = +1.2$ V.

The primary goal is to understand why each of these reactions predominates under different conditions. On the basis of the open cell potentials of each reaction versus the nominal hydrogen electrode, one would expect reaction 3, the oxidation of water, to predominate in all cases. This is, in fact, observed on all anode materials except smooth platinum, carbon, and a few others. As was mentioned earlier, current understanding suggests that, on these latter materials, organic acids are normally adsorbed on the electrode surface to a degree that prevents water from reaching it to be oxidized. However, at low current densities or in dilute solutions of acid, water electrolysis is observed on platinum.

Fig. 6.18 represents the electrode pair with details of the suggested concentration densities of ionic and molecular species, which is the key to the current hypothesis. Adjacent to the anode surface is a lipophilic layer formed by the adsorbed organic layer plus hydrocarbon products that have not yet escaped from the electrode surface. The integrity of this layer is a function of current density, which effects the production rate of these products, and its thickness is determined by the length of the acid molecules adsorbed on the surface.

Adjacent to the lipophilic layer on the anode is a low pH region. There is a corresponding high pH region at the cathode surface. Since less is going on at the cathode, formation of the high pH region is easier to comprehend. At the cathode the reaction is:

$$2H_2O + 2e^- \rightarrow H_2\uparrow + 2OH^-$$

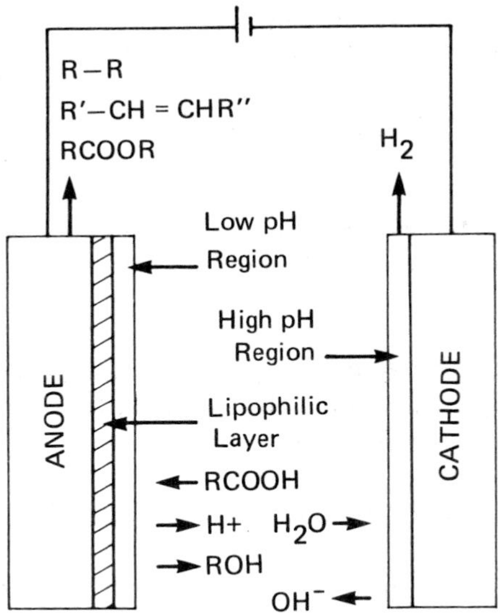

FIGURE 6.18 Model of the electrolysis process on a platinum anode.

That is, for every electron passed through the solution, a hydroxide ion is produced at the cathode. Since the only method available to receive hydroxide ions from the cathode surface is diffusion, a concentration gradient is established such that the diffusion flux is equal to the formation rate under steady state operation for a cell operating at neutral bulk pH and a cathode current density of 0.25 A/cm^2. Typical for this type of electrolysis, one would anticipate a hydroxide concentration at the cathode surface of 1.6 M in a well-stirred solution. Similarly, the anode, in order to maintain neutral change in the solution, must produce protons or consume organic acid anions at an equivalent rate. How much of each method occurs depends on the product mix as well as on the anion concentration and the current density. Normally, one net proton will be produced for each two electrons passed, giving a proton concentration adjacent to the electrode solution of 0.8 M under the same assumptions as those used for the cathode. This acid concentration is sufficiently high to effect the intraconversion of olefins and alcohols via a carbonium ion intermediate.

To answer the second question, that of why all of the organic acid molecules are not oxidized twice to form carbonium ion products if the second oxidation ($E = +1.9$ V) is easier than the first ($E = +2.2$ V), one must consider once again the lipophilic layer on the electrode. The low potential required for the second oxidation is achieved only when water is available near the electrode surface to stabilize the product. In the absence of water, the potential required for the second oxidation is hypothesized to be greater than or equal to the 2.2 V required for the first oxidation. This hypothesis forms the basis for an explanation of why alkane dimer is found with larger-chain acids and why olefin is the preferred carbonium ion product with this anode material.

The explanation for the lack of linear alkane formation at the carbon anode is based on the nature of carbon itself, which is known to contain active centers, the density of which is a function of the manner in which the carbon is made. It is assumed that the electrochemical reaction occurs only at these sites, which are usually relatively isolated on the electrode surface. Thus the actual current density at these sites is much higher than the apparent current density on the basis of the overall geometry of the anode. This view of the electrode reaction explains why oxygen formation is suppressed at lower apparent current densities on carbon than on platinum. It also provides two explanations of why alkane dimer formation is not normally observed. First, the sites are sufficiently isolated to disfavor a bimolecular reaction if, in fact, alkane dimer formation involves such a process; and second, it discourages the formation of a continuous lipophilic layer at the anode surface, which is presumed to be responsible for keeping the water required to stabilize carbonium ions formed in the second oxidation step away from the electrode surface.

Another point to be addressed is the amount and type of olefin produced. The model being presented assumes that conversion of olefin to carbonium ion is a reversible equilibrium process in the low-pH region and that the amount of olefin produced is a function of the escaping tendency of the olefin from the electrode surface. The electrochemical oxidation of the shorter acids produces more olefin (except acetic acid, for which olefin formation is not possible) than oxidation of the larger ones because of their higher vapor pressures. Table 6.5 lists the vapor pressures for various possible olefin products from the electrolysis of butyric, valeric, and caproic acids calculated at 35°C on the basis of the heat of vaporization as near this temperature as is readily available and the boiling point. The pentenes have vapor pressures of ~1000 mm Hg, the butenes 2000 mm Hg, and propylene 10,000 mm Hg, in accord with the typical olefin yields from caproic, valeric, and butyric acids of 20%, 40%, and 55%, respectively. The yields of the various olefin isomers from the electrolysis of valeric acid as reported by Koehl (1964) and confirmed in our laboratory of 50% 1-butene, 20% *trans*-2-butene, and 10% *cis*-2-butene are also in accord with the vapor pressures of these isomers. The mole fraction of each at the electrode surface, which is required along with the vapor pressure of the pure material to make an exact calculation of the isomeric composition, is probably based on the relative stability of each isomer, which would favor the *trans*-2-butene.

In summary, this hypothesis explains the five points raised earlier in the following manner:

1. The ratio of alkane dimer to olefin on platinum increases with the chain length of the acid electrolyzed because the lipophilic layer thus formed is thicker and is therefore more efficient at keeping water, which is required for carbonium ion stabilization, from the anode surface.

2 and 3. Alkane dimer is not normally found on carbon because electrolysis occurs only at active sites on the carbon anode; thus a continuous lipophilic layer is not formed. In the absence of a lipophilic layer, double oxidation to a carbonium ion occurs at a lower potential than radical formation.

TABLE 6.5 Vapor Pressure of Olefins at 35°C

Olefin	BP_{760}	ΔH_V *Cal/g*[1]	ΔH_V *Cal/mole*	*Vapor pressure at 35°C (mm Hg)*
Propylene	−47.7°C	104.6 (47.7°C)	4393	10,600
1-Butene	−6.3°C	86.8 (25°C)	4861	2600
cis-2-Butene	+3.7°C	86.8 (25°C)	4861	1870
trans-2-Butene	0.88°C	91.8 (25°C)	5141	2160
1-Pentene	30°C	85 est.	5950	890
2-Pentene	37.1°C	85 est.	5950	710

1. From Lange (1956).

4. The amount of olefin product is determined by the escaping tendency of the olefin, which increases exponentially with temperature.
5. Oxygen formation is suppressed at lower apparent current densities on carbon than on platinum because the entire area of the carbon electrode is not electroactive, resulting in higher actual current densities than are apparent.

In order to test this hypothesis, a number of experiments have been run. First, electrolysis experiments have been run at elevated pressures to test the assertion that olefin yield is a function of the escaping tendency of the olefin. Running at elevated total pressures increases the olefin partial pressure in the gas phase, which should result in a decrease in olefin yield. Tables 6.6, 6.7, and 6.8 present the effect of pressure on the yields of the various products for butyric, valeric, and caproic acids, respectively. Note that, in each case, a substantial reduction in olefin formation is achieved.

TABLE 6.6 Effect of Pressure on the Product Yields from the Electrolysis of Butyric Acid on Platinum

Pressure *Product*	*0 psig* *(% yield)*	*160 psig* *(% yield)*
Hexane	0	2.7
Isopropyl butyrate	53.4	67.0
Isopropanol	3.1	2.3
n-Propanol	1.0	0.9
Propylene	42.4	27.1

TABLE 6.7 Effect of Pressure on the Product Yields from the Electrolysis of Valeric Acid on Platinum

Pressure *Product*	*0 psig* *(% yield)*	*150 psig* *(% yield)*
n-Octane	42.9	63.3
sec-Butyl valerate	2.7	3.4
n-Butyl valerate	0.9	0
sec-Butanol	4.1	1.7
n-Butanol	1.4	1.3
Other aqueous	10.5	2.1
1-Butene	15.5	8.9
trans-2-Butene	5.5	2.1
cis-2-Butene	4.1	1.7
n-Butane	2.3	0.8
Other gaseous	10.0	4.2
Uncollected	0	10.5

TABLE 6.8 Effect of Pressure on the Product Yields from the Electrolysis of Caproic Acid on Platinum

Pressure *Product*	*0 psig* *(% yield)*[1]	*185 psig* *(% yield)*
Decane	76.1	87.8
Amyl caproates	0	0
2-Pentanol	3.3	1.0
1-Pentanol	0	0
Pentenes	20.6	11.2

1. Average of two runs.

Even more interesting is the observation that it is the alkane dimer production that increases with pressure. Table 6.9 gives the results of additional experiments on the effect of pressure on linear alkane dimer yield under a variety of experimental conditions. The reactions were all run in pairs with pressure as the only intentional variable except Run 75, which is included because it gave the highest linear alkane yield from valeric acid obtained to date. This observation is in accord with the hypothesis presented earlier in that the olefin retained at the electrode surface improves the integrity of the lipophilic layer, thereby favoring radical over carbonium ion products.

The significance of this finding can best be appreciated by quoting Lennart Eberson (1973), from a review article on the electrochemical behavior of carboxylic acids:

> Since the possibility of two competing pathways in the Kolbe reaction exists, the immediate problem to answer is whether one can influence the direction of a given process by proper manipulation of experimental variables so that either the radical or carbonium ion mechanism is favored. As we shall see, unidirectional selectivity can easily be achieved, in that one can always find conditions to favor the cationic mechanism; on the other hand, in many cases, because of structural features in the substrate, it may not be possible to realize the radical pathway even under the most favorable experimental conditions.

TABLE 6.9 Effect of Pressure on Linear Alkane Yield

Run Number	*Acid*	*Pressure*	*Percent Alkane*[1]
71	Valeric	160 psig	67%
72	Valeric	atm.	51%
73	Valeric	atm.	45%
74	Valeric	160 psig	59%
75	Valeric	160 psig	71%
76	Caproic	160 psig	80%
77	Caproic	atm.	68%
78	Caproic	160 psig	88%
79	Caproic	atm.	76%

1. Mole percent of acid consumed to form alkane dimer product.

In short, to date, Kolbe electrolysis has not been used commercially for the conversion of mixed and/or dilute organic acids to hydrocarbons because no technique was available whereby the yield of alkane dimer could be increased above the yields that have been observed for years, as given in Table 6.10 (Allen, 1958). By running electrolyses at elevated pressures, these values have been substantially exceeded in the current work.

Another experiment designed to demonstrate the reversibility of olefin formation was run using valeric acid in an electrolysis vessel initially pressurized with propylene. Figure 6.19 shows the distribution of liquid products from this experiment. The significant point is that *n*-propanol and *i*-propanol were identified in the product mix, an indication that olefins may be converted to alcohols under the reaction conditions.

In order to determine the effect of pH on the product distribution, a study was performed using a platinum anode run with 0.5 M of valeric acid at 35°C, 0.28 A/cm^2 without added supporting electrolyte. The results of those experiments are shown in Table 6.11. The amounts presented are based on the percent of the acid consumed that is required to form each product. In accord with the hypothesis, no systematic effect of pH on the product distribution was observed over the pH range 7–11, since the pH at the electrodes is predicted to be substantially different from and relatively independent of the pH in the bulk solution. It should be noted, however, that at pH near or below the pK's of these acids, a significant change in product distribution takes place, favoring ester formation.

Another important variable on the production distribution is temperature. The hypothesis would predict that olefin production would increase with increasing temperature because of the increase in vapor pressure. Fig. 6.20 is a plot of olefin product yield versus temperature for valeric acid at a carbon anode indicating the anticipated behavior. The plateau observed above 100°C may be due to the increase in pressure necessitated by operating temperatures above the boiling point of the solvent.

TABLE 6.10 Kolbe Electrolysis of Aliphatic Monobasic Carboxylates

		Percentage Yield		
Acid	*Structure*	*Paraffin*	*Olefin*	*Ester*
Acetic	CH_3COOH	85	2	2
Propionic	C_2H_5COOH	8	66	5
n-Butyric	C_3H_7COOH	14.5	53	10
iso-Butyric	$(CH_3)_2CHCOOH$	Trace	62	10
n-Valeric	C_4H_9COOH	50	18	4
iso-Valeric	$(CH_3)_2CHCH_2COOH$	43	42	5
Methyl ethyl acetic	$(CH_3)(C_2H_5)>CHCOOH$	10	42	10
Trimethyl acetic	$(CH_3)_3C\ COOH$	13	52	0
Caproic	$C_5H_{11}COOH$	75	7	1.5

From Allen (1958), p. 103.

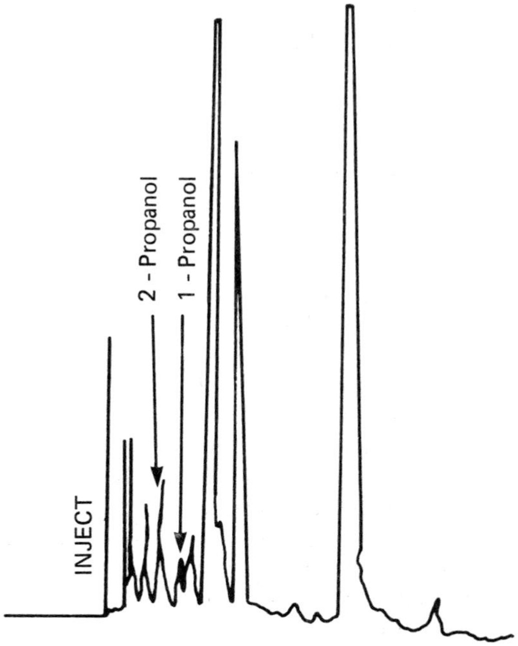

FIGURE 6.19 Gas chromatogram (liquid products).

V. PRELIMINARY PROCESS ECONOMICS

A. Description of Peat Process

A conceptual process design for the peat to mixed alcohol fuels plant is presented in Fig. 6.21. Peat is wet-harvested from the bog and pumped in a 3% solids slurry to the processing facility. The slurry is washed through a screen, where roots and stumps are separated. The peat slurry and small roots that pass through the screen are stored in a slurry tank. This slurry is concentrated to 8% total solids by a screening operation. This peat is then fed to the pretreatment reactor.

TABLE 6.11 Effect of pH on Product Distribution

Run No.	*pH*	*Percent Dimer*	*Percent Alkene*	*Percent Alcohol*	*Percent Ester*
4	7	32	25	18	7
6	9	34	30	15	6
7	11	28	34	27	6

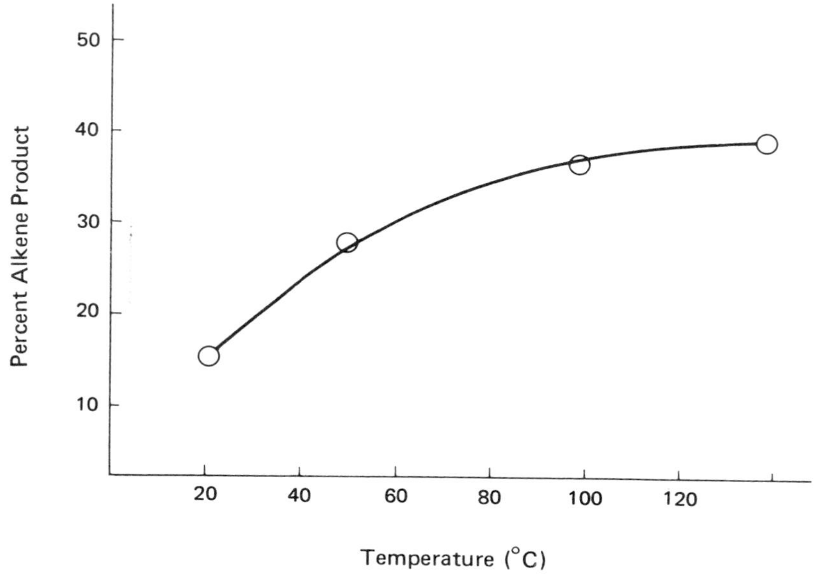

FIGURE 6.20 Effect of temperature on olefin formation (valeric acid on carbon anode).

1. Pretreatment Reactor (Alkaline Oxidation). The peat slurry is next processed through a plug flow staged reactor. Initially, peat is partially preheated through a heat exchanger, alkali is injected, and peat solubilization occurs. After solubilization is completed, oxygen is injected into the reactor, and the peat is further broken down into fermentable products. The products of the pretreatment are substituted single-ring aromatics, and vanillin is used as the model product for this analysis. The yield of fermentable dissolved solids is assumed to be 50% of the input peat solids, the rest remaining undissolved or being oxidized to CO_2. The heat released by the oxidation reaction is sufficient to raise the temperature from 250°F (120°C) after heat exchange to 480°F (250°C) in the oxidation reactor.

The unreacted residual peat solids are removed from the pretreated peat slurry by vacuum filtration or some other suitable means. The separated liquid stream is then sent to the fermentation tanks.

2. Fermentation Reactor. Anaerobic fermentation of the pretreated peat liquors occurs under mesophilic conditions (37°C). A fermenter incorporating cell recycle or an attached film technique is used to increase microorganism concentration and thus the conversion rate. When vanillin is assumed as the model pretreated peat product and valeric (pentanoic) acid as the average fermentation product, stoichiometric conversion is:

$$13C_8H_8O_3 + 33H_2O \rightarrow 17C_5H_{10}O_2 + 19CO_2$$

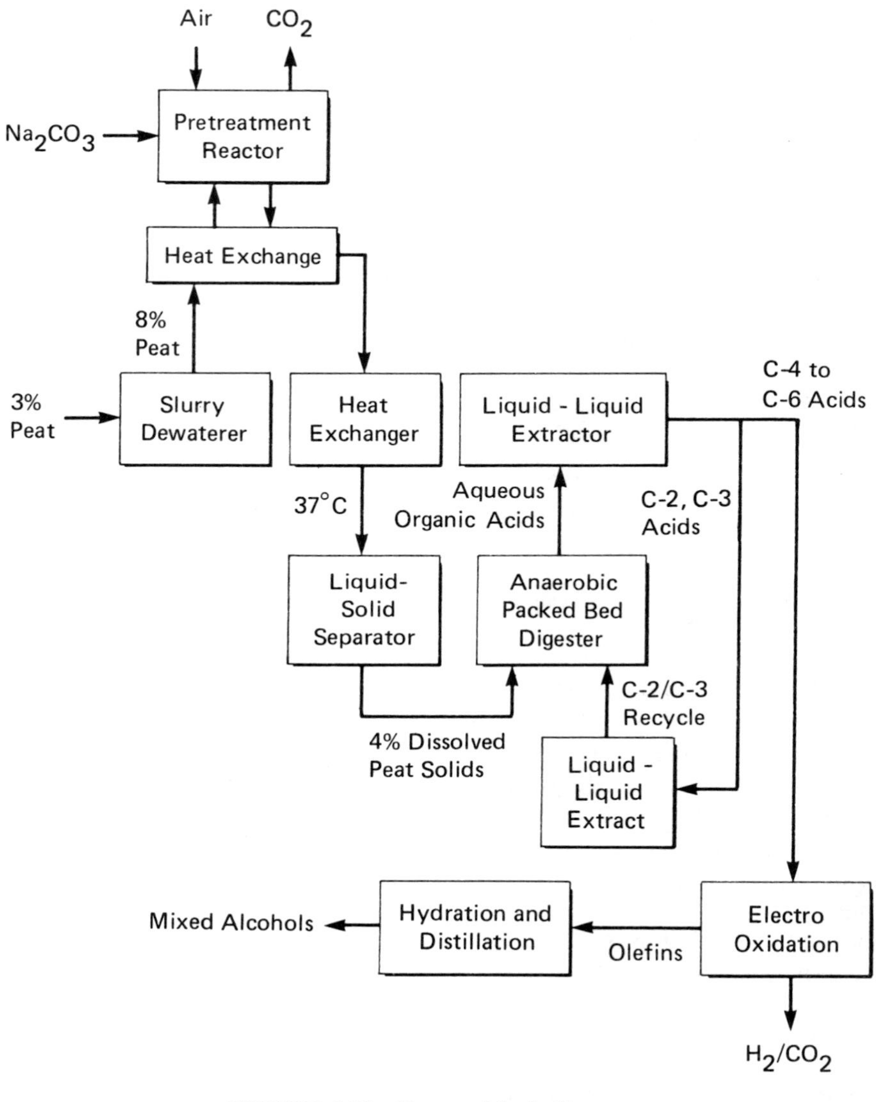

FIGURE 6.21 Process block diagram.

If 80% of the peat solids fed into the fermenter are converted to products, then 0.7 lb of valeric acid product is produced for every 1.0 lb of input pretreated peat. Water is consumed at the rate of 0.24 lb for every 1.0 lb of peat feed, and CO_2 is also produced at 0.34 lb for every 1.0 lb of feed.

The microbial conversion in the fermenters is assumed to follow pseudo–first order kinetics with a rate constant of 1.61 day^{-1}. This rate is reasonable for metabolism of a dissolved substrate. It results in 80% conversion with a 1-day hydraulic retention time in a fermenter that can be modeled as a batch or plug flow reactor.

3. Acid Removal by Liquid–Liquid Extraction. The organic acid product may be removed from the fermentation broth by using a series of two liquid–liquid extractors. In the first extractor, the fermentation broth is extracted with a hydrocarbon solvent such as kerosene, removing butyric and higher-molecular-weight organic acids. A second extractor is used to back-extract these acids into aqueous base. In this manner, the organic acids may be separated from other fermentation products and substrates and concentrated to levels required for the electrolytic oxidation. The extent of separation of acid product from pretreated peat feed, which has yet to be determined, is essential information for the final design of this operation. In this analysis, we assume that adequate separation is achieved even in the presence of high levels of substrate. The acetic and propionic acids remaining in the fermentation liquid must be extracted in a second series of liquid–liquid extractors containing a solvent such as trioctylphosphine oxide dissolved in kerosene that is suitable for removal of low-molecular-weight organic acids (Helsel, 1977). This stream of acetic and propionic acids can then be recycled back to the fermenters to help maintain steady state concentrations of these intermediates.

4. Electrolytic Oxidation of Organic Acids. Electrolytic oxidation of aqueous salts of organic acids may be used to convert the organic acids to a variety of products. For conversion to olefins, the electrolysis is to be performed on a platinum anode at 175°C. Products produced at the anode will be olefin, CO_2, and small quantities of linear alkanes. Hydrogen gas will be produced at the cathode. This hydrogen may be used to generate process electricity and steam.

The cell must be designed to convert 700 tons per day (tpd) of acids to olefins (6.23×10^6 gmoles/day). The electrolysis products will be 7.7×10^5 lb/day (3.5×10^5 kg) mixed olefins, 6.3×10^5 lb/day (2.9×10^5 kg) CO_2, and 2.7×10^4 lb/day (1.2×10^5 kg) H_2. The electrolytic oxidation of organic acids to olefins, CO_2, and H_2 is assumed to follow the stoichiometry:

$$C_5H_{10}O_2 \rightarrow C_4H_8 + CO_2 + H_2$$

The required voltage for the cell is 2.8 V, and a current density of 0.25 amp/cm^2 is used in this analysis.

5. Hydration of Olefins to Mixed Alcohols. Hydration of olefins to produce alcohols is a commercial process. Most industrial ethanol is produced by hydration of ethylene over acid catalysts such as Celite carrier impregnated with phosphoric acid (Considine, 1974; Kirk-Othmer, 1980). After hydration, a stream of about 50% ethanol is then distilled. The higher-weight olefins (propylene, butene, pentene) may also be hydrated by using this technology and then distilled. The following stoichiometry is used to represent the conversion of olefins to a mixed alcohol product:

$$C_4H_8 + H_2O \xrightarrow[\text{acid catalyst}]{} C_4H_{10}O$$

Details of this process step are not included here. The cost of this conversion is estimated on the basis of relative selling prices of ethylene and ethanol.

B. Equipment Size and Capital Cost Estimate

A preliminary process design is described for a commercial plant processing 2000 tpd of peat (dry basis). The peat is delivered to the plant at 3% total solids. Half the dry solids are converted to fermentable materials, and 80% of the material fed to the fermenters is converted to organic acids. Based on the stoichiometries of the conversions, 385 tpd of mixed olefins will be produced that can be hydrated to form 510 tpd of mixed alcohols.

An estimate of major equipment costs is made on the basis of total mass flows and rates of reaction. Capital and operating costs are obtained from traditional sources (Peters and Timmerhaus, 1980; Perry and Chilton, 1973; *Chemical Marketing Reporter,* 1982). Table 6.12 lists assumptions of process parameters.

1. Slurry Dewatering. The peat slurry, delivered to the plant at 3% solids, must be concentrated to 8% solids by water removal. The amount of water

TABLE 6.12 Assumed Operating Parameters

Parameter	*Value*
Solids content of delivered peat	3%
Solids content after dewatering	8%
Residence time for pretreatment	1 h
Conversion to fermentable material	50%
Rate constant for fermentation	1.61 day^{-1}
Retention time in fermenters	1 day
Conversion of fermentable portion	80%
Electrolysis cell voltage	2.8 V
Total peat feed (dry basis)	2000 tpd
Total olefin production	385 tpd
Total alcohol production	510 tpd
Heating value of olefin product	20,800 Btu/lb
Heating value of alcohol product	15,500 Btu/lb
Heating value of dried peat	8000 Btu/lb

to be removed is 42,000 tpd or 6800 gallons per minute (gpm). A fast rate filtration system is capable of removing 20 gpm/ft^2, indicating the requirement of a 350-ft^2 filtration unit.

2. Pretreatment Reactor. The pretreatment step requires a two-stage plug flow reactor. For 2000-tpd peat solids, the solubilization step requires 400 tpd of Na_2CO_3. Though base recycle is not detailed in this report, it is essential for reducing the cost of alkali in the final process design. The oxidation step is assumed to require 800 tpd of oxygen supplied as O_2 or as compressed air. The solubilization reactor is to be operated at 125–175°C and the oxidation at 250°C. Heat will be supplied by heat exchange and heat evolved in the oxidation reaction.

The size of the pretreatment reactor will be determined from the required residence times and rates of reaction at operating temperatures. A detailed engineering design to determine the reactor size will be performed when more precise operating conditions have been determined. A conservative estimate of 1-h residence time for the entire pretreatment will be used in this analysis. For a total flow rate through the reactor of 27,000 tpd (8% peat), a 3500-ft^3 pretreatment reactor is necessary.

Heat exchangers are required to cool the pretreated peat solution from 250°C to 37°C before feeding to the fermenters and to preheat the incoming peat slurry. Two sets of heat exchangers may be used. In the first, the pretreated peat at 250°C is heat-exchanged with incoming material at 10°C. The pretreated peat is cooled to 146°C, and the incoming slurry is heated to 120°C. A second set of heat exchangers is used to cool the pretreated peat from 146°C to 37°C with cooling water at 15°C. Single-pass counterflow heat exchangers are used for both applications. The necessary contact area may be calculated from the design equation:

$$q = UA\Delta T$$

where q is heat flow in Btu per hour; A is contact area in square feet; ΔT is the log mean temperature driving force; and U is the overall heat transfer coefficient, assumed to be 100 Btu/h·ft^2·°F. The required contact area for the first heat exchanger is 30,000 ft^2, and that for the second heat exchanger is 40,000 ft^2.

3. Liquid–Solid Separator. The effluent from the pretreatment step contains undissolved solids, which should be removed before feeding the digester, particularly if a high-rate digestion process such as an anaerobic packed bed filter system is used. A centrifuge or vacuum filter could be used for removing the residual solids. The size of the unit can be estimated from an assumed liquid throughput capacity of 10 gph/ft^2. The liquid throughput is 26,000 tpd (253,000 gallons per hour (gph)), requiring a vacuum filtration area of 25,000 ft^2. The energy requirement of this step is about 0.134 hp/ft^2 area, totalling 3350 hp, or about 5×10^8 Btu/day (assuming 33% efficiency).

4. Digester. The size of the digester is dependent on assumed values for slurry concentration, retention time, and mass throughput. Various digester designs promoting more efficient conversion, and thus reducing the capital cost, should be considered. These designs include the anaerobic contact process (using cell recycle) and the anaerobic packed bed filter. The increased capital cost of more efficient systems is more than compensated for by reduction in the required size of the digesters. An anaerobic packed bed filter digester, using stone as the packing, is assumed in this analysis. With an assumed rate constant of 1.61 day^{-1} and a 1-day hydraulic retention time, 80% conversion is achieved. For a mass flow of 28,600 tpd, 900,000 cubic feet of digester void volume is required. When 50% void volume is assumed in the packed digester, 1,800,000 cubic feet is the total size. If digesters are constructed 30 ft in diameter and 50 ft high, 50 are necessary to meet the required volume. Somewhat larger digesters may be built if they are designed as horizontal rather than vertical tubes. The cost of the packed digesters is estimated at $2.50 per cubic foot.

5. Liquid–Liquid Extractors. Removal of butyric, valeric, and caproic acids from the fermenters is accomplished by liquid–liquid extraction. Design equations for these extractors have been derived and presented previously (Sanderson et al., 1979). Scaling up from the mass flows assumed in previous estimates, 43,000 cubic feet of liquid–liquid extractors for higher acid removal are needed.

A second series of liquid–liquid extractors will be necessary to recycle the acetic and propionic acids. These extractors will use a hydrocarbon solvent containing 20% trioctylphosphine oxide, which has a partition coefficient for all organic acids of about 4. Acetic and propionic acids will be concentrated by tenfold before being returned to the digesters. The remaining liquid stream will be returned to the peat bog. Another set of 50,000 cubic feet extractors is assumed for this step. A more rigorous calculation of extractor design will be performed in subsequent analyses. A cost of $50 per cubic foot is assumed for the liquid–liquid extractors.

6. Electrolysis Cell. The electrolysis cell for the process must convert 700 tpd (6.23×10^6 gmoles) of acids to olefins. The required current is 2 electrons per mole of acid reacted, or 1.39×10^7 amp. At a current density of 0.25 amp/cm^2, 5570 m^2 of anode area is required. The electrode surface is platinum.

Hydrogen gas is also produced at the cathode at the rate of 27,000 lb/

day. This can be used in a molten carbonate fuel cell, which operates at 650°C and 60% efficiency, to produce electricity and process steam. Fuel cell cost is \$200/KW, and electrolysis cell cost is \$1000/m^2.

7. Hydration of Olefins. Capital costs of this process step include a catalytic reactor and distillation unit. Based on relative market prices for ethylene and ethanol, the cost of hydration has been estimated to be \$0.12 per pound of alcohol produced, including capital and processing costs, other materials, and a reasonable return on investment.

C. Estimation of Unit Fuel Cost

The operating parameters and assumed costs outlined in Tables 6.12 and 6.13 are used in this analysis. The plant capital costs are presented in Table 6.14 for production of olefins from peat. The cost of hydrating the olefins will be added on to the cost of producing mixed olefins. Annual operating costs are estimated in Table 6.15. Sodium carbonate costs may be reduced by incorporating a base recycle scheme into the process. The labor requirement is 125 people (at \$6/h). The gross annual operating cost is \$31,730,000, with electricity accounting for 40% of the total.

The average product cost is calculated from the gross annual expense including return on investment and divided by the gross production to obtain the unit cost. The cost routine used is a discounted cash flow routine for a period of 10 years. The plant is depreciated at a constant rate of 10% per year for 10 years. The gross annual operating expense (G) is calculated as follows:

$$G = OC + D + \frac{(CI/AC) - D}{0.5}$$

TABLE 6.13 Equipment and Material Costs

Item	*Cost*
Slurry dewaterer	\$70/ft^2
Pretreatment reactor	25/ft^3
Heat exchangers	7.00/ft^2
Liquid–solid separator	9.00/ft^2
Packed digester	2.50/ft^3
Liquid–liquid extractors	50.00/ft^3
Electrolysis cell	1000.00/m^2
Molten fuel cell	200.00/KW
DC rectifier/transformer	140.00/KW
Electricity	0.05/KWH
Raw peat (dry basis)	3.00/ton
Sodium carbonate	60/ton

TABLE 6.14 2000-TPD Facility Capital Costs

Equipment costs	
Slurry preparation	$ 24,000
Pretreatment reactor	88,000
Heat exchangers	490,000
Liquid/Solids separator	225,000
Digesters	4,500,000
Liquid–liquid extractors	4,650,000
Electrolysis cell	5,570,000
Molten fuel cell	2,450,000
DC rectifier/transformer	5,446,000
Pumps, piping, etc. (5% of above)	1,172,000
Total equipment costs	24,615,000
Supporting facilities (10%)	2,462,000
Total capital investment	27,077,000
Contractor's overhead and profit (10%)	2,708,000
Engineering and design (5%)	1,354,000
Subtotal plant investment	31,139,000
Contingency (15%)	4,671,000
Total plant investment	35,810,000
Interest during construction (18%)	6,446,000
Startup (9%)	3,223,000
Working capital (2%)	716,000
Total capital requirements	$46,195,000

TABLE 6.15 Annual Operating Costs

Raw materials	
Peat ($3/ton)	$ 2,190,000
Sodium carbonate ($60/ton)	8,760,000
Nutrients	500,000
Other chemicals	1,000,000
Utilities	
Electric (5¢/KWH)	12,850,000
Labor	
Operating	1,500,000
Maintenance	500,000
Supervision	300,000
Administration and overhead	1,380,000
Supplies	
Operating	500,000
Maintenance	500,000
Local taxes and insurance	1,750,000
Gross operating cost	$31,730,000

where OC is the gross annual operating cost, D is the annual depreciation, CI is the total capital requirement, and AC is the annual compounding factor:

$$AC = \sum_{j=1}^{10} \frac{1}{(1 + r)^j}$$

The annual production for the facility is 2.81×10^8 lb per year of mixed olefins, which are hydrated to form 3.72×10^8 lb per year of mixed alcohols. The tax rate on profits is 50%.

The selling price as a function of the rate of annual return on investment is plotted in Figs. 6.22 and 6.23. At 30% annual return on investment, the required selling price of the mixed olefins is \$0.21/lb and of the mixed alcohol fuel is \$1.88/gal. This is equivalent to \$17.84/MM Btu for the mixed alcohol fuel and \$10.10/MM Btu for the mixed olefin product.

D. Process Energy Balance

The process, as described, has a favorable energy balance; that is, the energy contained in the product olefin is more than the energy required in

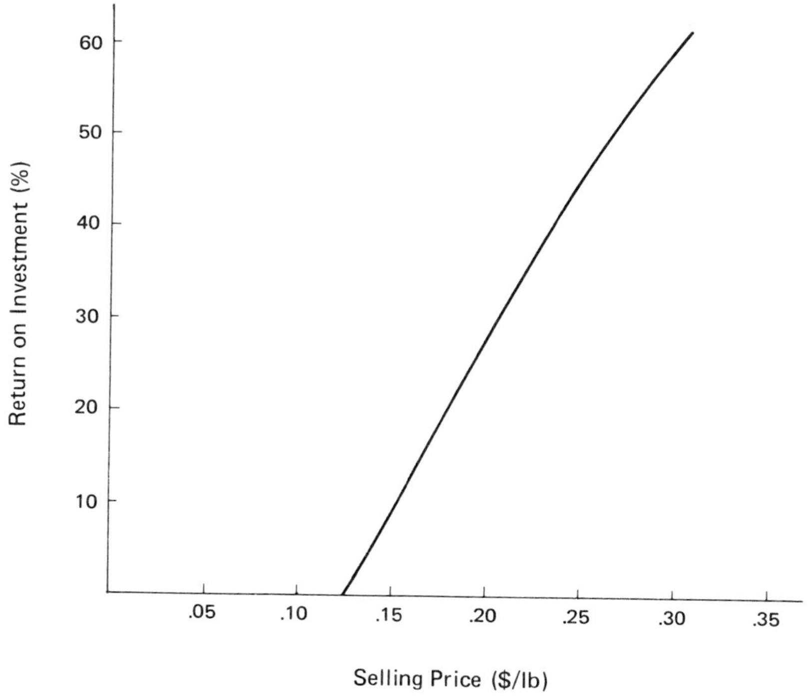

FIGURE 6.22 Cost of mixed olefin production.

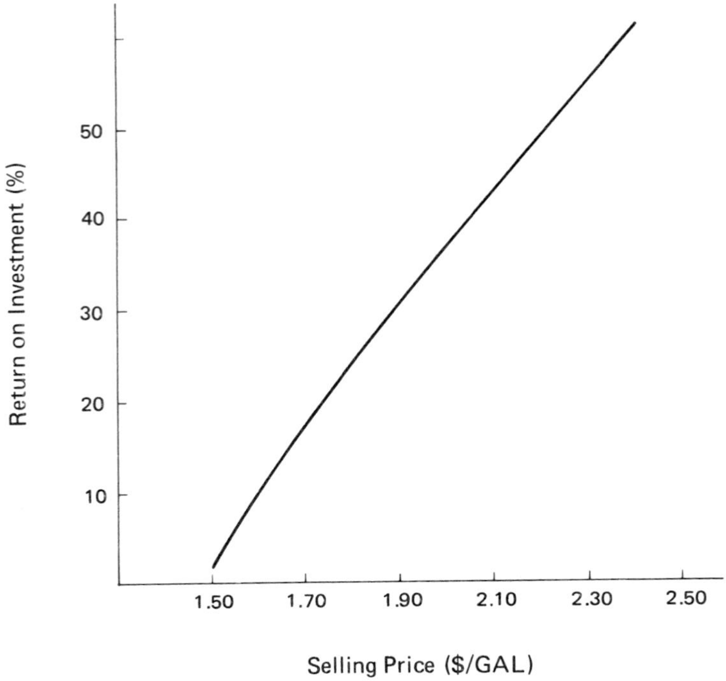

FIGURE 6.23 Cost of mixed alcohol production.

the process. The energy balance has been performed for the production of olefins from peat. The addition of a hydration step will reduce the value of the ratio of output/process energy.

The primary energy requirements and production, on an annual basis, are listed in Table 6.16. For all electric energy, the conversion 10,000 Btu/KWH is used. The major energy requirement is for electricity for the electrolytic oxidation of organic acids to olefins. The calculated ratio of

TABLE 6.16 Energy Production and Requirements Peat to Olefins (Annual Basis)

Description	*Required (Btu)*	*Produced (Btu)*
Pretreatment (air compression)	5×10^{11}	
Liquid/Solid Separator	2×10^{11}	
Electrolysis	3.4×10^{12}	
Molten Fuel Cell		1.1×10^{12}
Olefins		5.8×10^{12}
Totals	4.1×10^{12}	6.9×10^{12}

$$\text{Energy Balance: } \frac{\text{output}}{\text{process energy}} = 1.7/1.0$$

energy contained in the olefin to required process energy (not including the energy content of the wet peat) is 1.7/1.0.

ACKNOWLEDGMENTS

This work was carried out under U.S. Department of Energy Contract No. DE-AC02-81ER10914-1 under the sponsorship of Dr. Ryszard Gajewski, Director of Advanced Energy Projects, Office of Basic Energy Sciences.

REFERENCES

Aftring, R. P., and Taylor, B. F. (1981) *Arch. Microbiol. 130,* 101–104.
Allen, M. J. (1958) *Organic Electrode Processes,* Reinhold, New York.
Bache, R., and Pfennig, N. (1981) *Arch. Microbiol. 130,* 255–261.
Brown, A. C., and Walker, J. (1891) *Annalen der Chemie 261,* 107–128.
Buivid, M. G., Wise, D. L., Rader, A. M., McCarty, P. L., and Owen, W. F. (1980) *Resource Recovery and Conservation 5,* 117–138.
Chemical Marketing Reporter (1982) Schnell Publishing Co., Inc., New York.
Clark, F. M., and Fina, L. R. (1952) *Arch. Biochem. Biophys. 36,* 26–32.
Considine, D. M. (1974) *Chemical and Process Technology Encyclopedia* pp. 423–427, McGraw-Hill, New York.
Conway, B. E., and Dzieciuch, M. (1963a) *Can. J. Chem. 41,* 21.
Conway, B. E., and Dzieciuch, M. (1963b) *Can. J. Chem. 41,* 38.
Conway, B. E., and Dzieciuch, M. (1963c) *Can. J. Chem. 41,* 55.
Corey, E. J., Bauld, N. L., LaLinde, R. T., Casanova, J., Jr., and Kaiser, E. T. (1957) *J. Amer. Chem. Soc. 79,* 3182.
Dickinson, T., and Wynne-Jones, W. F. K. (1962a) *Trans. Farraday Soc. 58,* 382.
Dickinson, T., and Wynne-Jones, W. F. K. (1962b) *Trans. Farraday Soc. 58,* 388.
Dickinson, T., and Wynne-Jones, W. F. K. (1962c) *Trans. Farraday Soc. 58,* 400.
Dynatech (1981) Report No. 2115: Peat Biogasification Development Program, DOE Contract No. DE-AC 01-79ET14696, Cambridge, Mass.
Eberson, L. (1973) in *Organic Electrochemistry* (Baizer, M., ed.), Marcel Dekker, New York.
Evans, W. C. (1977) *Nature 270,* 17–22.
Ferry, J. G., and Wolfe, R. S. (1976) *Arch. Microbiol. 107,* 33–40.
Fichter, F. (1939) *Trans. Electrochem. Soc. 75,* 309–322.
Flynn, B. L., Jr. (1979) *Chem. Eng. Prog. 75,* April, 66–69.
Glasstone, S., and Hickling, A. (1934) *J. Chem. Soc.* 1878–1888.
Healy, J. B., Jr., and Young, L. Y. (1979) *Appl. Environ. Microbiol. 38,* 84–89.
Healy, J. B., Jr., Young, L. Y., and Reinhard, M. (1980) *Appl. Environ. Microbiol. 39,* 436–444.
Helsel, R. W. (1977) *Chemical Engineering Progress 73,* May, 55–59.
Hofer, H., and Moest, M. (1902) *Ann. Chem. 323,* 284.
Keith, C. L., Bridges, R. L., Fina, L. R., Iverson, K. L., and Cloran, J. A. (1978) *Arch. Microbiol. 118,* 173–176.
Kirk-Othmer (1980) *Encyclopedia of Chemical Technology* (3rd ed.), John Wiley, New York.

Koehl, W. J., Jr. (1964) *J. Amer. Chem. Soc. 86,* 4686.

Kolbe, H. (1849) *Ann. Chem. 69,* 257.

Lange, N. A. (1956) *Handbook of Chemistry* 9th ed., Handbook Publishers, Sandusky, Ohio.

Levy, P. F., Sanderson, J. E., Ashare, E., de Riel, S. R., and Wise, D. L. (1981a) Dynatech Report No. 2147: Liquid Fuels Production from Biomass, DOE Contract No. XB-0-9291-1, Cambridge, Mass.

Levy, P. F., Sanderson, J. E., Kispert, R. G., and Wise, D. L. (1981b) *Enzyme Microbiol. Technol. 3,* 207–215.

Mensinger, M. C. (1980) Wet Carbonization of Peat: State-of-the-Art Review. IGT Symposium, Arlington, Va., Dec. 1–3.

Oki, T., Ishikawa, H., and Okubo, K. (1978) *Mokuzai Gakkaishi 24,* 40–414.

Othmer, D. F. (1978) *Combustion. 50,* 44–47.

Perry, R. H., and Chilton, C. H. (1973) *Chemical Engineers' Handbook* (5th ed.), McGraw-Hill, New York.

Peters, M. S., and Timmerhaus, K. D. (1980) *Plant Design and Economics for Chemical Engineers,* McGraw-Hill, New York.

Petersen, J. (1900) *Zeitschr. für Physik Chemie 33,* 99, 295, 698.

Petersen, J. (1906) *Zeitschr. für Elektrochem. 12,* 22.

Rader, A. M. (1977) Testimony before U.S. House of Representatives Sub-Committee on Environment, Energy, and Natural Resources, Sept. 29.

Randall, T. L., and Knopp, P. V. (1980) *J. Water Poll. Cont. 52,* 2117–2130.

Ruoff, C. F., Ashare, E., Sanderson, J. E., and Wise, D. L. (1980) Continued Development of a Peat Biogasification Process. IGT Symposium, Arlington, Va., Dec. 1–3.

Sanderson, J. E., Garcia-Martinez, D. V., George, G. S., Dillon, J. J., Molyneaux, M. S., Barnard, G. W., and Wise, D. L. (1979) Liquid Fuels Production from Biomass. Dynatech Report No. 1931, Contract No. EG-77-C-02-4388-8, Cambridge, Mass.

Schall (1896) *Z. Electrochem. 3,* 83.

Soper, E. K., and Osborn, C. C. (1922) The occurrence and uses of peat in the United States. U.S. Geological Survey, Bulletin 728, Washington, D. C.

Taylor, B. F., Campbell, W. L., and Chinoy, I. (1970) *J. Bacteriol. 102,* 430–437.

Teletzke, G. H. (1964) *Chem. Eng. Prog. 60,* 33–38.

Tibbetts, T. E. (1968) Peat resources of the world—A Review, Proceedings of the Third International Peat Congress, Quebec, Canada.

Utley, J. H. P. (1974) in *Techniques of Chemistry* Vol. V, Part 1, *Technique of Electroorganic Synthesis* (Weinberg, N. L., ed.) p. 793, John Wiley, New York.

Waksman, S. A. (1930) *Amer. J. Sci. 19,* 32–54.

Walling, C. (1957) *Free Radicals in Solution* p. 581, John Wiley, New York.

Weedon, B. C. L. (1952) *Quart. Rev.* (London) *6,* 380.

Williams, R. J., and Evans, W. C. (1975) *Biochem. J. 148,* 1–10.

Zimmerman, F. J., and Diddams, D. G. (1960) *Tappi 43,* 710–715.

CHAPTER

7

Formation of Acetate Using Homoacetate Fermenting Anaerobic Bacteria

Lars G. Ljungdahl

I. INTRODUCTION

Acetic acid is an important industrial feedstock chemical that is produced from mineral oil or natural gas. During 1978, the production of acetate was 1.27×10^9 kg in the United States (*Standard and Poor's Industrial Surveys*, 1979). With the possibility of a shortage of oil and gas, an interest has developed in producing acetate from renewable biological resources by fermentations of carbohydrates and other "biomass" materials. This interest has focused on the "homoacetate fermenting or acetogenic bacteria," which are able to convert, for example, glucose and xylose almost stoichiometrically to acetate (reactions 1 and 2):

$$C_6H_{12}O_6 \rightarrow 3CH_3COOH \quad (1)$$

$$2C_5H_{10}O_5 \rightarrow 5CH_3COOH \quad (2)$$

The fermentation of sugars to acetate is a complex process. One mol of hexose is metabolized by the "Embden-Meyerhof" pathway to yield 2 mol of pyruvate, which are further metabolized to 2 mol of acetate formed from carbons 2 and 3 of the pyruvate and to 2 mol of one-carbon units formed from the carboxyl groups. These one-carbon units can be considered to be equivalent to carbon dioxide, and they are reduced to the third mol of acetate. The fermentation may then be viewed as a two-step process according to reactions (3) and (4), as was postulated by Barker (1944) and Wood (1952a):

$$C_6H_{12}O_6 + 2H_2O \rightarrow 2CH_3COOH + 2CO_2 + 8H^+ + 8e \quad (3)$$

$$2CO_2 + 8H^+ + 8e \rightarrow CH_3COOH + 2H_2O \quad (4)$$

$$\text{Net reaction:} \quad C_6H_{12}O_6 \rightarrow 3CH_3COOH \quad (1)$$

The mechanism of formation of acetate from CO_2 (reaction 4) has been studied in some detail by using *Clostridium thermoaceticum*. The reduction of CO_2 to the methyl group of the acetate apparently occurs via formate and one-carbon derivatives of tetrahydrofolate to yield methyltetrahydrofolate. In the final steps, the methyl group of methyltetrahydrofolate and the carboxyl group of pyruvate yield acetate in a reaction that is considered to be a transcarboxylation (Schulman et al., 1973) and that is catalyzed by, at least, a five-component enzyme system (Drake et al., 1981). Recently, it has been shown that carbon monoxide, in addition to the carboxyl group of pyruvate, can also serve as a precursor to the carboxyl group of acetate (Hu et al., 1982). The synthesis of acetate from carbon dioxide in acetogenic bacteria was recently reviewed (Ljungdahl and Wood, 1982).

It is evident from reactions (3) and (4) that carbon dioxide serves as an acceptor of electrons generated during fermentations of sugars by acetogenic bacteria. Most, if not all, of the acetogens may also use molecular hydrogen and very likely carbon monoxide as sources of electrons. This was noticed in 1932 by Fischer et al. (1932), using a bacterial population from sewage sludge. They postulated that acetate is synthesized according to reaction (5) and that carbon monoxide yields hydrogen as shown in reaction (6):

$$2CO_2 + 4H_2 \longrightarrow CH_3COOH + 2H_2O \quad (5)$$

$$CO + H_2O \longrightarrow CO_2 + H_2 \quad (6)$$

The first isolation of an acetogenic bacterium, *Clostridium aceticum*, was reported by Wieringa (1940), who showed that it forms acetate from carbon dioxide and hydrogen gas. Today, three additional bacteria have been described that carry out reaction (5): *Acetobacterium woodii* (Balch et al., 1977), *Acetogenium kivuii* (Leigh et al., 1981), and *Clostridium thermoautotrophicum* (Wiegel et al., 1981). These bacteria grow autotrophically and apparently generate their energy for growth as well as cell material by reducing carbon dioxide to acetate. Recent findings by J. G. Zeikus (personal communication) indicate that *C. thermoaceticum* also carries out reaction (5).

Acetogenic bacteria can also synthesize acetate from reduced one-carbon compounds such as carbon monoxide, formate, and methanol (reactions 7–9):

$$4CO + 2H_2O \longrightarrow CH_3COOH + 2CO_2 \quad (7)$$

$$4HCOOH \longrightarrow CH_3COOH + 2CO_2 + 2H_2O \quad (8)$$

$$4CH_3OH + 2CO_2 \longrightarrow 3CH_3COOH + 2H_2O \quad (9)$$

These reactions have been observed with *C. aceticum* (reaction 8) (Braun et al., 1981), *Clostridium formicoaceticum* (reaction 9) (Braun et al., 1981), *A. woodii* (reactions 8 and 9) (Balch et al., 1977; Bache and Pfennig, 1981), *C. thermoautotrophicum* (reactions 7 and 9) (Wiegel et al., 1981; Wiegel, 1982), and *C. thermoaceticum* (reaction 7) (Lynd et al., 1982). Alternatives to reactions (7) and (8) would be reactions (10) and (11):

$$2CO + 2H_2 \longrightarrow CH_3COOH \quad (10)$$

$$2HCOOH + 2H_2 \longrightarrow CH_3COOH + 2H_2O \quad (11)$$

Thus it appears to be possible to manufacture acetate from several one-carbon compounds with molecular hydrogen as an electron donor and using acetogenic bacteria as catalysts. An obvious possibility would be to use synthetic gas (syngas) as a substrate (reaction 10). This gas produced from coal and water contains a one-to-one mixture of carbon monoxide and hydrogen.

In addition to fermenting sugars and growing on free one-carbon compounds, the acetogenic bacteria may also utilize methoxyl groups of methoxylated aromatic compounds such as vanillic acid, syringic acid, ferulic acid, and sinapic acid. This was recently demonstrated with *A. woodii* (Bache and Pfennig, 1981), and it points to the possibility of partially converting lignin, which contains a substantial amount of methoxyl groups, to acetate by using acetogenic bacteria.

The last few years have witnessed an explosion of our knowledge of the acetogenic bacteria. For almost 30 years, the only such bacterium investigated was *C. thermoaceticum* (Ljungdahl and Wood, 1982). Lately, *C. formicoaceticum* (El Ghazzawi, 1967; Andreesen et al., 1970), *A. woodii* (Balch et al., 1977), *C. thermoautotrophicum* (Wiegel et al., 1981), and *Acetogenium kivuii* (Leigh et al., 1981) were discovered, and *C. aceticum* (Braun et al., 1981) was revived. Previously, a few sugars and a mixture of CO_2 and H_2 were considered as the substrates for acetogenes. Now, as is evident from the above discussion, it has been found that these bacteria can use a variety of substrates, which are converted to acetate. The aim of this chapter is to review the physiology of the acetogenic bacteria, including metabolic pathways and enzyme studies.

II. PHYSIOLOGY OF ACETOGENIC BACTERIA

A. Acetogenic Bacterial Species

The acetogenic bacteria to be included in this discussion are *Clostridium aceticum* (Wieringa, 1940; Braun et al., 1981), *Clostridium thermoaceticum* (Fontaine et al., 1942), *Clostridium formicoaceticum* (El Ghazzawi, 1967; Andreesen et al., 1970), *Acetobacterium woodii* (Balch et al., 1977), *Clostridium thermoautotrophicum* (Wiegel et al., 1981), and *Acetogenium kivuii* (Leigh et al., 1981). The properties of these bacteria are listed in Table 7.1. They have in common that they are anaerobic and rod-shaped, that they ferment 1 mol of fructose to 3 mol of acetate, and that they convert a variety of other substrates including one-carbon compounds to acetate as essentially the only product. The substrate spectra of the acetogenic bacteria are shown in Table 7.2.

It must be pointed out that, in addition to the bacteria to be considered in this paper, several other bacteria are able to synthesize acetate from CO_2 (Ljungdahl and Wood, 1982). They differ from the above-mentioned bacteria in that they form other products in addition to acetate, utilize other types of substrates, use a different pathway, or have not yet been well characterized. Notable among these bacteria are *Clostridium cylindrosporum* (Barker and Beck, 1942), *Clostridium acidiurici* (Barker and Beck, 1942), *Clostridium purinolyticum* (Dürre, 1981; Dürre et al., 1981), *Peptococcus* (*Diplococcus*) *glycinophilus* (Cardon and Barker, 1946), *Eubacterium limosum* (*Butyribacterium rettgeri*) (Barker and Haas, 1944; Genthner et al., 1981), *Butyribacterium methylotrophicum* (Zeikus et al., 1980), and a *Clostridium* that is physiologically similar to but morphologically different from *Clostridium aceticum* (Ohwaki and Hungate, 1977).

C. cylindrosporum, C. acidiurici, and *C. purinolyticum* ferment purines to ammonia CO_2 and acetate. Part of the acetate is formed from CO_2 (Karlsson and Barker, 1949; Schulman et al., 1972) via a pathway involving glycine and serine, which is different from the pathway of acetate synthesis with *C. thermoaceticum* (Ljungdahl and Wood, 1982; Waber and Wood, 1979; Dürre and Andreesen, 1982). *P. glycinophilus* grows on glycine, which is converted to ammonia, CO_2, and acetate. The acetate is partly formed from CO_2 (Schulman et al., 1972; Barker et al., 1948), and it can be assumed on the basis of enzyme studies (Sagers and Gunsalus, 1961; Klein and Sagers, 1966) that acetate is synthesized via a pathway similar to that used by the purine-fermenting clostridia.

E. limosum (Genthner et al., 1981) and *B. methylotrophicum* (Zeikus et al., 1980) appear to be similar, and it has been suggested that they are different strains of the same organism. They have similar guanine and cytosine content (49% and 49.8%, respectively), cell size, and temperature range. However, *B. methylotrophicum* forms spores, a feature that has not been demonstrated for *E. limosum*. Both bacteria ferment glucose, fructose, and lactate, with acetate and butyrate as the main products. Methanol and

TABLE 7.1 Properties of Acetogenic Bacteria

Property	*Clostridium aceticum*	*Clostridium thermoaceticum*	*Clostridium formicoaceticum*	*Acetobacterium woodii*	*Clostridium thermoautotrophicum*	*Acetogenium kivui*
Cell size (μm)	5 × 0.8–1.0	2.8 × 0.4	5–12 × 1.2–2	2 × 1	3–6 × 0.8–1.0	2–7.5 × 0.7
Spore formation	Yes	Yes	Yes	No	Yes	No
Flagellum	Peritrichous	Peritrichous	Peritrichous	1–5 Subterminal	3–8 Peritrichous	None
DNA composition G + C content (%)	33	53.6, 54	34.2, 34	39	53–55	38
Temperature						
Range (°C)	20–55	45–65	22–44	<40	36–70	50–72
Optimum (°C)	30	55–60	37	30	56–60	66
pH						
Range	7–9.5	6–8.5			4.5–7.6	5.3–7.3
Optimum	8.3	>7	7.2–7.8	6.7–7.5	5.7	6.4
Reference	Wieringa (1940), Braun et al. (1981)	Fontaine et al. (1942), Ljungdahl and Andreesen (1976), Matteuzzi et al. (1978)	El Ghazzawi (1967), Andreesen et al. (1970), Ljungdahl and Andreesen (1976), Matteuzzi et al. (1978)	Balch et al. (1977), Bache and Pfennig (1981), Mayer et al. (1977)	Wiegel et al. (1981)	Leigh et al. (1981)

TABLE 7.2 Substrate Spectra for Acetogenic Bacteria[1]

Substrate	*Clostridium aceticum*	*Clostridium thermoaceticum*	*Clostridium formicoaceticum*	*Acetobacterium woodii*	*Clostridium thermoautotrophicum*	*Acetogenium kivui*
Fructose	+[2]	+[3]	+[6]	+[7]	+[9]	+[11]
Glucose	—	+[3]	—	+[7]	+[9]	+[11]
Galactose	—	—	—	—	+[9]	—
Mannose	—	—	—	—	—	+[11]
Xylose	—	+[3]	—	—	+[9]	n.d.
Ribose	+[2]	—	+[6]	—	—	—
Mannitol	—	—	+[2]	—	n.d.	—
Gluconate	n.d.	—	+[6]	n.d.	n.d.	—
Mannonate	n.d.	n.d.	+[6]	n.d.	n.d.	n.d.
Galactonate	n.d.	n.d.	+[6]	n.d.	n.d.	n.d.
Glucoronate	n.d.	—	+[6]	—	n.d.	n.d.
Galacto-uronate	n.d.	n.d.	+[6]	—	n.d.	—
2-Keto-3-deoxy-gluconate	n.d.	n.d.	+[6]	n.d.	n.d.	n.d.
Glutamate	+[2]	—	+[6]	—	—	n.d.
Fumarate	+[2]	—	+[6]	—	n.d.	—
Malate	+[2]	—	+[6]	—	—	—
Serine	+[2]	n.d.	—	—	n.d.	—
Lactate	—	—	+[6]	+[7]	—	n.d.
Pyruvate	+[2]	+[4]	+[6]	+[8]	—	+[11]
Ethylene glycol	+[2]	n.d.	—	n.d.	—	n.d.
Ethanol	+[2]	n.d.	+[2]	—	—	—

TABLE 7.2 (Continued)

Substrate	*Clostridium aceticum*	*Clostridium thermoaceticum*	*Clostridium formicoaceticum*	*Acetobacterium woodii*	*Clostridium thermoautotrophicum*	*Acetogenium kivui*
Glycerol	—	—	+[6]	+[8]	—	—
Glycerate		n.d.	+[6]	+[7]	+[9]	—
CO_2/H_2	+[2]	+[12]	—	+[7]	+[9]	+[11]
CO	n.d.	+[5]	n.d.	n.d.	+[10]	n.d.
HCOOH	+[2]	n.d.	—	+[7]	+[9]	+[11]
CH_3OH/CO_2	—	n.d.	+[2]	+[8]	+[9]	—
Methoxyl groups	n.d.	n.d.	n.d.	+[8]	n.d.	n.d.
Methyl formate	n.d.	n.d.	n.d.	+[8]	n.d.	n.d.
Methylamine/ CO_2	n.d.	n.d.	n.d.	n.d.	+[10]	n.d.

1. A + indicates that the substrate supports growth, a — indicates the substrate is not utilized, and n.d. means that there is no reference indicating whether the substrate has been tested or not. Compounds tested and not utilized by any of the acetogenic bacteria are not included; among such compounds are disaccharides, starch, and cellulose.
2. From Braun et al. (1981).
3. From Fontaine et al. (1942).
4. From Barker and Kamen (1945).
5. From Lynd et al. (1982).
6. From Andreesen et al. (1970).
7. From Balch et al. (1977).
8. From Bache and Pfennig (1981).
9. From Wiegel et al. (1981).
10. From Wiegel (1982).
11. From Leigh et al. (1981).
12. From J. G. Zeikus (personal communication).

H_2/CO_2 are also utilized as energy sources. Growth on methanol seems to require the presence of CO_2 and acetate. The methanol is converted by *E. limosum* to acetate, butyrate, and a small amount of caproate, whereas with *B. methylotrophicum,* butyrate is the main product. With H_2/CO_2 as the energy source, *B. methylotrophicum* produces only acetate, and *E. limosum* forms acetate and a small amount of butyrate.

E. limosum grows with CO as the sole source of energy (Genthner and Bryant, 1982). Acetate and CO_2 are the major products. The original isolate of *B. methylotrophicum* is not able to use carbon monoxide as the sole energy source. However, when this strain is grown in the presence of acetate, methanol, and bicarbonate with CO as the gas phase, the CO is consumed. Using such growth conditions, Lynd et al. (1982) have been able to select a strain that grows on CO without the presence of methanol and acetate. The strain appears to be stable and has retained its capacity to grow on CO for 8 months. Acetate and CO_2 are the products from CO.

B. methylotrophicum contains corrinoid levels from 0.35 to 7.9 $\mu g\ mg^{-1}$ cell dry weight when cells are grown on glucose or methanol, respectively (Zeikus et al., 1980). Such high levels of corrinoids are found in the acetogenic bacteria (Ljungdahl et al., 1966; Tanner et al., 1978; Clark et al., 1982), in which corrinoids are proposed to be fundamental for the synthesis of acetate. It is proposed that acetate synthesis in *B. methylotrophicum* and *E. limosum* occurs via a pathway similar to that used by the acetogenic bacteria. The difference between the types of bacteria would then be that *E. limosum* and *B. methylotrophicum* are able to produce butyrate in addition to acetate, whereas the acetogenic bacteria form only acetate. Of course, one should perhaps include *E. limosum* and *B. methylotrophicum* among the acetogenes.

1. Clostridium Aceticum. *C. aceticum* was isolated in 1936 by Wieringa (1936, 1940), and it was the first pure culture obtained of an acetogenic bacterium, with the capacity to form acetate from CO_2 and molecular hydrogen according to reaction (5). Wieringa (1940) noticed that when CO_2 is supplied in the form of $NaHCO_3$ and the gas phase is pure hydrogen, the acetate formation is, according to reaction (12),

$$3NaHCO_3 + 4H_2 \longrightarrow CH_3\,COONa + Na_2CO_3 + 4H_2O \qquad (12)$$

Under such circumstances, the medium becomes alkaline in spite of acetate production. This shift in pH is prevented by supplying CO_2 in the gas phase. Wieringa (1940) also observed that a low redox potential is advantageous and included 0.1% Na_2S in the medium. Higher concentrations (0.5%) of Na_2S are inhibiting.

In 1948, Karlsson et al. (1948) developed for *C. aceticum* a synthetic medium containing minerals, a vitamin-free casein hydrolyzate, biotin, pyridoxamine, pantothenic acid, and sodium thioglycolate (to replace Na_2S). After this work, the bacterium was considered to be lost until 1981, when a spore preparation of the original strain was discovered in a culture collection of H. A. Barker. From this preparation, the bacterium has been revived. It has now been well described by Braun et al. (1981), who have deposited it in the German Collection of Microorganisms in Göttingen under DSM number 1496.

C. aceticum grows chemolithotrophically on H_2/CO_2 and organotrophically on fructose, ribose, glutamate, fumarate, malate, pyruvate, formate, and ethanol (Table 7.2) at 30°C and at an optimum pH between 8 and 8.5. Interestingly, chemolithotrophic growth occurs at pH 8.5 but not at pH 7.5. At pH 7.5, fructose is rapidly fermented to acetate. Braun (1979) and Braun and Gottschalk (1981) have investigated whether *C. aceticum* can simultaneously utilize H_2/CO_2 and fructose. Their work revealed that H_2 is an inhibitor of the fructose fermentation at pH 8.5, but not at pH 7.5, and that some H_2 is evolved from fructose during the stationary phase when the fermentation is under an atmosphere of N_2. They also found that hydrogenase activity in fructose-grown cells (Braun et al., 1981) is about one tenth of that in H_2/CO_2-grown cells.

2. Clostridium Thermoaceticum. *C. thermoaceticum* (DSM 521 and ATCC 39073), isolated by Fontaine et al. (1942), is thermophilic and grows optimally between 55° and 60°C and at pH 7–8 under an atmosphere of 100% CO_2. It ferments fructose, glucose, and xylose to acetate (reactions 1 and 2) and pyruvate to acetate and CO_2 (reaction 13):

$$4CH_3COCOOH + 2H_2O \longrightarrow 5CH_3COOH + 2CO_2 \qquad (13)$$

Recently, it has been found that CO, and perhaps also H_2/CO_2, can support growth and be reduced to acetate (Lynd et al., 1982; J. G. Zeikus, personal communication).

Theoretically, according to reactions (1) and (2), glucose, fructose, and xylose are 100% converted to acetate. In reality, 85% of the sugars are converted to acetate, and about 5% enter the cells (Fontaine et al., 1942; Andreesen et al., 1973). The remaining 10% of the sugars are most likely decomposition products due to caramelizing (Fontaine et al., 1942), which occurs at incubations at thermophilic temperatures (Wiegel et al., 1979).

An apparent "diauxie" type of fermentation is exhibited by *C. thermoaceticum* when the medium contains mixtures of xylose, fructose, and glucose (Andreesen et al., 1973). Thus xylose is fermented faster than either fructose or glucose; and with a mixture of fructose and glucose, the fructose is almost completely gone before glucose is fermented. This diauxie phenomenon may affect the use of *C. thermoaceticum* in industrial fermentations if mixed substrates are used.

The original medium used by Fontaine et al. (1942) for *C. thermoaceticum* consisted of a phosphate buffer, magnesium sulfate, calcium chloride, tryptone, glucose, liver extract, and calcium carbonate. This medium was improved by removing the liver extract and calcium carbonate and by adding yeast extract, sodium bicarbonate, and sodium thioglycolate (Wood, 1952; Barker and Kamen, 1945) or Na_2S (Poston et al., 1966) and ferrous sulfate (Poston et al., 1966) as well as boiled and filtered tomato juice (Ljungdahl and Wood, 1965). Eventually, it was found that the tomato juice could be replaced by a solution containing molybdate, tungstate, selenite, manganese chloride, zinc chloride, nickel chloride, boric acid, aluminum potassium sulfate, and copper chloride (Andreesen et al., 1973). The rationale for adding many of these metals is now obvious. Cobalt is a constituent of corrinoids, which take part in the final steps of acetate synthesis from methyltetrahydrofolate and pyruvate (Drake et al., 1981; Ljungdahl and Andreesen, 1976; Poston et al., 1966; Ghambeer et al., 1971); formate dehydrogenase contains tungsten, selenium, and iron (Ljungdahl and Andreesen, 1976; Saiki et al., 1981; Yamamoto et al., 1982, 1983); and carbon monoxide dehydrogenase has nickel (Drake et al., 1980; Diekert and Thauer, 1980) plus, as recently was demonstrated, zinc and iron (Ragsdale et al., 1982, 1983).

C. thermoaceticum is now routinely grown in a rich medium (Ljungdahl and Andreesen, 1978) containing glucose, yeast extract, tryptone, Na-thioglycolate, bicarbonate, phosphate, and minerals as shown in Table 7.3. The medium is made up in three solutions, which are autoclaved separately before being combined and gassed with 100% CO_2 until a pH of 8.5 or lower is obtained. It is then placed in an incubator maintained at 58°C and allowed to reach a temperature of 55°C before inoculation. During this time and during growth, a slow constant stream of CO_2 is bubbled through the medium. Depending on the size of the inoculum, the fermentation ceases within 24–48 h when the glucose is completely fermented. The medium then contains about 0.25 M of acetate and a cell mass between 10 and 17 g/l. J. R. Andreesen (personal communication) has developed a synthetic medium for *C. thermoaceticum*. In this medium, yeast extract and tryptone have been replaced by lysine, methionine, nicotinic acid amide, and biotin.

It is noted in Table 7.1 that *C. thermoaceticum* has an optimum pH above 7 and does not grow below pH 6. Schwartz and Keller (1982) have pointed out that recovery of acetic acid from a fermentation would be easier than recovery of acetate. For this and other reasons, they have selected from a wild-type *C. thermoaceticum* strain (ATCC 39073) an acetate-tolerant strain derived by ethyl methane sulfonate mutagenesis. From this strain, using low pH and high acetic acid concentration as selective pressures, they have successfully obtained a culture of *C. thermoaceticum* that can grow and produce acetic acid at pH 4.5. With this culture (ATCC 31490), acetic acid is produced to a concentration of 4.5 g/l.

C. thermoaceticum has been the subject of intensive investigations to evaluate the pathway of the synthesis of acetate from CO_2. This work, which has led to the present concept of acetate synthesis, has been reviewed twice (Ljungdahl and Wood, 1969, 1982). Some details and recent findings of the pathway and of properties of the enzymes involved will be discussed later.

3. Clostridium Formicoaceticum. El Ghazzawi (1967) isolated from mud an anaerobic mesophilic bacterium with several properties similar to *C. aceticum* (Wieringa, 1936, 1940). However, it was unable to use molecular hydrogen to reduce CO_2, and fructose was fermented to acetate and a smaller amount of formate. Andreesen et al. (1970) isolated several strains of a similar bacterium. The name *C. formicoaceticum* (type strain ATCC 27076; DSM 92) was adopted to reflect the formation of formate as a product. However, formate is formed only at the end of a fermentation.

C. formicoaceticum ferments fructose, ribose, several aldonic and uronic acids, glutamate, fumarate, malate, lactate, pyruvate, ethanol, glycerol, glycerate, and a CH_3OH/CO_2 mixture (Table 7.2). The fermentation of fructose occurs via the Embden-Meyerhof pathway (Andreesen et al., 1970; Linke, 1969a; O'Brien and Ljungdahl, 1972), which yields pyruvate. The pyruvate and CO_2 are therefore most likely converted to acetate according to the pathway postulated to occur in *C. thermoaceticum* (Linke, 1969b; O'Brien and Ljungdahl, 1972). The fermentation of fructose by *C. formicoaceticum* is stimulated by 0.5–1% sodium bicarbonate or by formate in the medium (Andreesen et al., 1970). However, a gas phase of 100% CO_2, which is very beneficial for growth of *C. thermoaceticum,* is inhibitory for *C. formicoaceticum*. Therefore *C. formicoaceticum* is preferably grown under a gas phase of N_2.

To ferment aldonic and perhaps also uronic acids, *C. formicoaceticum* uses a modified Entner-Doudoroff pathway (Andreesen and Gottschalk, 1969). The first step of the breakdown of gluconate is by dehydration to form 2-keto-3-deoxygluconate, which is phosphorylated to yield 2-keto-3-deoxy-6-phosphogluconate. The latter is then cleaved to yield pyruvate and glyceraldehyde-3-phosphate, which are fermented by steps common to glycolysis. Gluconate dehydrase, 2-keto-3-deoxygluconate kinase, and 2-keto-3-deoxy-6-phosphogluconate aldolase have been found in *C. formicoaceticum* and partly characterized (Andreesen and Gottschalk, 1969).

The metabolism of L-malate and fumarate in *C. formicoaceticum* has been investigated by Dorn (1976) and Dorn et al. (1978). The fermentations of fumarate and malate are according to reactions (14) and (15), respectively:

$$3 \text{ fumarate} + 2H_2O \longrightarrow 2 \text{ succinate} + \text{acetate} + 2CO_2 \quad (14)$$

$$3 \text{ malate} \longrightarrow 0.16 \text{ succinate} + 4.2 \text{ acetate} + 2.9CO_2 + 0.08H_2O \quad (15)$$

When fumarate is used as a substrate in the presence of formate, formate functions as an electron donor, and more succinate is produced than is shown in reaction (14). Molar growth yields were determined for the two substrates, and they were found to be about equal, although a much lower growth yield was to be expected with fumarate than with malate. It was suggested (Dorn et al., 1978) that reduction of fumarate to succinate is coupled with electron transport phosphorylation. This suggestion is supported by the presence in *C. formicoaceticum* of a *b*-type cytochrome and menaquinone (Gottwald et al., 1975), which may function as electron carriers in a fumarate-reducing system.

Several recipes for media for *C. formicoaceticum* have been described (El Ghazzawi, 1967; Andreesen et al., 1970; Linke, 1969a; Gottwald et al., 1975). The medium listed in Table 7.3 for *C. thermoaceticum* also supports growth of *C. formicoaceticum,* provided that the glucose is replaced with fructose and that the gas phase is N_2 instead of CO_2. Common with *C. thermoaceticum, C. formicoaceticum* requires trace metals in the medium. Nutritional studies have shown that the formation of formate dehydrogenase in *C. formicoaceticum* depends on selenium and tungsten (Andreesen et al., 1974; Leonhardt and Andreesen, 1977) and that nickel is required for carbon monoxide dehydrogenase (Diekert and Thauer, 1978). Cobalt is also needed for corrinoid formation.

4. Acetobacterium Woodii. The type strain (ATCC 29683; DSM 1030) of *A. woodii* was isolated from black sediment from a pond at Woods Hole,

TABLE 7.3 Medium for *Clostridium thermoaceticum*

Solution	*Ingredient*	*Amount (g)*	*Volume (ml)*	*Final Concentration*[1] *(mM)*
A	Glucose	18		100
	Water		150	
B	Yeast extract (Difco)	5		
	Tryptone (Difco)	5		
	$(NH_4)_2SO_4$	1		7.6
	$MgSO_4 \cdot 7H_2O$	0.25		1
	$Fe(NH_4)_2(SO_4)_2 \cdot 6H_2O$	0.039		0.1
	$Co(NO_3)_2 \cdot 6H_2O$	0.03		0.1
	$Na_2WO_4 \cdot 2H_2O$	0.0033		0.01
	$Na_2MoO_4 \cdot 2H_2O$	0.0024		0.01
	$ZnCl_2$	0.0014		0.01
	Na_2SeO_3	0.0002		0.001
	$NiCl_2 \cdot 6H_2O$	0.0002		0.001
	Sodium thioglycolate (Difco)	0.5		4.4
	Water		550	
C	$HaHCO_3$	16.8		200
	K_2HPO_4	7		40
	KH_2PO_4	5.5		40
	Water		300	

1. Concentration after mixing of solutions A, B, and C.

Massachusetts (Balch et al., 1977). Other strains have been obtained from Crystal Lake, Urbana, Illinois (Leigh et al., 1981), and from waste water lagoons in New Zealand (strains NZva16 and NZva24) (Bache and Pfennig, 1981). It is an anaerobic non-spore-forming rod, which grows at a neutral pH at around 30°C. It utilizes a number of one-carbon substrates, including CO_2/H_2, HCOOH, CH_3OH/CO_2, methyl formate, and methoxyl groups of aromatic compounds. All strains grow on fructose and probably also on glycerate, but results differ with regard to growth on glucose, lactate, glycerol, and pyruvate (Balch et al., 1977; Bache and Pfennig, 1981). Acetate is the only or the main product; traces of succinate may be formed from organic substrates (Balch et al., 1977). The pathway of acetate synthesis from CO_2 appears to be similar to that of *C. thermoaceticum* (Tanner et al., 1978).

Braun and Gottschalk (1981) observed that, when *A. woodii* grows under a N_2 gas phase on fructose, glucose, or lactate with less than 0.04% bicarbonate in the medium, molecular hydrogen evolves. With a H_2 gas phase instead of N_2, the fermentation was completely inhibited. They also reported that hydrogenase activity in H_2/CO_2-grown cells was similar to that in fructose-grown cells. These results open the possibility that molecular hydrogen is an intermediate in the fermentation of fructose and in the synthesis of acetate from CO_2.

Experiments conducted by Winter and Wolfe (1980) support this possibility. A co-culture of *A. woodii* with *Methanobacterium* strain AZ ferments fructose according to reaction (16):

$$C_6H_{12}O_6 \rightarrow 2CH_3COOH + CH_4 + CO_2 \qquad (16)$$

A. woodii does not produce methane, and *Methanobacterium* strain AZ utilizes only H_2/CO_2, which is converted to methane. Consequently, the fermentation as shown in reaction (16) is best explained by postulating a fermentation of fructose by *A. woodii* to acetate, CO_2, and molecular hydrogen (reaction 17) and a consequent formation of methane by the *Methanobacterium* species (reaction 18):

$$C_6H_{12}O_6 + 2H_2O \rightarrow 2CH_3COOH + 2CO_2 + 4H_2 \qquad (17)$$

$$CO_2 + 4H_2 \rightarrow CH_4 + 2H_2O \qquad (18)$$

$$\text{Net reaction:} \quad C_6H_{12}O_6 \rightarrow 2CH_3COOH + CH_4 + CO_2 \qquad (16)$$

One may wonder if other acetogenic bacteria in addition to *A. woodii* and *C. aceticum* (Braun and Gottschalk, 1981) produce hydrogen and if molecular hydrogen formation is a general phenomenon in acetogenes during heterotrophic growth. It is interesting that *C. thermoaceticum* possesses a hydrogenase, as was recently shown by Drake (1982). Previous attempts to detect hydrogenase in *C. thermoaceticum* have not been successful, and it was believed that this bacterium lacked hydrogenase and that electrons were forced to enter the CO_2-reducing pathway leading to acetate without involvement of molecular hydrogen.

A. woodii can grow on methanol in the presence of CO_2 according to reaction (9) (Bache and Pfennig, 1981). A surprising observation is that the New Zealand strains and the type strain also in the presence of CO_2 can grow on methoxyl groups derived from methoxylated aromatic acids such as vanillic acid. This means that *A. woodii* is able to cleave a phenylether bond to form a C-1 fragment, most likely methanol, and the corresponding hydroxy derivative of the aromatic acid. The C-1 fragment is subsequently converted to acetate, whereas the aromatic product is not further degraded. When aromatic acids contain two (syringic acid) or three (3,4,5-trimethoxybenzoic acid) methoxyl groups, all groups are fermented. A fermentation balance using syringic acid is shown in reactions (19–21) (Bache and Pfennig, 1981):

$$2C_9H_{10}O_5 + 4H_2O \rightarrow 2C_7H_6O_5 + 4CH_3OH \quad (19)$$

$$4CH_3OH + 2CO_2 \rightarrow 3CH_3COOH + 2H_2O \quad (20)$$

Net reaction:

$$2\ \text{syringic acid} + 2CO_2 + 2H_2O \rightarrow 2\ \text{gallic acid} + 3CH_3COOH \quad (21)$$

An additional remarkable property of the *A. woodii* strains was discovered by Bache and Pfennig (1981). They observed that the double bond of phenylacrylate derivatives is reduced and functions as an electron acceptor. Thus in fermentations of ferulic acid, which contains one methoxyl group and one acrylic side chain, hydrocaffeic acid with one propionic side chain and acetate are formed. The normal molar ratio of acetate over methoxyl groups (reaction 21) was then decreased from 0.75 to 0.5.

5. Clostridium Thermoautotrophicum. Wiegel et al. (1981) recently isolated an ubiquitous thermophilic bacterium that grows chemolithotrophically on CO_2/H_2 and chemoorganotrophically on fructose, glucose, galactose, and glycerate. Two of the strains grow on xylose. This bacterium also grows on reduced one-carbon compounds, including CO, formate, CH_3OH/CO_2, and CH_3NH_2/CO_2 (Wiegel et al., 1981; Wiegel, 1982). With all substrates, acetate is the only product of significance. The name *C. thermoautotrophicum* was adopted to reflect the fact that the bacterium is an anaerobic spore former that grows autotrophically at high temperatures (V_{max} = 70°C). The type strain is deposited in the German Collection of Microorganisms in Göttingen under the number DSM 1974.

C. thermoautotrophicum differs from *C. thermoaceticum* in that it grows without adaptation on several one-carbon compounds (Table 7.2) and that it has a lower pH range (Table 7.1). Optimum pH for *C. thermoautotrophicum* is 5.7, whereas that for *C. thermoaceticum* is over 7. Thus *C. thermoautotrophicum* would perhaps be a better choice from which to select an acidophilic acetic acid producer than *C. thermoaceticum,* the choice of Schwartz and Keller (1982).

The medium for *C. thermoautotrophicum* is that described by Braun et al. (1979), who worked out a procedure for the isolation of acetogenic bacteria. Yeast extracts appears to be necessary for high growth yields.

The synthesis of acetate from CO_2 apparently occurs via the pathway proposed for *C. thermoaceticum*. This is suggested by the presence in *C. thermoautotrophicum* of formate dehydrogenase, carbon monoxide dehydrogenase, the tetrahydrofolate enzymes, and high levels of corrinoids (Clark et al., 1982). Extracts from *C. thermoautotrophicum* also catalyze a pyruvate-dependent acetate formation from methyltetrahydrofolate.

Wiegel (1982; personal communication) has carried out fermentations using CH_3OH/CO_2, ^{14}CO, $^{14}CO/CO_2$, or $^{14}CO/CO_2/H_2$ at pH 5.7 and 60°C. The fastest growth rate was obtained with a methanol concentration of 0.5%; however, growth was observed also with 6.5%. Clearly, *C. thermoautotrophicum* is quite tolerant to methanol. With ^{14}CO as the only carbon substrate, Wiegel showed that a total synthesis of acetate from CO occurs. Thus the acetate formed had a ^{14}C-specific activity per mol that was twice that of the original ^{14}CO. He also showed, with a mixture of $^{14}CO/H_2$, that the hydrogen was used as a source of electrons, which indicates the feasibility of acetate formation according to reaction (10). In the fermentation with $^{14}CO/CO_2$ and $^{14}CO/CO_2/H_2$, the specific activities of the produced acetate were lower than when ^{14}CO was the lone carbon source. The results may indicate either that CO_2 may directly be used for acetate synthesis or that an isotope equilibrium between CO and CO_2 occurs or both. An isotope exchange between CO and CO_2 may be catalyzed by carbon monoxide dehydrogenase. Drake et al. (1981) and Hu et al. (1982) have shown that enzymes from *C. thermoaceticum* catalyze a synthesis of acetyl CoA from CO and methyltetrahydrofolate according to reaction (22):

$$CO + CH_3H_4\text{folate} + \text{CoA} \xrightarrow{\text{ATP}} CH_3COCoA + H_4\text{folate} \quad (22)$$

It may be assumed that a similar reaction occurs also in *C. thermoautotrophicum*. One would then expect CO to be the precursor of the carboxyl group and CO_2 to be the precursor of the methyl group of acetate. The latter group would be formed from CO_2 via formate- and tetrahydrofolate-dependent reactions (Ljungdahl and Wood, 1982; Clark et al., 1982). The results obtained by J. Wiegel (personal communication) seem to support these suggestions.

6. Acetogenium Kivui. The number of hydrogen-oxidizing acetogenic bacteria was recently enlarged by the isolation of *A. kivui* (Leigh et al., 1981). It is a non-spore-forming, nonmotile, thermophilic, and anaerobic rod. It grows optimally at pH 6.4 and at 66°C and forms acetate from H_2/CO_2, formate, fructose, glucose, mannose, and pyruvate. A defined mineral medium is sufficient. However, yeast extract or trypticase increases the cell yield. The medium must be reduced by cysteine or sulfide, which cannot be replaced by thioglycolate or dithiotreitol. The type strain of *A. kivui* has been deposited in the American Type Culture Collection (ATCC 33488) and in the German Collection (DSM 2030).

B. Pathway of Acetate Synthesis from CO_2

With the risk of being considered biased, I would like to state that what we know about synthesis of acetate from CO_2 has come from work with *C. thermoaceticum*. There are two good reasons for this. The first reason is simply that the only homoacetate-fermenting organism that was available from 1948 to 1967 was *C. thermoaceticum*. The second reason is Harland G. Wood, who has the vision, curiosity, and enthusiasm to find and examine seemingly odd but important biological phenomena, one of them being how *C. thermoaceticum* carries out a synthesis of acetate from CO_2. Consequently, this section will mostly deal with the pathway of acetate synthesis in *C. thermoaceticum*. However, it can be quite safely said that evidence so far obtained indicates the same pathway with minor modifications also in the other acetogenic bacteria.

The general concept of the fermentation of glucose and the synthesis of acetate in *C. thermoaceticum* is shown in Fig. 7.1. It is a rather complicated fermentation; for the sake of discussion, the fermentation will be divided up in sections as follows:

1. Hexose (glucose) → pyruvate.
2. Pyruvate → acetate + CO_2 with formation of reduced ferredoxin (left side in Fig. 7.1).
3. Formate dehydrogenase.
4. Carbon monoxide dehydrogenase.
5. Tetrahydrofolate enzymes.
6. Acetate formation from methyltetrahydrofolate, and pyruvate or carbon monoxide.

1. From Hexose to Pyruvate. All acetogenic bacteria discussed in this chapter carry out a homoacetate fermentation of fructose (Table 7.2); some, in addition, ferment glucose and galactose. *C. thermoaceticum* and some strains of *C. thermoautotrophicum* ferment xylose. Ribose is fermented by *C. formicoaceticum* and *C. aceticum*. The fermentation pathway of xylose in *C. thermoaceticum* has not been investigated. Isotope tracer studies using $^{14}CO_2$ with *C. thermoaceticum* (Ljungdahl and Wood, 1965) indicate the presence of transketolase and transaldolase. However, these enzymes, which may be involved in the fermentation of xylose, have not been demonstrated in *C. thermoaceticum*.

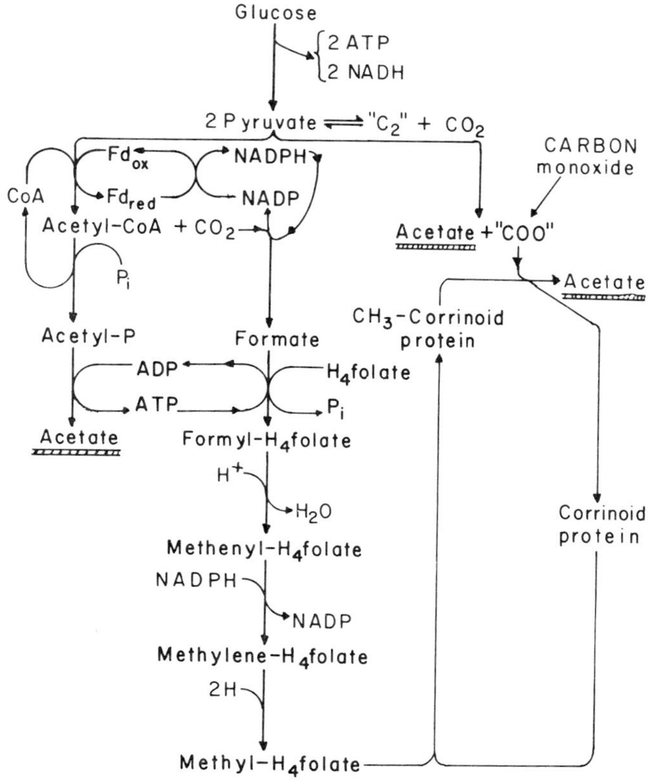

FIGURE 7.1 Proposed pathway of fermentation of glucose to 3 mol of acetate by *C. thermoaceticum.*

The fermentation of glucose in *C. thermoaceticum* proceeds via the Embden-Meyerhof pathway. This has been ascertained with the use of 1-^{14}C, 2-^{14}C, and 3,4-^{14}C-glucose (Wood, 1952b; Lentz, 1956). According to this pathway, the glucose is split between carbons 3 and 4 to yield two C_3 units, which are converted to pyruvate. Further metabolism of the pyruvate leads to acetate from glucose carbons 1, 2, and 5, 6, whereas carbons 3 and 4 yield one-carbon units, which are converted to acetate. None of the enzymes of the Embden-Meyerhof pathway has been proven in *C. thermoaceticum,* with the exception of glyceraldehyde phosphate dehydrogenase, which is NAD-specific (Thauer, 1972). The net result of the conversion of glucose to pyruvate is shown in reaction (23):

$$C_6H_{12}O_6 + 2NAD^+ + 2ADP + 2P_i \rightarrow 2CH_3COCOOH + 2NADH + 2H^+ + 2ATP \quad (23)$$

Fermentations of ^{14}C-labeled fructose with *A. woodii* (Winter and Wolfe, 1980) and with *C. formicoaceticum* (Linke, 1969a; O'Brien and Ljungdahl, 1972; Andreesen and Gottschalk, 1969) yielded results indicating that the Embden-Meyerhof pathway is also used in these organisms. In addition, all the enzymes of this pathway have been demonstrated in *C. formicoaceticum* (Andreesen et al., 1970).

2. Pyruvate Metabolism. According to Fig. 7.1 pyruvate is metabolized by two different mechanisms. On the right-hand side of Fig. 7.1, it can be seen that carbons 2 and 3 of pyruvate are converted to acetate and carbon 1 to a "COO" unit, which in a reaction with methyltetrahydrofolate, most likely mediated by a corrinoid–protein enzyme system, forms acetate (Drake et al., 1981). This latter reaction has been referred to as a "transcarboxylation" (Schulman et al., 1973), and it will be considered later. According to the left-hand side of Fig. 7.1, part of the pyruvate is metabolized via a pyruvate–ferredoxin oxidoreductase, transacetylase, and acetate kinase enzyme system according to reaction (24), in which Fd designates ferredoxin:

$$CH_3COCOOH + Fd_{ox} + ADP + P_i \xrightarrow{CoA} CH_3COOH + CO_2 + Fd_{red} + ATP \qquad (24)$$

The components needed to catalyze reaction (24) have been purified. Two different ferredoxins were noted by J. LeGall (personal communication, 1975) in *C. thermoaceticum*. These have been purified. Ferredoxin I (Yang et al., 1977) has one $[Fe_4S_4]$ cluster, whereas Ferredoxin II (Elliott and Ljungdahl, 1982a) has two such clusters. Ferredoxin I, which has been sequenced (Elliott et al., 1982), functions in the pyruvate–ferredoxin oxidoreductase reaction (Yang et al., 1977), and the resulting reduced ferredoxin is used for the reduction of $NADP^+$ by a reduced ferredoxin–NADP reductase (Thauer, 1972; Yang et al., 1977). NADPH may then be used for the reduction of CO_2 to formate.

Extracts of *C. thermoaceticum* catalyze an exchange reaction between CO_2 and pyruvate according to reaction (25) (Ljungdahl and Wood, 1963):

$$CH_3COCOOH + {}^{14}CO_2 \rightarrow CH_3CO^{14}COOH + CO_2 \qquad (25)$$

This reaction depends on catalytic amounts of CoA, thiamine pyrophosphate, and divalent cations such as Mn^{2+} or Mg^{2+}. An oxygen-sensitive protein with M_r around 250,000 catalyzes the reaction. When it is incubated with [2-^{14}C-pyruvate] and thiamine pyrophosphate, radioactive hydroxyethyl–thiamine pyrophosphate is formed (Schaupp and Ljungdahl, 1971). Recently, Drake et al. (1981) isolated a similar protein fraction (F_4), which also catalyzes a pyruvate–CO_2 exchange (reaction 25) and a cleavage of pyruvate to form acetyl CoA when flavin adenine dinucleotide (FAD) is added as an electron acceptor. Whether one of the ferredoxins would work as an electron acceptor instead of FAD has apparently not been tested.

Drake et al. (1981) has also isolated phosphotransacetylase from *C. thermoaceticum* (fraction F_1). This enzyme is stimulated by Mn^{2+}. Schaupp and Ljungdahl (1974) have studied the acetate kinase in *C. thermoaceticum*. This enzyme is apparently allosterically affected by ATP, a result that indicates that the acetate synthesis is controlled.

3. Formate Dehydrogenase. The first step to form the methyl group of acetate from CO_2 is the reduction of CO_2 to formate (Lentz and Wood, 1955). *C. thermoaceticum* contains an NADP-dependent formate dehydrogenase, which catalyzes the reversible reaction (26) (Thauer, 1972; Ljungdahl and Wood, 1963; Andreesen and Ljungdahl, 1974; Yamamoto et al., 1983):

$$CO_2 + NADPH \rightarrow HCOO^- + NADP^+$$

or (26)

$$HCO_3^- + NADPH + H^+ \rightarrow HCOO^- + NADP^+ + H_2O$$

It is not known whether CO_2 or HCO_3^- is the substrate. However, Thauer et al. (1975) showed that the ferredoxin-dependent formate dehydrogenase of *Clostridium pasteurianum* utilizes CO_2 as the substrate.

Nutritional studies (Andreesen et al., 1973) showed that the formation of formate dehydrogenase in *C. thermoaceticum* depends on the presence of either tungstate or molybdate, and selenite as well as ferrous salts in the medium. The formation of formate dehydrogenases in other anaerobic bacteria (Ljungdahl, 1980), including *C. formicoaceticum* (Andreesen et al., 1974) and *A. woodii* (Schobert, 1977), is similarly affected by these chemicals. With the use of ^{75}Se-selenite and ^{185}W-tungstate, the incorporation of selenium and tungsten into the formate dehydrogenase of *C. thermoaceticum* has been demonstrated (Ljungdahl and Andreesen, 1976; Andreesen and Ljungdahl, 1973).

Several attempts have been made to purify the formate dehydrogenase from *C. thermoaceticum*. (Saiki et al., 1981; Ljungdahl and Andreesen, 1978; Andreesen and Ljungdahl, 1974). They were only partially successful, probably owing to an irreversible inactivation of the enzyme by O_2. An apparent $K_i = 7.6 \times 10^{-6}$ M for O_2 has been observed (Yamamoto et al., 1982, 1983). Recently, the enzyme has been obtained as a homogenous protein (Yamamoto et al., 1983). It has $M_r = 340{,}000$ and consists of two each of two different subunits, giving the composition $\alpha_2\beta_2$. The larger subunit (α) and the smaller subunit (β) have $M_r = 96{,}000$ and 76,000, respectively. The enzyme with a specific activity of 1100 μmol min^{-1} mg^{-1}, measured in the direction of formate oxidation, most likely contains 2 g-at of W, 2 g-at of Se, and about 36 g-at of Fe per mol. The selenium is in the larger subunits in the form of selenocysteine. The enzyme was grown in a medium containing both tungsten and molybdenum. However, the pure enzyme has less than 0.1 g-at of Mo per mol, and molybdenum is clearly not an important component of the enzyme. The formate dehydrogenase from *C. thermoaceticum* represents the first example of a naturally occurring active tungsten enzyme.

The purified enzyme catalyzes the reduction of CO_2 with NADPH, without any NADPH-regenerating system (Yamamoto et al., 1983), which have been used in previous studies to demonstrate the reversibility of reaction (26) (Thauer, 1972). An apparent $K_{eq} = [HCOO^-][NADP^+]/[HCO_3^-][NADPH] = 1.48 \times 10^{-3}$ has been obtained at pH 7.5 and 55°C. Andreesen and Ljungdahl (1974), with a partially purified preparation of formate dehydrogenase from *C. thermoaceticum,* observed that it catalyzed an oxidation of NADPH with the flavin nucleotides FMN and FAD at a rate equal to or better than that of the formate dehydrogenase activity. The purified enzyme also catalyzes these reactions, but at a rate less than 1% of that of the formate dehydrogenase activity. The previous high activities with FMN and FAD were clearly due to the presence of contaminating enzymes in the formate dehydrogenase preparations.

4. Carbon Monoxide Dehydrogenase. Carbon monoxide dehydrogenase catalyzes reaction (27):

$$CO + H_2O \rightarrow CO_2 + 2H^+ + 2e \qquad (27)$$

The bacterial oxidation of CO was noted almost 80 years ago (see the review by Schlegel and Meyer, 1981). Rather recently, Diekert and Thauer (1978) reported that *C. thermoaceticum* and *C. formicoaceticum* oxidize CO and that this occurred with growing cultures fermenting glucose or fructose. With ^{14}CO, most of the label was recovered in CO_2, and a smaller amount in acetate. Growth was not observed on CO alone; the CO was co-metabolized in fermentations of carbohydrates or pyruvate. However, Lynd et al. (1982) have reported that *C. thermoaceticum* can grow on CO. That CO alone can support growth has been demonstrated with *E. limosum* (Genthner, 1982) and *Butyribacterium methylotrophicum* (Lynd et al., 1982).

Cell extracts of the two acetogenic bacteria oxidized CO with methyl viologen, benzyl viologen, triquat, and methylene blue as electron acceptors, whereas NAD, NADP, and *C. pasteurianum* ferredoxin did not work. Drake et al. (1980), with a partly purified preparation of the *C. thermoaceticum* enzyme, reported that *C. thermoaceticum* ferredoxin (most likely ferredoxin I (Yang et al., 1977)), *Desulfovibrio vulgaris* cytochrome C_3, FMN, and membrane-bound *b*-type cytochrome (Gottwald et al., 1975) from *C. thermoaceticum* also worked as electron acceptors.

In the absence of CO, the enzyme was reversibly inactivated by CN^-. This inactivation was not seen in the presence of CO. Methyl iodide and other alkyl halides in the presence of CO also inactivated the enzyme. This inactivation was reversed by light treatment, which led to a speculation that the enzyme was perhaps Vitamin B_{12}-dependent (Diekert and Thauer,

1978). However, Drake et al. (1980), with the partly purified preparation of CO dehydrogenase, did not observe the inhibition by alkyl halides. On the other hand, Ragsdale et al. (1983) who recently purified the CO dehydrogenase from *C. thermoaceticum* to homogeneity, found inhibition with methyl iodide, which was not reversible by light. The methyl iodide reaction and its relation to the unknown chromophore of the CO dehydrogenase will perhaps be explained when more is known about the chromophore.

There is a strong possibility that this chromophore contains nickel. Diekert and Thauer (1980), Drake et al. (1980), and Diekert and Ritter (1982) have presented evidence for this, and Drake et al. (1980), using ^{63}Ni, found it to be incorporated into the protein fraction having CO dehydrogenase activity. Nickel was also found in the homogenous preparation of the enzyme, and more recent studies show that nickel is redox active and binds CO (Ragsdale et al., 1982, 1983).

Ragsdale et al. (1983) have determined several properties of the homogeneous CO dehydrogenase that has a specific activity of 700 using methyl viologen as acceptor. Its M_r is apparently 440,000. It consists of three each of two different subunits with M_r = 76,000 and 72,000. Thus the composition is $\alpha_3\beta_3$. The hexameric enzyme contains 3 Ni, 6 Zn, and 36 Fe per molecule. The product of oxidation of CO is CO_2. Free formate is not formed, and there is no indication that formate is a tightly bound but dissociable intermediate. Pyruvate, Acetyl CoA, CoA, and ATP do not effect the apparent K_m for methyl viologen or the apparent V_{max} of the pure enzyme.

Hu et al. (1982) and Drake et al. (1981) have found that CO, in addition to the carboxyl group of pyruvate (Schulman et al., 1973), is a precursor of the carboxyl group of acetate. Carbon monoxide dehydrogenase is a part of the enzyme system that forms the acetate. These reactions will be discussed later.

5. Tetrahydrofolate Enzymes. Methyltetrahydrofolate and methylcorrinoids are precursors of the methyl group of acetate (Drake et al., 1981; Ljungdahl et al., 1966; Poston et al., 1966; Ghambeer et al., 1971). The evidence for this has been discussed at length in a recent review (Ljungdahl and Wood, 1982). The formation of methyltetrahydrofolate from formate is catalyzed by a sequence of reactions involving formyltetrahydrofolate synthetase (reaction 28), methenyltetrahydrofolate cyclohydrolase (reaction 29), methylenetetrahydrofolate dehydrogenase (reaction 30), and methylenetetrahydrofolate reductase (reaction 31):

$$\text{HCOOH} + \text{ATP} + \text{H}_4\text{folate} \rightarrow \text{HCO–H}_4\text{folate} + \text{ADP} + \text{P}_i \quad (28)$$

$$\text{HCO–H}_4\text{folate} + \text{H}^+ \longrightarrow \text{CH–H}_4\text{folate}^+ + \text{H}_2\text{O} \quad (29)$$

$$\text{CH–H}_4\text{folate}^+ + \text{NADPH} \longrightarrow \text{CH}_2\text{–H}_4\text{folate} + \text{NADP}^+ \quad (30)$$

$$\text{CH}_2\text{–H}_4\text{folate} + \text{XH}_2 \longrightarrow \text{CH}_3\text{–H}_4\text{folate} + \text{X} \quad (31)$$

Extracts of the acetogenic bacteria *C. thermoaceticum* (Andreesen et al., 1973), *C. formicoaceticum* (O'Brien and Ljungdahl, 1972), *A. woodii* (Tanner et al., 1978), and *C. thermoautotrophicum* (Clark et al., 1982) have high activities of these enzymes (Table 7.4). This indicates the importance of the tetrahydrofolate enzymes in the acetogenic bacteria.

Formyltetrahydrofolate synthetases have been purified from *C. thermoaceticum* (Ljungdahl et al., 1970) and *C. formicoaceticum* (O'Brien et al., 1976). The enzymes are very similar. They have the same $M_r = 240{,}000$ and consist of four identical subunits. Many properties of the *C. thermoaceticum* enzyme have been studied, most recently the quarternary structure, using electron microscopy (Mayer et al., 1982), and the feature that a tyrosine residue is located in the binding site of tetrahydrofolate (Elliott and Ljungdahl, 1982b).

A single protein from *C. thermoaceticum* catalyzes both the conversion of formyltetrahydrofolate to methenyltetrahydrofolate (reaction 29) and the methylenetetrahydrofolate (reaction 30) (Ljungdahl et al., 1980). These activities in *C. formicoaceticum* and *A. woodii* are catalyzed by separate proteins (Ljungdahl et al., 1980; Clark and Ljungdahl, 1982). The enzyme that catalyzes reaction (31) has recently been purified from *C. formicoaceticum* (J. E. Clark, personal communication). The nature of electron acceptor X is still unknown; however, rubredoxin functions as an acceptor in the reaction.

6. Acetate Formation from Methyltetrahydrofolate, Pyruvate, and Carbon Monoxide. The most impressive recent work on the acetate synthesis from CO_2 is that published by Drake et al. (1981). They found that, in *C. thermoaceticum,* the five protein fractions F_1, F_2, F_3, F_4, and ferredoxin are required for the formation of acetyl-phosphate from pyruvate and methyltetrahydrofolate (reaction 32):

$$CH_3COCOOH + 2HPO_4^{2-} + CH_3\text{–}H_4\text{folate} \rightarrow 2CHCOPO_4^{2-} + H_4\text{folate} + H_2O \quad (32)$$

Fractions F_1, F_2, and F_4 were purified to homogeneity. As was discussed earlier, F_1 has been identified as phosphotransacetylase, and F_4 as pyruvate–ferredoxin oxidoreductase. The F_2 fraction is considered a methyltransferase. It is a dimer with $M_r = 58{,}800$. It may transfer the methyl group of methyltetrahydrofolate to a corrinoid protein to form a Co-methylcorrinoid protein. Fraction F_3 contains a mixture of proteins, one of them almost certainly the corrinoid protein. A second protein in the F_3 fraction is carbon monoxide dehydrogenase. To complete reaction (32), all five protein components are required as well as CoA, ATP, and TPP. When CoA was added in substrate amounts, the requirement for fraction F_1 disappeared, and instead of acetyl-phosphate the product was acetyl-CoA. This is, of course, according to the assignment of fraction F_1 as phosphotransacetylase.

In a later paper, Hu et al. (1982) reported that fractions F_2 and F_3

TABLE 7.4 Levels of Tetrahydrofolate Enzymes in Extracts of Homoacetate-Fermenting Bacteria

Organism	*Growth Condition*	*Formyl–H_4folate Synthetase Specific Activity*[1]	*Methenyl–H_4folate Cyclohydrolase Specific Activity*[1]	*Methylene–H_4folate Dehydrogenase Specific Activity*[1,2]	*Methylene–H_4folate Reductase Specific Activity*[1]	*Reference*
C. thermoaceticum	Glucose + CO_2	7–18	0.7–1.9	1–2.6	0.011–0.055	Andreesen et al. (1973)
C. formicoaceticum	Fructose + CO_2	8.7–10	2–4	8.1–39.2	0.016	O'Brien and Ljungdahl (1972)
A. woodii	H_2 + CO_2	9.0–12.7	0.59–1.07	0.59–1.27	n.d.	Tanner et al. (1978)
A. woodii	Fructose + N_2	8.8–9.2	0.24–0.36	0.36–0.81	n.d.	Tanner et al. (1978)
C. thermoautotrophicum	H_2 + CO_2	10.8	0.49	2.03	n.d.	Clark et al. (1982)
C. thermoautotrophicum	Glucose + CO_2	11.1	0.63	0.82	n.d.	Clark et al. (1982)
C. thermoautotrophicum	Methanol + CO_2	7.7	0.94	3.10	n.d.	Clark et al. (1982)

Abbreviation: n.d. = not determined.
1. Specific activities are given in μmol of substrate converted min^{-1} mg^{-1} or protein.
2. Methylene–H_4folate dehydrogenase in *C. thermoaceticum* and *C. thermoautotrophicum* is NADP-dependent, whereas in *C. formicoaceticum* and *A. woodii,* this enzyme is NAD-dependent.

catalyze the synthesis of acetyl-CoA from CO and methyltetrahydrofolate (reaction 33):

$$CO + CH_3\text{–}H_4\text{folate} + CoA \xrightarrow{ATP} CH_3COCoA + H_4\text{folate} \quad (33)$$

CO and CoA are the precursors of the COCoA group of acetyl CoA, whereas the methyl group of methyltetrahydrofolate yields the methyl group of acetate.

Fraction F_3 alone catalyzes a reversible exchange between the carbonyl group of acetyl CoA and Co (Hu et al., 1982). This was shown by formation of ^{14}CO from $CH_3{}^{14}COCoA$. The exact mechanism of this exchange reaction is not yet known in detail. Hu et al. (1982) suggested that the carbon monoxide dehydrogenase and the corrinoid protein present in fraction F_3 are responsible for this exchange.

Since both the carboxyl group of pyruvate (Schulman et al., 1973; Drake et al., 1981) and CO (Drake et al., 1981; Hu et al., 1982) are precursors to the carboxyl group of acetate, it is logical to speculate that these compounds give rise to a common C_1 intermediate. This intermediate subsequently forms the carboxyl group of acetate. At the present time, the nature of the C_1 intermediate is not known. Hu et al. (1982) and Drake et al. (1981) favor it to be at the oxidation level of formate and have designated it [HCOOH]. However, the possibility may also exist that it can be more oxidized and be designated [COO]. This suggestion is based on the facts that CO_2 is the product of CO in the CO dehydrogenase reaction and that the carboxyl group of pyruvate exchanges with CO_2 and not with formate.

C. Electron Transfers and Energy Metabolism in Acetogenic Bacteria

1. Electron Transfer Components in Acetogenic Bacteria. The isolation of bacteria that grow chemolithotrophically by the reduction of CO_2 with H_2 to yield acetate indicates that acetate synthesis must be coupled with the generation of ATP. Indirect evidence also exists for this conclusion. The growth yields of *C. thermoaceticum* on glucose, fructose, and xylose are higher than were expected on the basis of phosphorylation obtained during glycolysis and pyruvate conversion to acetate and CO_2 (Andreesen et al., 1973). It was suggested that phosphorylation is coupled with electron transfer to reduce CO_2 to acetate. However, direct evidence does not exist for electron transport phosphorylation.

The acetogenic bacteria have several electron transfer proteins. *C. thermoaceticum* contains two different ferredoxins (Yang et al., 1977; Elliott and Ljungdahl, 1982a), two rubredoxins (Yang et al., 1980), cytochrome *b*, and menaquinone (Gottwald et al., 1975). *C. formicoaceticum* also has cytochrome *b* and menaquinone (Gottwald et al., 1975), and S. W. Ragsdale (personal communication) has recently isolated ferredoxin, rubredoxin, and flavodoxin from this bacterium.

It has already been discussed that *C. thermoaceticum* ferredoxin I functions in the final steps of acetate synthesis, in which the methyl group of methyltetrahydrofolate or a methylcorrinoid together with the carboxyl group of pyruvate forms acetate (Drake et al., 1981; Poston and Stadtman, 1967). It also functions with CO dehydrogenase (Drake et al., 1980). Whether or not ferredoxin II functions in these reactions is not known. Ferredoxin II appears to be associated with the cell membrane, whereas ferredoxin I is cytoplasmic (Elliott and Ljungdahl, 1982a). Both ferredoxins have similar oxidation–reduction potentials. However, compartmentalization and the fact that ferredoxin I has one iron cluster and ferredoxin II has two clusters suggest that the ferredoxins have different roles.

A function for rubredoxins in the metabolism of acetogenic bacteria is yet to be found. The two rubredoxins of *C. thermoaceticum* are irreversibly reduced by NADH and NADPH in a reaction that is catalyzed by a FMN-containing protein. This protein is referred to as rubredoxin reductase (Ljungdahl et al., 1981).

Cytochrome *b* from *C. thermoaceticum* has been shown to be reduced by CO in the CO dehydrogenase reaction (Drake et al., 1980). In *C. formicoaceticum*, the cytochrome *b* clearly functions in the reduction of fumarate to succinate (Dorn et al., 1978), and it is believed that this reduction is coupled with electron transport phosphorylation.

It is evident from the above discussion that electron transfer reactions are playing important roles in acetogenic bacteria. It can be predicted that these reactions are involved in the regulations of fermentations of carbohydrates to acetate and in fixation of CO_2 or CO to acetate. If the acetogenic bacteria are to be used in industrial processes, an understanding of electron transport and energy generation in these bacteria would be essential.

2. Hydrogenase and H_2 Cycling. The need for hydrogenase in acetogenic bacteria during growth on a CO_2/H_2 mixture is obvious, and hydrogenase has been demonstrated in *C. aceticum* (Braun and Gottschalk, 1981), *A. woodii* (Braun and Gottschalk, 1981), and *C. thermoaceticum* (Drake, 1982).

With *C. aceticum* and *A. woodii*, it has been shown that hydrogen is evolved during fermentations of fructose, glucose, and lactate (Braun and Gottschalk, 1981). The experiments conducted by Winter and Wolfe (1980), using *A. woodii* in co-culture with a *Methanobacterium*, strongly indicated the formation of hydrogen during fermentation of fructose. In these experiments, interspecies hydrogen transfer was clearly demonstrated.

Recently, it was proposed that an H_2 cycling system is present in *Desulfovibrio gigas* (Odom and Peck, 1981; LeGall et al., 1982). In such a system, H_2 is generated inside the cell by a cytoplasmic hydrogenase.

The molecular hydrogen passes through the cell membrane to the outside, where a second hydrogenase oxidizes it to protons. The oxidation generates electrons, which are transferred back into the cell and used to reduce SO_4^{2-}. The protons generated on the outside of the membrane may also enter the cell through a reverse ATPase enzyme system with concomitant formation of ATP.

The reduction of CO_2 to acetate by acetogenic bacteria is similar to that of the reduction of sulfate by sulfate-reducing bacteria. It is known that sulfate reduction involves ATP generation (LeGall et al., 1982; Peck, 1966). It does not seem too farfetched to postulate that an H_2 cycling mechanism may occur also in the acetogenic bacteria. Such a mechanism, if it occurs, would explain the high growth yield during fermentations of organic substrates. It may also explain energy generation during growth on CO_2/H_2.

III. CONCLUSION

Many bacteria produce acetate, but only the homoacetate-fermenting bacteria form acetate as a sole product in fermentations of carbohydrates. Thus these bacteria could be useful for industrial production of acetate from renewable sources. Unfortunately, they have limitations. All of the bacteria discussed in this paper ferment fructose; a few ferment glucose; only one, *C. thermoaceticum,* ferments xylose; and none of them ferment disaccharides, starch, cellulose, or xylan. The most available materials for fermentation are glucose and xylose, which can be obtained by hydrolysis from the polysaccharides. Since *C. thermoaceticum* is the only bacterium that ferments both glucose and xylose, it seems to be the bacterium of choice. *C. thermoaceticum* could perhaps be further improved by genetic means so that it could ferment disaccharides such as cellobiose and sucrose and perhaps also starch or cellulose.

Alternative substrates to carbohydrates would be one-carbon compounds, which also are converted to acetate by the acetogenic bacteria. Most of them grow on a mixture of CO_2 and H_2 and also on more reduced one-carbon compounds such as CO, formate, and methanol. Probably the most interesting substrate would be a mixture of CO and H_2 (syngas), which can be formed from carbon and water.

We have some appreciation of the pathway of acetate synthesis from studies using *C. thermoaceticum.* It has been discussed at some length in this paper. However, the recent discoveries that the acetogenic bacteria can grow on diverse one-carbon substrates have led to speculations that the present accepted pathway may have to be modified. For instance, there are the possibilities that carbon monoxide is an intermediate and that hydrogen is formed and reutilized. The latter possibility is quite exciting, since it appears to point toward a new concept of energy generation in anaerobic bacteria.

ACKNOWLEDGMENTS

Work on acetogenic bacteria at the University of Georgia is being supported by U.S. Public Health Service Grant AM27323 from the National Institute of Arthritis, Diabetes, Digestive and Kidney Diseases and by Contract DEAS09-79 ER10449 from the U.S. Department of Energy. Thanks are due to Jan Andreesen, Joan Clark, Shiu-Mei Hsu, Harry D. Peck, Jr., Steve Ragsdale, Jürgen Wiegel, Isamu Yamamoto, and J. G. Zeikus for valuable discussions and for supplying unpublished data.

REFERENCES

Andreesen, J. R., El Ghazzawi, E., and Gottschalk, G. (1974) *Arch. Microbiol. 96,* 103.

Andreesen, J. R., and Gottschalk, G. (1969) *Arch. Microbiol. 69,* 160.

Andreesen, J. R., Gottschalk, G., and Schlegel, H. G. (1970) *Arch. Microbiol. 72,* 154.

Andreesen, J. R., and Ljungdahl, L. G. (1973) *J. Bacteriol. 116,* 867.

Andreesen, J. R., and Ljungdahl, L. G. (1974) *J. Bacteriol. 120,* 6.

Andreesen, J. R., Schaupp, A., Neurauter, C., Brown, A., and Ljungdahl, L. G. (1973) *J. Bacteriol. 114,* 743.

Bache, R., and Pfennig, N. (1981) *Arch. Microbiol. 130,* 255.

Balch, W. E., Schoberth, S., Tanner, R. S., and Wolfe, R. S. (1977) *Int. J. Syst. Bacteriol. 27,* 355.

Barker, H. A. (1944) *Proc. Nat. Acad. Sci. U.S.A. 30,* 88.

Barker, H. A., and Beck, J. V. (1942) *J. Bacteriol. 43,* 291.

Barker, H. A., and Haas, V. (1944) *J. Bacteriol. 47,* 301.

Barker, H. A., and Kamen, M. D. (1945) *Proc. Nat. Acad. Sci. U.S.A. 31,* 219.

Barker, H. A., Volcani, B. E., and Cardon, B. P. (1948) *J. Biol. Chem. 176,* 803.

Braun, K. (1979) "Untersuchungen zum autotrophen, heterotrophen und mixotrophen Wachstum von *Acetobacterium woodii* and *Clostridium aceticum,*" Ph.D. dissertation, University of Göttingen, West Germany.

Braun, K., and Gottschalk, G. (1981) *Arch. Microbiol. 128,* 294.

Braun, M., Mayer, F., and Gottschalk, G. (1981) *Arch. Microbiol. 128,* 288.

Braun, M., Schoberth, S., and Gottschalk, G. (1979) *Arch. Microbiol. 120,* 201.

Cardon, B. P., and Barker, H. A. (1946) *J. Bacteriol. 52,* 699.

Clark, J. E., and Ljungdahl, L. G. (1982) *J. Biol. Chem. 257,* 3833.

Clark, J. E., Ragsdale, S. W., Ljungdahl, L. G., and Wiegel, J. (1982) *J. Bacteriol. 151,* 507.

Diekert, G., and Ritter, M. (1982) *J. Bacteriol. 151,* 1043.

Diekert, G. B., and Thauer, R. K. (1978) *J. Bacteriol. 136,* 597.

Diekert, G., and Thauer, R. K. (1980) *FEMS Microbiol. Lett. 7,* 187.

Dorn, M. (1976) "Vergärung von Fumarat and L-Malat durch *Clostridium formicoaceticum,*" Ph.D. dissertation, University of Göttingen, West Germany.

Dorn, M., Andreesen, J. R., and Gottschalk, G. (1978) *J. Bacteriol. 133,* 26.
Drake, H. L. (1982) *J. Bacteriol. 150,* 702.
Drake, H. L., Hu, S.-I., and Wood, H. G. (1980) *J. Biol. Chem. 255,* 7174.
Drake, H. L., Hu, S.-I., and Wood, H. G. (1981) *J. Biol. Chem. 256,* 11,137.
Dürre, P. (1981) "Selenabhängige Vergärung von Purinen und Glycin durch *Clostridium purinolyticum* and *Peptococcus glycinophilus,*" Ph.D. dissertation. University of Göttingen, West Germany.
Dürre, P., Andersch, W., and Andreesen, J. R. (1981) *Int. J. Syst. Bacteriol. 31,* 184.
Dürre, P., and Andreesen, J. R. (1982) *Abstracts of the Annual Meeting of the American Society for Microbiology,* K38, American Society for Microbiology, Washington, D.C.
El Ghazzawi, E. (1967) *Arch. Mikrobiol. 57,* 1.
Elliott, J. I., and Ljungdahl, L. G. (1982a) *J. Bacteriol. 151,* 328.
Elliott, J. I., and Ljungdahl, L. G. (1982b) *Arch. Biochem. Biophys. 215,* 245.
Elliott, J. I., Yang, S.-S., Ljungdahl, L. G., Travis, J., and Reilly, C. F. (1982) *Biochemistry 21,* 3294.
Fischer, F., Lieske, R., and Winzer, K. (1932) *Biochem. Zeitschr. 245,* 2.
Fontaine, F. E., Peterson, W. H., McCoy, E., Johnson, M. J., and Ritter, G. J. (1942) *J. Bacteriol. 43,* 701.
Genthner, B. R. S., and Bryant, M. P. (1982) *Appl. Environ. Microbiol. 43,* 70.
Genthner, B. R. S., Davis, C. L., and Bryant, M. P. (1981) *Appl. Environ. Microbiol. 42,* 12.
Ghambeer, R. K., Wood, H. G., Schulman, M., and Ljungdahl, L. (1971) *Arch. Biochem. Biophys. 143,* 471.
Gottwald, M., Andreesen, J. R., LeGall, J., and Ljungdahl, L. G. (1975) *J. Bacteriol. 122,* 325.
Hu, S.-I., Drake, H. L., and Wood, H. G. (1982) *J. Bacteriol. 149,* 440.
Karlsson, J. L., and Barker, H. A. (1949) *J. Biol. Chem. 178,* 891.
Karlsson, J. L., Volcani, B. E., and Barker, H. A. (1948) *J. Bacteriol. 56,* 781.
Klein, S. M., and Sagers, R. D. (1966) *J. Biol. Chem. 241,* 197.
LeGall, J., Moura, J. J. G., Peck, H. D., Jr., and Xavier, A. V. (1982) in *Iron Sulfur Proteins* Vol. IV (Spiro, T., ed.) p. 177, John Wiley, New York.
Leigh, J. A., Mayer, F., and Wolfe, R. S. (1981) *Arch. Microbiol. 129,* 275.
Lentz, K. E. (1956) "Studies on the synthesis of acetate by *Clostridium thermoaceticum,*" Ph.D. dissertation, Western Reserve University, Cleveland, Ohio.
Lentz, K., and Wood, H. G. (1955) *J. Biol. Chem. 215,* 645.
Leonhardt, U., and Andreesen, J. R. (1977) *Arch. Microbiol. 115,* 277.
Linke, H. A. B. (1969a) *Zentralbl. Bakteriol. Parasitenk, Infektionskr. Hyg. Abt. 2, 123,* 369.
Linke, H. A. B. (1969b) *Arch. Mikrobiol. 64,* 203.
Ljungdahl, L. G. (1980) in *Molybdenum and Molybdenum Containing Enzymes* (Coughlan, M., ed.) p. 463, Pergamon Press, New York.
Ljungdahl, L. G., and Andreesen, J. R. (1976) in *Microbial Production and Utilization of Gases* (Schlegel, H. G., Gottschalk, G., and Pfennig, N., eds.) p. 151, E. Goltze, Göttingen, West Germany.
Ljungdahl, L. G., and Andreesen, J. R. (1978) *Meth. Enzymol. 53,* 360.
Ljungdahl, L. G., Brewer, J. M., Neese, S. H., and Fairwell, T. (1970) *J. Biol. Chem. 245,* 4791.

Ljungdahl, L. G., Bryant, F., Carreira, L., Saiki, T., and Wiegel, J. (1981) in *Trends in the Biological Fermentations for Fuels and Chemicals* (Hollaender, A., ed.) p. 397, Plenum, New York.
Ljungdahl, L., Irion, E., and Wood, H. G. (1966) *Fed. Proc. 25,* 1642.
Ljungdahl, L. G., O'Brien, W. E., Moore, M. R., and Liu, M.-T. (1980) *Meth. Enzymol. 66,* 599.
Ljungdahl, L., and Wood, H. G. (1963) *Bacteriol. Proc.* Abstr. P58.
Ljungdahl, L., and Wood, H. G. (1965) *J. Bacteriol. 89,* 1055.
Ljungdahl, L. G., and Wood, H. G. (1969) *Ann. Rev. Microbiol. 23,* 515.
Ljungdahl, L. G., and Wood, H. G. (1982) in *B12* Vol. II (Dolphin, D., ed.) p. 165, John Wiley, New York.
Lynd, L., Kerby, R., and Zeikus, J. G. (1982) *J. Bacteriol. 149,* 255.
Matteuzzi, D., Hollaus, F., and Biavati, B. (1978) *Int. J. Syst. Bacteriol. 28,* 528.
Mayer, F., Elliott, J. I., Sherod, D., and Ljungdahl, L. G. (1982) *J. Eur. Biochem. 124,* 397.
Mayer, F., Lurz, R., and Schobert, S. (1977) *Arch. Microbiol. 115,* 207.
O'Brien, W. E., Brewer, J. M., and Ljungdahl, L. G. (1976) in *Enzymes and Proteins from Thermophilic Microorganisms* (Zuber, H., ed.) *Experientia Supplementum* Vol. XXVI, p. 249, Birkhäuser Verlag, Basel.
O'Brien, W. E., and Ljungdahl, L. G. (1972) *J. Bacteriol. 109,* 626.
Odom, J. M., and Peck, H. D., Jr. (1981) *FEMS Lett. 12,* 47.
Ohwaki, K., and Hungate, R. E. (1977) *Appl. Environ. Microbiol. 33,* 1270.
Peck, H. D., Jr. (1966) *Biochem. Biophys. Res. Comm. 22,* 112.
Poston, J. M., Kuratomi, K., and Stadtman, E. R. (1966) *J. Biol. Chem. 241,* 4209.
Poston, J. M., and Stadtman, E. R. (1967) *Biochem. Biophys. Res. Comm. 26,* 550.
Ragsdale, S. W., Clark, J. E., Ljungdahl, L. G., Lundie, L. L., and Drake, H. L. (1983) *J. Biol. Chem. 258,* 2364.
Ragsdale, S. W., Ljungdahl, L. G., and DerVartanian, D. V. (1982) *Biochem. Biophys. Res. Commun. 108,* 658.
Sagers, R. D., and Gunsalus, I. C. (1961) *J. Bacteriol. 81,* 541.
Saiki, T., Shackleford, G., and Ljungdahl, L. G. (1981) in *Selenium in Biology and Medicine* (Spallholz, J. E., Martin, J. L., and Ganther, H. E., eds.) p. 220, AVI Publishing Company, Westport, Conn.
Schaupp, A., and Ljungdahl, L. (1971) *Bacteriol. Proc. Abstr.* P258.
Schaupp, A., and Ljungdahl, L. G. (1974) *Arch. Microbiol. 100,* 121.
Schlegel, H. G., and Meyer, D. (1981) in *Microbial Growth on* C_1*-compounds* (Dalton, H., ed.) p. 105, Heydon and Son, London.
Schobert, S. (1977) *Arch. Microbiol. 114,* 143.
Schulman, M., Ghambeer, R. K., Ljungdahl, L. G., and Wood, H. G. (1973) *J. Biol. Chem. 248,* 6255.
Schulman, M., Parker, D., Ljungdahl, L. G., and Wood, H. G. (1972) *J. Bacteriol. 109,* 633.
Schwartz, R. D., and Keller, F. A., Jr. (1982) *Appl. Environ. Microbiol. 43,* 117.
Standard and Poor's Industry Surveys (1979) October 25.
Tanner, R. S., Wolfe, R. S., and Ljungdahl, L. G. (1978) *J. Bacteriol. 134,* 668.
Thauer, R. K. (1972) *FEBS Lett. 27,* 111.
Thauer, R. K., Kaufer, B., and Fuchs, G. (1975) *Eur. J. Biochem. 55,* 111.
Waber, L. J., and Wood, H. G. (1979) *J. Bacteriol. 140,* 468.

Wiegel, J. (1982) *Abstracts of the Annual Meeting of the American Society for Microbiology* I107, American Society for Microbiology, Washington, D.C.
Wiegel, J., Braun, M., and Gottschalk, G. (1981) *Curr. Microbiol. 5,* 255.
Wiegel, J., Ljungdahl, L. G., and Rawson, J. R. (1979) *J. Bacteriol. 139,* 800.
Wieringa, K. T. (1936) *Antonie van Leeuwenhoek 3,* 263.
Wieringa, K. T. (1940) *Antonie van Leeuwenhoek 6,* 251.
Winter, J. U., and Wolfe, R. S. (1980) *Arch. Microbiol. 124,* 73.
Wood, H. G. (1952a) *J. Biol. Chem. 194,* 905.
Wood, H. G. (1952b) *J. Biol. Chem. 199,* 579.
Yamamoto, I., Liu, S.-M., Saiki, T., and Ljungdahl, L. G. (1982) *Fed. Proc. 41,* 888, Abstr. 3615.
Yamamoto, I., Saiki, T., Liu, S.-M., and Ljungdahl, L. G. (1983) *J. Biol. Chem. 258,* 1821.
Yang, S.-S., Ljungdahl, L. G., DerVartanian, D. V., and Watt, G. D. (1980) *Biochim. Biophys. Acta 590,* 24.
Yang, S.-S., Ljungdahl, L. G., and LeGall, J. (1977) *J. Bacteriol. 130,* 1084.
Zeikus, J. G., Lynd, L. E., Thompson, T. E., Krzycki, J. A., Weimer, P. J., and Hegge, P. W. (1980) *Curr. J. Microbiol. 3,* 381.

CHAPTER

8

An Overview of Process Technology for the Production of Liquid Fuels and Chemical Feedstocks via Fermentation

Janice A. Phillips
Arthur E. Humphrey

I. INTRODUCTION

Lignocellulose is the major renewable alternative to the crude oil that forms the basis for our present fuel and chemicals infrastructure. Like crude oil, for which it is a potential substitute, and coal, its leading nonrenewable competitor, lignocellulose is a heterogeneous feedstock. The design of technically and economically viable processes for the conversion of this material must recognize not only the problems that the complexity of composition and structure of lignocellulose present, but also the potential.

Lignocellulose consists of three main components: cellulose, hemicellulose, and lignin. Cellulose, a polymer of β-D-1,4-linked anhydrous glucose units, comprises 40–60% of the cell wall material of trees and plants. The individual cellulose molecules are linked together to form elementary fibrils, which in turn are aggregated, via intermolecular hydrogen bonding, into larger subunit structures referred to as microfibrils. These microfibrils contain alternating sequences of highly ordered (or crystalline) and randomly oriented (or amorphous) phases and are imbedded in a matrix of hemicellulose (Chang et al., 1981; Fan et al., 1980; Rånby, 1969; Rowland, 1975; Shafizadeh and McGinniss, 1971; Wenzl, 1970; Wood, 1970). This latter carbohydrate component, comprising 20–50% of the plant dry weight, is a branched polymer of pentose sugars. The principal five-carbon monomeric units of this polypentose are xylose, arabinose, galactose, and rhamnose. The composition of a specific hemicellulose fraction depends on the source from which it is isolated; the hemicellulose of hardwoods consists mainly of xylose, in the form of glucuronoxylan, with trace amounts of galactose, arabinose, and mannose, while the hemicellulose of softwoods consists mainly of mannose, in the form of glucomannan (Timmell, 1964; Wenzl, 1970; Whistler and Richards, 1970). The cellulosic and hemicellulosic fractions are encrusted in an amorphous layer of lignin. This latter component, which accounts for the remaining 10–20% of the plant material, is a complex three-dimensional polymer formed by carbon–carbon or ether bonds between phenylpropane units; its composition is also a function of the source from which it is isolated (Goheen et al., 1963; Reddy and Forney, 1978; Sarkanen and Ludwig, 1971).

On a generalized basis, the idea of converting lignocellulosic materials to fuels and chemicals is not conceptually different from the present practice of refining crude oil. The first step, as shown in Fig. 8.1, is the fractionation or separation of the feedstock into its three main components. With further processing, each of these fractions can then be converted into fuels, chemicals, or feed and foodstuffs. The schemes in Figs. 8.1–8.3 are complex illustrations of the theoretical manner in which lignocellulose can be envisioned as a substitute for petroleum-based fuels and chemicals. Depending on the form in which it is extracted from the raw material, lignin can be

- burned to generate process energy (as is typically practiced in the pulp and paper industry);
- converted (via pyrolysis, hydrogenolysis, sulfonation, oxidation, etc.) to low-molecular-weight chemicals such as vanillin, dimethylsulfoxide, phenol and phenol derivatives, benzene and benzene derivatives, catechols, and methyl mercaptan;
- used in solution systems as a dispersant, emulsion stabilizer, complexing agent, precipitant, or coagulant;
- used in the manufacture of various polymeric materials such as thermosetting resins;
- used as a matrix for adsorption and ion exchange;
- or converted through carbonization and pyrolysis to active carbon or carbon or graphite filters (Falkehag, 1975; Glasser, 1981; Goheen et al., 1963; Goldstein, 1975, 1976; *Chemical & Engineering News*, 1975).

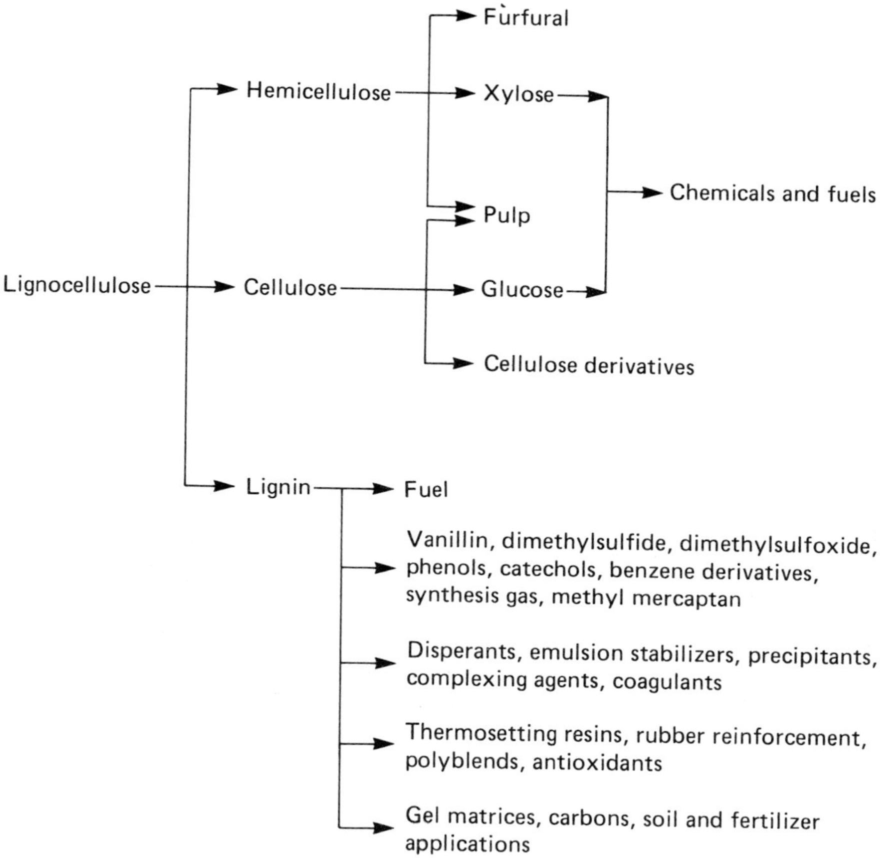

FIGURE 8.1 Proposed separation and use scheme for lignocellulose.

The cellulosic fraction, depending on its purity, strength, and fiber characteristics, can be marketed as pulp or converted to cellulose derivatives such as rayon, cellophane, cellulose acetate, carboxymethyl cellulose, and cellulose butyrate. Hemicellulose can be left as a component of the pulp to enhance the mechanical performance of the fibers during subsequent paper-making operations or can be completely degraded to furfural. From the standpoint of fuels and chemicals production, however, greater utility can be made of both the cellulose and hemicellulose if they are hydrolyzed to the soluble sugars glucose and xylose, respectively. As shown in Figs. 8.2 and 8.3, these sugars can then serve as feedstocks for a variety of fermentations from which commodity chemicals such as ethanol, acetic acid, acetone, butanol, 2,3-butanediol, and lactic acid can be produced; these chemicals can then serve as intermediates for the petrochemical, agricultural chemical, plastics, and pharmaceuticals industries. Both glucose and xylose can be fermented to single cell protein, a possible animal feed

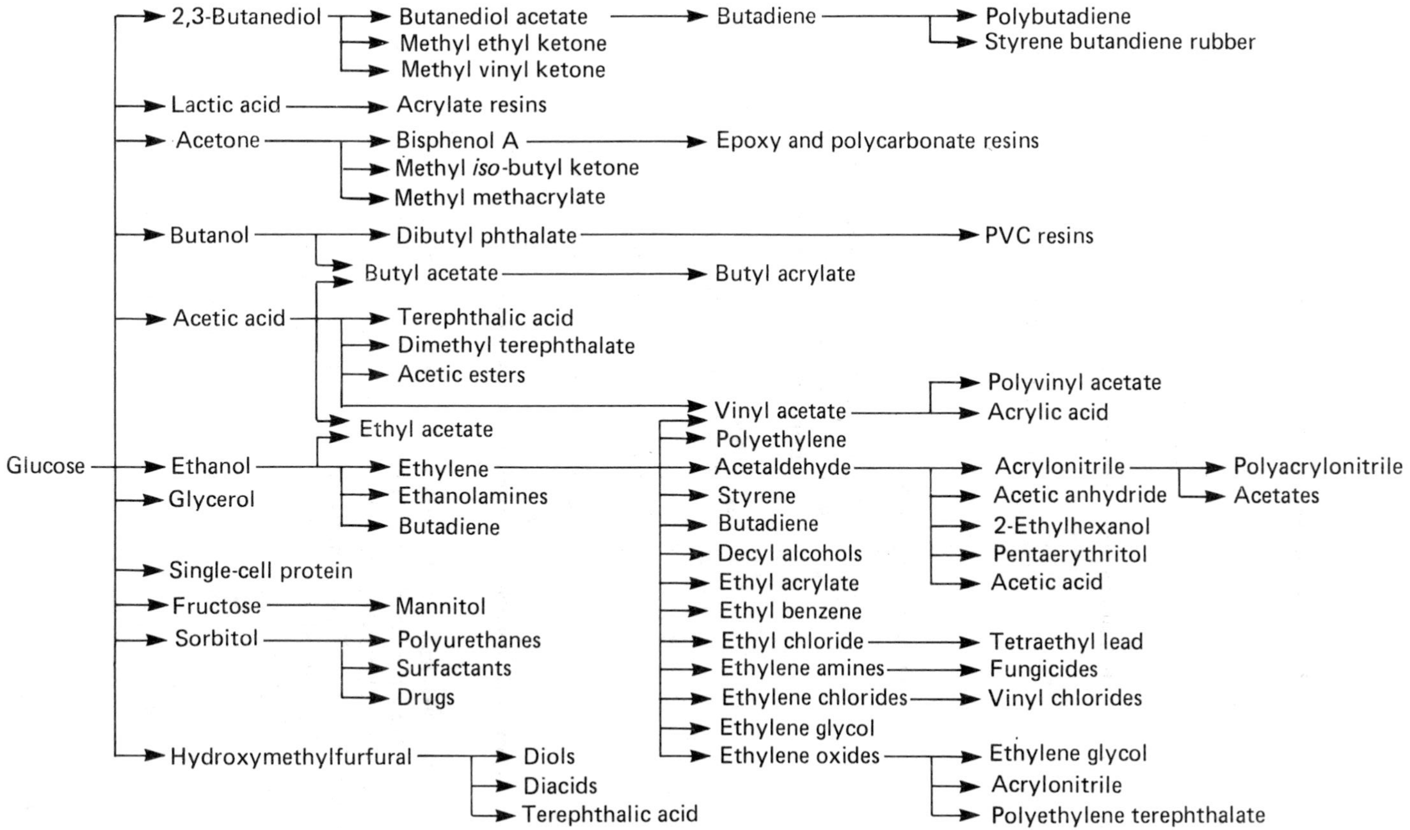

FIGURE 8.2 Chemicals and foods obtainable from glucose.

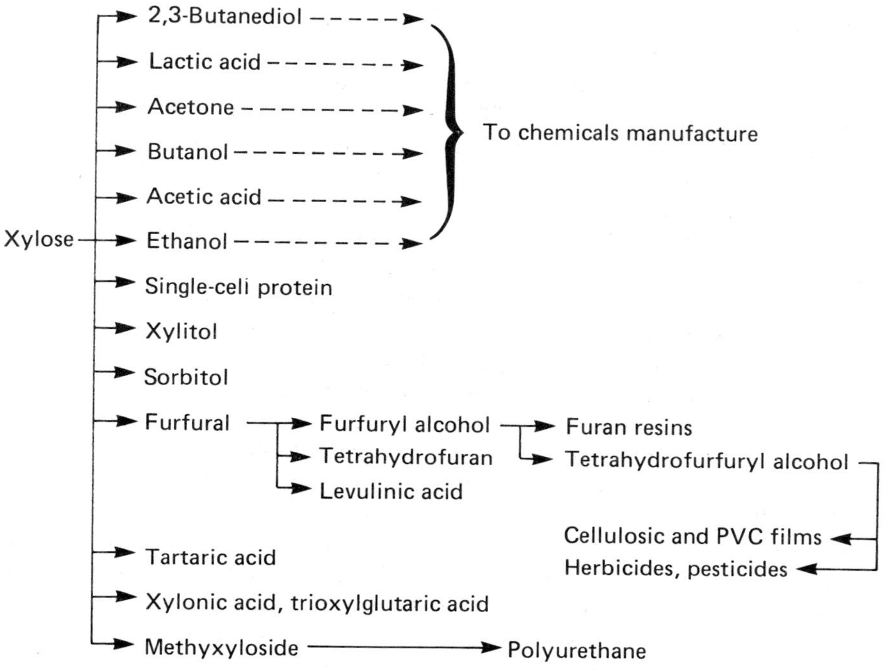

FIGURE 8.3 Chemicals and foods obtainable from xylose.

material. Glucose can be enzymatically isomerized to fructose, a major sweetener in the food-processing industry; chemically converted to sorbitol; or hydrolyzed to hydroxymethyl-furfural. Xylose can be hydrogenated to xylitol, a potentially attractive but as yet unexploited sweetening agent, or derivatized to compounds such as xylonic and trioxyglutaric acid and methylxyloside, which can be used in the manufacture of binders, sequestering agents, and polyurethanes. The economic viability of these suggested uses will depend on the competition from currently available sources of food, petroleum, and petroleum-derived compounds. Obviously, the cost of producing fuels and chemicals by fermentation routes will depend on the cost associated with the raw material and the efficiency of the conversion steps employed.

Sources and characteristics of lignocellulosic raw materials can be found in U.S. Department of Agriculture (1978), Anderson (1977), Office of Solid Waste (1975), Bartley et al. (1981), Burt (1978), Office of Technology Assessment (1980a, b), Sloneker (1976), Stephens and Heichel (1975), Stone (1976), Tillman (1977), and Tyner (1981).

The nature and composition of the raw material will influence the types of steps selected for the conversion facility, particularly for the initial step of fractionation. Lignocellulose will be available either as tree or grass crops specifically grown for feedstock purposes or as residues from the agricultural and forestry industries.

The current surface production rate of conventional forest crops that have been bred for their wood quality, the characteristic of primary importance in commercial timber production, is estimated at 2–4 tons/acre/year. A significant increase in this productivity could be achieved through the monoculture of selected tree species developed for short-rotation production, such as the hybrid poplar. This hardwood is an ideal candidate for energy farming: it grows rapidly (5–7 ft. annually); it is easily established with cuttings; and it is self-regenerating; after an initial 4 years of growth, the tree can be harvested biannually, new growth occurring by coppicing (resprouting from stumps). Average sustained yields of 10–15 dry tons of wood/acre/year with an average composition of 51% cellulose, 23% hemicellulose, 24% lignin, and 2% extractives have been predicted (Schoenmaker, 1980). The hybrid poplar can be grown on marginal cropland or noncommercial timberland generally considered unsuitable for agricultural purposes.

The alternative source of lignocellulose is organic solid waste. Approximately 1.3×10^9 dry tons of this material are generated annually in the United States. Most, as shown in Table 8.1, are residues from the agricultural and forest industries, manure from feedlots, and urban refuse (or municipal solids wastes (MSW)). Typical analyses of several of the agricultural wastes are given in Table 8.2; an average composition of MSW is given in Table 8.3. In general, these materials are highly variable in

TABLE 8.1 Quantities and Sources of Organic Wastes (10^6 tons/year)

Source	*Amount*
Agricultural crops and food wastes	400
Manure	260
Urban refuse (MSW)	220
Logging and wood manufacturing wastes	200
Industrial wastes	76
Municipal sewage solids	25
Miscellaneous organic wastes	90
Total	1271

TABLE 8.2 Compositional Analysis of Agricultural Residues

Residue	*Percent Cellulose*	*Percent Hemicellulose*	*Percent Lignin*	*Percent Extractives*	*Percent Ash*
Sugarcane bagasse	33	30	19	6	2
Wheat straw	30–33	22–28	14–18	3–7	9–11
Rice straw	32–37	19–24	9–13	4–5	12–18
Corn stover	34–36	16–24	15–19	2–6	2–4

TABLE 8.3 Average Composition of Municipal Solids Waste

Type	*Percent of Total*	*Percent Cellulose*
Paper products	35–45	80
Food wastes	12–15	20–30
Yard wastes	15–18	20–30
Plastics	1–5	0
Cloth, rubber, leather, and synthetics	1–2	0
Metal	8–10	0
Glass	6–10	0
Wood	0–12	40–50
Inerts	1–10	0

composition. The bulk of this waste material is not readily available as a feedstock; it is widely dispersed and, consequently, difficult to collect. Manure, the collection of which is easily centralized, is available as a relatively dilute material containing less than 10% solids. In addition, although classified as a waste product, much of this material has alternative uses. Most crop residues are plowed back into the soil to prevent erosion. Many of the wood residues from timber harvesting and lumber manufacturing are used for the manufacture of pulp, particle board, and hardboard.

Prices for these biomass materials are difficult to estimate. Projections for the cost of tree crops range from $10 to $50 per dry ton. Although most of the organic wastes presently have no production cost attached to them, charges for collection and transportation of these residues to a centralized processing location and the cost of their replacement in current agriculture and forestry practices could result in charges of $20 to $60 per dry ton.

II. CONVERSION STEPS

An integrated process for the production of fuels and chemicals via fermentation will include four major steps: (1) pretreatment, (2) hydrolysis, (3) fermentation, and (4) product recovery. These individual process steps may be carried out sequentially or combined to increase operating efficiency and reduce capital investment.

A. Pretreatment

The primary purpose of pretreatment is to alter the compositional and structural characteristics of the feedstock (Cowling and Brown, 1969) that hinder the access of hydrolytic agents to the carbohydrate by removing lignin and disrupting the hydrogen-bonding network between cellulose molecules. The pretreatment should be conducted under conditions that result in minimum degradation of the hemicellulose and cellulose and extract the lignin fraction in its most valuable form. In theory, a distinct separation of the individual wood components should be included as an objective, although this does not seem to be easily achievable in practice.

Pretreatments can generally be classified on the basis of their primary effect on the raw material as either physical or chemical. Physical pretreatments result in drastic reductions in macroscopic particle dimensions and changes in microscopic molecular configuration with little or no chemical modification of the lignin, hemicellulose, or cellulose. Chemical treatments, on the other hand, can result in extensive modification and even degradation of these components; they effect no changes in the gross particle size but can disrupt both the intracrystalline and intercrystalline hydrogen bonding of cellulose.

The principal physical pretreatments are milling and radiation treatment. Milling—ball milling, two-roll or compression milling, hammer milling, colloid milling, cryomilling, disc refining, and so on—decreases the particle size, crystallinity (Fan et al., 1981a, b; Millett et al., 1979; Ryu et al., 1982; Sasaki et al., 1979), and degree of polymerization (Ryu et al., 1982; Tassinari et al., 1980) and increases the surface area (Fan et al., 1981b; Ryu et al., 1982) and the bulk density of the raw material. Consequently, milling has been shown to be generally effective in increasing the susceptibility of a wide variety of materials to hydrolysis by acids and enzymes (Andren et al., 1976a; Mandels et al., 1974; Millet et al., 1979). The changes in physical properties and extent of enhancement in hydrolytic susceptibility obtained are dependent not only on the particular milling operation chosen (Fan et al., 1981b; Mandels et al., 1974; Pew and Weyna, 1962; Tassinari et al., 1980) and the operating conditions (Millett et al., 1975; Tassinari and Macy, 1977; Tassinari et al., 1980, 1982), but also on the characteristics of the feedstock (Millett et al., 1979; Ryu et al., 1982). For example, hardwoods are more responsive than softwoods to vibratory ball milling; among the hardwoods, the increase in susceptibility to acids and enzymes following milling is species-specific (Millett et al., 1975). Similar trends have been noted among agricultural residues, municipal and pulp mill wastes, and chemical pulps (Andren et al., 1976a). Most grinding operations are conducted on relatively dry feedstocks with moisture contents of less than 10%. The presence of water reduces the effectiveness of the milling; water exerts a plasticizing effect, reducing the rate at which particle size is reduced with little significant change in the crystallinity of the substrate. Optimum moisture content must be defined on the basis of the substrate being processed and the milling conditions (Tassinari et al., 1982). Since most lignocellulosic materials have an initial moisture content in excess of 50% at the time of harvest or collection, drying of the substrate prior to milling will be necessary. The effectiveness of wet milling can be significantly enhanced if simultaneous hydrolysis of the substrate is performed (Kelsey and Shafizadeh, 1980).

The operating conditions to be considered as relevant to the process result will obviously be specific to the type of mill employed. Time of milling (Fan et al., 1981b; Millett et al., 1979), speed, and size, material of construction, and quantity of balls used (Kelsey and Shafizadeh, 1980) are important in ball milling. Roller clearance, roll speed, roll speed ratio, and number of passes affect energy requirements and efficiency of pretreatment in compression milling operation (Tassinari and Macy, 1977; Tassinari et al., 1982).

Attempts have been made to relate the enzymatic susceptibility of milled lignocellulosic materials to the changes effected in their physical structure. Although rates of enzymatic hydrolysis vary with crystallinity, surface area, and degree of polymerization, for milled substrates the rates are most dependent on crystallinity index; several empirical and semiempirical models have been developed to describe this relationship (Fan et al., 1981a; Ryu et al., 1982).

Milling as a pretreatment option has two major advantages: it significantly reduces the bulk density of the substrate, with the result that higher solids concentrations can be handled in subsequent operations (Tassinari et al., 1980), and most milling operations are developed commercial entities, scalable for design purposes (Gracheck et al., 1981; Tassinari et al., 1980). However, the hydrolytic susceptibility of a milled substrate is a linear function of the time and extent of milling, and most mills are batch processes. Consequently, milling is both an energy-intensive and a capital-intensive pretreatment. Estimates of the power requirements for several milling techniques as applied to different substrates and the costs based on sugar yield in enzymatic hydrolysis are summarized in Table 8.4.

Gamma and electron beam irradiation reduces the degree of polymerization and crystallinity and increases the "solubility" of lignocellulose (Dunlap et al., 1976; Fan et al., 1981a, b; Kamakura and Kaetzu, 1978, 1982; Millett et al., 1976). The source of radiation is unimportant. A dosage range of 1×10^8 to 5×10^8 rads has been shown to be optimal for a wide range of materials including Solka Floc, wheat and rice straw, chaff, sawdust, cotton linters, sulfite pulp, extracted spruce wood, and newspaper. At these dosages, about 15% of the carbohydrate is destroyed (Millett et al., 1976). The high operating cost associated with irradiation (estimates exceed $100/ton) precludes its consideration as a viable pretreatment.

For easily degraded substrates such as newspaper, milling alone is a sufficient pretreatment (Mandels et al., 1972, 1974; Mandels and Sternberg, 1976; Nystrom, 1975). However, for native lignocellulosic materials such as poplar and the agricultural residues, the structural changes resulting from milling are not sufficient; some degree of delignification must be effected before significant increases in hydrolytic susceptibility—particularly to enzymes—occur. Chemical pretreatments with agents such as NaOH, NH_3, amines, H_2SO_4, HCl, SO_2, hypochlorite, ammonium bisulfite, steam, and polyhydroxy alcohols are more effective than physical pretreatments in achieving this. Most of the agents listed are nonselective in their action, catalyzing simultaneously the disruption of the crystalline cellulose structure

TABLE 8.4 Power Requirements and Operating Costs for Milling

Substrate	*Process*	*Power Requirement (kWh/ton)*	*Cost of Milling* $/ODT	*Cost of Milling* ¢/lb Sugar[1]	*Reference*
Newspaper	Batch compression	600	24.00[2]	2.2	Tassinari et al. (1980)
	Continuous compression milling	420	16.80[2]	1.8	
MSW	Hammer milling		1.90	0.4	Gracheck et al. (1981)
	Ball milling		9.50	1.9	
	Continuous vibratory rod milling		6.00–8.00	1.0–1.4	
Poplar	Two-stage double disc attrition milling	600	24.00[2]		Holtzapple (1981)
	Attrition milling	250	10.00[2]		
	Two-stage hammer milling	45	1.80[2]		
	Hammer milling	30–85	1.20–3.40[2]		
	Two-stage wood hog	30	1.20[2]		
	Wood planer	20	0.80[2]		

1. Estimated from cited conversion efficiencies.
2. Estimated on the basis of cited energy requirements and electricity at $0.04/kWh.

and removal of the lignin. In addition, most chemical agents, especially the acids and bases, degrade both lignin and carbohydrate solubilized during the pretreatment. Consequently, yields from chemical pretreatments are low (in most cases, only 40–70% of the raw material is recovered in the solid phase), and the potential by-product value of lignin and carbohydrates extracted is limited.

Bases, such as sodium hydroxide and ammonia, exert a strong swelling action on wood. Unlike the intercrystalline swelling action of water, which is reversible upon dehydration of the substrate, the swelling with bases is an intracrystalline phenomenon affecting both amorphous and crystalline regions of the cellulose and resulting in irreversible changes in the basic cellulose structure (Millett et al., 1976; Howsman and Sisson, 1954). These bases hydrolyze—through saponification or ammonolysis—the intermolecular bonds of the 4-O-methyl glucuronic acid moieties attached to the xylan side chains that function as cross-links restricting the swelling of wood in H_2O (Tarkow and Feist, 1969). Although the action of both bases is similar, NaOH is more effective than NH_3; at a given concentration, the rate of reaction is faster with NaOH. The level of alkali required is related to the combined acetyl and carbonyl content of the lignocellulosic material; maximum effects, in terms of ruminant digestibility, are obtained at NaOH concentrations of 7–10% (based on weight of the lignocellulose) for poplar, wheat straw, and sugar maple (Millett et al., 1975; Tarkow and Feist, 1969). The reaction is typically conducted at 20°–90°C for reaction times in excess of 1 h. The extent of swelling in wood is limited by the constraining network of lignin, a fact suggesting that delignification prior to alkaline treatment may be beneficial. Treatment with NaOH results in loss of lignin, cellulose, and hemicellulose, the extent depending on the alkali concentration, reaction time and temperature, particle size, and nature of the material being treated (Nelson and Schuerch, 1956, 1957). The delignification reaction catalyzed by alkali is first-order with respect to lignocellulose concentration; the alkaline extraction of the hemicellulose is a diffusion-limited process in which the rate of diffusion of hemicellulose sugars out of the wood matrix is controlled by the degree of swelling of and reaction of alkali with cellulose (Nelson and Schuerch, 1957). Data for the alkaline pretreatment of wheat straw, illustrating the loss of lignin and carbohydrate from and enzymatic susceptibility of the pretreated residue, are summarized in Table 8.5 (Detroy et al., 1981). There is little or no degradation of the hemicellulose sugars extracted with alkali, although irreversible recombination with lignin and cellulose can reduce the yield (Nelson and Schuerch, 1957; Wenzl, 1970). At the low temperatures utilized in this pretreatment, lignin is most probably solubilized as phenols and 1,2 glycols (Wenzl, 1970). Although an effective procedure for swelling of and extraction of an undegraded pentosan fraction from lignocellulose, alkaline pretreatment suffers from two major disadvantages. First, the rate of hemicellulose extraction is severely constrained by the presence of lignin, which functions as a cage to limit the swelling

TABLE 8.5 NaOH Pretreatment of Wheat Straw (Detroy et al., 1981)

	Percent Loss			*Percent Conversion to*
Treatment Conditions	*Hemicellulose*	*Cellulose*	*Lignin*	*Glucose**
3% NaOH (based on straw), 25°C, 1 hr	32	13	17	51
3% NaOH (based on straw), 90°C, 1 hr	60	9	35	75

* 10 IU cellulase/gm wheat straw, 45°C, 6 hr.

of cellulose; consequently, the rate of delignification, which is relatively slow at temperatures less than 100°C, controls the overall reaction (Wenzl, 1970). Second, a significant portion of the alkali—as much as 40–50%—is consumed during the pretreatment (Wenzl, 1970), a factor that will have significant impact on the pretreatment economics. Losses of NH_3, through sorption within or reaction with the lignocellulose would not have as drastic an economic repercussion; in biologically based conversion processes, the retained ammonia could serve as a source of nitrogen in subsequent fermentations.

Acids catalyze a more rapid extraction or hydrolysis of the pentosans. Acid pretreatment can be accomplished via autohydrolysis (Lora and Wayman, 1978; MacDonald and Mathews, 1979; Millett et al., 1976; Neese et al., 1977) or reaction with dilute solutions of sulfuric or hydrochloric acid (Knappert et al., 1980, 1981; Kouklos and Valkanas, 1982). Autohydrolysis is a high-temperature steam treatment in which the organic acids formed from the acetyl groups present in the wood catalyze the hydrolysis of hemicellulose. Typical autohydrolysis conditions are temperatures of 130–220°C and reaction times of 4–120 min. Approximately 20–35% of the wood is solubilized during treatment; the autohydrolysis liquors contain primarily solubilized and degraded pentoses and small amounts of lignin derivatives (Aronovsky and Gortner, 1930). Autohydrolysis pretreatment has been used to increase the enzymatic susceptibility of aspen (Macdonald and Mathews, 1979); enhance the solubility of lignin from aspen, poplar, and eucalyptus in organic extractants (Lora and Wayman, 1978; Wayman and Lora, 1979); and improve the digestibility of manure (Neese et al., 1977). The acids catalyze some decomposition of the solubilized sugar; in complex substrates such as manure, caramelization catalyzed by the ammonia, amines, and amino acids present in the feedstock occurs.

Dilute acid addition can be used to reduce either the temperature or the reaction times in an effort to minimize these degradation reactions. Since acids catalyze both the hydrolysis and degradation of hemicellulose and cellulose, dilute acid pretreatment is conducted at conditions that favor hemicellulose solubilization without glucose formation, that is, at temperatures less than 220°C, acid concentrations less than 1.0% by weight, and reaction times on the order of 12 seconds (Knaeppert et al., 1980, 1981). The response of the substrate to this pretreatment is variable (Knaeppert et al., 1980). The acid reduces the degree of polymerization of the substrate, although increases in enzymatic susceptibility do not correlate well with changes in DP. Twofold to fourfold increases in the enzymatic susceptibility of oak, corn stover, newsprint, and poplar have been effected by this treatment. Yields vary with reaction conditions over the range of 45–98%, typically averaging 70–80% (Knaeppert et al., 1981).

At the treatment conditions employed with acids and alkali to extract hemicellulose in an undegraded form, little decomposition and solubilization of the lignin occur. The more severe conditions of reaction time, temperature, and chemical concentration employed in pulping processes based on NaOH, H_2SO_4, sodium chlorite, ammonium bisulfite, SO_2, and the like are necessary to remove the bulk of lignin from wood. However, such processes can severely degrade the cellulose and hemicellulose and yield an extensively modified lignin product (Glasser, 1981; Wenzl, 1970); the chemical costs associated with them preclude their consideration for lignocellulose conversion. Although complete lignin removal is neither necessary nor sufficient to achieve a high degree of substrate susceptibility (Holtzapple, 1981), if the lignin is to be processed for its by-product value, the maximum quantity of lignin with the highest potential value should be extracted. It has been shown that lignin can be extracted from wood without extensive modification of its structure with aqueous organic solvent mixtures. A number of mono- and polyhydroxy-alcohols have been investigated for this purpose, and a summary of these systems is given in Table 8.6. Solvent delignification is a complex process involving breakdown of the lignin–carbohydrate complex, solvation of the breakdown products, and repolymerization and/or redeposition of the breakdown products on the solids (Hansen, 1981; Kleinert, 1974; Lora and Wayman, 1978: Wayman and Lora, 1979). It is a two-step reaction. Step one, referred to as "bulk delignification," is an initial, rapid depolymerization of the bulk of the lignin and blocking of the repolymerization of the depolymerized lignin. This initial reaction is independent of the type of solvent, the solvent-to-water ratio, or the chemical additives used in the delignification itself (Hansen, 1981; Kleinert, 1974) and dependent on the type of wood (hardwoods delignify faster than softwoods), moisture content, aging, and prior pretreatment of the wood (Kleinert, 1974, 1975; Wayman and Lora, 1979). Step two, "residual delignification," is characterized by a slower rate of removal of lignin modified by self-condensation or grafting onto the carbohydrates. It has been suggested that the organic solvent swells the wood structure, the extent of swelling increasing with the polarity of the solvent (Hansen, 1981), and acts as a solvent for the lignin (Aronovsky and Gortner, 1936). However, no relationship between the solubility of the free lignin in a solvent and the degree of delignification has been developed (Hansen, 1981). The most effective delignifying agents, such as the aromatic

TABLE 8.6 Summary of Solvent Delignification Systems Investigated

Substrate	*Solvent and Additives*	*Reaction Temperature (°C)*	*Liquid/Wood*	*Time at Temperature*	*Reference*
Aspen sawdust and chips	Methanol, ethanol, *n*- and *iso*-propanol, *n*-, *iso*- and *tert*-butanol and amyl alcohol, dioxane, ethylene glycol, glycerol, D-glucose, D-mannitol, urea, *n*-butanol + urea/H_2O	157–186	15/1	4 h	Aronovsky and Gortner (1936)
Paper birch, basswood, aspen, red oak, white oak, silver maple, red pine, jack pine, black spruce, white spruce, tamarack, balsam fir	*n*-Butanol/H_2O	158	16/1	6 h	McMillen et al. (1938)
Beech, pine sapwood, pine	Phenol/H_2O + H_2SO_4, CH_3COOH, HCl, H_3PO_4, CH_3OSO_3H, oxalic acid	110–170	4/1	0–6 h	Schweers (1974)
Spruce and poplar sawdust	Ethanol/H_2O	185	10/1	0–60 min	Kleinert (1974)

TABLE 8.6 (Continued)

Substrate	*Solvent and Additives*	*Reaction Temperature (°C)*	*Liquid/Wood*	*Time at Temperature*	*Reference*
Southern yellow pine meal	*n*-Butanol/H_2O	174–205	15/1	4–12 h	Bowers and April (1977)
Autohydrolyzed aspen meal and chips, poplar, eucalyptus	Dioxane/H_2O	175–215	unspecified	0–120 min	Lora and Wayman (1978)
Hybrid poplar, southern hardwoods, eucalyptus, spruce autohydrolyzed in presence of 2-naphthol	Dioxane/H_2O	155–195	unspecified	0–300 min	Wayman and Lora (1979)
Southern yellow pine meal	*n*-Butanol, phenol/H_2O	175–205	15/1	0.5–12 h	April et al. (1979)
Cottonwood wafers	Ethanol/H_2O + H_2SO_4, $NaHSO_4$, oxalic acid	155–175 with "preconditioning" at 100°C for 3 h	unspecified	unspecified	Sarkanen (1980)
Poplar, aspen, MSW	Ethanol, butanol/H_2O + NaOH, H_2SO_4, NH_3, $FeCl_3$, Na_2CO_3, $NH_4H_2PO_4$	160–190	10/1	0.5–2.0 h	Holtzapple (1981)

alcohols, act as free radical scavengers in the presence of water, reducing or eliminating secondary condensation reactions (April et al., 1979; Kleinert, 1975). The aqueous phase functions as a catalyst for the hydrolysis reactions and as a solvent for the solubilized carbohydrates (Aronovsky and Gortner, 1936; Kleinert, 1974). Extensive delignification—accomplished in solvent–water systems without chemical addition at reaction times in excess of 2 h—is accompanied by almost complete extraction of the pentosans and some removal of the cellulose as shown in Table 8.7. Reduction in reaction times and temperature can be achieved by the addition of acidic and basic additives to the delignifying liquor (Holtzapple, 1981; Sarkanen, 1980; Schweers, 1974). Acidic "catalysts"[1] increase the extent of pentosan removal (Holtzapple, 1981; Kleinert, 1974); basic "catalysts" prevent or reduce the hydrolysis of pentosans (Aronovsky and Gortner, 1936; Holtzapple, 1981). As shown in Fig. 8.4, hemicellulose removal is directly related to the extent of delignification for a wide range of treatment conditions. Extraction of the hemicellulose via autohydrolysis prior to solvent delignification increases the delignification rate and yields a relatively pure pentose stream (Bharoocha, 1980; Lora and Wayman, 1978; Wayman and Lora, 1979).

The fractionation of components typically achieved via solvent delignification is outlined in Fig. 8.5 (Myerly et al., 1981). The characteristics of the lignin extracted by this pretreatment are summarized in Table 8.8. The solid phase, that is, the cellulosic fraction, is of pulp-grade quality (Katzen et al., 1980; Myerly et al., 1981). The extent of solvent recovery depends on the solvent used. Recoveries in excess of 90% have been reported for ethanol and butanol (April et al., 1979; Bailey, 1942). Only 70% of phenol is directly recoverable as a result of its reaction with lignin (April et al., 1979; Schweers, 1974); but phenol makeup can be regenerated by processing of this solvent (Kreiger, 1982). As shown in Fig. 8.6, butanol extraction of poplar chips increases their susceptibility to enzymatic hydrolysis. However, solvent delignification does not result in extensive modification of the crystalline cellulose structure (Holtzapple, 1981). Unless some swelling of the intracrystalline and intercrystalline regions of cellulose is performed, for example, by swelling in NaOH as shown in Fig. 8.7, the enhancement in enzymatic susceptibility is limited.

The steam explosion of lignocellulose (Solar Energy Research Institute, 1980; Bungay, 1981) is attractive as a pretreatment technique because, in a single step, the physical alterations of macroscopic and microscopic structure and removal of lignin necessary to enhance the hydrolytic susceptibility of the substrate can be effected, and a high quality, relatively pure lignin produced. The pretreatment, based on a modification of the Masonite process, is conducted in a high-pressure gun; wood is rapidly heated with high-pressure (500–600 psi) steam, held at temperature for times varying between 20 and 120 seconds, and then rapidly decompressed through the gun nozzle.

---

[1] It is not known to what extent these "catalysts" are consumed in the reaction.

The steam catalyzes the autohydrolysis of the pentosans to xylose and furfural; lignin freed during the pretreatment is recoverable in nearly native form by washing of the solids. The physical disruption of the biomass occurs as a result of the shear forces to which it is exposed during decompression. Yields, lignin recovery, and pentosan degradation are affected by the residence time and temperature in the gun. Extensive thermal degradation of the pentosan, through hydrolysis to furfural or condensation with the lignin (Jurasek, 1979), occurs. After extraction, the steam-exploded material is a suitable substrate for enzyme production and enzymatic hydrolysis (Jurasek, 1979; Saddler et al., 1982). A modification of this process, referred to as freeze-explosion, in which liquid ammonia is used instead of steam, has recently been developed for the pretreatment of agricultural residues (Dale and Moreira, 1982).

B. Hydrolysis of Cellulose and Hemicellulose

The solid phase remaining after pretreatment can be hydrolyzed to soluble sugars with either acids or enzymes.

1. Acid Hydrolysis. The acid hydrolysis of cellulose and hemicellulose can be accomplished with dilute or concentrated acids (Ladisch, 1979; Wenzl, 1970).

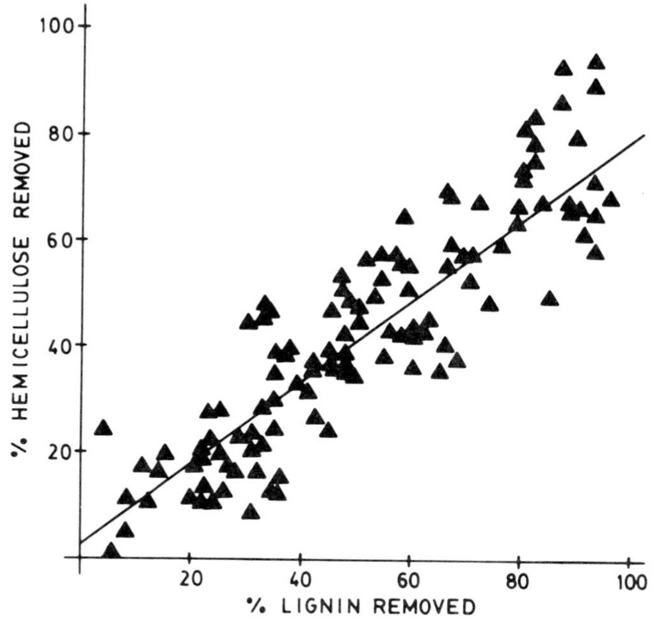

FIGURE 8.4 Correlation of hemicellulose removal with lignin removal.

TABLE 8.7 Yields from Solvent Delignification

Substrate	Solvent	Temperature (°C)	Time (h)	Percent Wood Yield	Percent Original Component Remaining: Lignin	Cellulose	Pentosans	Reference
Aspen Sawdust	Water	186	4	63.4	57.2	74.8	14.8	Aronovsky and Gortner (1936)
	Water/methanol	158	4	70.2	41.6	100.7	49.0	
	Water/ethanol	165	4.5	62.5	32.6	94.5	34.2	
	Water/*n*-propanol	172	4	60.0	23.3	94.5	27.7	
	Water/*iso*-propanol	162	4	66.1	38.0	96.7	38.1	
	Water/*n*-butanol	174	4	54.1	12.8	91.2	22.6	
	Water/*iso*-butanol	174	4	57.1	19.5	92.2	21.9	
	Water/*tert*-butanol	157	4	68.1	44.0	95.2	41.3	
	Water/*n*-amyl alcohol	177	4	51.5	18.0	83.6	10.3	
	Water/*iso*-amyl alcohol	177	4	52.6	15.9	85.3	11.0	
	Water/*tert*-amyl alcohol	159	4	60.2	29.3	81.2	17.4	
	Water/dioxane	176	4	63.4	34.9	95.2	32.9	
	Water/ethylene glycol	—	4	53.2	12.1	85.0	11.0	
	Water/glycerol	—	4	54.9	28.0	80.3	9.7	
Paper birch	Water/*n*-butanol	158	6	45.3	8.4	87.0	19.0	McMillen et al. (1938)
Basswood				43.6	11.5	80.5	16.3	
Aspen				54.2	13.3	92.9	18.0	
Red oak				45.4	11.2	90.0	17.3	
White oak				43.1	9.6	80.6	19.2	
Silver maple				46.3	10.3	87.7	16.4	
Red pine				58.3	39.9	84.8	52.8	
Jack pine				59.5	40.1	86.2	27.8	

TABLE 8.7 (Continued)

Substrate	*Solvent*	*Temper-ature (°C)*	*Time (h)*	*Percent Wood Yield*	*Percent Original Component Remaining*			*Reference*
					Lignin	*Cellu-lose*	*Pento-sans*	
Black spruce				58.4	37.8	85.5	30.2	
White spruce				60.0	42.2	89.2	29.5	
Tamarack				57.2	49.0	88.7	21.9	
Balsam fir				68.4	55.2	85.9	41.9	
Southern yellow pine	Water/*n*-butanol	174	4	90.0	80.4	—	—	Bowers and April (1977)
			4	92.6	82.0	—	—	
			4	84.7	80.0	—	—	
		205	4	80.0	76.2	—	—	
		174	8	76.7	72.4	—	—	
		205	8	73.3	62.4	—	—	
		174	12	74.0	62.3	—	—	
		205	12	70.7	53.4	—	—	
Southern yellow pine	Water/phenol	205	2	38.0	10.7	—	—	April et al. (1979)
		175	4	46.0	21.5	—	—	
				45.4	21.2	—	—	
				45.0	20.6	—	—	
		205	4	36.0	11.5	—	—	
				36.0	10.4	—	—	
				36.0	10.5	—	—	
				36.2	10.3	—	—	
		205	12	29.0	9.3	—	—	
Hybrid poplar chips	Water/*n*-butanol	160	12	92.5	89.8	93.6	93.3	Holtzapple (1981)

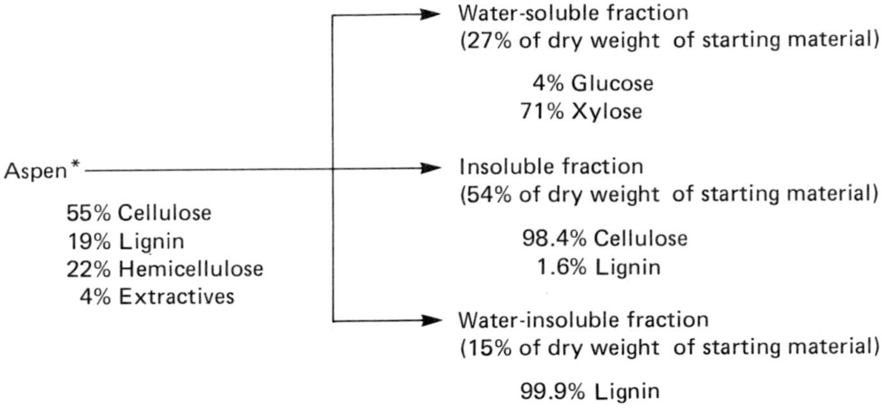

*Wiley-milled to 4 mm; treated with 50% aqueous ethanol mixture (12:1 solvent:H_2O ratio) at 200°C; two extractions for 45 min each; fresh solvent charge to each extraction.

FIGURE 8.5 Typical separation of lignocellulosic components via solvent delignification (from Myerly et al., 1981).

TABLE 8.8 Chemical Characteristics of Lignins

Characteristic	*Lignosulfonates*	*Kraft Lignin*	*Organosolv Lignin*
Elemental composition			
% C	53	66	63
% H	5.4	5.8	5.5
% S	6.5	1.6	—
Contaminants	Miscellaneous carbohydrate degradation products	None	None
Functionality			
Phenolic OH	1.9	4.0	4.8
Aliphatic OH	7.5	9.5	5.0
SO_3H	16.0	—	
SH	—	3.3	
OCH_3	12.5	14.0	19.0
Solubility	Water	Alkali	Organic solvents
Molecular weight	400–150,000	2000	700
Predominant interunit bond	Aryl-alkyl ethers (β-O-4)	C-C bonds (polystyrene type) involving side-chains, aromatic rings, and dialkyl ethers	C-C bonds (hydrolysis lignin type) between side-chains, aromatic rings, and diaryl ethers

From Glasser (1981).

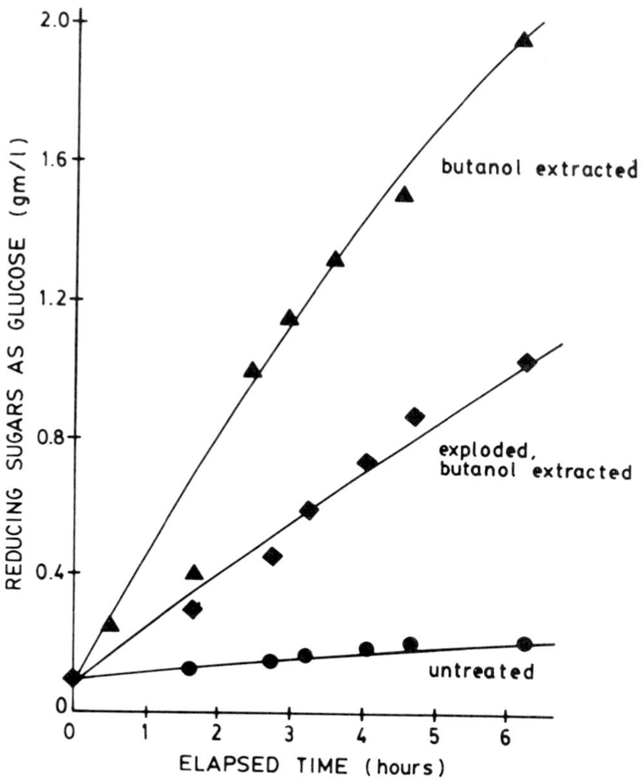

FIGURE 8.6 Hydrolysis of solvent delignified poplar chips by extracellular cellulase of *Thermomonospora* sp. YX (25 g of chips per liter; chips wet-milled in blender before hydrolysis; enzyme activity = 1.0 mg of glucose released from filter paper per ml enzyme·hour).

Dilute acid hydrolysis is conducted at acid concentrations of less than 2%, at temperatures between 160° and 250°C, and at reaction times of less than 1 h. The reaction of cellulose with dilute acid is a four-step sequence (Wenzl, 1970), as depicted in Fig. 8.8, consisting of (1) a rapid conversion of native cellulose to a stable hydrocellulose fraction, (2) a first-order cleavage of the hydrocellulose to soluble polysaccharides, (3) a rapid hydrolysis of the soluble polysaccharides to simple sugars, and (4) a first-order decomposition of the sugar to a degraded product. Since both the conversion of native cellulose to hydrocellulose and the hydrolysis of the soluble polysaccharides to simple sugars are rapid, this sequence has been simplified to two first-order reactions consisting of hydrolysis of the cellulose to simple sugars followed by degradation of the sugars (Wenzl, 1970). An analogous reaction sequence has been used to describe hemicellulose degradation. The relevant kinetic equations for cellulose hydrolysis are summarized in

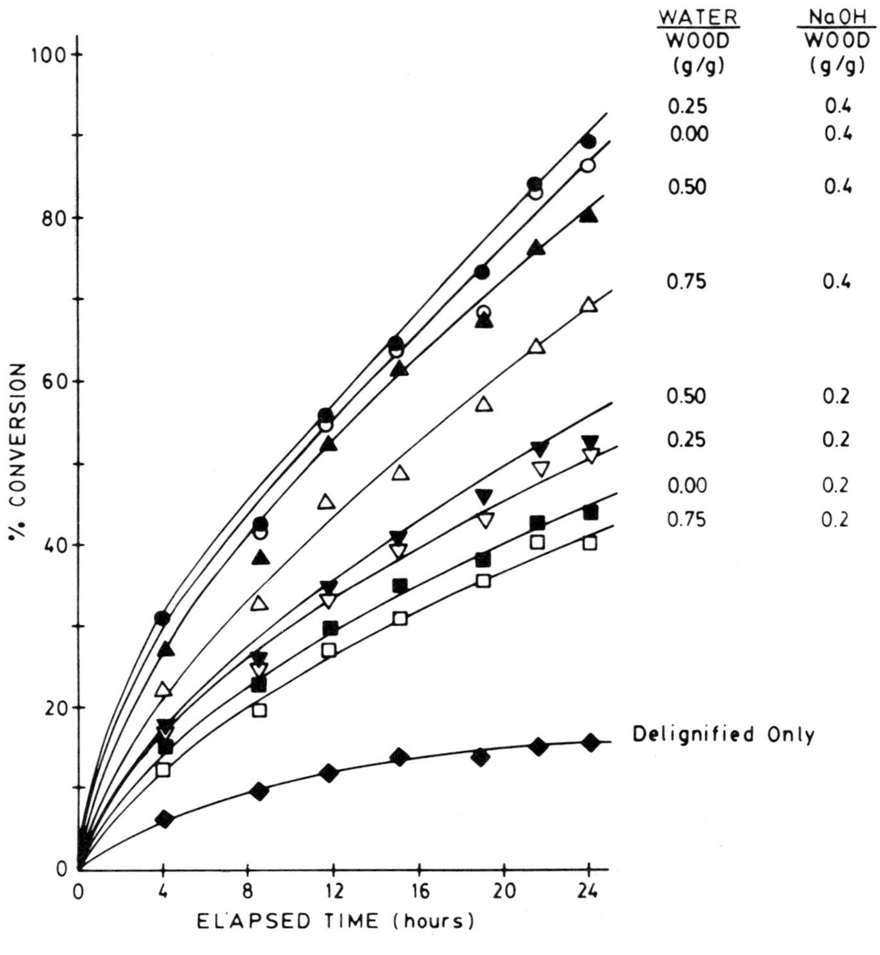

FIGURE 8.7 Saccharification of swollen, delignified wood (delignification conditions: 1:1 30% NH_4OH:ethanol, 188°C, 2 h, 10:1 liquid:wood; swelling conditions: aqueous NaOH:ethanol, at water:wood and NaOH:wood ratios shown, 165°C, 0.5 h).

Fig. 8.8. (An alternative kinetic model, incorporating similar features but based on a different relationship of the rate constants to acid concentration, has been developed for the dilute acid hydrolysis of ryegrass straw (Grant et al., 1977).) The individual rate constants are functions of the acid concentration and temperature. Values of the kinetic parameters for the hydrolysis of the cellulosic fraction of Douglas fir (Saeman, 1945), Solka Floc (Thompson and Grethlein, 1979), kraft paper (Fagan et al., 1971), and barley straw (C. G. Sinclair and R. Quintero-Ramirez, unpublished manuscript) with sulfuric acid are summarized in Table 8.9. The values of these parameters are

$$\text{Native cellulose} \xrightarrow{(1)} \text{Hydrocellulose} \xrightarrow{(2)} \text{Soluble polysaccharide} \xrightarrow{(3)} \text{Simple sugars} \xrightarrow{(4)} \text{Degradation products}$$

$$\underset{(C_x)}{\text{Native cellulose}} \xrightarrow{k_1} \underset{(C_1)}{\text{Simple sugars}} \xrightarrow{k_2} \underset{(C_0)}{\text{Degradation products}}$$

$$\frac{d[C_x]}{dt} = -k_1 [C_x]$$

$$\frac{d[C_1]}{dt} = y_1 k_1 [C_x] - k_2 [C_1]$$

$$\frac{d[C_0]}{dt} = y_2 k_2 [C_1]$$

$$k_1 = K_1 (A)^m \exp(-E_1/RT)$$

$$k_2 = K_2 (A)^n \exp(-E_2/RT)$$

FIGURE 8.8 Reaction scheme for dilute acid hydrolysis of cellulose.

dependent on the type of acid used, the rate of the hydrolysis of the native cellulose of Solka Floc increasing at a given molar concentration of acid in the order $HCl > H_2SO_4 > SO_2 > H_3PO_4$ (Wenzl, 1970). The hemicellulose fraction is more rapidly hydrolyzed than the cellulosic fraction, and the soluble sugars formed from hemicellulose are more rapidly degraded, as shown by the data in Table 8.10. The maximum yield of soluble sugars is consequently limited by the loss to degradation products such as furfural, hydroxymethyl furfural, and levulinic, formic, and acetic acids; as shown in Fig. 8.9, yields of soluble sugars of no more than 65–70% are obtainable under isothermal conditions. Reaction times of substantially less than 6

TABLE 8.9 Dilute Acid Hydrolysis Kinetic Parameters Cellulosic Fraction

Parameter	*Douglas Fir*[1]	*Solka Floc*[2]	*Kraft Paper*[3]	*Barley Straw*[4]
K_1 (min^{-1})	1.73×10^{19}	1.22×10^{19}	28×10^{19}	2.77×10^{16}
K_2 (min^{-1})	2.38×10^{14}	3.79×10^{14}	4.9×10^{14}	1.15×10^{14}
E_1 (cal/g mole)	42,900	42,500	45,100	36,329
E_2 (cal/g mole)	32,800	32,700	32,800	32,229
m	1.34	1.16	1.78	1.314
n	1.02	0.69	0.555	1.105

1. From Saeman (1945).
2. From Thompson and Grethlein (1979).
3. From Fagan et al. (1971).
4. From C. G. Sinclair and R. Quintero-Ramirez (unpublished manuscript).

TABLE 8.10 Rate Constants for the Acid Catalyzed Degradation of Soluble Sugars

Sugar	k_2 (min^{-1})
D-Glucose	0.0242
D-Galactose	0.0263
D-Mannose	0.0358
D-Arabinose	0.0421
D-Xylose	0.0721

From Wenzl (1970).

seconds are necessary to achieve these maximum conversions; at longer times, greater amounts of soluble sugar are decomposed (see Fig. 8.10). Reaction times of 0.22 min (13 seconds) have been demonstrated in laboratory-scale reactors (Thompson and Grethlein, 1979) but will probably be difficult to achieve in full-scale operation (Mendelsohn and Wettstein, 1981). Dilute acid hydrolysis has formed the basis of several commercial processes, all based on H_2SO_4 (Faith, 1945; Gilbert et al., 1952; Harris et al., 1945;

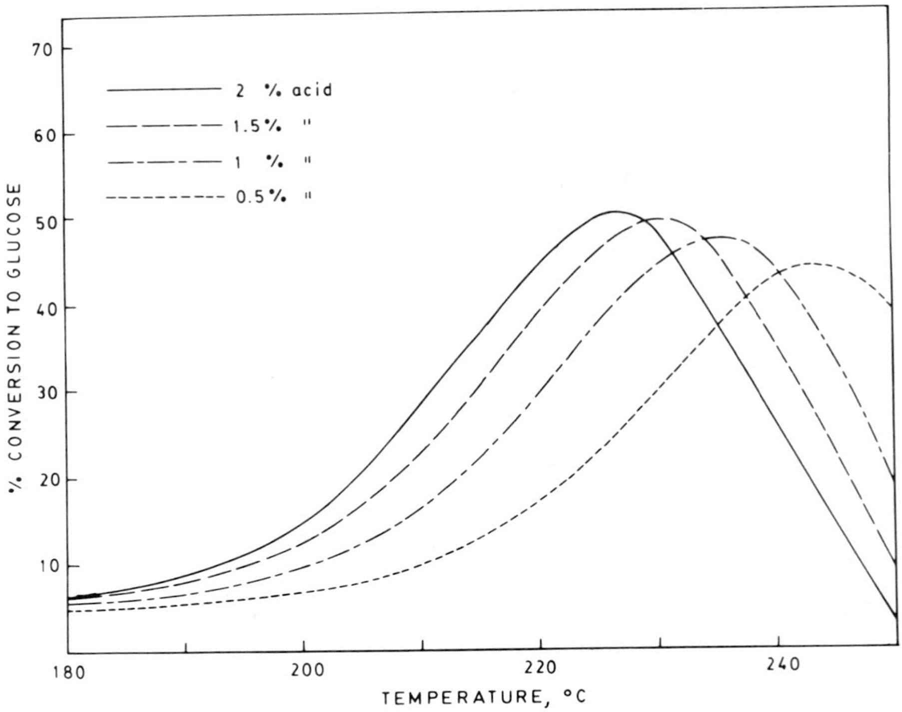

FIGURE 8.9 Conversion of cellulose to glucose by acid hydrolysis as a function of reaction temperature and acid concentration using Saeman's constants for sulfuric acid.

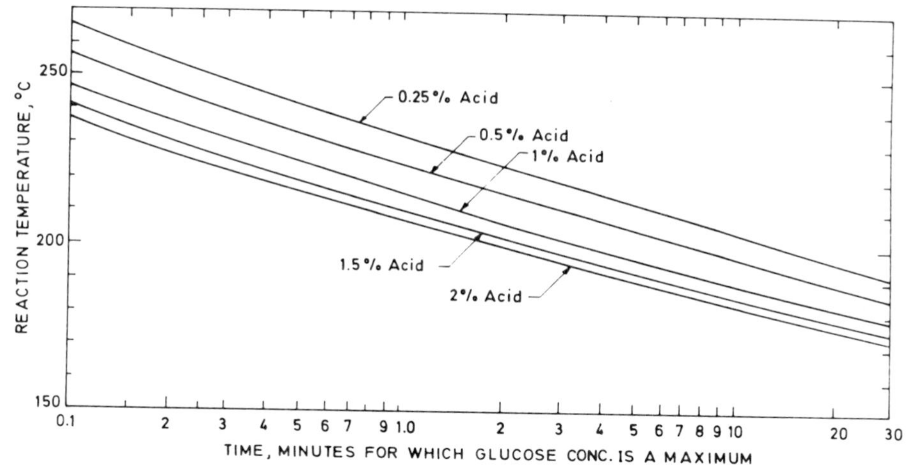

FIGURE 8.10 Time required for maximum conversion of cellulose to glucose as a function of reaction temperature and acid concentration.

Mendelsohn and Wettstein, 1981; Sherrard and Kressman, 1945; Wenzl, 1970). Sugar yields range between 55% and 72% of the hemicellulose and cellulose; sugar concentrations in the product streams prior to evaporative concentration vary between 4% and 6%.

Concentrated acid hydrolysis is conducted at more moderate temperatures (20–100°C), at high acid concentrations (>35% HCl; >60% H_2SO_4), and for times varying between 10 min and 6 h. Under properly controlled reaction conditions, concentrated acid hydrolysis converts cellulose and hemicellulose into water-soluble oligosaccharides without appreciable decomposition of the sugar. The hydrolysis is a two-stage reaction proceeding as shown below through the formation of an acid–cellulose addition complex prior to the hydrolysis to glucose (Wenzl, 1970):

$$\text{Native cellulose} \rightarrow \text{Acid–cellulose complex (swollen, soluble)} \rightarrow \text{Glucose} \rightarrow \text{Decomposition products}$$

The addition complex differs from the native cellulose in its swelling, solubility, and hydrolysis characteristics. The acid concentration required to achieve maximum swelling and solubility of this complex depends on the reaction temperature, moisture content of the wood, and acid-to-cellulose ratio. The swelling reaction in H_2SO_4 is not affected by the hemicellulose, but the extent of swelling is constrained by the lignin network. Subsequent hydrolysis of the intermediate is a first-order reaction and can be carried out after dilution of the reaction mixture to an acid concentration of <10%. The overall hydrolysis of wood with concentrated acid is controlled by the diffusion of acid into and dissolved carbohydrate out of the wood particle and is therefore dependent on the particle size of the substrate, the reaction temperature, the mode of contact between acid and solid, and the extent of lignification of the substrate. At pilot-scale, production of a 10–12% glucose solution, corresponding to 85–90% of the theoretical yield, from agricultural residues has been demonstrated (Dunning and Lathrop, 1945).

The dissolution of cellulose, resulting in the formation of an extensively swollen cellulose structure, can also be accomplished with quaternary ammonium bases, aprotic solvents, and metal complexes (Turbak et al., 1980). The dissolved cellulose, freed from the lignin and hemicellulose, can be reprecipitated in an amorphous form and subsequently hydrolyzed to soluble sugar with acid or enzyme. This concept has been refined into the process scheme outlined in Fig. 8.11, which consists of (1) dilute acid hydrolysis

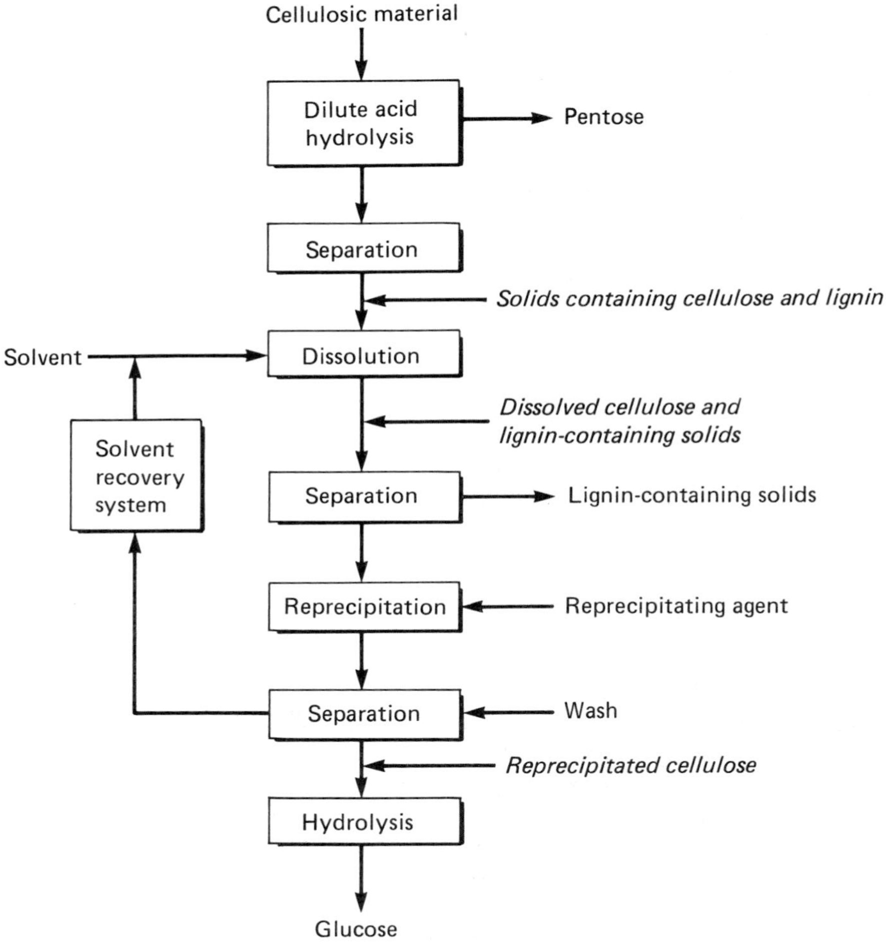

FIGURE 8.11 Process for the dissolution of cellulose.

of the cellulosic material to remove pentosans, (2) treatment with a solvent that breaks the lignin–cellulose bond and dissolves the cellulose, (3) separation of the undissolved lignin solids from the cellulose, (4) reprecipitation of the cellulose as an amorphous solid with the appropriate agent, (5) separation of the reprecipitated cellulose from the solvent, and (6) final hydrolysis of the reprecipitated cellulose to glucose (Tsao and Chou, 1981; Tsao et al., 1981). Cadoxen (Hsu et al., 1980; Ladisch et al., 1978; Tsao et al., 1981) (a mixture of 4.5–7% cadmium oxide or hydroxide with 25–30% ethylenediamine) and 60–90% H_2SO_4 (Tsao and Chou, 1981) have been identified as suitable solvents. Water and organic solvents such as methanol are used as reprecipitating agents with Cadoxen and H_2SO_4, respectively. Yields of cellulose as high as 97% after reprecipitation have been reported. The amorphous cellulose, which will recrystallize on standing, is extremely susceptible to enzymatic hydrolysis; conversions in excess of 80% of the cellulose to glucose in less than 10 h have been demonstrated for AVICEL, bagasse, cornstalks, tall fescue, and orchard grass treated by this technique (Hsu et al., 1980; Ladisch et al., 1978). The major claims for this process have been the high recovery and easy regenerability of the solvents (Tsao and Chou, 1981; Tsao et al., 1981). The technique has been applied only to pure cellulosic substrates and agricultural wastes; no woods have been tested.

2. Enzymatic Hydrolysis. The alternative to acid hydrolysis is enzymatic hydrolysis. The specificity of the enzymatic reaction eliminates the problem of sugar degradation. Unfortunately, because of the differences in the natures of the catalytic agents involved, enzymatic hydrolysis proceeds at a significantly slower rate than acid hydrolysis; the size of the enzymes that catalyze the hydrolysis of cellulose and hemicellulose relative to the size of the pores within the substrate restricts their access to the reaction sites.

The complex of enzymes that hydrolyzes the β-1,4-glucan linkages of cellulose is referred to as cellulase. In general, these enzymes are classified as β-1,4-glucan-glucanohydrolases. If the product of the attack by the enzyme is glucose, the enzyme is specifically designated as a β-1,4-glucan-glucohydrolase. If, however, the product is cellobiose, the enzyme is classified as a β-1,4-glucan-cellobiohydrolase. Those cellulases that attack the substrate in a random manner are given the additional designation of endo-glucanases; those that cleave glucoside units from the terminal end of the polymer chain are denoted as exo-glucanases. A summary of this general nomenclature is given in Table 8.11. This rather systematic classification is often confused by the use of generic terms such as AVICELase, filter paperase, and CMCase, which reflect the activity of an enzyme against a particular substrate rather than the inherent nature of the reaction catalyzed by the enzyme.

The exact nature and mode of action of the components of the cellulase system are matters of continuing research and controversy. The original

TABLE 8.11 General System of Cellulase Nomenclature Based on Mode of Enzyme Attack and Reaction Product

Reaction Product	*Mode of Attack on Substrate*	
	Random	*Endwise*
Glucose	Endo-β-1,4-glucan-glucohydrolase	Exo-β-1,4-glucan-glucohydrolase
Cellobiose	Endo-β-1,4-glucan-cellobiohydrolase	Exo-β-1,4-glucan-cellobiohydrolase

theory of multiple enzyme action was advanced by Reese et al. (1950) to explain the fact that, while certain cellulolytic fungi grew readily on cellulosic substrates, the action of the cell-free enzymes was usually limited to the hydrolysis of soluble cellulose derivatives or swollen cellulose. Reese proposed the existence of a factor designated as C_1, which was required, in conjunction with a C_x enzyme, for the hydrolysis of native cellulosic materials. Only the C_x enzyme was present in those cellulolytic systems incapable of hydrolyzing the crystalline forms of cellulose. The overall pattern for enzymatic hydrolysis outlined by Reese involved the conversion of native cellulose to reactive cellulose by the C_1 component, followed by the attack of this reactive cellulose by the C_x component in a random fashion to produce glucose and soluble oligosaccharides. The soluble oligosaccharides were then hydrolyzed to glucose by a β-glucosidase.

Refinements of this preliminary theory of enzymatic cellulose hydrolysis are based on the understanding of the action of purified cellulase components (Emert et al., 1974; Gritzali and Brown, 1979; Halliwell, 1962, 1965; Halliwell and Riaz, 1970; Halliwell and Vincent, 1977; Kanda et al., 1979; King, 1965; Li et al., 1965; Reese et al., 1957; Selby, 1969; Selby and Maitland, 1967; Wood, 1968, 1972; Wood and McCrae, 1972, 1975, 1977, 1979). Gilligan and Reese (1954) and Reese and Gilligan (1953) identified multiple C_x components with different relative activities toward swollen and soluble substrates. The purified C_x fractions have subsequently been designated as endo-β-1,4-glucanases and exo-β-1,4-glucanases. The endo-glucanases, the randomly acting C_x components, hydrolyze high-molecular-weight glucans; the major products of the reaction are cellobiose and cellotriose. The rate of hydrolysis of the glucans by these endo-glucanase components increases with the degree of polymerization (DP), within the limits of substrate solubility. Endo-glucanase action is associated with a rapid change in solution fluidity and a relatively slow increase in reducing end group production. The exo-glucanases, the endwise acting C_x components, successively remove the glucose from the nonreducing end of the glucan polymer, resulting in a rapid increase in reducing end groups.

Postulates concerning the action of the C_1 component against cellulose, proposed since the original C_1–C_x concept, have depicted C_1 as being (1) a random C_x component, (2) a C_x enzyme small enough to penetrate the crystalline areas of the cellulose lattice, or (3) a protein that more tightly bonds to cellulose than cellulose bonds to itself, thus opening up the micellar structure. Each of these hypotheses has subsequently been experimentally disproven. A true understanding of the function of this enzyme has been clouded by its association with C_x components, even in what were considered to be purified preparations. From data on purified C_1 preparations from which the C_x contaminant has been removed, the C_1 component has been identified as an exo-glucanase (Halliwell, 1975; Halliwell and Riaz, 1970; Halliwell and Vincent, 1977; Kanda et al., 1979; King, 1965; Li et al., 1965). In its purest form, the enzyme is incapable of solubilizing native cellulose. Cellobiose is the major product when the C_1 enzyme is contacted with swollen cellulose for an extended period. Consequently, C_1 is now understood to be a β-1,4-glucan-cellobiohydrolase. It is not solely responsible for initiating the attack on crystalline cellulose; the accompanying action of the C_x glucanases is required to accomplish this hydrolysis effectively. Cellulose hydrolysis is thus presently thought to be a sequential attack in which the randomly acting C_x glucanases initiate the attack and the new chain ends that are produced are then hydrolyzed instantly by the C_1 component in order to prevent reformation of the glycosidic linkage.

In addition to the endo-glucanase and exo-glucanase activities, the cellulase systems of most organisms include a β-glucosidase, sometimes referred to as cellobiase. The primary action of this enzyme is the hydrolysis of cellobiose. In most cases, this enzyme has been shown to have activity against higher oligosaccharides, but the activity decreases markedly as the degree of polymerization increases. In addition, β-glucosidase is less specific about the nature of the glucan linkage it will attack; it will hydrolyze β-1,1, β-1,2, β-1,3, and β-1,6 bonds.

The generalized scheme of multiple-enzyme action of the cellulase system is summarized in Fig. 8.12.

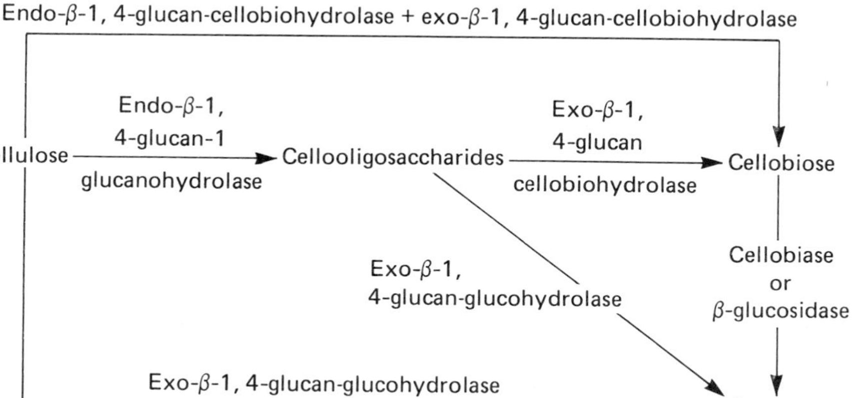

FIGURE 8.12 Generalized mechanism of enzymatic cellulose hydrolysis.

A somewhat analogous system of enzymes is responsible for the hydrolysis of hemicellulose (Dekker and Richards, 1976; Reilly, 1981). Referred to as "hemicellulases" or "xylanases," this enzyme complex consists of three major types of hydrolases: (1) endo-xylanases, enzymes that attack the interior of the xylan molecule; (2) exo-xylanases, enzymes that produce xylose by endwise attack on xylan; and (3) β-xylosidase, an enzyme analogous in action to β-glucosidase, which catalyzes the formation of xylose from xylobiose and other short chain xylooligosaccharides. The endo-xylanases are further subclassified on the basis of their activity at L-arabinosyl-initiated branch points of the xylan backbone and on whether the products of their action are xylose and xylobiose or xylooligosaccharides.

Cellulases are formed by many bacteria, fungi, higher plants, and some invertebrate animals. However, investigations into cellulase production have been restricted primarily to the microbial sources. In addition, since the present focus of cellulase work is the maximization of soluble sugar production from native cellulose, research efforts have been directed primarily at isolating organisms that produce cellulase complexes capable of rapidly degrading these native materials. Consequently, the number of organisms actively being investigated for cellulase production is small in comparison with the number of organisms that have been identified as cellulase producers. The various species of bacteria, fungi, and actinomycetes that degrade cellulose have been isolated most commonly from soil, rotting wood, or exposed textile materials. Comprehensive listings of the microorganisms that exhibit cellulolytic activity have been supplied by Siu (1951) and Gascoigne and Gascoigne (1960). The predominant cellulose-decomposing fungi are represented by only a very few species within a wide listing of genera; primary among the fungal isolates capable of degrading cellulose are *Trichoderma, Aspergillus, Penicillium, Sporotrichum, Fusarium, Stachybotrys,* and *Basidomycetes*. Enzyme production by fungi is usually adaptive; the enzyme complex is induced or enhanced by the presence of the substrate. Cellulose itself, however, is not the inducer; low levels of the soluble sugars resulting from cellulose hydrolysis activate cellulase synthesis. The cellulases of fungi are usually secreted, from living cells, into the culture medium.

Bacteria capable of disintegrating cellulose fall into one of six genera: *Clostridium, Cellulomonas, Cellvibrio, Cytophaga, Sporocytophaga,* and *Corynebacterium*. There are anaerobic cellulose-decomposing bacteria as well as aerobic species. Cellulase production by bacteria is constitutive, that is, the cellulase enzymes are produced whether or not cellulose or a cellooligosaccharide, required by fungi for induction of cellulase synthesis, is present in the microbial environment. Levels of cellulase produced by bacteria are typically lower than fungal levels, and in some cases, cellulases are intracellular.

The members of the actinomycetes that are responsible for wood decomposition belong primarily to the genera *Actinomyces, Streptomyces,* and *Thermomonospora*. Taxonomically, these organisms exhibit properties common to both the fungi and the bacteria. Cellulases produced by various species of actinomycetes are excreted into the culture fluid during fermentation.

No broad quantitative comparison of the cellulolytic activities of fungal, bacterial, and actinomycetal cultures has been published. Although indications are that such a comparison has been conducted in various laboratories, only data for selected species have been reported. Differences in media used for enzyme production, substrates used in the assay procedures, and conditions under which the activity measurements were made make a comparison of enzymatic activities measured in different laboratories difficult.

Of the many microbial sources of cellulolytic enzymes, the fungus *Trichoderma reesei* has been the most extensively researched (Andren et al., 1975, 1976a, b; Andreotti et al., 1978, 1980; Brown and Halsted, 1975; Brown and Zainudeer, 1977; Cinq-Mars and Howell, 1977; Cuskey et al., 1980; Dwivedi and Ghose, 1979; Gallo et al., 1979; Ghose, 1969; Griffin et al., 1974; Katz and Reese, 1968; Mandels, 1979; Mandels et al., 1971; Montenecourt et al., 1979, 1981; Nagai et al., 1976; Noo Mangat and Howell, 1978; Nystrom and Allen, 1976; Nystrom and DiLuca, 1978; Peiterson, 1975, 1977; Peiterson et al., 1977; Reese and Mandels, 1980; Reese and Ryu, 1980; Shin et al., 1978; Spano, 1976; Spano et al., 1978; Sternberg, 1976a, b; Sternberg and Dorval, 1979; Sternberg and Mandels, 1979, 1980; Tangnu et al., 1981). The original wild-type strain has been subjected to mutation by a number of investigators to increase dramatically its enzyme production capacity and alter the characteristics of its enzyme system (Cuskey et al., 1981). A summary of the mutant strains developed worldwide is given in Fig. 8.13. The enzyme production characteristics of some of these mutants in batch culture are summarized in Table 8.12. The endo-enzymes, exo-enzymes, and cellobiase of the cellulase system produced by *Trichoderma reesei* are extracellular. Cellulase production in all mutant strains is induced by cellulose, lactose (Andreotti et al., 1980), and sophorose (Loewenberg and Chapman, 1977; Nisizawa et al., 1971; Sternberg and Mandels, 1979, 1980) and repressed by glucose and cellobiose (Allen and Mortensen, 1981; Hulme and Stranko, 1970; Reese et al., 1969; Shin et al., 1978). Cellulose hydrolysis syrups, containing β-dimers and trimers of glucose formed by the transferase activity of the cellulases, are also suitable substrates for enzyme production; enzyme titers of 10.7 IU/ml (filter paper activity) have been produced (Allen and Mortensen, 1981). The highest yields of cellulase are usually produced on cellulose. The enzyme level is proportional to the initial cellulose concentration and dependent on the nature of the cellulose; the highest yields—approximately 15 IU/ml—are reported on AVICEL, Solka Floc, and two roll-milled cotton. The productivity in batch culture also depends on the manner in which the environmental conditions in the fermentor are controlled. The best productivities—as high as 100 IU/ml/h—have been reported in fermentations for which both the pH and the temperature were adjusted over the batch time (Tangnu et al., 1981). *T.*

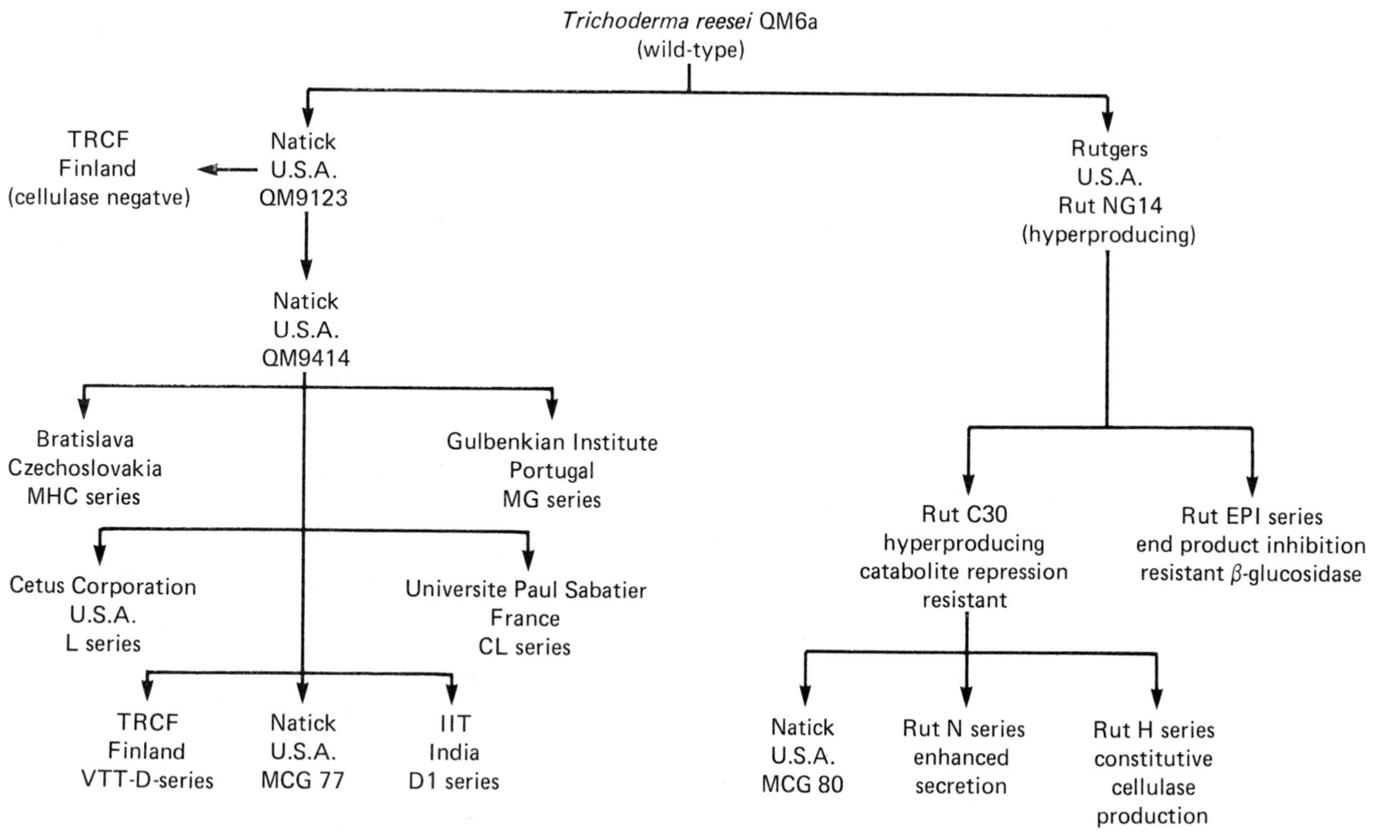

FIGURE 8.13 Worldwide programs for mutation of *Trichoderma reesei*.

reesei has been shown to produce xylanases when grown on oat straw (L. Viikari, M. Linko, and T.-M. Enari, unpublished results) and Solka Floc (Tangnu et al., 1981).

The cellulase preparations from *T. reesei* have been used in the hydrolysis of a wide variety of lignocellulosic materials (Andren et al., 1975, 1976a, b; Cinq-Mars and Howell, 1977; Dwivedi and Ghose, 1979; Ghose, 1969; Katz and Reese, 1968; Mandels, 1979; Noo Mangat and Howell, 1978; Reese and Mandels, 1980; Reese and Ryu, 1980; Spano, 1976; Spano et al., 1976). The saccharification yields vary with the extent of pretreatment of the substrate and the enzyme loading. Cellulase preparations from *T. reesei* have been used to convert as much as 70–80% of milled poplar shavings and newspaper to reducing sugars in less than 30 h. With more recalcitrant sources of cellulose such as bagasse, less than 50% hydrolysis of the substrate is obtained (Mandels, 1979). Optimum conditions for the hydrolysis are pH 4.8–5.0, 50°C, and cellulase loadings of 10–20 IU/g initial substrate. Some components of the complex are inactivated at these conditions (Reese and Mandels, 1980), and the enzymes have been shown to be shear-sensitive (Reese and Ryu, 1980). Supplementation of *T. reesei* cellulase with β-glucosidase from an alternate microbial source is required if glucose is the desired sugar product. Relatively high concentrations of cellobiose (approximately 50% of the total sugars produced) accumulate in the hydrolysis liquor either because *T. reesei* has poor β-glucosidase activity or because its β-glucosidase is strongly inhibited (Sternberg, 1976b; Sternberg et al., 1977). Consequently, β-glucosidase addition reduces the cellulase requirements (Mandels et al., 1981). The role of xylanases in the hydrolysis of pretreated cellulosics and in the presence of cellulases has not been explored.

Other microorganisms, such as the thermophiles *Thermomonospora* sp. (Ferchak et al., 1980; Hägerdal et al., 1978, 1979, 1980; Moreira et al., 1981a, b; Phillips and Humphrey, 1980) and *Thielavia terrestris* (Skinner and Tokuyama, 1978; *Chemical & Engineering News,* 1978), have been suggested as alternative sources of cellulases.[2] Both produce enzyme systems that are capable of degrading crystalline cellulose and are extremely stable at higher hydrolysis temperatures—50–70°C. Their immediate potential as cellulase producers is limited, however, by a number of factors including low absolute titers of cellulase. As shown by the data in Table 8.13, both wild-type strains of *Thermomonospora* sp. YX and *Thielavia terrestris* produce absolute levels of cellulase activity comparable to the early mutant strain QM9414 of *T. reesei.* The specific activity of the wild-type strain of *Thermomonospora* sp. is higher than that of *T. reesei* QM9414; insufficient data are available to make a comparison of specific activities with *Thielavia.*

---

[2] The organism referred to here as *Thermomonospora* sp. has been mislabeled in other publications as *Thermoactinomyces*. It has been reclassified, on the basis of the heat resistance of the spores and its amino acid composition, as *Thermomonospora* sp. (H. Lechvalier, personal communication to J. Forro, General Electric Co., 1979).

TABLE 8.12 Enzyme Production of Mutant Strains of *Trichoderma reesei*

	Substrate		Enzyme Activity (IU/ml)							
Strain	*Type*	*Concentration (percent)*	*Filter Paper*	C_1	C_x	*β-Glucosidase*	*Extracellular Protein (g/l)*	*Fermentation Time (h)*	*Fermentation Conditions*	*Reference*
QM9414	Solka Floc SW40	1	2.5–3.5					110–15	pH ≥ 3.0, 29°C, 0.05–0.1 vvm, 220 RPM, 0–0.2% Tween 80	Nystrom and Allen (1976)
		1	1.8		25		3.0	160	pH ≥ 3.5, 27°C, 0.1–0.2 vvm, 200–250 RPM, 0.1% Tween 80	Andreotti et al. (1978)
		0.94 2.69 5.11	0.9 1.7–3.0 2.3					130–160	pH ≥ 3.25, $T = 33°C$ ≤ 24 hrs, $T = 29°C$ ≥ 25 hrs, 0.2% Tween 80	Nystrom and DiLuca (1978)
	Solka Floc NIB40	2.55 5.04 8.0	2.2–3.1 3.25–4.6 2.75						$T = 29°C$	Gallo et al. (1979)
	Two-roll-milled cotton	6	9.0	13–15			13.6	332	pH ≥ 3.0, 27°C, 0.2 vvm, 300–500 RPM, 0.1% Tween 80, no urea	

TABLE 8.12 (Continued)

	Substrate		Enzyme Activity (IU/ml)							
Strain	*Type*	*Concentration (percent)*	*Filter Paper*	C_1	C_x	*β-Glucosidase*	*Extracellular Protein (g/l)*	*Fermentation Time (h)*	*Fermentation Conditions*	*Reference*
NG14	Solka Floc BW400	2	2.8–3.6					164		
	Two-roll-milled cotton	6	14.5	13–15			21.2	332		
MCG77	Solka Floc BW200	2	2.9–3.9					92		
	Two-roll-milled cotton	6	10.5	13–15			16.2	332		
Rut C30	Solka Floc BW200	5	14.4	1.0	3.48	26	20	192	pH ≥ 5.0, 25°C, 0.2% Tween 80, no urea	Tangnu et al. (1981)

TABLE 8.13 Comparison of Cellulase Activities of *Trichoderma reesei* QM9414, *Thermomonospora* sp. YX, and *Thielavia terrestris*

Characteristic	Trichoderma reesei *QM9414*	Thermomonospora *sp. YX*	Thielavia terrestris
Temperature (°C)	50	65	60
Protein (mg/ml)	1.36	0.17	—
Filter paper activity			
(IU/ml)	0.54	0.36	0.93
(IU/mg protein)	0.4	2.1	
C_x Activity			
(IU/ml)	15.5	13.0	16.5
(IU/mg protein)	11.4	76.5	
Cotton activity			
(IU/ml)	2.1	0.36	
(IU/mg protein)	1.53	2.1	

Cultivation times for both of these microorganisms are significantly shorter than that for *T. reesei*—12–18 h to maximum enzyme production for *Thermomonospora* sp. and 40–48 h for *Thielavia terrestris*.

C. Fermentation

The soluble hexoses and pentoses produced by acid or enzymatic hydrolysis of pretreated lignocellulose are suitable substrates for a variety of homo- and hetero-fermentations by bacteria, yeast, and fungi, which have been extensively reviewed (Eveleigh, 1981; Gong et al., 1981b; Prescott and Dunn, 1949; Rose, 1978; Rosenberg, 1980). Because of ethanol's utility as a fuel and versatility as a chemical feedstock, most current research investigations are concentrated on improving the efficiency of microbial processes by which it is produced.

1. Direct Fermentation of Lignocellulose-Derived Sugars to Ethanol. Yeast (that is, *Saccharomyces cerevisiae* or *Saccharomyces carlsbergensis*) is still the organism of choice for the production of alcohol from glucose, primarily because the product concentration obtainable in this fermentation is high (10–12% ethanol). However, the use of these particular organisms in conversion processes based on total biomass utilization is limited by their inability to ferment aldopentoses, such as xylose, to alcohol. In yeast capable of metabolizing pentoses, xylose utilization occurs via the sequential catabolic steps shown in Fig. 8.14 (Gong et al., 1981d; P. Y. Wang and Schneider, 1980). The relationship between this pathway and the inability of *S. cerevisiae* and *S. carlsbergensis* to assimilate xylose or xylitol has not been defined. Growth on and production of ethanol from the intermediate D-xylulose by

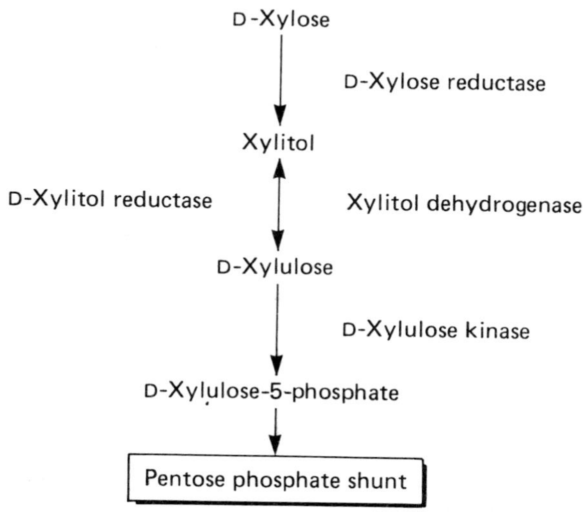

FIGURE 8.14 Steps in xylose metabolism in yeast.

S. cerevisiae and *S. carlsbergensis* has been demonstrated, with low concentrations of ethanol (<0.5% in 1% xylulose, <1% in 5% xylulose) being produced under microaerophilic and anaerobic conditions (Ueng et al., 1981; P. Y. Wang and Schneider, 1980; P. Y. Wang et al., 1980a, b); some strains produce the polyols arabitol and xylitol as by-products. Of the nonxylose utilizers screened, *Schizosaccharomyces pombe* produces the best yields of ethanol with insignificant production of polyol products (Ueng et al., 1981). The conversion of xylose to ethanol by yeast has been mediated by the addition to the fermentation of xylose isomerase, which catalyzes the conversion of xylose to xylulose. Ethanol yields ranging from 10% to 85% of theoretical based on the initial amount of xylose added to the culture have been achieved in laboratory-scale demonstrations using soluble and immobilized xylose isomerase (commercially available as glucose isomerase) and the xylulose-fermenting yeasts *Schizosaccharomyces pombe, Kluyveromyces lactis, Saccharomyces cerevisiae,* and *Candida tropicalis* (Gong et al., 1981a; Jeffries, 1981a and 1981b; P. Y. Wang et al., 1980a). The rates and yields are limited by factors such as the viability of the yeast under the conditions of cultivation, the hydrolysis of the isomerase by proteases produced by the yeast, the accumulation of substances in the culture medium that are inhibitory to both the yeast and the enzyme, the diversion of substrate to xylitol production, the incompatibility of the conditions required for optimum growth and ethanol production by the yeast and those required for optimal activity of the enzyme, and the equilibrium concentration of xylulose produced by the isomerizing enzyme.

Three yeasts have been specifically identified as being capable of effecting the direct conversion of xylose to ethanol. *Pachysolen tannophilus* ferments xylose, at a yield of 52% of theoretical, under microaerophilic conditions (Schneider et al., 1981). At the oxygen tensions favoring ethanol production, the generation times are >4 h. This organism will neither grow nor ferment at alcohol concentrations exceeding 4–5%, and under some cultural conditions, it converts a fraction of the D-xylose to the unfermentable alditol, xylitol (H. Schneider, Y. K. Chan, R. Maleska, I. A. Veliky, and P. Y. Wang, submitted for publication). In batch cultures, ethanol concentrations of no greater than 0.5%, obtained after 3 days of cultivation, have been reported; enhanced levels and rates of production have been achieved through recycling and immobilization. *Candida* sp. XF217 produces ethanol from xylose under aerobic and anaerobic conditions, with significantly greater levels of ethanol production induced by the presence of air (Gong et al., 1981c, d); *Candida tropicalis* ATCC 1369, on the other hand, ferments xylose to ethanol only under aerobic conditions (Jeffries, 1981a).

The bacterium *Zymomonas mobilis* has been proposed as an alternative candidate for the production of ethanol (Arcuri, 1982; Arcuri et al., 1980; Cromie and Doelle, 1980; Dawes et al., 1966; Grote et al., 1980; K. J. Lee et al., 1979, 1980, 1981a, b; J. H. Lee et al., 1981; Lyness and Doelle, 1980, 1981; Rogers et al., 1979, 1980a, b). Originally isolated from the fermented beverages of Europe and the tropical regions of America, Asia, and Africa, this anaerobe uses the Entner-Doudoroff pathway, a pathway typically used by aerobic organisms, to catabolize glucose (Dawes et al., 1966; Raps and DeMoss, 1962; Swings and DeLey, 1977). *Zymomonas* has several advantages over the commonly utilized yeast strains: (1) it has higher volumetric ethanol productivities (120 g/l/h for *Zymomonas* versus 30 g/l/h for *Saccharomyces*) in continuous culture employing recycle; (2) it has high ethanol yields (1.8–1.9 mole ethanol/mole glucose utilized); (3) it tolerates high (10–25%) sugar concentrations; and (4) it produces lower quantities of biomass. Its tolerance to ethanol is comparable to that of the best yeast strains. However, the organism has higher maintenance energy requirements, the maintenance coefficient being a function of the initial sugar concentration. Currently available wild-type and mutant strains of *Z. mobilis* can ferment only glucose, fructose, and sucrose; they are incapable, as are yeast, of metabolizing pentoses and the insoluble carbohydrates starch and cellulose. The kinetics of growth and ethanol production by the two wild-type strains ATCC 10988 (referred to as ZM1) and ZM4 (isolated from sugarcane juice), a more ethanol tolerant strain, have been studied in batch and continuous culture (K. J. Lee et al., 1979, 1980; Rogers et al., 1980a, b); the batch culture results as a function of initial glucose concentration are summarized in Table 8.14. Yield and rate data for two mutant strains—ZM7 derived from ATCC 10988 and ZM10 derived from NCIB 11199—have been obtained for the fermentation of sucrose (Lyness and Doelle, 1980) and for strains ZM1 and ZM4 for the fermentation of glucose (K. J. Lee et al., 1981b) in batch culture as a function of temperature. Growth and ethanol production on two practical substrates—molasses (Van Vuuren and Meyer, 1982) and

TABLE 8.14 Summary of Yield and Rate Data for Growth and Ethanol Production by *Zymomonas mobilis* strains ZM1 and ZM4 in Batch Culture on Glucose

Strain	ZM1 (ATCC 10988)				ZM4 (isolated from sugarcane)				
Fermentation mode	Batch				Batch				
Temperature	30°C				30°C				
pH	5.0				5.0				
Glucose concentration (%)	10	15	20	25	10	15	20	25	30
Specific growth rate, (h^{-1})	0.212	0.165	0.146	0.133	0.35	0.27	0.22	0.18	0.13
Specific ethanol productivity, q_p (g/g/h)	2.50	2.52	2.49	2.53	5.2	4.2	5.1	5.4	4.3
Specific glucose uptake rate, q_s (g/g/h)	5.47	5.22	5.15	5.45	10.9	8.9	10.5	11.3	8.7
Cell yield, $y_{x/s}$ (g/g)	0.038	0.036	0.028	0.019	0.032	0.030	0.018	0.015	0.015
Ethanol yield, $y_{p/s}$ (g/g)	0.491	0.490	0.496	0.472	0.48	0.47	0.49	0.48	0.49
Ethanol yield (percent of theoretical)	96.3	96.1	97.2	92.5					
Maximum ethanol concentration (g/l)		77	100	102		78	105	117	127

From Cromie and Doelle (1980), K. J. Lee et al. (1980), and Rogers et al. (1979).

sugarcane chips (Shalita et al., 1981)—have been studied. Ethanol production by strain 10988 decreased with increasing molasses concentration, possibly as the result of an inhibitor in the substrate, and at high sugar concentrations (>10%) was significantly less than production by *S. cerevisiae*. Ethanol production on sugarcane chips by strain CP4 was comparable to that obtained with *S. cerevisiae*. Fermentations have also been conducted in membrane (Rogers et al., 1980a), immobilized cell (Arcuri, 1982; Arcuri et al., 1980; Grote et al., 1980), and vacuum fermentation systems (J. H. Lee et al., 1981).

2. Direct Fermentation of Cellulose to Ethanol. The biochemical conversion of cellulose could be simplified by integration of the saccharification of the biomass and the fermentation of the soluble sugars into a single process step. This can be accomplished in one of two ways. First, by the addition of cellulolytic and hemicellulolytic enzymes to the culture of a noncellulolytic alcohol-producing organism. For feasible operation, the cultural conditions of the organism must be compatible with those established as optimal for the activity of the enzyme. Possible combinations include *Saccharomyces cerevisiae* (which grows at 32°C, pH 4.2) or *Zymomonas mobilis* (which grows at 30°C, pH 5.0) supplemented with cellulases from *Trichoderma reesei* (with optimal activity at 50°C, pH 4.5), *Clostridium thermohydrosulfuricum* (which grows at 60–65°C, pH 6.0–7.0) supplemented with cellulases/hemicellulases from *Thermomonospora* sp. (with optimal activity at 55–65°C, pH 6.0–7.0), and *Thermoanaerobacter ethanolicus* (which grows at 37–78°C, pH 5.7–8.6) supplemented with cellulases from *Thielavia terrestris* (with optimal activity at 50–70°C, pH 3.5–5) or *Thermomonospora* sp. Besides potentially reducing the capital investment, this scheme provides a mechanism for minimizing or eliminating the inhibitory effect that cellobiose, which accumulates in hydrolysis mixtures owing to low or instable β-glucosidase, exerts on the enzymes and that normally limits the rate and extent of enzymatic hydrolysis.

The alternative process is the mixed thermophilic culture of cellulolytic and noncellulolytic microorganisms such as *Clostridium thermocellum* (McBee, 1950; Ng et al., 1977) and *Clostridium thermosaccharolyticum* (Avgerinos and Wang, 1980; Ng et al., 1977) or *Clostridium thermohydrosulfuricum* (Zeikus et al., 1980). *Cl. thermocellum* produces extracellular enzymes that hydrolyze both the cellulosic and hemicellulosic fractions of the biomass to soluble sugars, but the microorganism ferments only glucose. Both *Cl. thermosaccharolyticum* and *Cl. thermohydrosulfuricum* will ferment the accumulated pentoses. In laboratory experiments on the mixed cultures of *Cl. thermocellum* and *Cl. thermosaccharolyticum,* product mixtures containing ethanol and acetic and lactic acids in the weight ratio of 1/0.35/0.1 and 1/0.5/0.25 were obtained at conversion efficiencies of 0.57 and 0.32 grams of solvents per gram of cellulose from Solka Floc and corn stover, respectively (Cooney et al., 1978). The major limitation to this technology is the low concentrations of products in the fermentation broth (2.5% mixed solvents, <2% ethanol).

D. Recovery Systems

The problem of low product concentration and its impact on recovery costs is encountered in most fermentation systems. However, it is particularly acute in systems for the biological production of fuels and chemicals. As shown in Figs. 8.15 and 8.16, the energy required for recovery of *n*-butanol and ethanol by conventional distillation technology increases significantly as the concentration of either solvent in the feed to the distillation section decreases. At ethanol and butanol feed concentrations less than 6–8% and 3–4% by weight, respectively, the net energy recovery from the process is less than optimal. A number of strategies have been investigated to reduce the energy requirements for distillation, including improved heat integration through vapor recompression, vacuum distillation, the use of extractants such as salts to modify the vapor–liquid equilibrium (Furter and Cook, 1967), supercritical CO_2 extraction, and the use of selected adsorbents for dehydration of azeotropic mixtures (Hong et al., 1982; Ladisch and Dyck, 1979). The low concentrations of products in the fermentation broth are,

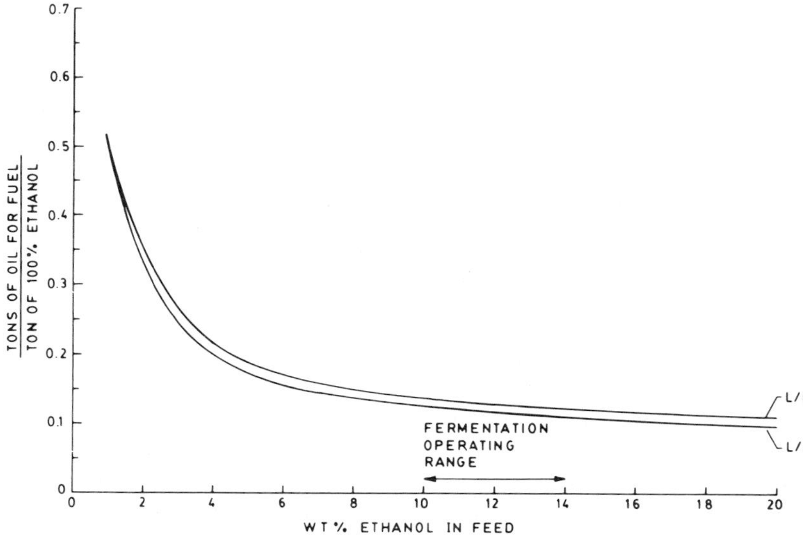

FIGURE 8.15 Energy requirements for the purification of ethanol by distillation as a function of the feed ethanol concentration and reflux ratio (atmospheric distillation to obtain 92 wt. % ethanol product; minimum reflux ratio = 1.27).

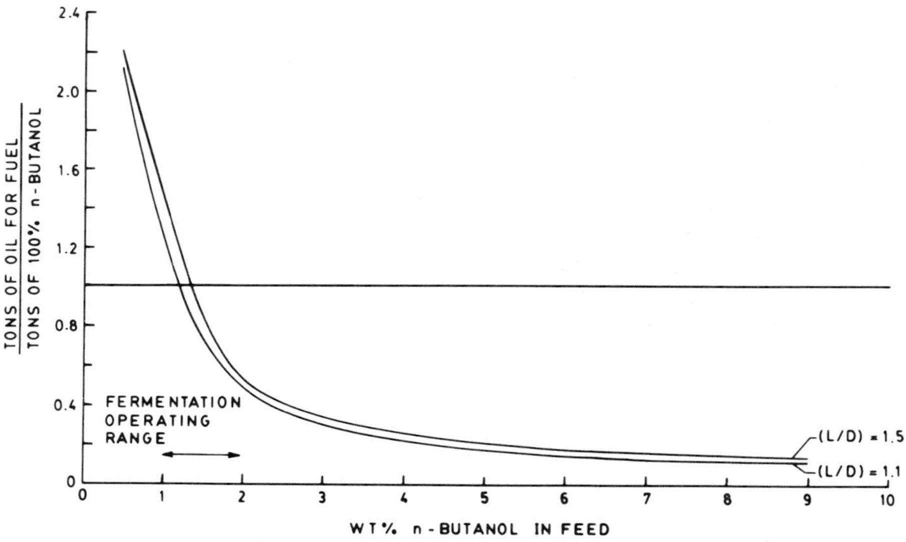

FIGURE 8.16 Energy requirements for the purification of *n*-butanol by distillation as a function of the feed ethanol concentration and reflux ratio (atmospheric distillation to obtain 48 wt. % *n*-butanol product; minimum reflux ratio = 1.0).

in most cases, a direct result of the inhibitory and toxic effects of the solvents on the microorganisms used to produce them. Elimination of the inhibition could be accomplished by continuous extraction of the product from the fermentation broth. If this removal was done so that the product was obtained from the fermentor in more concentrated form, then downstream energy requirements for product recovery could be reduced. Two modifications to standard fermentation technology have been proposed to accomplish this: vacuum fermentation (Boeckler, 1948; Cysewski and Wilke, 1977; Ramalingham and Finn, 1977) and extractive fermentation (Hashimoto, 1979; Hernandez-Mena et al., 1980; Ribaud, 1980).

In the vacuum system, the ethanol fermentation is conducted under a reduced pressure, usually 35–75 mm Hg if yeasts are used that grow optimally between 32° and 37°C. The concentration of ethanol in the exit gas stream of the fermentor is increased as a result of its higher volatility at these pressures. Economically and energetically, the system is limited by the vast quantities of CO_2 that must be handled by the compressors (Maiorella and Wilke, 1980). Total energy requirements are 14% greater than those required in conventional distillation; ethanol productivity, as a result of the removal of end product inhibition, yeast recycle, and continuous operation, is 80 g/l·h, or 40-fold higher than the productivity of conventional batch fermentation (Maiorella and Wilke, 1980). An alternative strategy—the FLASHFERM process (Maiorella et al., 1979)—has been proposed to reduce the compression requirements. In this process, the fermentation is conducted under atmospheric conditions, and the CO_2 gas generated is vented from the fermentor. The ethanol is stripped from the liquid phase by recirculating a stream of whole broth through an external flash chamber maintained at reduced pressures. The quantity of CO_2 then being handled is reduced to that which is soluble in the fermentation broth. Ethanol productivities of 80 g/l·h can be maintained. If the process is coupled to conventional atmospheric distillation, energy requirements are 12% greater than conventional operation; however, if it is coupled to vacuum distillation, an energy savings of 49% can be realized.

In extractive fermentation, the dilute solvent is stripped from the fermentation broth by continuous contacting of the liquor with an organic solvent into which the product can be absorbed (Akhnana, 1981; Hashimoto, 1979; Hernandez-Mena et al., 1980; Minier and Goma, 1981, 1982; Ribaud, 1980; H. Y. Wang et al., 1981). The extractant used should be selective for the product being recovered, immiscible in water, nontoxic to the microorganism, cheap in relation to the cost of the product, and volatile in relation to the product. The selection of an extractant that satisfies all of these criteria is difficult and represents the major obstacle to commercialization of this process. The use of polypropylene glycol P1200 (Ribaud, 1980) and 1-dodecanol (Minier and Goma, 1982) to extract ethanol and of dibutyl phthalate to extract *n*-butanol from the broths of *Saccharomyces cerevisiae* and *Cl. acetobutylicum* fermentations, respectively, has been shown to be effective in increasing both the amount of sugar utilized and the quantities of solvents produced.

III. PROCESS DESIGNS AND ECONOMICS

Several integrated processes based on the technologies outlined have been proposed. They include:

1. The acid hydrolysis scheme outlined by Grethlein (1978) and shown in Fig. 8.17. In this process, a slurry of newsprint is hydrolyzed at 260°F with 1% H_2SO_4 in a short residence time plug flow reactor to yield, after neutralization, a hydrolyzate containing glucose, hydroxymethylfurfural (HMF), xylose, furfural, and small quantities of acetic acid and $CaSO_4$. The quantities of HMF and furfural in the sugar stream are reduced by flash vaporization of the reactor effluent prior to neutralization. The furfural is recovered by distillation, and the residual solids from the hydrolysis are used to generate process steam.
2. The acid hydrolysis process being developed and piloted at Purdue University shown in Fig. 8.18 (Gong et al., 1981a; Ladisch and Dyck, 1979; Ladisch et al., 1978; Tsao and Chou, 1981; Tsao et al., 1978; Tsao et al., 1981). In this process, the hemicellulose fraction of ground corn stover is removed by dilute acid hydrolysis at 90–120°C for 1–4

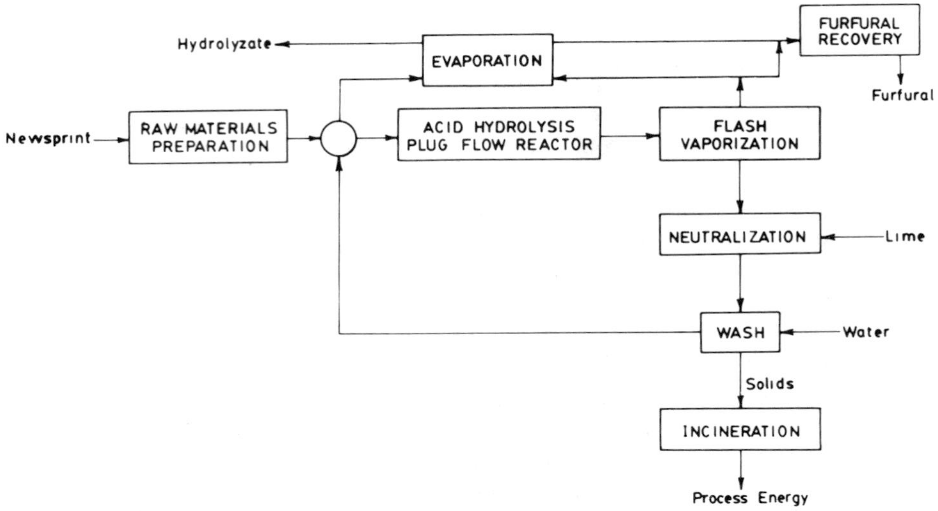

FIGURE 8.17 Outline of Grethlein/Dartmouth acid hydrolysis process.

h. The solubilized pentoses are removed by filtration, and the solids are swollen and solubilized in concentrated sulfuric acid at room temperature. After addition of water to reduce the acid concentration to 8%, the hydrolysis of cellulose to glucose is completed at 90–100°C. After filtration and neutralization of the sugar syrups, the hexoses and pentoses are fermented in a two-stage recycle reactor consisting of a

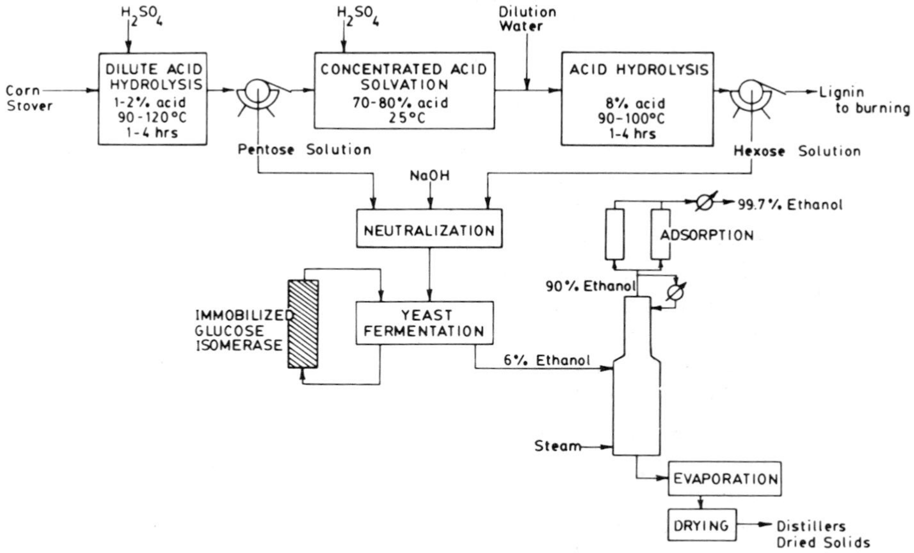

FIGURE 8.18 Outline of Tsao/Purdue acid hydrolysis process.

main fermentation vessel and an external column of immobilized glucose isomerase for conversion of the xylose to xylulose. The ethanol product is concentrated to approximately 90 wt. % by distillation; further purification is achieved by selective adsorption of the ethanol from the distillate vapor using cornmeal.

3. The enzymatic hydrolysis scheme developed by Natick (Allen, 1976; Spano et al., 1978), shown in Fig. 8.19, for the conversion of urban waste, wheat straw, or poplar to ethanol. The process consists of production of cellulases by aerobic fermentation of *Trichoderma reesei* on the raw cellulosic material, pretreatment of the cellulosic substrate by milling, enzymatic hydrolysis of the pretreated substrate to produce a 10% sugar solution, and fermentation of the sugar solution by *Saccharomyces cerevisiae* to produce ethanol. The by-products are spent solids from the cellulose fermentation and enzymatic hydrolysis steps.
4. The enzymatic hydrolysis scheme proposed by Wilke for the conversion of newsprint and corn stover (Wilke and Mitra, 1975; Wilke and Yang, 1975a, b; Wilke et al., 1976a, b, 1981), outlined in Fig. 8.20. This process involves (a) the pretreatment of the cellulosic material to enhance the susceptibility of the substrate to enzymatic hydrolysis; (b) countercurrent contacting of the pretreated substrate with the product sugar stream to adsorb enzyme onto the solids; (c) the production of makeup enzyme by continuous aerobic fermentation of *Trichoderma reesei* on product sugars and/or cellulose; (d) the enzymatic hydrolysis of the pretreated substrate to produce a sugar solution; (e) the concentration of the sugar solution in multieffect evaporators; (f) continuous anaerobic fermentation of the concentrated sugar solution by *Saccharomyces cerevisiae* to

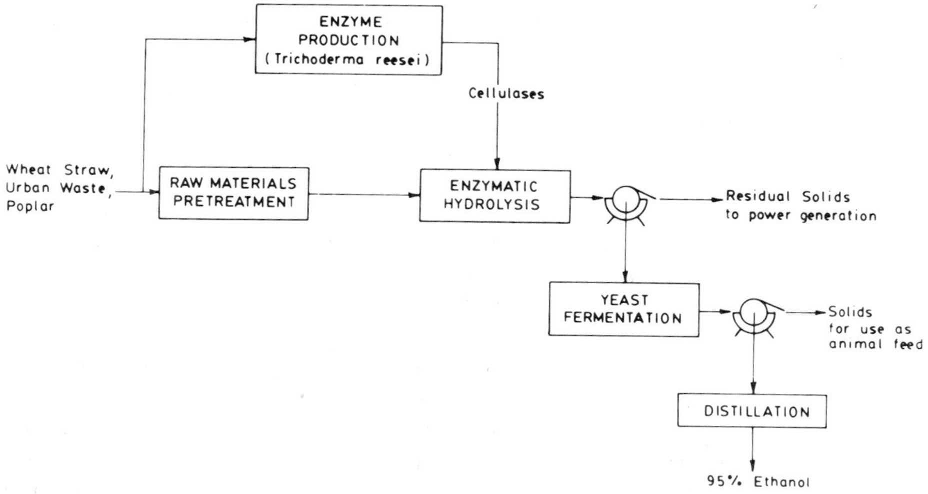

FIGURE 8.19 Outline of Natick enzymatic hydrolysis process.

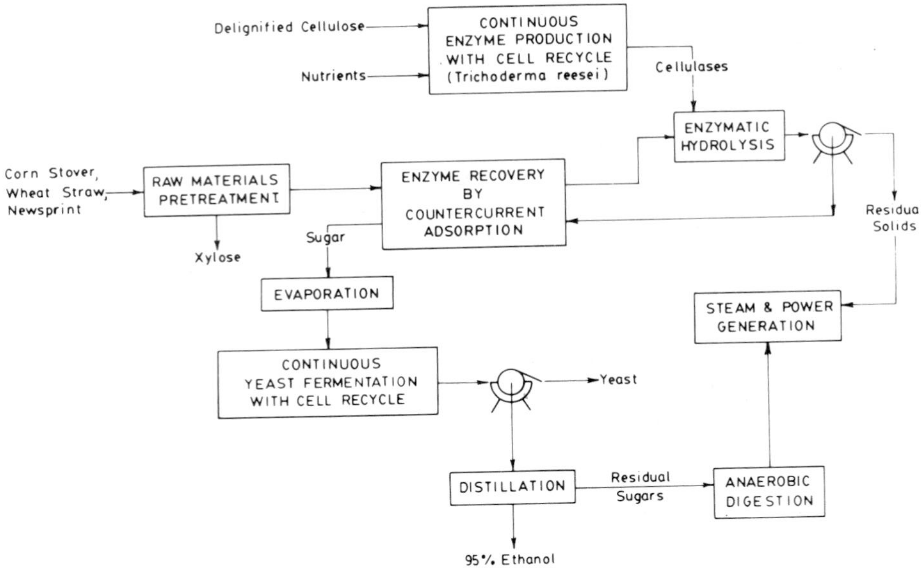

FIGURE 8.20 Outline of Wilke enzymatic hydrolysis process.

produce ethanol; and (g) concentration of ethanol by distillation. The enzyme production step is a two-stage cell growth/enzyme induction fermentation with continuous recycle of cell mass. By-products are spent solids cakes from the enzyme production and hydrolysis steps and yeast cakes produced in the ethanol fermentation step and from the residual sugars recovered during ethanol distillation.

5. The enzymatic hydrolysis process being jointly developed by research teams at Lehigh University, University of Pennsylvania, Hahnemann Medical College, and General Electric Co., outlined in Fig. 8.21, for the production of ethanol from hybrid poplars (Nolan, 1981). The process steps include (a) pretreatment of the poplar by solvent pulping with aqueous alcohol mixtures to isolate a high-quality lignin by-product and hemicellulose molasses; (b) cellulase production by the thermophilic actinomycetes *Thermomonospora* sp.; (c) hydrolysis of the cellulosic fraction of the pretreated wood to soluble sugars; (d) continuous anaerobic fermentation of these sugars to ethanol with *Saccharomyces cerevisiae;* and (e) recovery of the ethanol by extractive salt distillation.
6. The combined saccharification and fermentation originally conceived at Gulf, shown in Fig. 8.22, for the production of ethanol by *Saccharomyces cerevisiae* supplemented with *T. reesei* enzymes (Emert and Huff, 1978; Emert and Katzen, 1979, 1980; Emert et al., 1974; Gauss et al., 1976; Grachek et al., 1981; Horton et al., 1980; Huff and Yata, 1976). The feedstock is a mixture of municipal solids waste, sawmill waste, and pulp mill waste. Both the alcohol and enzyme production

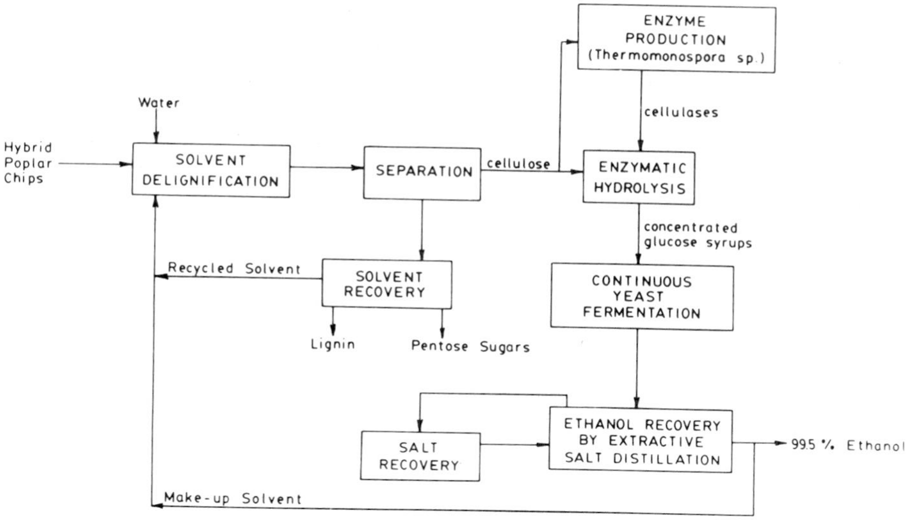

FIGURE 8.21 Outline of Lehigh/Penn/GE enzymatic hydrolysis process.

steps are continuous processes. The beer produced is about 3.5% ethanol and is recovered by distillation.

7. The continuous mixed culture fermentation developed at MIT for the production of ethanol and lactic and acetic acids from corn stover via the co-culture of *Cl. thermocellum* and *Cl. thermosaccharolyticum* (Avgerinos and Wang, 1980).

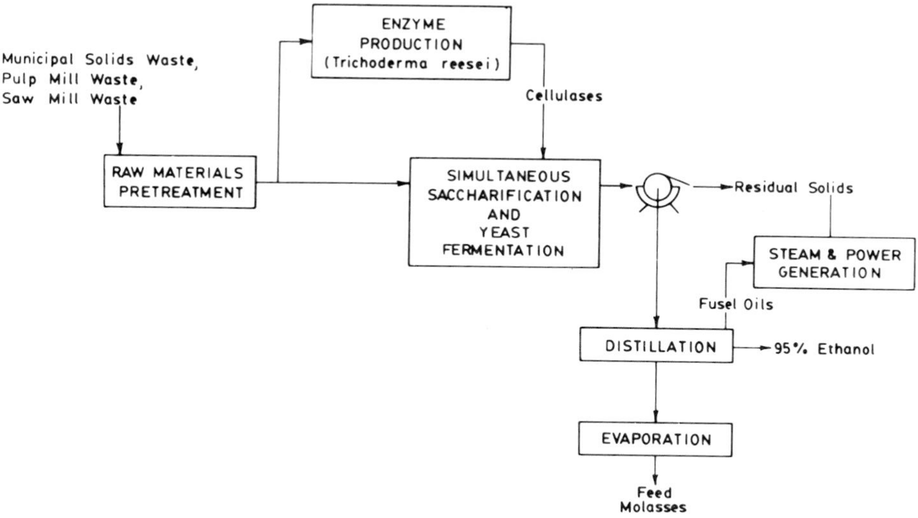

FIGURE 8.22 Outline of Emert/Gulf simultaneous saccharification–fermentation process.

Various attempts have been made to estimate the capital and operating expenses associated with these process schemes (Allen and Mortensen, 1981; Avgerinos and Wang, 1980; Cysewski and Wilke, 1978; Emert and Katzen, 1979; Nolan, 1981; Spano et al., 1978; Wilke et al., 1976a, 1981). The manufacturing costs calculated are based on varying assumptions of feedstock and utility costs, equity, interest rates, and return on investment; consequently, comparison of the information is difficult. However, from the costs calculated it is obvious that

1. the economics of acid hydrolysis depend strongly on the recoverability of the acids used;
2. the major cost of the enzymatic hydrolysis processes is associated with the enzyme production step; and
3. the competitive position of any process for the conversion of cellulosic materials is highly dependent on the feedstock costs, the recoverability of any pretreatment chemicals used, the sugar yields obtainable from the hydrolysis, and the product recovery techniques employed.

These conclusions pinpoint the areas of needed research.

REFERENCES

Akhnana, L. Y. (1981) "Alcohol Fermentation Involving Liquid–Liquid Extraction," M.S. thesis, University of Manchester Institute of Science and Technology, Manchester, England.

Allen, A. L. (1976) *AIChE Symp. Ser. 72* (158), 115–118.

Allen, A. L., and Mortensen, R. E. (1981) *Biotech. Bioeng. 23,* 2641–2645.

Anderson, L. L. (1977) in *Fuels from Waste* (Anderson, L. L., and Tillman, D. A., eds.), Academic Press, New York.

Andren, R. K., Mandels, M. H., and Medeiros, J. E. (1975) *Appl. Poly. Symp. 28,* 205–219.

Andren, R. K., Erickson, R. J., and Medeiros, J. E. (1976a) *Biotech. Bioeng. Symp. 6,* 177–203.

Andren, R. K., Mandels, M., and Medeiros, J. E. (1976b) *Proc. Biochem. 11,* 2–11.

Andreotti, R. E., Mandels, M., and Roche, C. (1978) Proceedings of the Bioconversion Symposium, pp. 249–267, IIT, Delhi.

Andreotti, R. E., Medeiros, J. E., Roche, C. R., and Mandels, M. (1980) presented at Second International Course-cum-Symposium on Bioconversion and Biochemical Engineering, IIT February 17–March 6, IIT, Delhi.

April, G. C., Kamal, M. M., Reddy, J. A., Bowers, G. H., and Hansen, S. M. (1979) *TAPPI 62*(5), 83–85.

Arcuri, E. J. (1982) *Biotech. Bioeng. 24,* 595–604.

Arcuri, E. J., Worden, R. M., and Shumate, S. E., II (1980) *Biotech. Lett. 2*(11), 499–504.

Aronovsky, S. S., and Gortner, R. A. (1930) *Ind. Eng. Chem. 22*(3), 264–274.
Aronovsky, S. S., and Gortner, R. A. (1936) *Ind. Eng. Chem. 28*(11), 1270–1276.
Avgerinos, G. C., and Wang, D. I. C. (1980) in *Annual Reports on Fermentation Processes* Vol. IV (Tsao, G. T., ed.) Academic Press, New York.
Bailey, A. J. (1942) *Ind. Eng. Chem. 34*(4), 483–485.
Bartley, D. A., Vigil, S. A., and Tchobanoglous, G. (1981) *Biotech. Bioeng. Symp. 10,* 67–80.
Bharoocha, R. (1980) "Chemicals from Wood by Organic Solvent Delignification," M.S. thesis, University of Alabama, Tuscaloosa.
Boeckler, B. C. (1948) U.S. Patent 2,440,925.
Bowers, G. H., and April, G. C. (1977) *TAPPI 60*(8), 102–104.
Brown, D. E., and Halsted, D. J. (1975) *Biotech. Bioeng. 17,* 1199–1210.
Brown, D. E., and Zainudeen, M. A. (1977) *Biotech. Bioeng. 19,* 941–958.
Bungay, H. R. (1981) *Energy, The Biomass Options,* John Wiley, New York.
Burt, J. P. (1975) *Agricultural Waste Management Field Manual,* Soil Conservation Service, U.S. Department of Agriculture, Washington, D.C.
Castanon, M., and Wilke, C. R. (1980) *Biotech. Bioeng. 22,* 1037–1053.
Chang, M. M., Chou, T. Y. C., and Tsao, G. T. *Adv. Biochem. Eng. 20,* 16–42.
Chemical & Engineering News (1975) "Wood Draws Attention as Plastics Feedstock," *53*(16), 13–14.
Chemical & Engineering News (1976) "Costs Prohibit Cellulosics Use as Feedstock," *54*(16), 12.
Chemical & Engineering News (1978) "SRI Discovers High-Temperature Cellulase," *56,* 22.
Chemical & Engineering News (1979) "Continuous Cellulose-to-Glucose Process," *57,* 19–20.
Chemical & Engineering News (1981) "USDA Process Converts Xylose to Ethanol," *59,* 54–55.
Cinq-Mars, G. V., and Howell, J. (1977) *Biotech. Bioeng. 19,* 377–385.
Cooney, C. L., Wang, D. I. C., Wang, S.-D., Gordon, J., and Jimenez, M. (1978) *Biotech. Bioeng. Symp. 8,* 103–114.
Cowling, E. B., and Brown, W. (1969) in *Cellulases and Their Applications* (Hajny, G. J., and Reese, E. T., eds.) *Advances in Chemistry Series* Vol. 95, American Chemical Society, Washington, D.C.
Cromie, S., and Doelle, H. W. (1980) *Biotech. Lett. 2,* 357–362.
Cuskey, S. M., Schambart, D. H. J., Chase, T., Jr., Montenecourt, B. S., and Eveleigh, D. E. (1980) *Dev. Ind. Microbiol. 21,* 471–480.
Cuskey, S. M., Frein, E. M., Montenecourt, B. S., and Eveleigh, D. E. (1981) presented at FEMS Symposium on Overproduction of Microbial Products, Czechoslovakia.
Cysewski, G. R., and Wilke, C. R. (1977) *Biotech. Bioeng. 19,* 1125–1143.
Cysewski, G. R., and Wilke, C. R. (1978) *Biotech. Bioeng. 20,* 1421–1444.
Dale, B. E., and Moreira, M. J. (1982) presented at Fourth Symposium on Biotechnology in Energy Production and Conservation, Gatlinburg, Tennessee, May 11–14.
Dawes, E. A., Ribbons, D. W., and Large, P. J. (1966) *Biochem. J. 98,* 795–803.
Dekker, R. F. H., and Richards, G. N. (1976) *Adv. Carb. Chem. 32,* 277–352.
Detroy, R. W., Lindenfelsen, L. A., Sommer, S., and Orton, W. L. (1981) *Biotech. Bioeng. 23,* 1527–1535.

Dunlap, C. E., Thomson, J., and Chiang, L. C. (1976) *AIChE Symp. 72*(158), 58–63.
Dunning, J. W., and Lathrop, E. C. (1945) *Ind. Eng. Chem. 37*(1), 24–29.
Dwivedi, C. P., and Ghose, T. K. (1979) *J. Ferment. Technol. 57*(1), 15–24.
Emert, G. H., and Huff, G. (1978) testimony before Special Senate Hearing on Alcohol Fuels, January 31.
Emert, G. H., and Katzen, R. E. (1979) presented at 72nd Annual AIChE Meeting, San Francisco, Calif., November 29.
Emert, G. H., Gum, E. K., Jr., Lang, J. A., Liu, T. H., and Brown, R. D., Jr. (1974) in *Food Related Enzymes* (Whitaker, J. R., ed.) *Advances in Chemistry Series* Vol. 136, American Chemical Society, Washington, D.C.
Emert, G. H., Katzen, R., Frederickson, R. E., and Kaupisch, K. F. (1980) *Chem. Eng. Prog. 76,* 47–52.
Eveleigh, D. E. (1981) *Scientific American 245*(3), 154–179.
Fagan, R. D., Grethlein, H. E., Converse, A. O., and Porteus, A. (1971) *Environ. Sci. Tech. 5*(6), 545–547.
Faith, W. L. (1945) *Ind. Eng. Chem. 37*(1), 9–11.
Falkehag, S. I. (1975) *Appl. Poly. Symp. 28,* 247–257.
Fan, L. T., Lee, Y.-H, and Beardmore, D. H. (1980) *Adv. Biochem. Eng. 14,* 101–117.
Fan, L. T., Lee, Y.-H., and Beardmore, D. H. (1981a) *Biotech. Bioeng. 23,* 419–424.
Fan, L. T., Gharpuay, M. M., and Lee, Y.-H. (1981b) *Biotech. Bioeng. Symp. 11,* 29–45.
Ferchak, J. D., Hägerdal, B., and Pye, E. K. (1980) *Biotech. Bioeng. 22,* 1527–1542.
Furter, W. F., and Cook, R. A. (1967) *Int. J. Heat Mass Transfer 10,* 23–36.
Gallo, B. J., Andreotti, R., Roche, C., Ryu, D., and Mandels, M. (1979) *Biotech. Bioeng. Symp. 8,* 89–102.
Gascoigne, J. A., and Gascoigne, M. M. (1960) *Biological Degradation of Cellulose,* Butterworths, London.
Gauss, W. F., Suzuki, S., and Takagi, M. (1976) U.S. Patent 3,990,944.
Ghose, T. K. (1969) *Biotech. Bioeng. 11,* 239–261.
Gilbert, N., Hobbs, S. A., and Levine, J. D. (1952) *Ind. Eng. Chem. 44*(7), 1712–1720.
Gilligan, W., and Reese, E. T. (1954) *Can. J. Microbiol. 1,* 90–107.
Glasser, W. G. (1981) *Forest Prod. J. 31*(3), 24–29.
Goheen, D. W., Glennie, D. W., and Hoyt, C. H. (1963) *Kirk-Othmer Encyclopedia of Chemical Technology 12,* 361–381.
Goldstein, I. S. (1975) *Science 189*(4206), 847–852.
Goldstein, I. S. (1976) *Biotech. Bioeng. Symp. 6,* 293–301.
Gong, C.-S., Chen, L.-F., Flickinger, M. C., Chiang, L.-C., and Tsao, G. T. (1981a) *Appl. Environ. Microbiol. 41*(2), 430–436.
Gong, C.-S., Chen, L.-F., Flickinger, M. C., and Tsao, G. T. (1981b) *Adv. Biochem. Eng. 20,* 93–118.
Gong, C.-S., Ladisch, M. R., and Tsao, G. T. (1981c) *Biotech. Lett. 3*(11), 657–662.
Gong, C.-S., McCracken, L. D., and Tsao, G. T. (1981d) *Biotech. Lett. 3*(5), 245–250.

Gracheck, S. J., Rivers, D. B., Woodford, L. C., Giddings, K. E., and Emert, G. H. (1981) *Biotech. Bioeng. Symp. 11,* 47–65.

Grant, G. W., Han, Y. W., Anderson, A. W., and Frey, K. L. (1977) *Dev. Ind. Microbiol. 18,* 599–611.

Grethlein, H. (1978) *Biotech. Bioeng. 20,* 503–525.

Griffin, H. L., Sloneker, J. H., and Inglett, G. E. (1974) *Appl. Microbiol. 27*(6), 1061–1066.

Gritzali, M., and Brown, R. D., Jr. (1979) in *Hydrolysis of Cellulose: Mechanisms of Enzymatic and Acid Catalysis* (Brown, R. D., Jr., and Jurasek, L., eds.) *Advances in Chemistry Series* Vol. 181, American Chemical Society, Washington, D.C.

Grote, W., Lee, K. J., and Rogers, P. L. (1980) *Biotech. Lett. 2*(11), 481–486.

Hägerdal, B., Ferchak, J., and Pye, E. K. (1978) *Appl. Environ. Microbiol. 36,* 606–612.

Hägerdal, B., Harris, H., and Pye, E. K. (1979) *Biotech. Bioeng. 21,* 345–355.

Hägerdal, B., Ferchak, J. D., and Pye, E. K. (1980) *Biotech. Bioeng. 22,* 1515–1526.

Halliwell, G. (1962) *Biochem. J. 85,* 67–72.

Halliwell, G. (1975) *Proceedings of the Symposium on Enzymatic Hydrolysis of Cellulose,* pp. 319–336. Aulanko, Finland.

Halliwell, G., and Riaz, M. (1970) *Biochem. J. 116,* 35–42.

Halliwell, G., and Vincent, R. (1977) *Proceedings of the Bioconversion Symposium,* pp. 413–434, IIT, Delhi.

Hansen, S. (1981) "The Role of Solvent in Aqueous-Organic Solvent Delignification of Southern Yellow Pine," Ph.D. thesis, University of Alabama, Tuscaloosa.

Harris, E. E., Beglinger, E., Hajny, G. J., and Sherrard, E. C. (1945) *Ind. Eng. Chem. 37*(1), 12–23.

Hashimoto, Y. (1979) "Optimization of an Extractive Fermentation for the Production of Butanol," M.S. thesis, University of Pennsylvania, Philadelphia.

Hernandez-Mena, R., Ribaud, J. A., and Humphrey, A. E. (1980) Poster No. F-7.3.8, Sixth International Fermentation Symposium, London, Ontario, July 20–25.

Holtzapple, M. T. (1981) "The Pretreatment and Enzymatic Saccharification of Poplar Wood," Ph.D. thesis, University of Pennsylvania, Philadelphia.

Hong, J., Voloch, M., Ladisch, M. R., and Tsao, G. T. (1982) *Biotech. Bioeng. 24,* 725–730.

Horton, G. L., Rivers, D. B., and Emert, G. H. (1980) *Ind. Eng. Chem. Prod. Res. Dev. 19,* 422–429.

Howsman, J. A., and Sisson, W. A. (1954) *Cellulose and Cellulose Derivatives,* Interscience Publishers, New York.

Hsu, T. A., Ladisch, M. R., and Tsao, G. T. (1980) *CHEMTECH 10,* 315–319.

Huff, G. F., and Yata, N. (1976) U.S. Patent 3,990,945.

Hulme, M. A., and Stranko, D. W. (1970) *Nature 226,* 469–470.

Jeffries, T. W. (1981a) *Biotech. Lett. 3*(5), 213–218.

Jeffries, T. W. (1981b) *Biotech. Bioeng. Symp. 11,* 315–324.

Jurasek, L. (1979) *Dev. Ind. Microbiol. 20,* 177–183.

Kamakura, M., and Kaetsu, I. (1978) *Biotech. Bioeng. 20,* 1309–1315.

Kamakura, M., and Kaetsu, I. (1982) *Biotech. Bioeng. 24,* 991–997.

Kanda, T., Nakakubo, S., Wakabayashi, K., and Nisizawa, K. (1979) in *Hydrolysis of Cellulose: Mechanisms of Enzymatic and Acid Catalysis* (Brown, R. D., Jr., and Jurasek, L., eds.) *Advances in Chemistry Series* Vol. 181, American Chemical Society, Washington, D.C.

Katz, M., and Reese, E. T. (1968) *Appl. Microbiol. 16*(2), 419–420.

Katzen, R., Frederickson, R., and Brush, B. F. (1980) *Chem. Eng. Prog. 76*(2), 62–67.

Kelsey, R. G., and Shafizadeh, F. (1980) *Biotech. Bioeng. 22,* 1025–1036.

King, K. W. (1965) *J. Ferment. Technol. 43*(2), 79–94.

Kleinert, T. N. (1974) *TAPPI 57*(8), 99–102.

Kleinert, T. N. (1975) *TAPPI 58*(8), 170–176.

Knappert, D., Grethlein, H., and Converse, A. (1980) *Biotech. Bioeng. 22,* 1449–1463.

Knappert, D., Grethlein, H., and Converse, A. (1981) *Biotech. Bioeng. Symp. 11,* 67–77.

Kouklos, E. G., and Valkanas, G. N. (1982) *Ind. Eng. Chem. Prod. Res. Dev. 21*(2), 309–314.

Krieger, J. (1982) *Chemical & Engineering News 60*(12), 38.

Ladisch, M. R. (1979) *Proc. Biochem. 14,* 21–23, 25.

Ladisch, M. R., and Dyck, K. (1979) *Science 205,* 898–900.

Ladisch, M. R., Ladisch, C. M., and Tsao, G. T. (1978) *Science 120,* 743–745.

Lee, K. J., Tribe, D. E., and Rogers, P. L. (1979) *Biotech. Lett. 1,* 421–426.

Lee, K. J., Skotnicki, M. L., Tribe, D. E., and Rogers, P. L. (1980) *Biotech. Lett. 2,* 339–344.

Lee, K. J., Skotnicki, M. L., Tribe, D. E., and Rogers, P. L. (1981a) *Biotech. Lett. 3*(5), 207–212.

Lee, K. J., Skotnicki, M. L., Tribe, D. E., and Rogers, P. L. (1981b) *Biotech. Lett. 3*(6), 291–296.

Lee, J. H., Woodward, J. C., Pagan, R. J., and Rogers, P. L. (1981) *Biotech. Lett. 3*(4), 177–182.

Li, L. H., Flora, R. M., and King, K. W. (1965) *Arch. Biochem. Biophys. 111,* 439–447.

Loewenberg, J. R., and Chapman, C. M. (1977) *Arch. Microbiol. 113,* 61–64.

Lora, J. H., and Wayman, M. (1978) *TAPPI 61*(6), 47–50.

Lyness, E., and Doelle, H. W. (1980) *Biotech. Lett. 2*(12), 549–554.

Lyness, E., and Doelle, H. W. (1981) *Biotech. Lett. 3*(5), 257–260.

Macdonald, D. G., and Mathews, J. F. (1979) *Biotech. Bioeng. 21,* 1091–1096.

Maiorella, B., and Wilke, C. R. (1980) *Biotech. Bioeng. 22,* 1749–1751.

Maiorella, B., Blanch, H. W., and Wilke, C. R. (1979) presented at 72nd National AIChE Meeting, November 29, San Francisco, Calif.

Mandels, M. (1979) Proceedings of the Third Annual Biomass Energy Systems Conference, pp. 281–290, Golden, Col.

Mandels, M., and Sternberg, D. (1976) *J. Ferment. Technol. 54,* 267–286.

Mandels, M., Weber, J., and Parizek, R. (1971) *Appl. Microbiol. 21,* 152–154.

Mandels, M., Hontz, L., and Brandt, D. (1972) *Proceedings of the Army Science Conference,* pp. 16–31, West Point, N.Y.

Mandels, M., Hontz, L., and Nystrom, J. (1974) *Biotech. Bioeng. 16,* 1471–1493.

Mandels, M., Medeiros, J. E., Andreotti, R. E., and Bissett, F. H. (1981) *Biotech. Bioeng. 23,* 2009–2026.

McBee, R. H. (1950) *Bacteriol. Rev. 14,* 51–63.

McMillen, J. M., Gortner, R. A., Schmitz, H., and Bailey, A. J. (1938) *Ind. Eng. Chem. 30*(12), 1407–1409.

Mendelsohn, H. R., and Wettstein, P. (1981) *Chem. Eng. 88*(12), 62–63, 65.

Millett, M. A., Baker, A. J., and Satter, L. D. (1975) *Biotech. Bioeng. Symp. 5,* 193–219.

Millett, M. A., Baker, A. J., and Satter, L. D. (1976) *Biotech. Bioeng. Symp. 6,* 125–153.

Millett, M. A., Effland, M. J., and Caulfield, D. F. (1979) in *Hydrolysis of Cellulose: Mechanisms of Enzymatic and Acid Catalysis* (Brown, R. D., Jr., and Jurasek, L., eds.) *Advances in Chemistry Series* Vol. 181, American Chemical Society, Washington, D.C.

Minier, M., and Goma, G. (1981) *Biotech. Lett. 3*(8), 405–408.

Minier, M., and Goma, G. (1981) *Biotech. Bioeng. 24,* 1565–1579.

Montenecourt, B. S., Schamhart, D. H. J., Cuskey, S. M., and Eveleigh, D. E. (1979) *Proceedings of the Third Annual Biomass Energy Systems Conference* pp. 85–90, Golden, Col.

Montenecourt, B. S., Frein, E. M., Cuskey, S. M., and Eveleigh, D. E. (1981) presented at the Second World Congress of Chemical Engineers, October 4–9, Montreal.

Moreira, A. R., Phillips, J. A., and Humphrey, A. E. (1981a) *Biotech. Bioeng. 23,* 1325–1338.

Moreira, A. R., Phillips, J. A., and Humphrey, A. E. (1981b) *Biotech. Bioeng. 23*(6), 1339–1348.

Myerly, R. C., Nicholson, M. D., Katzen, R., and Taylor, J. M. (1981) *CHEMTECH 10* (March 1981), 186–192.

Nagai, S., Onodera, M., and Aiba, S. (1976) *Eur. J. Appl. Microbiol. 3,* 9–18.

Neese, N., Wallick, J., and Harper, J. M. (1977) *Biotech. Bioeng. 19,* 323–336.

Nelson, R., and Schuerch, C. (1956) *J. Poly. Sci. 22,* 435–448.

Nelson, R., and Schuerch, C. (1957) *TAPPI 40*(6), 419–426.

Ng, T. K., Weimer, P. J., and Zeikus, J. G. (1977) *Arch. Microbiol. 114,* 1–7.

Nisizawa, T., Suzuki, H., and Nisizawa, K. (1971) *J. Biochem. 70,* 387–393.

Nolan, E. J. (1981) "A Systems Analysis for the Production of Selected Fuels and Chemicals from Biomass," Ph.D. thesis, University of Pennsylvania, Philadelphia.

Noo Mangat, M., and Howell, J. A., (1978) *AIChE Symp. 74*(172), 77–81.

Nystrom, J. (1975) *Biotech. Bioeng. Symp. 5,* 221–224.

Nystrom, J. M., and Allen, A. L. (1976) *Biotech. Bioeng. Symp. 6,* 55–74.

Nystrom, J. M., and DiLuca, P. H. (1978) *Proceedings of the Bioconversion Symposium,* pp. 293–304, IIT, Delhi.

Office of Solid Waste (1975) *A Technical, Environmental and Economic Evaluation of the "Wet Processing System for the Recovery and Disposal of Municipal Solid Waste,"* Environmental Protection Agency, Washington, D.C.

Office of Technology Assessment (1980a) *Energy from Biological Processes,* U.S. Congress, Washington, D.C.

Office of Technology Assessment (1980b) *Energy from Biological Processes* Vol. II, *Technical and Environmental Analyses,* U.S. Congress, Washington, D.C.

Peitersen, N. (1975) *Biotech. Bioeng. 17,* 361–374.

Peitersen, N. (1977) *Biotech. Bioeng. 19,* 337–348.

Peitersen, N., Medeiros, J., and Mandels, M. (1977) *Biotech. Bioeng. 19,* 1091–1094.

Pew, J. C., and Weyna, P. (1962) *TAPPI 45*(3), 247–256.
Phillips, J. A., and Humphrey, A. E. (1980) presented at Second International Course-cum-Symposium on Bioconversion and Biochemical Engineering, Feb. 17–March 6, IIT, Delhi.
Prescott, S. C., and Dunn, C. G. (1949) *Industrial Microbiology,* McGraw-Hill, New York.
Ramalingham, A., and Finn, R. K. (1977) *Biotech. Bioeng. 19,* 583–589.
Rånby, B. (1969) in *Cellulases and Their Applications* (Hajny, G. J., and Reese, E. T., eds.) *Advances in Chemistry Series* Vol. 95, American Chemical Society, Washington, D.C.
Raps, S., and DeMoss, R. D. (1962) *J. Bacteriol. 84,* 115–118.
Reddy, C. A., and Forney, L. (1978) *Dev. Ind. Microbiol. 19,* 27–34.
Reese, E. T., and Gilligan, W. (1953) *Arch. Biochem. Biophys. 45,* 74–82.
Reese, E. T., and Mandels, M. (1980) *Biotech. Bioeng. 22,* 323–335.
Reese, E. T., and Ryu, D. Y. (1980) *Enzyme Microbiol. Technol. 2,* 239–240.
Reese, E. T., Siu, R. H. G., and Levinson, H. S. (1950) *J. Bacteriol. 59,* 485–487.
Reese, E. T., Segal, L., and Tripp, V. W. (1957) *Text. Res. J. 27,* 626–632.
Reese, E. T., Lola, J. E., and Parrish, F. W. (1969) *J. Bacteriol. 100*(3), 1151–1154.
Reilly, P. J. (1981) in *Trends in the Biology of Fermentations for Fuels and Chemicals* (Hollaender, A., ed.), Plenum Press, New York.
Ribaud, J. (1980) "Feasibility Study on the Use of Extractive Fermentation to Enhance Ethanol Production in the Yeast Fermentation," M.S. thesis, University of Pennsylvania, Philadelphia.
Rogers, P. L., Lee, K. J., and Tribe, D. E. (1979) *Biotech. Lett. 1,* 165–170.
Rogers, P. L., Lee, K. J., Skotnicki, M. L., and Tribe, D. (1980a) presented at Second International Course-cum-Symposium on Bioconversion and Biochemical Engineering, February 17–March 6, IIT, Delhi.
Rogers, P. L., Lee, K. J., and Tribe, D. E. (1980b) *Proc. Biochem. 15,* 7–11.
Rose, A. H., ed. (1978) *Primary Products of Metabolism* Vol. II, *Economic Microbiology,* Academic Press, New York.
Rosenberg, S. L. (1980) *Enzyme Microbiol. Technol. 2,* 185–193.
Rowland, S. P. (1975) *Biotechnol. Bioeng. Symp. 5,* 183–191.
Ryu, D. D. Y., Lee, S. B., Tassinari, T., and Macy, C. (1982) *Biotech. Bioeng. 24,* 1047–1067.
Saddler, J. N., Brownell, H. H., Clermont, L. P., and Levitin, N. (1982) *Biotech. Bioeng. 24,* 1389–1402.
Saeman, J. (1945) *Ind. Eng. Chem. 37,* 43–52.
Sarkanen, K. V. (1980) in *Progress in Biomass Conversion* (Sarkanen, K. V., and Tillman, D. A., eds.) Academic Press, New York.
Sarkanen, K. V., and Ludwig, C. H. (1971) *Lignins: Occurrence, Formation, Structure and Reactions,* John Wiley, New York.
Sasaki, T., Tanaka, T., Nanbu, N., Sato, Y., and Kainuma, K. (1979) *Biotech. Bioeng. 21,* 1031–1042.
Schneider, H., Wang, P. Y., Chan, Y. K., and Maleszka, R. (1981) *Biotech. Lett. 3*(2), 89–92.
Schoenmaker, S. (1980) *Alternative Sources of Energy 45,* 18–20.
Schweers, W. H. M. (1974) *CHEMTECH 4* (August), 490–493.
Selby, K. (1969) in *Cellulases and Their Applications* (Hajny, G. J., and Reese,

E. T., eds.) *Advances in Chemistry Series* Vol. 95, American Chemical Society, Washington, D.C.

Selby, K., and Maitland, C. C. (1967) *Biochem. J. 104,* 716–724.

Shafizadeh, F., and McGinniss, G. D. (1971) *Adv. Carb. Chem. Biochem. 26,* 297–349.

Shalita, Z. P., White, M. D., Katz, M., Zur, M., and Mizrahi, A. (1981) *Biotech. Lett. 3*(12), 729–733.

Sherrard, E. C., and Kressman, F. W. (1945) *Ind. Eng. Chem. 37*(1), 4–8.

Shin, S. B., Kitagawa, Y., Suga, K., and Schikawa, K. (1978) *J. Ferment. Technol. 56*(4), 396–402.

Siu, R. G. H. (1951) *Microbial Decomposition of Cellulose,* Reinhold, New York.

Skinner, W. A., and Tokuyama, F. (1978) U.S. Patent 4,081,328.

Sloneker, J. H. (1976) *Biotech. Bioeng. Symp. 6,* 235–250.

Solar Energy Research Institute (1980) *Biomass Refining Newsletter* (Summer), Golden, Col.

Spano, L. A. (1976) *Proceedings of the IGT Symposium on Clean Fuels from Biomass, Sewage, Urban Refuse and Agricultural Wastes* pp. 325–348, Orlando, Fla.

Spano, L. A., Medeiros, J., and Mandels, M. (1976) *Resource Recovery and Conservation 1,* 279–294.

Spano, L., Allen, A., Tassinari, T., and Mandels, M. (1978) *Proceedings of the Second Annual Symposium on Fuels from Biomass,* pp. 671–684, DOE, Troy, N.Y.

Stephens, G. R., and Heichel, G. H. (1975) *Biotech. Bioeng. Symp. 5,* 27–42.

Sternberg, D. (1976a) *Biotech. Bioeng. Symp. 6,* 35–53.

Sternberg, D. (1976b) *Appl. Environ. Microbiol. 31*(5), 648–654.

Sternberg, D., and Dorval, S. (1979) *Biotech. Bioeng. 18,* 181–191.

Sternberg, D., and Mandels, G. R. (1979) *J. Bacteriol. 139,* 761–769.

Sternberg, D., and Mandels, G. R. (1980) *J. Bacteriol. 144,* 1197–1199.

Sternberg, D., Vijayakumar, P., and Reese, E. T. (1977) *Can. J. Microbiol. 23,* 139–147.

Stone, N. A. (1976) *Biotech. Bioeng. Symp. 6,* 223–234.

Swings, J., and DeLey, J. (1977) *Bacteriol. Rev. 41,* 1–46.

Tangnu, S. K., Blanch, H. W., and Wilke, C. R. (1981) *Biotech. Bioeng. 23,* 1837–1849.

Tarkow, H., and Feist, W. C. (1969) in *Cellulases and Their Applications* (Hajny, G. J., and Reese, E. T., eds.) *Advances in Chemistry Series* Vol. 95, American Chemical Society, Washington, D.C.

Tassinari, T., and Macy, C. (1977) *Biotech. Bioeng. 19,* 1321–1330.

Tassinari, T., Macy, C., Spano, L., and Ryu, D. D. Y. (1980) *Biotech. Bioeng. 22,* 1689–1705.

Tasssinari, T. H., Macy, C. F., and Spano, L. A. (1982) *Biotech. Bioeng. 24,* 1495–1505.

Thompson, D. R., and Grethlein, H. E. (1979) *Ind. Eng. Chem. Prod. Res. Dev. 18*(3), 166–169.

Tillman, D. A. (1977) in *Fuels from Waste* (Anderson, L. L. and Tillman, D. A., eds.), Academic Press, New York.

Timell, T. E. (1964) *Adv. Carb. Chem. 19,* 247–302.

Tsao, G. T., and Chou, T. Y. (1981) U.S. Patent 4,266,981.

Tsao, G. T., Ladisch, M., Ladisch, C., Hsu, T. A., Dale, B., and Chou, T. (1978) in *Annual Reports on Fermentation Processes* Vol. II (Perlman, D., ed.), Academic Press, New York.
Tsao, G. T., Ladisch, M. R., Ladisch, C. M., and Hsu, T.-A. (1981) U.S. Patent 4,281,063.
Turbak, A. F., Hammer, R. B., Davies, R. E., and Hergert, H. L. (1980) *CHEMTECH 10,* 51–57.
Tyner, W. E. (1981) *Biotech. Bioeng. Symp. 10,* 81–90.
Ueng, P. P., Hunter, C. A., Gong, C.-S., and Tsao, G. T. (1981) *Biotech. Lett. 3*(6), 315–320. U.S. Department of Agriculture (1978) *Agricultural Statistics,* U.S. Government Printing Office, Washington, D.C.
Van Vuuren, H. J. J., and Meyer, L. (1982) *Biotech. Lett. 4*(4), 253–256.
Wang, H. Y., Robinson, F. M., and Lee, S. S. (1981) *Biotech. Bioeng. Symp. 11,* 555–566.
Wang, P. Y., and Schneider, H. (1980) *Can. J. Microbiol. 26,* 1165–1168.
Wang, P. Y., Johnson, B. F., and Schneider, H. (1980a) *Biotech. Lett. 2*(6), 273–278.
Wang, P. Y., Shopsis, C., and Schneider, H. (1980b) *Biochem. Biophys. Res. Comm. 94*(1), 248–254.
Wayman, M., and Lora, J. H. (1979) *TAPPI 62*(9), 113–114.
Wenzl, H. F. J. (1970) *The Chemical Technology of Wood,* Academic Press, New York.
Whistler, R. L., and Richards, E. L. (1970) in *The Carbohydrates,* (Pigman, W., and Hoston, D., eds.), Academic Press, New York.
Wilke, C. R., and Mitra, G. (1975) *Biotech. Bioeng. Symp. 5,* 253–274.
Wilke, C. R., and Yang, R. D. (1975a) *Appl. Poly. Symp. 28,* 175–188.
Wilke, C. R., and Yang, R. D. (1975b) in *Proceedings of the Symposium on Enzymatic Hydrolysis of Cellulose,* pp. 485–506, Aulanko, Finland.
Wilke, C. R., Cysewski, G. R., Yang, R. D., and von Stockar, U. (1976a) *Biotech. Bioeng. 18,* 1315–1323.
Wilke, C. R., Yang, R. D., and von Stockar, U. (1976b) *Biotech. Bioeng. Symp. 6,* 155–175.
Wilke, C. R., Yang, R. D., Sciamanna, A. F., and Freitas, R. P. (1981) *Biotech. Bioeng. 23,* 163–183.
Wood, T. M. (1968) *Biochem. J. 109,* 217–227.
Wood, T. M. (1970) *World Review of Nutrition and Dietetics 12,* 227–265.
Wood, T. M. (1972) *Proceedings of the Fourth International Fermentation Symposium: Fermentation Technology Today,* pp. 711–718, Osaka, Japan.
Wood, T. M., and McCrae, S. I. (1972) *Biochem. J. 128,* 1183–1192.
Wood, T. M., and McCrae, S. I. (1975) *Proceedings of the Symposium on Enzymatic Hydrolysis of Cellulose,* pp. 231–254, Aulanko, Finland.
Wood, T. M., and McCrae, S. I. (1977) *Proceedings of the Bioconversion Symposium,* pp. 1–10, IIT, Delhi.
Wood, T. M., and McCrae, S. I. (1979) in *Hydrolysis of Cellulose: Mechanisms of Enzymatic and Acid Catalysis,* (Brown, R. D., Jr., and Jurasek, L., eds.) *Advances in Chemistry Series* Vol. 181, American Chemical Society, Washington, D.C.
Zeikus, J. G., Ben-Bassat, A., and Hegge, P. W. (1980) *J. Bacteriol. 143,* 432–440.

CHAPTER

9

Applications of Oxidative and Reductive Biocatalysis

David R. Light
James R. Swartz

I. INTRODUCTION

In the following presentation, we address the field of the biocatalysis of oxidative and reductive reactions. This chapter is not intended to be a comprehensive review; such a work could fill an entire volume. Instead, we would like to provide an overview of the area, with particular emphasis on the basic enzymatic mechanisms involved as well as some of the problems involved in industrial applications. These become especially relevant as the emerging power of recombinant DNA technology is applied to commercial biocatalysis.

The use of enzymes and microorganisms for oxidative and reductive chemical transformations is well established. An enzymatic approach offers regioselectivity and stereoselectivity and eliminates competing side reactions and the danger of overoxidation or overreduction. Excellent reviews cover the field. Reviews that address the use of microorganisms (Sih and Rosazza, 1976; Johnson, 1978; Fonken and Johnson, 1972) and enzymes (Hamilton, 1976; Jones, 1980) to accomplish oxidations and oxygenations of functional groups and classes of chemicals are recommended, as are reviews devoted to biotransformations of classes of specific compounds including steroids (Kieslich, 1980a, 1980b; Kieslich and Sebek, 1979), aromatic and aliphatic hydrocarbons (Abbott and Gledhill, 1971; Perry, 1977; McKenna and Kallio, 1965; Pirnik, 1977; Gibson, 1971; Cain, 1980), alkaloids (Vining, 1980), and drugs (Smith and Rosazza, 1975; Abbott, 1976; Rosazza and Smith, 1979). While such reviews present many useful examples and applications, they could be of even greater use to the design of biotechnological processes if the examples of oxidative transformations given also included the cofactor and coenzyme requirements as well as the mechanism of the enzymes involved. Unfortunately, the type of enzyme responsible for a microbial transformation is often unknown, and for many enzymes, the mechanism remains to be established.

Until recently, the development of bioconversion processes has been dependent on the availability of appropriate microorganisms or enzymes and on the ability of genetic selection to enhance the levels of the requisite enzymes, eliminate feedback inhibition, or eliminate competing side reactions. However, useful selection strategies are often unattainable, and, particularly in the area of oxidative enzymology, many potentially useful enzymes are found in organisms for which the design of selective strategies is difficult. For general synthetic use, the most useful enzyme will catalyze the modification of a particular functional group on a molecule but will be fairly nonspecific with respect to the binding tolerance for the remainder of the molecule (Hamilton, 1976). This explains the interest in the broad substrate specificity of the bacterial methane monooxygenases (Higgins, 1982; Higgins et al., 1980; Dalton, 1980; Colby et al., 1977), which catalyze the hydroxylation and epoxidation of a wide variety of substrates. However, an even broader range of oxygenations can be catalyzed by the cytochrome P450 enzymes in the mammalian liver (Wislocki et al., 1980; Ullrich, 1979; Lu and West, 1980). But while the use of mammalian liver microsomal enzyme systems for industrial processes has been proposed (Sofer, 1979), it is unlikely that they will be useful for large-scale reactions in their present form. Recombinant DNA technology offers the potential to clone and express in high yields enzymes and proteins from any source, including higher plants and animals as well as microorganisms that may otherwise have little potential to be developed into a useful strain for fermentation. Thus the range of potential catalysts that are now, in theory, available to the investigator interested in the development of oxidative processes has been significantly expanded.

As the range of biocatalysis increases, so does the importance of large-scale process development. Even with a well-designed catalytic system, factors such as capital and substrate costs, cofactor requirements, and substrate solubility may render a process uneconomical. These problems will become increasingly important as the field emerges and recombinant DNA technology begins to be applied to oxidative biocatalysis.

II. INFLUENCE OF ENZYME CHARACTERISTICS ON BIOCATALYTIC PROCESS DESIGN

When a biocatalytic process is examined, it is particularly important to be aware of the nature of the enzyme catalyst, the coenzyme requirements, and the mechanism for a given oxidative reaction. Each can greatly influence the process design.

A. Nature of Enzyme Catalyst

Whether enzymes are isolated and utilized or are cloned into a new organism, the total number of polypeptides required must be known. Enzymes that catalyze oxidative transformations, particularly the monooxygenases (Dalton, 1980; Griffith et al., 1978; Colby and Dalton, 1978; Gunsalus et al., 1974) may require additional electron transferring protein components to supply reducing equivalents to the active site of the monooxygenase. Often, the requirement for a particular ferredoxin or flavoprotein is quite exacting, and the possibilities for substituting an alternate reductant are limited. Thus for some microbial transformations, it may be necessary to clone and express more than one gene product in order to assemble all the protein components required for catalysis. In addition, an active enzyme may be multimeric in nature and may require specific conditions in order to associate. Finally, enzymes that are normally membrane-bound may require a hydrophobic environment in order to express activity, in contrast to enzymes that are freely soluble in aqueous media.

B. Cofactor and Coenzyme Requirements

If microbial transformations are desired and if the protein components are cloned and expressed, it is also necessary to consider whether or not the host organism can supply in sufficient quantity the necessary cofactors and coenzymes required for catalysis. Cofactor and coenzyme requirements must also be examined with the use of isolated enzymes. This consideration is especially pertinent when enzymes that require unusual cofactors and coenzymes are evaluated, for example, methoxatin (Salisbury et al., 1979), 6-hydroxyflavins (Mayhew et al., 1974), or factor 420 (a derivative of 7,8-didemethyl-8-hydroxy-5-deaza FMN) (Jacobson et al., 1982; Eirich et al., 1978), or when more common cofactors are utilized in an unusual manner, for example, the natural covalent attachment of flavins to flavoenzymes (Ohta-Fukuyama et al., 1980; Edmondson et al., 1976; Nakagawa et al., 1975; Singer and Edmondson, 1974; Iwatsuki et al., 1980; Steenkamp et al., 1978; McIntire et al., 1980). To synthesize an active catalyst, the host organism must have the enzymatic capability of synthesizing the necessary coenzyme and of attaching the coenzyme to the apoenzyme.

C. Enzyme Mechanism

The enzyme mechanism will also impact on the final design of the process. Often it is possible to choose from several alternatives for the same resulting reaction; for example, if one is interested in the oxidation of toluene derivatives to benzylalcohols,

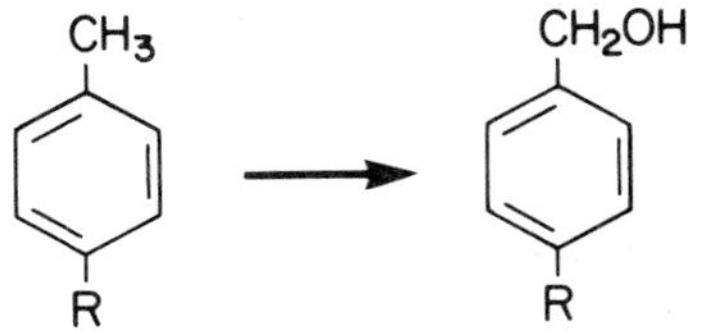

(1)

a number of potential enzyme systems exist. The methane monooxygenases (R = –H) from *Methyloccus capsulatus* (*Bath*) (Colby et al., 1977) or *Methylosinus trichosporium OB36* (Higgins et al., 1980), a microsomal cytochrome P450 (R = $-SCH_2CH_3$) from rat liver (Light et al., 1982; Waxman et al., 1982), and *p*-cresol methylhydroxylase (R = –OH) from *Pseudomonas putida* (McIntire et al., 1980, 1981) have all been shown to catalyze this type of reaction. The former two enzymes are different examples of monooxygenases (iron–sulfur containing and heme containing, respectively) that catalyze reactions with the following stoichiometry:

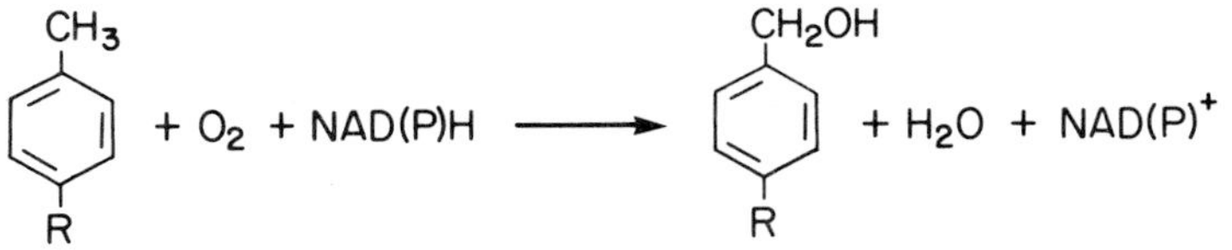

(2)

Thus, as will be discussed below, the monoxygenase oxygenation requires a source of two-electron equivalents and a mole of molecular oxygen for each cycle of catalysis. In contrast, the latter enzyme, *p*-cresol methylhydroxylase, is an example of nonpyridine-linked dehydrogenase that catalyzes a reaction with a much different stoichiometry:

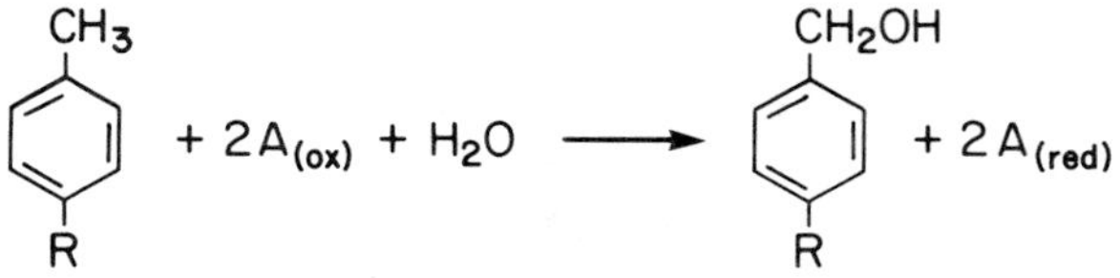

(3)

Here $A_{(ox)}$ and $A_{(red)}$ represent an oxidized and a reduced one-electron acceptor such as phenazine methosulfate (PMS), 2,6-dichlorophenolindophenol (DCPIP), or potassium ferricyanide. Among the considerations that will affect reactor design is that the latter enzyme may be run under anaerobic conditions, while the monooxygenase reaction must obviously be run in an aerated medium. One further example of the effect of enzyme mechanism on reactor design is provided by the consideration of the simple oxidation of β-D-glucose to δ-gluconolactone:

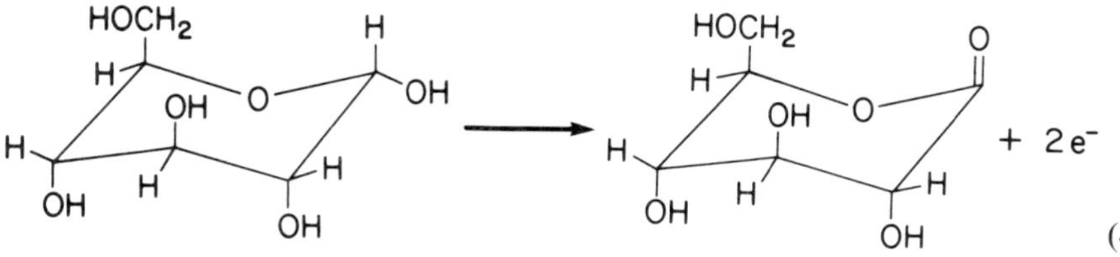

(4)

This oxidation is catalyzed by the FAD containing glucose oxidase from *Aspergillus niger* (Stankovich et al., 1978) and other sources (Barker and Shirley, 1980) (Eq. 5); by nonpyridine nucleotide–dependent methoxatin containing glucose dehydrogenases from *Acinetobacter calcoaceticus* (Duine et al., 1979), *Gluconobacter suboxydans,* and *Pseudomonas aeruginosa* (Ameyama et al., 1981; Matsushita et al., 1980) (Eq. 6); and by the pyridine nucleotide–dependent glucose dehydrogenase also from *Gluconobacter suboxydans* (Adachi et al., 1980) or other species (Barker and Shirley, 1980) (Eq. 7):

$$\text{glucose} + O_2 \longrightarrow \text{gluconolactone} + H_2O_2 \qquad (5)$$

$$\text{glucose} + 2\ A_{(ox)} \longrightarrow \text{gluconolactone} + 2\ A_{(red)} \qquad (6)$$

$$\text{glucose} + \text{NAD(P)}^+ \rightleftharpoons \text{gluconolactone} + \text{NAD(P)H} \qquad (7)$$

Again, constraints on process configuration will be experienced depending upon which of these three kinds of enzymes are used. Eqs. (5) and (6) are shown as being irreversible, since the oxidants are O_2 or one-electron acceptors with relatively high redox potentials,

$$E^{0'}\ (O_2) = +0.82\ \text{V}; \qquad E^{0'}\ (\text{ferricyanide}) = +0.36\ \text{V};$$
$$E^{0'}\ (\text{DCPIP}) = +0.22\ \text{V}; \qquad E^{0'}\ (\text{PMS}) = +0.08\ \text{V} \quad (\text{pH } 7,\ 30°\text{C})$$

compared with that of the glucose/gluconolactone couple ($E^{0'} = -0.45$ V). Glucose oxidase utilizes the relatively inexpensive oxidant, O_2, but generates reactive hydrogen peroxide, which must be removed from the reactor. The nonpyridine nucleotide-dependent dehydrogenase may be run anaerobically and generates a reduced species that may be regenerated electrochemically. Finally, Eq. (7) is written as a reversible reaction, reflecting the relatively low redox potential of the pyridine nucleotides ($E^{0'} = -0.32$ V). The pyridine nucleotide-dependent dehydrogenase may also be run anaerobically, and the reaction can be driven to completion by reoxidizing NADPH. However, the direct electrochemical reoxidation of NADPH is not yet practical. The topic of cofactor regeneration will be addressed in a separate section.

III. TYPES OF OXIDATIVE ENZYME MECHANISMS

Because the nature and mechanism of an oxidative enzyme impact so strongly on its practical application, this section is organized around the classification of enzyme types. Not all members of a particular class of enzyme are discussed, but rather representative examples. Thus the following classes of oxidative enzymes, some of which have already been mentioned, are discussed: dehydrogenases (pyridine nucleotide–dependent and non-pyridine nucleotide–dependent), oxidases, peroxidases and catalases, mono-oxygenases, and dioxygenases. This approach to the field of oxidative enzymology in biotechnological processes allows the convenient introduction of several important topics in places where they are most relevant to the type of enzyme under discussion. Sections that cover specific cofactors, coenzymes, deleterious effects of reactive intermediates of dioxygen reduction, phenol coupling, cooxidations, and plasmids coding for degradative oxidative enzymes are introduced at appropriate points. Finally, based on the premise that recombinant DNA technology will allow the utilization of enzymes from mammalian sources, comparisons between similar reactions carried out by enzymes from microbial sources and mammalian sources are made throughout.

A. Dehydrogenases

Dehydrogenases catalyze the often reversible oxidation or reduction of a given functional group without the direct participation of molecular oxygen (or of hydrogen peroxide). For example, in the dehydrogenation of a hydroxyl group to a carbonyl group,

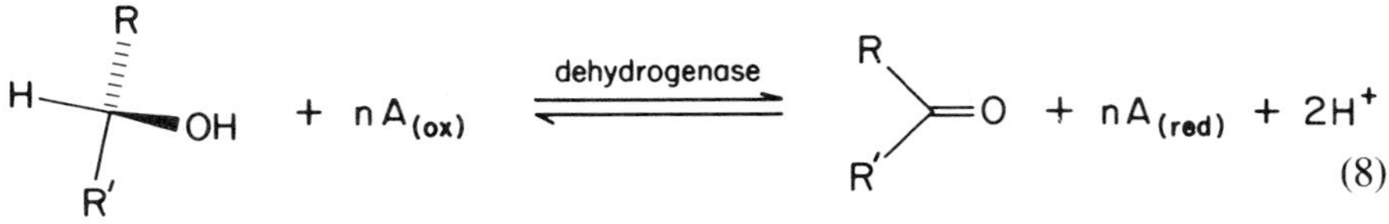

(8)

$A_{(ox)}$ may be any one of a number of electron acceptors, including one-electron-accepting dyes, cytochromes and ferredoxins ($n = 2$), or flavoproteins or two-electron-accepting dyes, quinones, or pyridine nucleotides ($n = 1$). It is convenient to consider the pyridine nucleotide-dependent dehydrogenases separately from other dehydrogenases.

1. Pyridine Nucleotide–Dependent Dehydrogenases. The pyridine nucleotide–dependent dehydrogenases may utilize either one or both of the relatively low-potential ($E^{0'} = -0.32$ V) obligate 2-electron redox cofactors, NAD^+ or $NADP^+$, which differ only by the phosphorylation of the 2′-hydroxyl of the adenosyl ribose. Hydride transfer occurs to and from carbon 4 of the nicotinamide ring:

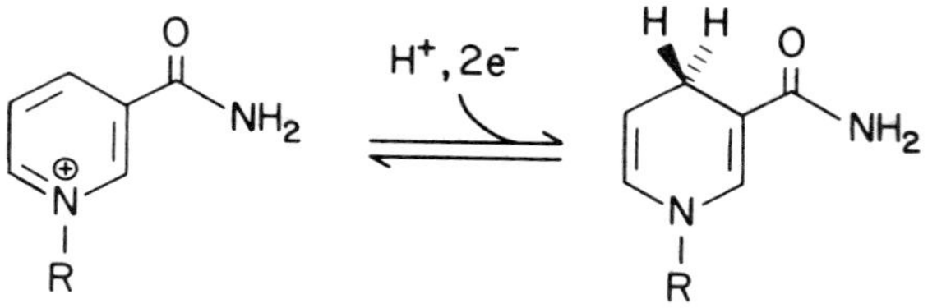

(9)

The three-dimensional structures of a number of pyridine nucleotide dependent dehydrogenases have been determined. These include lactate dehydrogenase (Adams et al., 1973; Holbrook et al., 1975), malate dehydrogenase (Tsernoglov et al., 1972; Banaszak and Bradshaw, 1975), liver alcohol dehydrogenase (Eklund et al., 1976; Samama et al., 1981; Cedergren-Zeppezauer et al., 1982; Eklund and Branden, 1979), and glyceraldehyde-3-phosphate dehydrogenase (Harris and Waters, 1976). The last-named enzyme is not a simple pyridine nucleotide–linked dehydrogenase but couples the oxidation of an aldehyde, glyceraldehyde-3-phosphate, to the formation of the high energy phosphate anhydride, 1,3-diphosphoglycerate:

$$\begin{array}{l} H{-}C{=}O \\ \quad | \\ H{-}C{-}OH \\ \quad | \\ \;CH_2OPO_3^{=} \end{array} + HPO_4^{=} + NAD^{+} \rightleftharpoons \begin{array}{l} O{=}C{-}OPO_3^{=} \\ \quad | \\ H{-}C{-}OH \\ \quad | \\ \;CH_2OPO_3^{=} \end{array} + H^{+} + NADH \qquad (10)$$

Pyridine nucleotide–dependent dehydrogenases not only function in the oxidation of alcohols, aldehydes, and amines, but can also catalyze the reduction of unsaturated carbon bonds. For example, 5α-reductase catalyzes the addition of hydride across a carbon–carbon double bond (Moore and Wilson, 1975). In this case, the substrate, androsterone, is a Michael acceptor, and the enzyme catalyzes the transfer of hydride from NADPH to C-5 of the steroid ring to form dihydrotestosterone as shown in Eq. (11):

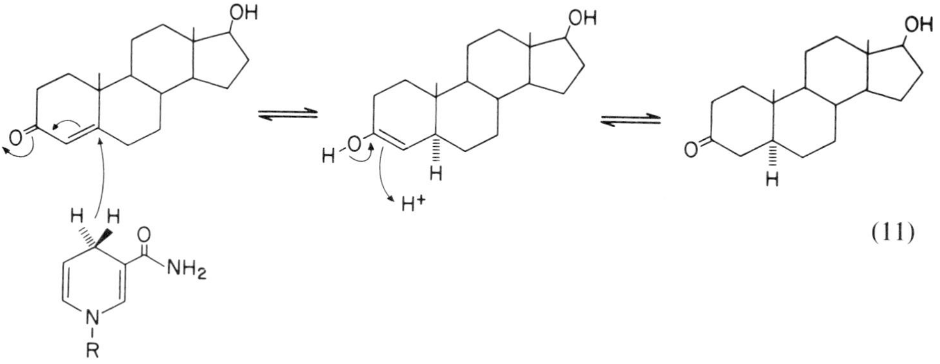

(11)

Although the hydride transfer step in Eq. (11) is shown as being reversible, it is not practical to utilize 5α-reductase in the direction of steroid oxidation (Moore and Wilson, 1975). The reduction of the double bond at carbon 7 of 5,7,22-cholestrienol is catalyzed by a similar NADPH-dependent enzyme (Koroly and Dempsey, 1981).

The pyridine nucleotide-linked dehydrogenases catalyze a variety of reactions, but there is a remarkable similarity in the coenzyme-binding domains of the enzymes that have been characterized (Ohlsson et al., 1974; Rossmann et al., 1974, 1975). Reduced pyridine nucleotides are the required electron source for a number of enzymes of interest to researchers in the area of biotechnology. Thus pyridine nucleotide–dependent dehydrogenases have received considerable attention because they offer a potential route to coenzyme regeneration. In addition, they catalyze a number of reactions of potential commercial utility.

As an example of this class of oxidative enzyme, some of the catalytic and structural features of the broadly specific mammalian liver alcohol dehydrogenase will be described. Mammalian liver alcohol dehydrogenase (LADH) is a metalloenzyme and a dimer of identical 40,000-dalton subunits. Each subunit contains one catalytic site zinc atom and one zinc atom involved only in stabilization of enzyme structure. The presence of a metal in the active site is not common to all pyridine nucleotide–linked dehydrogenases; however, a number of other alcohol and polyol dehydrogenases share this feature (Jornvall et al., 1981; Jeffery et al., 1981; Patel et al., 1981). The subunits of LADH are held together by hydrophobic interactions and cannot be dissociated in the native state. Portions of both subunits are required to form the hydrophobic substrate-binding "barrel" that leads the alcohol or aldehyde to the active site zinc. The pyridine nucleotide is bound by an entirely separate domain on each subunit. Thus a functional enzyme is 80,000 daltons and contains four zinc atoms, two substrate-binding sites, and two coenzyme-binding sites (Branden et al., 1975; Klinman, 1981). Although the possibility of cooperativity between the active sites has been explored (Dunn et al., 1979), the evidence does not support any form of cooperativity or "half-site" reactivity (Andersson and Pettersson, 1982). The combination of structural data and kinetic data has led to a rather thorough understanding of how this catalyst works (Klinman, 1981).

Depending on the substrate and the direction of the reaction, turnover of horse liver alcohol dehydrogenase is partially or completely limited by release of the coenzyme, NAD^+ or NADH (Klinman, 1981; Cook and Cleland, 1981). Although neither the rate of hydride transfer from the alcohol to the oxidized coenzyme nor the rate of hydride transfer from the reduced coenzyme to the aldehyde determines the turnover rate of the enzyme, the mechanism of this hydride transfer step has been extensively studied. Kinetic studies that involve both the determination of secondary isotope effects and the determination of equilibrium isotope effects have revealed the details of this hydride transfer step (Cook et al., 1981). The evidence suggests that, during the removal of the hydroxyl proton and formation of the alkoxide anion, alcohol dehydrogenase catalyzes the distortion of the pyridine ring of NAD^+ from a planar conformation into a boat conformation. This distortion gives the coenzyme carbonium ion character at carbon 4 and greatly facilitates the subsequent transfer of a hydride from carbon 1 of the alkoxide (Cook et al., 1981).

Liver alcohol dehydrogenase has broad substrate specificity, and its potential has been explored with numerous substrates. These include allylic alcohols and aldehydes (Pietruszko et al., 1973), α-haloalcohols (Kvassman et al., 1981), 2-substituted tetrahydrothiopyran-4-ones (J. Davies and Jones, 1979), bridged bicyclic alcohols and ketones (Irwin and Jones, 1976; Irwin et al., 1978), steroids (Pietruszko et al., 1966), and substituted 1, 5-pentane diols (Irwin and Jones, 1977). In the latter study, horse LADH was shown to catalyze the oxidation of 3-methylpentane-1, 5-diol with enantiotopic specificity to produce (−)-(3S)-3-methylvalerolactone of 90% optical purity when levels of NAD^+ are maintained by recycling NADH (Eq. 12):

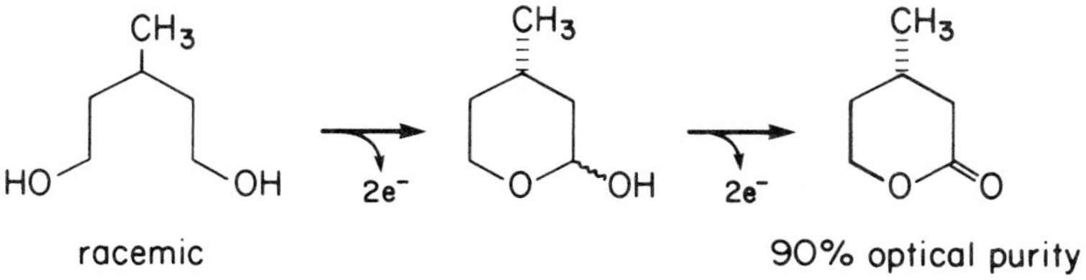

(12)

Note that this observation implies the ability to oxidize both the appropriate prochiral hydroxyethyl group and the intermediate hemiacetal.

Although alcohol dehydrogenases have not been used in traditional industrial processes, a number of processes have been developed that suggest areas of potential commercialization. Applications that have been patented or that have been demonstrated on a small scale include the production of chiral solvents from the appropriate ketones; immobilized enzyme electrodes for the selective detection of aldehydes and alcohols (Lamed et al., 1981); the reductive conversion of limonoate A-ring lactone to a nonbitter compound, 17-dehydrolimonoate A-ring lactone during the processing of orange and other citrus juices (Eq. 13) (Hasegawa et al., 1982); the resolution of ketone-

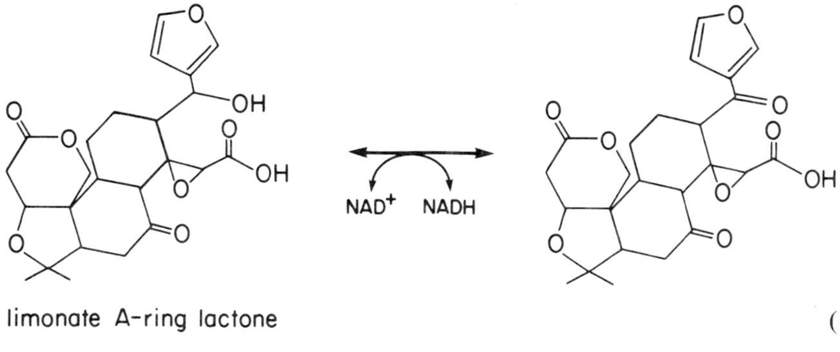

(13)

containing racemates by conversion to diastereomers (Sih and Rosazza, 1976); and the regioselective and stereoselective reduction of numerous steroids, alkaloids, prostaglandins, and miscellaneous carbonyl compounds (Sih and Rosazza, 1976; Fonken and Johnson, 1972; Kieslich, 1980a).

The problem of the low solubility of steroids in water has been addressed by the use of immobilized hydroxysteroid dehydrogenase in a two-phase water–organic solvent (ethylacetate or butylacetate) system (Carrea et al., 1979; Cremonesi et al., 1975). Additional utility may be gained by the alteration of the substrate specificity of yeast alcohol dehydrogenase by selection in the presence of allyl alcohol (Wills et al., 1982) or by the selection of natural variants specific for long-chain alkyl alcohols (Tassin and Vandecasteele, 1972) and secondary alcohols (Bryant and Ljungdahl, 1981; Patel et al., 1981; Niehaus et al., 1978).

2. Flavin-Containing Pyridine Nucleotide–Linked Dehydrogenases. Enzymes in this class catalyze the same reaction stoichiometry (Eq. 10) as the simple pyridine nucleotide–linked dehydrogenases. However, they contain the redox active flavin coenzyme (FAD or FMN) in their active sites. This additional coenzyme allows the movement of electrons from the obligate two-electron donor (NADH or NADPH) to either two-electron or one-electron acceptors (Eq. 14):

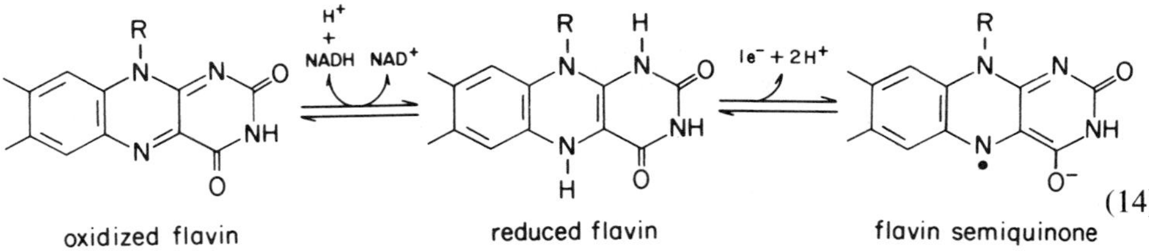

(14)

The one-electron acceptors that these pyridine nucleotide–linked flavoprotein dehydrogenases will accept as substrates include ferredoxins and cytochromes and thus allow these enzymes to function as electron transport catalysts. The ultimate electron acceptors include a variety of monooxygenases, which are described below, as well as the mechanistically less well-defined microsomal fatty acid desaturase (C. H. Williams, 1976). The examples of pyridine nucleotide–linked flavoprotein dehydrogenases that utilize two-electron acceptors as substrates all have active site disulfides in addition to the active site flavin. Reviews (C. H. Williams, 1976; Holmgren, 1980) are recommended for structural and mechanistic information concerning lipoamide dehydrogenase, which oxidizes reduced lipoamide to the disulfide form; glutathione reductase, which reduces the disulfide of oxidized glutathione; and thioredoxin reductase, which reduces the disulfide bond of the thioredoxin, a small electron transport protein active in electron transfer to ribonucleotide reductase, methionine sulfoxide reductase, and other redox enzymes. The structures of two of the intermediates of the reduction of oxidized glutathione by NADPH in the active site of crystalline glutathione reductase have been elucidated (Pai and Schulz, 1983). Thus it is possible to trace the transfer of electrons from the dihydronicotinamide portion of NADPH to the flavin of FAD; from the flavin to the active site disulfide; and from the resulting thiols to the disulfide of oxidized glutathione.

In this class of reactive disulfide–containing flavoprotein dehydrogenases is the recently characterized enzyme, mercuric reductase (Fox and Walsh, 1982), which catalyzes the following reaction:

$$\text{NADPH} + \text{RS-Hg-SR} \longrightarrow \text{NADP}^{+} + \text{Hg}^{\bullet} + 2\ \text{RSH} \qquad (15)$$

This enzyme is of potential biotechnological interest, since it will also reduce salts of gold and silver to colloidal gold and silver metal (Summers and Sugarman, 1974).

3. Dehydrogenases Not Dependent Upon Pyridine Nucleotides. A wide variety of enzymes catalyze the dehydrogenase reaction (shown for the dehydrogenation of an alcohol in Eq. (8)) and do not utilize pyridine nucleotides as the electron acceptor. Common coenzymes for this class of enzymes are, again, the flavins (FAD or FMN). For example, fatty acyl–CoA dehydrogenase contains noncovalently bound FAD and catalyzes the β-oxidation of the coenzyme A thioesters of saturated fatty acids coupled to the reduction of the protein electron acceptor, electron transfer flavoprotein (Crane and Beinert, 1956). A key intermediate in the reaction catalyzed by fatty acyl–CoA dehydrogenase is a charge transfer complex between the C-2 anion of the thioester substrate and oxidized FAD (Eq. 16) (Schmidt et al., 1981; McFarland et al., 1982):

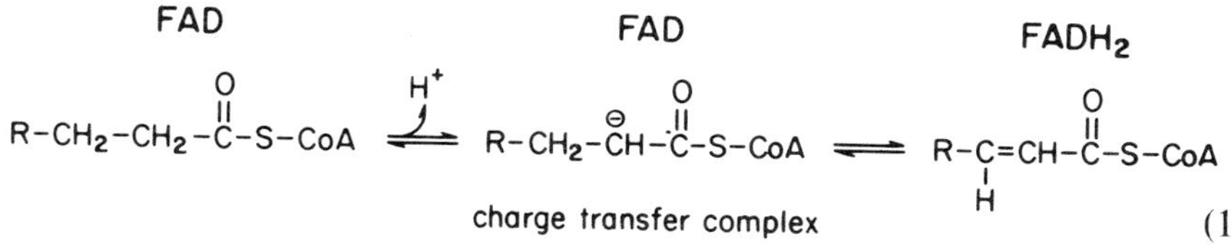

(16)

The bacterial enzyme, carbon monoxide dehydrogenase, contains three moles of FAD and couples the oxidation of carbon monoxide to quinone reduction (Kim and Hegeman, 1981). Another bacterial dehydrogenase, *p*-cresol methylhydroxylase, contains, in addition to a covalently bound FAD, a *c*-type cytochrome (McIntire et al., 1981) (Eq. 3). Succinate dehydrogenase also contains covalently bound FAD as well as iron–sulfur centers (Hatefi and Stiggall, 1976).

Additional coenzymes that have been described in nonpyridine nucleotide–linked dehydrogenases include heme and methoxatin. A heme containing amine dehydrogenase has been characterized that catalyzes the oxidative deamination of benzylamine to benzaldehyde in the presence of cytochrome *c* as an electron acceptor (Durham and Perry, 1978). The methoxatin-containing nonpyridine-linked dehydrogenases couple electron transfer from their respective substrates to a number of artificial electron acceptors that substitute for an as yet uncharacterized physiological acceptor. Methoxatin is a coenzyme for a methanol dehydrogenase purified from a strain of methylotrophic bacteria, *Pseudomonas* M27. When first observed, this coenzyme could not be classified with any of the known coenzymes (Anthony and Zatman, 1967), but it has subsequently been identified as an orthoquinone (Salisbury et al., 1979; Duine et al., 1980; Cruse et al., 1980):

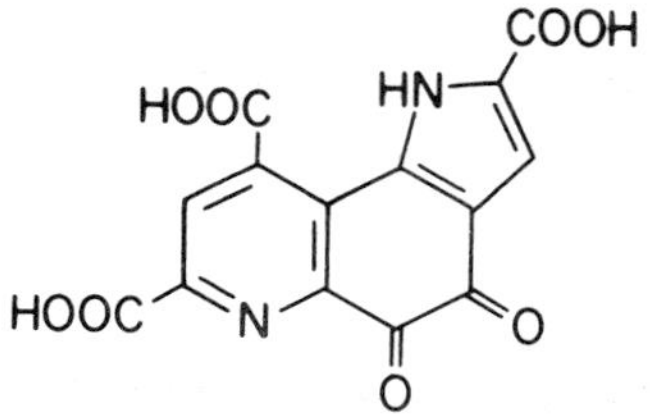

(17)

Methoxatin has also been identified as the coenzyme that is bound to a variety of dehydrogenases, including several methanol dehydrogenases (Ohta et al., 1981; Patel et al., 1978; Duine et al., 1978), an alcohol dehydrogenase (Duine and Frank, 1981a, 1981b), and several glucose dehydrogenases (Duine et al., 1979, 1980; Ameyama et al., 1981), and that has been implicated as being the prosthetic group covalently bound to methylamine dehydrogenase (De Beer et al., 1980). While a number of mechanisms for the role of methoxatin in methanol oxidation have been proposed (Higgins et al., 1981; Forrest et al., 1980), two observations need to be reconciled in a successful mechanism. By using electron spin resonance (ESR) spectroscopy measurements of an active sample of the methoxatin-containing methanol dehydrogenase from *Methylomonas methanica,* it was demonstrated that a substantial amount of the coenzyme existed in the semiquinone form and that this one-electron reduced form could be correlated with the active enzyme (Mincey et al., 1981). However, studies of methanol dehydrogenase from *Hyphomicrobium X* have led to the view that free radical species are not catalytically competent (De Beer et al., 1983). A second observation that must be explained is the fact that the purified enzyme requires ammonia in order to be activated (Higgins et al., 1981).

Hydrogenases may be grouped with the nonpyridine nucleotide–linked dehydrogenases (although an $NADP^+$-dependent hydrogenase has been characterized (Gray and Gest, 1965)). Methanogenic bacteria utilize an FAD-containing dehydrogenase to effect the oxidation of H_2 coupled to the reduction of the cofactor F_{420} (Jacobson et al., 1982). Additional hydrogenases link H_2 oxidation to ferredoxin reduction and are iron–sulfur enzymes (Schlegel and Schneider, 1978; Orme-Johnson, 1973; Erbes et al., 1975; Gitlitz and Krasna, 1975). Hydrogenases also catalyze an exchange reaction between hydrogen isotopes that does not require an additional electron acceptor (Eq. 18) (Krasna, 1979).

$$^{3}HH + H_2O \rightleftharpoons H_2 + {}^{3}HHO \tag{18}$$

Because enzymes can readily function in aqueous solutions, the utilization of the O_2-insensitive hydrogenase from *Alcaligenes eutrophus* has been suggested as a catalyst to affect the detritiation of water from nuclear power plants by taking advantage of the isotope effect on this exchange reaction (Klibanov and Huber, 1981).

B. Enzymes That Utilize Oxygen and Related Species

The dehydrogenases are capable of catalyzing a wide variety of redox reactions; however, they have a disadvantage when large-scale production is contemplated, since they require expensive cofactors. Enzymes that are capable of coupling the oxidation of a starting material directly to oxygen have the advantage of utilizing a cheap and available oxidant. The complete reduction of molecular oxygen to water requires a total of four electrons, and some biochemically relevant intermediates of the reduction of O_2 are shown:

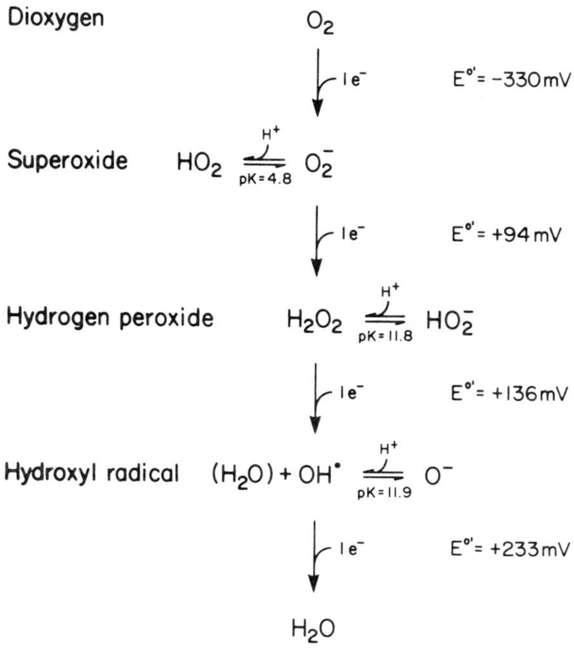

(19)

The remainder of the enzymes that will be discussed either utilize or generate one or more of the oxygen-derived species shown.

1. The Oxidases. The oxidases catalyze the two-electron reduction of molecular oxygen to hydrogen peroxide,

$$AH_2 + O_2 \rightarrow A + H_2O_2 \qquad (20)$$

coupled to the two-electron oxidation of a variety of substrates. Because they are capable of using this inexpensive oxidant directly, they are of considerable interest to researchers in the area of biotechnology.

a. Types of Oxidases. Flavin-containing oxidases and copper-containing oxidases have been described; however, the majority of oxidases are of the former type. The simplest flavoprotein oxidases contain one mole of flavin per mole of monomeric protein as in the case of glucose oxidase (Stankovich et al., 1978) and cyclohexylamine oxidase from *Pseudomonas* (Tokeida et al., 1977); however, methanol oxidase from yeast is isolated as an octamer of eight identical subunits with one FAD per subunit (Courderc and Baratti, 1980; Mincey et al., 1980). Bovine liver monoamine oxidase does not contain iron as had once been thought and has recently been shown to be a simple flavoenzyme (Ichinose et al., 1982). Pig liver glycolate oxidase contains the FMN level of the unusual flavin, 6-hydroxyflavin (Mayhew et al., 1974). The simple flavoprotein oxidases have been reviewed by Bright and Porter (1975).

The catalytic mechanism of the flavoprotein oxidases can be divided into two half-reactions. In the first half-reaction, the flavin coenzyme is reduced by the substrate as in the left-hand side of Eq. (14) (but with the substitution of the appropriate substrate for NADH). In the case of L-α-hydroxy acid oxidase and D-amino acid oxidase, this reductive half-reaction has been shown to involve the initial formation of an α-carbanion intermediate (Cromartie and Walsh, 1975; Marcotte and Walsh, 1978; Bruice, 1980; Walsh, 1980). The second half-reaction is the oxidation of the reduced coenzyme by molecular oxygen and is thought to involve the transient formation of the 4α-hydroperoxy flavin intermediate (Bruice, 1980; Walsh, 1980; Ronchi et al., 1982).

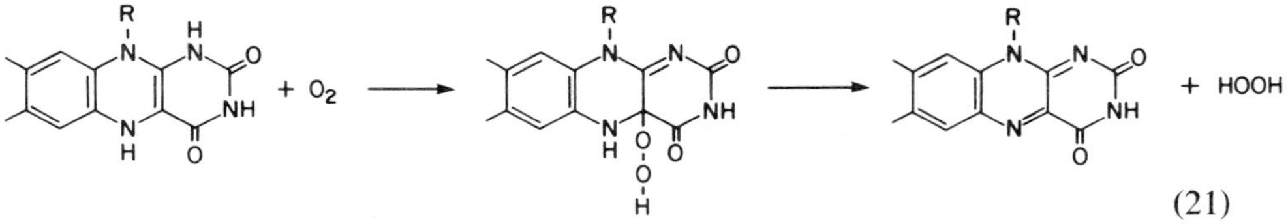

(21)

This general mechanism is not consistent with the fact that the majority of flavins in yeast methanol oxidase exist in semiquinone form; however, these flavin radicals do not appear to be catalytically functional (Mincey et al., 1980). By far the best-characterized of the flavoprotein oxidases is pig kidney D-amino acid oxidase, for which primary sequence information has recently been obtained (Ronchi et al., 1982).

Examples of copper-containing oxidases that reduce O_2 to H_2O_2 include plasma amine oxidase and galactose oxidase. Plasma amine oxidase catalyzes a reaction which is identical to that of liver monoamine oxidase and cyclohexylamine oxidase:

$$R_1-\underset{R_2}{\overset{H}{C}}-NH_3^+ + O_2 + H_2O \longrightarrow \begin{matrix}R_1\\R_2\end{matrix}\!\!>C{=}O + H_2O_2 + NH_4^+ \qquad (22)$$

In contrast to these simple flavoenzymes, plasma amine oxidase contains Cu^{2+} and a second as yet unidentified coenzyme, which, at least in the bovine enzyme, has been demonstrated not to be pyridoxal phosphate (K. A. Berg and Abeles, 1980; Malmstrom et al., 1975). Galactose oxidase, on the other hand, contains simply one mole of copper (Cu^{2+} or Cu^+) per mole of enzyme (Malmstrom et al., 1975). An example of a much more complicated oxidase is xanthine oxidase, which contains flavin (FAD), molybdenum, and iron. Xanthine oxidase is a flavin- and metal-containing pyridine nucleotide-linked dehydrogenase that is converted to an O_2-dependent oxidase upon modification of a single thiol group on the enzyme (Kaminski and Jezewska, 1982). The mechanism of xanthine oxidase differs from that of the simple flavoprotein oxidases described below; and it is, in fact, a hydroxylase that incorporates oxygen from water into the final product (Bray, 1975). In addition, the overall stoichiometry does not always correspond to that shown in Eq. (20), since during turnover, the O_2-dependent form of xanthine oxidase generates a mixture of hydrogen peroxide (H_2O_2) and superoxide (O_2^-) in a ratio that is dependent upon pH (Komai et al., 1969; Fridovich, 1970). Finally, discussion of an unusual group of oxidases, the blue copper oxidases, will be included in the monooxygenase section below. In contrast to the oxidases discussed in this section, these oxidases reduce O_2 to H_2O rather than to H_2O_2.

The oxidases that have received the most interest from an industrial standpoint are cholesterol oxidase and glucose oxidase. Cholesterol oxidase oxidizes cholesterol to cholest-4-ene-3-one:

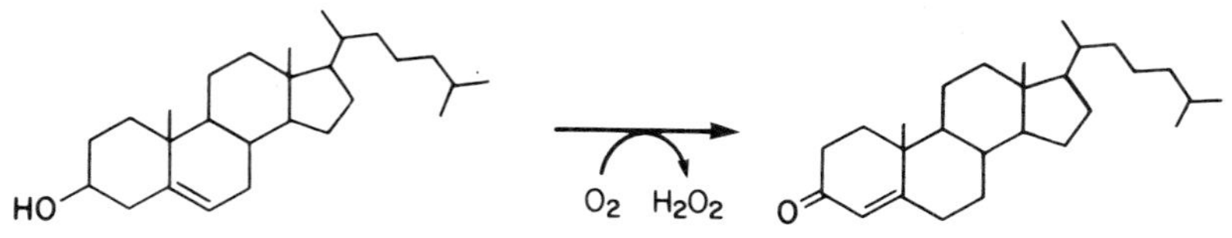

(23)

As shown, this reaction represents a Δ5, Δ4 isomerization in addition to the 3β-alcohol oxidation. In the case of cholesterol oxidase from *Nocardia rhodochrous,* these two activities have been shown to copurify and are probably catalyzed by one enzyme (Cheetham, 1979).

Cholesterol oxidase is used in the clinical assay of cholesterol (Cheetham, 1979), and the details of its production from various microbial sources have been reviewed (M. G. Halpern, 1981). In addition, the large-scale production of steroids using free and immobilized cells that contain high levels of the enzyme has been reported (Duarte and Lilly, 1980; Buckland et al., 1975). Since cholesterol is sparingly soluble in water, a suspension of *Nocardia* sp. cells in a two-phase system of water and carbon tetrachloride was employed. This system had the added benefit of the higher solubility of O_2 in the inorganic solvent when compared to aqueous solutions. Under the conditions reported, it was possible to perform seven 10-hour runs with high yields before about 50% of the initial activity was lost (Buckland et al., 1975).

Two types of glucose oxidase activities have been reported as shown:

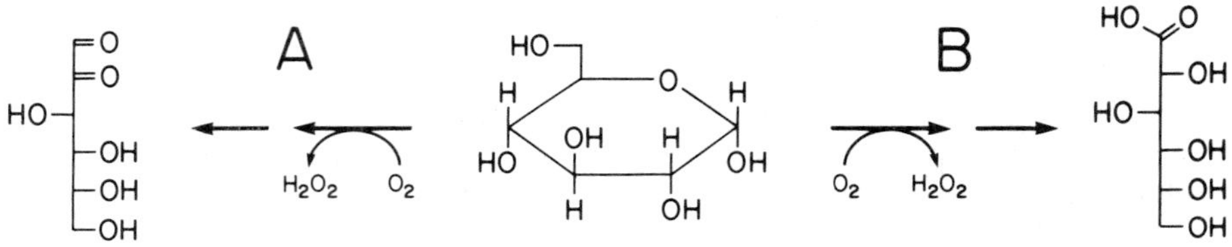

(24)

"Glucose oxidase" usually indicates glucose-1-oxidase (reaction B in Eq. 24). This flavoenzyme has been best characterized as isolated from *Aspergillus niger* (Stankovich et al., 1978), where it was discovered (D. Muller, 1928); however, its occurrence in other sources has been reviewed (Barker and Shirley, 1980). Glucose-1-oxidase is used in products designed for clinical assays of glucose (Barker and Shirley, 1980). Glucose-1-oxidase activity results in the production of the commercially important acid, D-gluconic acid, via the intermediate gluconolactone. The other oxidase enzyme that catalyzes reaction A in Eq. (24) has as yet been undefined with respect to its coenzyme requirement. This enzyme, glucose-2-oxidase, has been identified by its activity in the basidiomycete *Oudemansiella mucida* (Volc et al., 1978). There has been some interest in this enzyme recently as an alternative to glucose isomerase for the formation of fructose from glucose. In the alternative process, glucose-2-oxidase is employed to form the glucosone, D-arabino-2-hexosulose, from glucose and O_2 (reaction A in Eq. 24). The dicarbonyl intermediate is reduced chemically at the aldehyde carbon-1 to generate the desired product, fructose (Neidleman et al., 1979).

b. Deleterious Effects of Hydrogen Peroxide. A number of oxidases have received considerable study as immobilized catalysts; these include methanol oxidase (Baratti et al., 1978), D-amino acid oxidase (Brodelius et al., 1980), L-amino acid oxidase (Allen and Fink, 1980), and glucose-1-oxidase (Prenosil, 1978; Carter et al., 1980; Buchholz and Jaworek, 1979; Lilly et al., 1979; Mosbach, 1982; Alberti and Klibanov, 1982). Almost without exception, the major drawback to the use of these enzymes as production-scale catalysts results from the buildup of the product, hydrogen peroxide, and the inactivation of the enzyme by high levels of this agent. Although alternative proposals for the inactivation of glucose-1-oxidase during catalytic turnover have been put forward (Bourdillon et al., 1982), most efforts have been directed to circumventing the deleterious effects of H_2O_2, since it is well known to deactivate enzymes (Cho and Bailey, 1977) and damage cells (Ibrahim and Schlegel, 1980a). The coimmobilization of an oxidase with catalase has been utilized in order to remove H_2O_2 continuously via the disproportionation of H_2O_2 to O_2 and H_2O (Prenosil, 1978; Brodelius et al., 1980; Buchholz and Jaworek, 1979; Lilly et al., 1979; Carter et al., 1980). However, catalase itself is deactivated in the presence of H_2O_2 (Ibrahim and Schlegel, 1980b); and, in fact, the rate of catalase inactivation is nearly double the rate of glucose oxidase inactivation in a coimmobilization system (Buchholz and Jaworek, 1979), so that high ratios of catalase to oxidase must be employed. As an alternative to the enzymatic disproportionation of H_2O_2, inorganic catalysts have been employed to effect this reaction. Manganese oxide entrapped with whole cells (Brodelius et al., 1980; Mosbach, 1982) and coimmobilized ruthenium salts (Carter et al., 1980; Allen and Fink, 1980) have both proven to be more effective and longer-lived than catalase. In considering a process, the added cost of these metal catalysts must be taken into account, especially in light of the report of progressive poisoning of a ruthenium-based catalyst by gluconic acid, when it was coimmobilized with glucose-1-oxidase. Perhaps the most novel solution to this general problem is to eliminate O_2 completely as the oxidant and thus eliminate H_2O_2 buildup. Benzoquinone will replace O_2 in the oxidative half-reaction (Eq. 21) of glucose-1-oxidase and is reduced to the industrially important hydroquinone (Alberti and Klibanov, 1982):

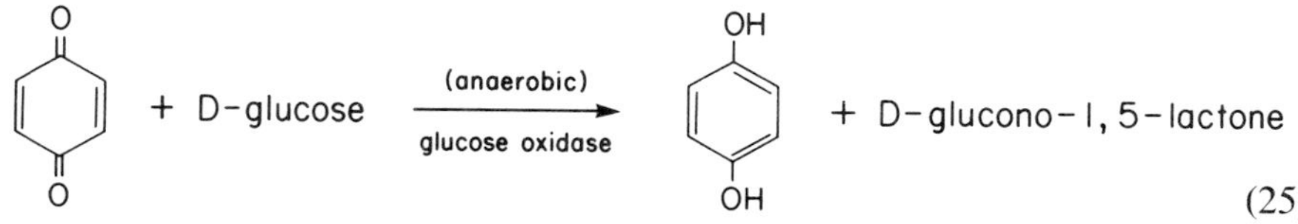

(25)

Essentially, glucose-1-oxidase in Eq. (25) is acting as a dehydrogenase, and the major problem now becomes the removal of the "cofactor" hydroquinone from the product.

As we will see, the problem of H_2O_2 reactivity is not unique to the oxidases that generate H_2O_2 as an obligate product but will be important in the consideration of the peroxidases that utilize hydrogen peroxide as a substrate.

2. Hydroperoxidases. A class of oxidative enzymes that offer a unique solution to both the problem of cofactor (NAD(P)H) regeneration posed by the dehydrogenases and monooxygenases (see below) and the problem of O_2 transfer rates and O_2 solubility posed by the oxidases, dioxygenases, and monooxygenases exists in a group of hemoproteins that catalyze the hydroperoxidase reaction:

$$\text{reduced substrate} + \text{ROOH} \longrightarrow \text{ROH} + \text{HO}^- + \text{oxidized substrate} \quad (26)$$

Thus the oxidant is hydrogen peroxide or an alkylperoxide, and the list of reductants includes hydrogen peroxide, chloride, bromide, iodide, phenols, glutathione, and ascorbic acid, among others. The advantage of the hydroperoxidase stoichiometry is that it involves neither reduced pyridine nucleotide nor O_2. Since the solubility of hydrogen peroxide in water (completely miscible) exceeds that of O_2 (0.2 mM), the size of a reactor required for a hydrogen peroxide–based process might be expected to be less than the size of a reactor required for a molecular oxygen–based process. However, high concentrations of H_2O_2 are extremely destructive to enzymes, as was discussed above; in practice, the maximum concentration of H_2O_2 that could be used in a fermentor is 0.01–0.1 mM (Ibrahim and Schlegel, 1980a, b). Although the use of hydrogen peroxide as an oxidant will pose difficulties until the challenge of enzyme inactivation is met, the potential usefulness of the kinds of reactions that the peroxidases catalyze will ensure continued interest in their development.

All peroxidative enzymes are hemoproteins and thus contain a coenzyme that is derived from iron porphyrin and may be bound to the protein covalently or noncovalently:

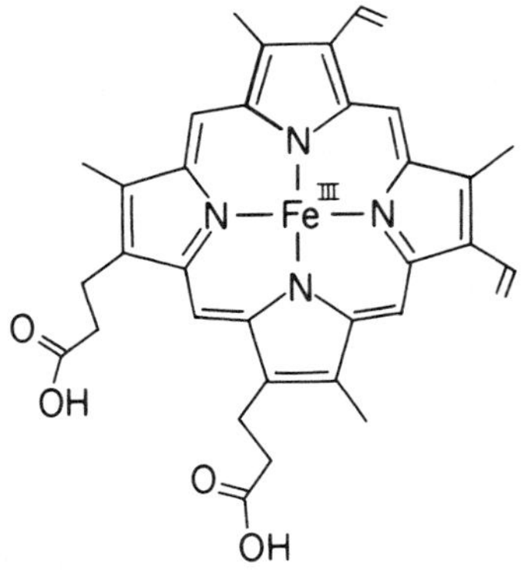

(27)

iron porphyrin (protoporphyrin IX)

Classically, the enzyme intermediates formed during turnover of the peroxidases correspond to unique spectral intermediates. Much effort has gone into the elucidation of the exact chemical nature of these spectral intermediates, and a very generalized reaction scheme may be written as follows:

$$Fe^{III} + H_2O_2 \longrightarrow \underset{\substack{\text{compound I} \\ \text{(olive green)}}}{Fe(H_2O_2)}$$

$$Fe(H_2O_2) \xrightarrow[AH \quad A^{\bullet}]{} \underset{\substack{\text{compound II} \\ \text{(red)}}}{Fe^{IV}} \xrightarrow[AH \quad A^{\bullet}]{} Fe^{III} \qquad (28)$$

The overall reaction sequence shown in Eq. (28) results in the formation of two moles of H_2O.

The symbols Fe^{III}, $Fe(H_2O_2)$, and Fe^{IV} are a shorthand notation for the complete iron–porphyrin coenzyme. It is essential to keep this in mind, since the porphyrin portion of the coenzyme is obviously not merely a passive support for the active site iron. Porphyrin radicals are evoked during discussion of the peroxidase mechanism (Castro, 1980), and such radical species have been identified. For example, in the case of horseradish peroxidase compound I, endor spectroscopy has confirmed the presence of a porphyrin π-cation radical (Roberts et al., 1981). For convenience, the general scheme depicted in Eq. (28) will be used in the following discussion, and the reader is referred to reviews for detailed elaboration of the peroxidase mechanism (Yamazaki, 1974; Castro, 1980; Morrison and Schonbaum, 1976).

a. Catalases. Most hemoproteins will catalyze the disproportionation of hydrogen peroxide to oxygen and water:

$$2\,H_2O_2 \longrightarrow O_2 + 2\,H_2O \qquad (29)$$

However, catalases are hemoproteins that have evolved such that they catalyze this reaction at least 10^4 times faster than other hemoproteins and have high turnover numbers of $2 \times 10^5\ s^{-1}$ per heme. Compund II is not an intermediate in the catalase reaction, and a second molecule of hydrogen peroxide serves as the reductant and converts the initial hydrogen peroxide adduct, compound I, directly to the initial Fe^{III} resting state enzyme form. However, compound II may be formed from catalase compound I by the action of organic hydroperoxides and phenol (Castro, 1980). As was discussed above, the application of catalase as an agent for H_2O_2 removal in bioreactors is severely limited by its propensity to inactivate during normal turnover.

b. Peroxidases and Phenol Coupling. As is shown in Eq. (28), the peroxidase reaction may formally be shown to result in the formation of two radical species, which can be expected to react subsequently with each other. When phenol and phenol derivatives serve as the reductant, a variety of products can be formed owing to the delocalization of the unpaired electron in the reacting phenol group:

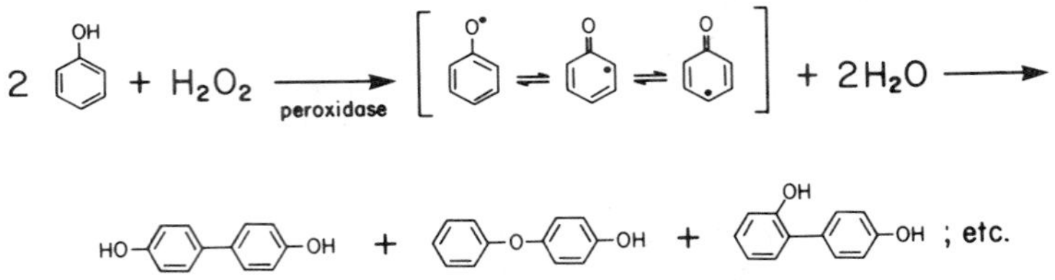

(30)

The potential product mixture from substituted phenols and polyphenols can be quite complex and will reflect a combination of the intrinsic reactivity of all of the possible aryloxy radicals as well as the biochemical environment on the peroxidase enzyme surface (Brown, 1967). The existence of a wide variety of polyphenols and alkaloids in nature may indicate the existence of specific peroxidases for the enzymic coupling of phenols; however, such peroxidases have not been characterized. The blue copper oxidases discussed below also have the capability to couple phenols (Brown, 1967).

A process to synthesize the important monomer 4,4′-dihydroxybiphenyl from phenol using horseradish peroxidase and H_2O_2 has been developed (Schwartz and Hutchinson, 1981). However, attempts to use peroxidases as synthetic agents for the specific oxidative phenolic coupling and hydroxylation of alkaloids has resulted is disappointingly low regiospecificity and low stereospecificity of the products formed (Vining, 1980; Holland, 1981). This lack of specificity has been used to advantage in the development of a process to remove potentially carcinogenic phenols and aromatic amines from industrical waste water. Thus a combination of horseradish peroxidase and H_2O_2 added to mixtures of aromatic amines including benzidine, naphthylamines, 4-aminobiphenyl, and *p*-phenylazoaniline will polymerize and cross-link these compounds via phenol coupling into a material that can be sedimented or filtered and removed at efficiencies approaching 100% (*McGraw-Hill's Biotechnology Newswatch,* 1982; Klibanov et al., 1980; Klibanov and Morris, 1981).

Finally, certain steroids can act as the reductant in the peroxidase reaction (Eq. 28) and are hydroxylated as a result. These steriods must have a 1,3-dicarbonyl group; the reaction of 16-ketoprogesterone is an example:

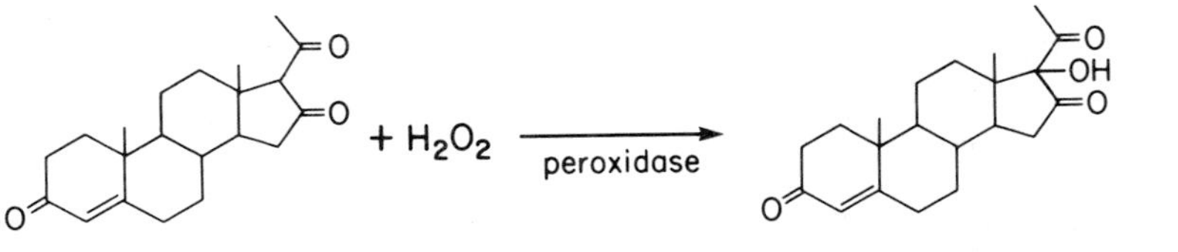

(31)

16-ketoprogesterone 17-hydroxy, 16-ketoprogesterone

A process utilizing a variety of 16-ketosteroids and horseradish peroxidase has been developed (Neidleman et al., 1967a, b).

c. Chloroperoxidase and Biochemical Halogenation. Haloperoxidases catalyze the peroxidative formation of carbon–halogen bounds; for example, the standard assay of chloroperoxidase utilizes the chlorination of the monochloroketone, monochlorodimedone (Hollenberg and Hager, 1978):

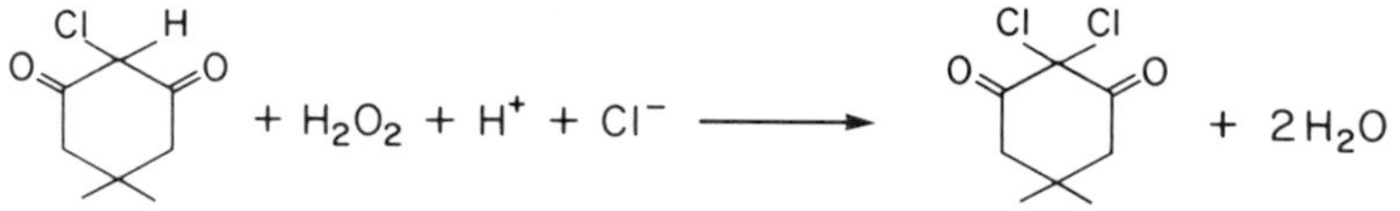

(32)

The halogenating species is formed by the oxidation of the appropriate halide ion by compound I (Eq. 28) of the appropriate peroxidase. The halide specificity depends on the peroxidase; thus chloroperoxidase catalyzes the oxidation of all halides except F^-, lactoperoxidase catalyzes the oxidation of Br^- and I^-, and horseradish peroxidase catalyzes only the peroxidative oxidation of I^- (Morrison and Schonbaum, 1976). These different reactivities probably reflect protein-mediated variations of the redox potential of the heme coenzyme, and horseradish peroxidase can be converted to an effective chlorinating agent by reaction with chlorite ion, $NaClO_2$, in the absence of H_2O_2 and Cl^- (Hollenberg et al., 1974). The halogenating species involved in the haloperoxidase reaction is most likely an enzyme-bound halogenium ion (—X^+) or hypohalite ion (—OX^+) (Morrison and Schonbaum, 1976; Hollenberg et al., 1974). In contrast, chlorination by the antimicrobial peroxidase, myeloperoxidase, does not require an acceptor such as monochlorodimedone during catalytic turnover, since this enzyme generates free hypochlorous acid (Harrison and Schultz, 1976):

$$H_2O_2 + Cl^- + H^+ \xrightarrow[\text{myeloperoxidase}]{} HOCl + H_2O \tag{33}$$

Myeloperoxidase can also oxidize I^- and Br^- (Morrison and Schonbaum, 1976), and halogenation simply involves the reaction of the hypohalous acid and any species in solution capable of undergoing halogenation.

Probably the best characterized haloperoxidase is the monomeric heme protein, chloroperoxidase from *Caldariomyces fumago* (Hollenberg and Hager, 1978), for which preliminary X-ray crystallographic data have recently been obtained (Rubin et al., 1982). Chloroperoxidase is able to function to some extent in nonaqueous solvents (Cooney and Hueter, 1974), and it has been demonstrated to have potential for chlorinating and brominating a variety of steriods as well as compounds containing a thiazole ring (Neidleman et al., 1969). Perhaps the biggest drawback for the use of this enzyme is the pH optimum for its halogenating activity, which is quite low ($pH < 3.0$) (Hollenberg and Hager, 1978; Cooney and Hueter, 1974; Neidleman et al., 1969). A bromoperoxidase has been partially purified from a marine algae, *Bonnemaisonia hamifera,* that is capable of catalyzing peroxidative brominations at neutral pH (Theiler et al., 1978). A process has been patented for the production of propylene oxide from peroxidase-generated propylene halohydrin (Neidleman et al., 1981).

3. Oxygenases. By far, most oxidative processes based on microbial fermentation or other applications of biotechnology are dependent on enzymes that fall into the classification of oxygenases (Sih and Rosazza, 1976; Hamilton, 1976; Fonken and Johnson, 1972; Johnson, 1978). Oxygenases are enzymes that catalyze the incorporation of one atom (monooxygenases) or both atoms (dioxygenases) of dioxygen into the oxidized product. As is the case with many microbial processes, the oxygenase enzymes responsible for a given transformation have often not been purified; and in many cases, much is still to be learned about the mechanistic details of a given reaction. Two areas that represent the greatest use of oxygenase-based processes are the oxidative transformation of hydrocarbons and the oxidative transformation of steroids. There has been much interest in the production of fatty acids and dicarboxylic acids from hydrocarbons (Abbott and Gledhill, 1971; Miall, 1980), and patented examples include the use of bacteria (*Corynebacterium*) (Dahlstrom and Jachnig, 1973, 1974; Minato et al., 1974) and yeast (*Candida tropicalis*) (Yamada et al., 1978, 1979). The ability of microorganisms to oxidatively split aromatic rings (Perry, 1977; McKenna and Kallio, 1965; Gibson, 1971) has figured in a process to produce aromatic hydroxy acids from naphthalene, phenanthrene, and anthracene (Hill, 1967). Oxygenase chemistry is predominant in the area of steroid conversions (Kieslich, 1980a, 1980b; Kieslich and Sebek, 1979), and examples include 11β-hydroxylation (Hanisch et al., 1980; Takeda et al., 1962; Shull et al., 1953; Kieslich and Raspe, 1964), 4α-hydroxylation (Kieslich and Raspe, 1964), and 16α-hydroxylation (Iida and Iizuka, 1981; Chun et al., 1981).

An understanding of the mechanism of oxygenase enzymes will lead to a more effective utilization of these important catalysts in industrial processes. This is especially true because genetic engineering methods will allow the possibility of using oxygenases from a variety of sources including mammalian and plant enzymes. The nature of the products derived from the biochemical oxygenation of a given molecule will depend upon the source of the oxygenase. An often cited example is the oxygenation of aromatic hydrocarbons by mammalian systems versus bacterial systems (Gibson, 1971; Cain, 1980; May, 1979):

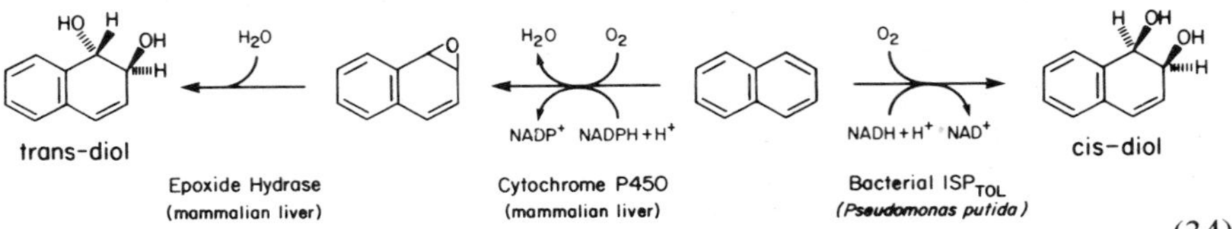

(34)

The *cis*-diol product of naphthalene oxygenation results from the mode of oxygen activation by the NADH-dependent toluene dioxygenase system of *Pseudomones putida* (Subramanian et al., 1981), whereas *trans*-diol is a consequence of the two-step process of cytochrome P450 monooxygenase–dependent epoxidation and hydrolysis of the resultant epoxide by epoxide hydrase (Wislocki et al., 1980). A further example is the observation that, in contrast to mammalian systems, bacterial oxidative degradation of the steriod side chain does not involve a progesterone-derived intermediate (Martin, 1977; Kieslich, 1980b):

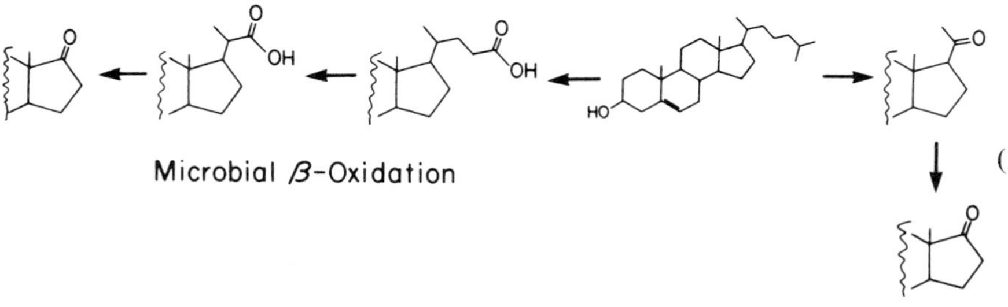

Microbial β-Oxidation

(35)

Mammalian Cytochrome P450 Dependent Oxygenation

The mammalian enzymes responsible for 17-ketosteroid formation are both cytochrome P450–dependent monooxygenases, the side chain cleavage cytochrome P450 (Light and Orme-Johnson, 1981; Wilson and Harding, 1970) and the 17α-hydroxylase (Nakajin et al., 1981), while the bacterial side chain removal is not dependent upon oxygenases at all, but rather involves an anaerobic mechanism of dehydrogenation and water addition similar to the β-oxidation of fatty acids (Martin, 1977; Sih et al., 1968).

The subject of microbial cooxidation or cometabolism is tied to the mechanism of oxygenase enzymes. Microbes are often able to oxygenate compounds that do not serve to support growth of the organism if these compounds are presented together with substrates that are metabolizable (Perry, 1979; Quayle, 1980). In addition to molecular oxygen, most monooxygenases and some dioxygenases also required a source of electrons in the form of the reduced pyridine nucleotide cofactor. The required cooxidant can be viewed as providing the reducing equivalents necessary to regenerate NADH or NADPH during turnover of the oxygenase. Thus processes based on the use of these enzymes will also require consideration of the topic of cofactor regeneration discussed below. The remainder of this section touches upon the wide variety of oxygenases and the even wider variety of reactions that they catalyze.

a. The Monooxygenases. The majority of monooxygenases catalyze the incorporation of one atom of dioxygen into the substrate with the concomitant oxidation of one mole of NADH or NADPH:

$$AH + O_2 + NAD(P)H + H^+ \longrightarrow AOH + H_2O + NAD(P)^+ \tag{36}$$

These monooxygenases may utilize heme, nonheme iron, copper, pteridine, or flavin as the coenzyme and, in addition, often do not interact directly with the reduced pyridine nucleotide. Instead, ancillary electron transfer proteins are required to deliver electrons to the active site of the monooxygenase.

Flavin- and Pterin-Dependent Monooxygenases. Flavoprotein monooxygenases are directly reduceable by NAD(P)H, while pterin-dependent monooxygenases require an additional enzyme to provide tetrahydropteridine. These enzymes are discussed together, however, since in each case, the oxygenating species may be an enzyme-bound organic hydroperoxide. Excellent reviews of the flavin- and pterin-dependent monooxygenases are recommended (Flashner and Massey, 1974; Kaufman and Fisher, 1974; Massey and Hemmerich, 1975). The flavin monooxygenases have been called "the gentle oxygen addition or insertion reagents of biochemistry" (Bruice, 1980), since they do not appear to be capable of catalyzing the insertion of oxygen into isolated carbon–carbon double bonds or aliphatic carbon–hydrogen bonds. The mechanism of active oxygen generation by the flavoprotein monooxygenases is composed of two parts. In the first half, the oxygenase functions as an NAD(P)H oxidase, in that the active site flavin is reduced by NAD(P)H and the reduced flavin reacts with O_2 to form an oxygen adduct, which stop-flow spectroscopy indicates is a 4a-hydroperoxy intermediate (Eq. 21) (Entsch et al., 1976). Rather than break down to oxidized flavin and H_2O_2, however, this enzyme-bound hydroperoxide serves to oxygenate the substrate and produces a 4a-hydroxy intermediate, which loses H_2O to regenerate the oxidized flavin (Bruice, 1980; Walsh, 1980; Massey and Hemmerich, 1975):

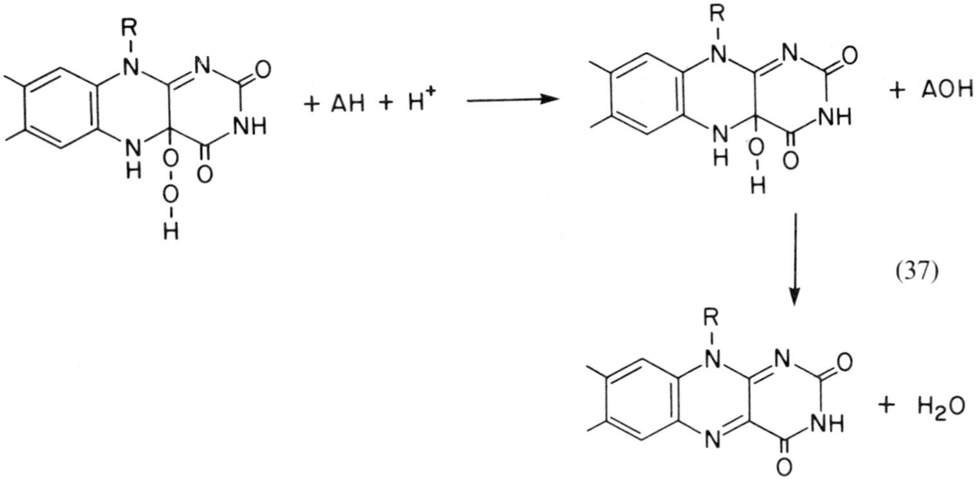

(37)

The 4a-hydroperoxide intermediate or related species is capable of hydroxylating electron-rich aromatic rings in the case of the aromatic hydroxylases (Flashner and Massey, 1974; Massey and Hemmerich, 1975, 1978). The aromatic hydroxylases include the well-characterized *p*-hydroxybenzoate hydroxylase (Husain and Massey, 1979; Entsch et al., 1980; Husain et al., 1978) and salicylate hydroxylase, which is purified from a pseudomonad able to grow on salicylate (Kamin et al., 1978). This oxygenated intermediate is also capable of oxidizing nitrogen (secondary and tertiary amines, hydrazines, etc.) and sulfur (sulfides, disulfides, thiols, thioamides, etc.), and this activity is catalyzed by the mammalian liver microsomal FAD-containing monooxygenase (Ziegler, 1980; Beaty and Ballou, 1981a, b; Prough et al., 1981; Poulsen and Ziegler, 1979; Light et al., 1982). Finally, oxygenated flavoprotein monooxygenases are capable of acting as Baeyer-Villiger agents in the lactonization of ketones:

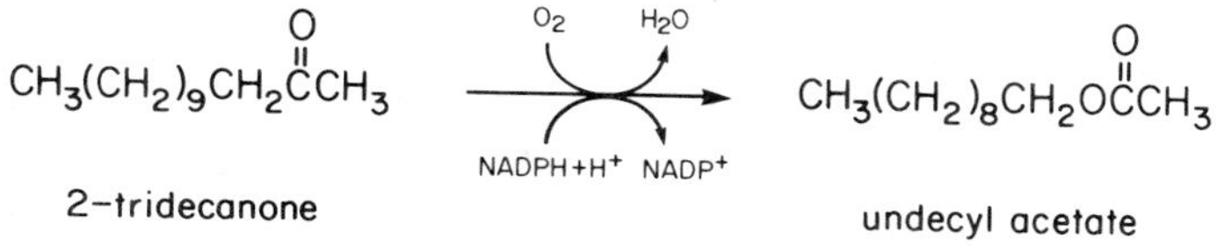

(38)

Thus a flavoprotein ketone monooxygenase has been purified (Britton and Markovetz, 1977) from *Pseudomonas cepacia* that can utilize 2-tridecanone as a sole carbon source (Forney et al., 1967). Cyclohexanone monooxygenase, which catalyzes a similar reaction, has been better characterized (Ryerson et al., 1982) and is found in microorganisms that are capable of growth on cyclohexane (Anderson et al., 1980).

The known examples of pterin-dependent monooxygenases are all mammalian aromatic amino acid hydroxylases (Kaufman and Fisher, 1974; Massey and Hemmerich, 1975), the most well-characterized example of which is liver phenylalanine hydroxylase. During turnover, the pteridine cofactor, biopterin, cycles between tetrahydrobiopterin and quinoid dihydrobiopterin:

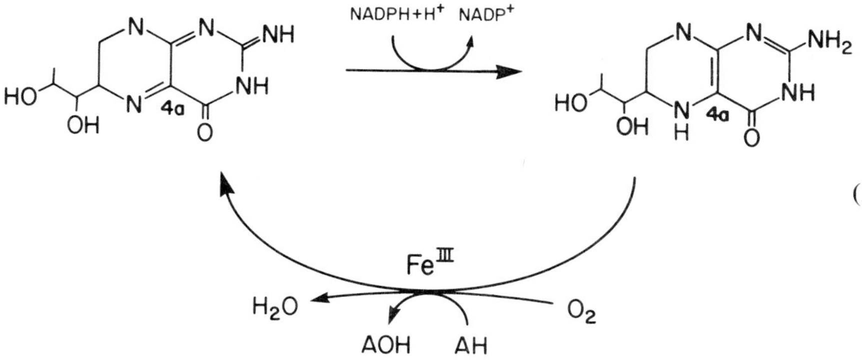

(39)

However, unlike the flavoprotein monooxygenases, which contain a tightly bound flavin, the pteridine cofactor dissociates from the monooxygenase between hydroxylation events, and a second enzyme, dihydropteridine reductase, is required to provide the reduced tetrahydrobioprotein required for catalysis. The obvious structural similarity of tetrahydrobiopterin to reduced flavin, coupled with studies of phenylalanine activity with a variety of pterin analogs (Bailey and Ayling, 1980; Kaufman, 1979; Lazarus et al., 1981), has led to the suggestion that the initial site of oxygen attachment during turnover is the 4a position of the pterin cofactor. However, although there is spectral evidence for the release of a 4a-hydroxy adduct from phenylalanine hydroxylase after the hydroxylation event, a 4a-hydroperoxy adduct has not been observed (Lazarus et al., 1981). In addition, phenylalanine hydroxylase is absolutely dependent on the presence of one mole of iron per subunit for full activity (Shiman and Jefferson, 1982; Gottschall et al., 1982). Thus it is still uncertain whether the pterin-dependent monooxygenases are more similar to the flavin monooxygenases or to the iron-dependent monooxygenases discussed below.

Cytochrome P450-Dependent Monooxygenases. The cytochrome P450–dependent monooxygenases have received by far the greatest amount of interest because of the impressive reactivity of the oxygenating intermediate that they generate. The identity of this intermediate and the detailed mechanism of oxygenation are still under investigation. The coenzyme common to cytochrome P450 enzymes is protoporphyrin IX (Eq. 27) and they derive their name from a spectral intermediate unique to hemoproteins, which is generated from the complexation of carbon monoxide to the reduced heme and which absorbs at 450 nm. Since cytochrome P450 monooxygenases are unable to accept electrons directly from NAD(P)H, they require additional electron transfer proteins (flavoproteins and ferredoxins) to provide electrons in order to catalyze the monooxygenase reaction (Eq. 36).

The inability of reduced pyridine nucleotides to reduce cytochrome P450 is a result of the fact that the heme cofactor is an obligate one-electron acceptor, in contrast to flavins and pterins, which can accept two electrons. Thus during catalysis of the monooxygenase reaction (Eq. 36), the two required electrons must be introduced to the active site iron one at a time with the intervening addition of oxygen. This sequential addition of electrons forms part of the well-known cytochrome P450 catalytic cycle:

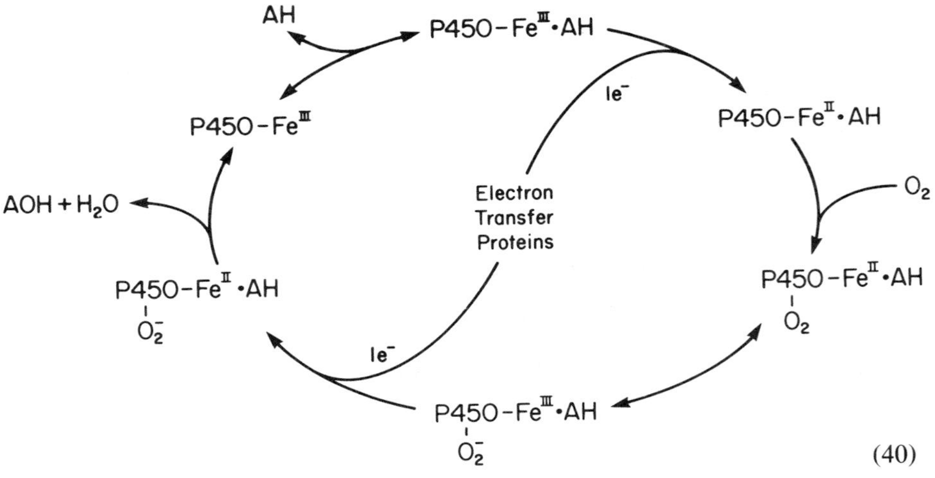

(40)

It is well-documented that the binding of substrate to a cytochrome P450 enzyme can increase the redox potential of the heme by as much as 100 mV (Gunsalus et al., 1974; Light and Orme-Johnson, 1981) and thus greatly facilitate heme reduction by the first electron. While this first electron can be supplied by artificial electron donors, the transfer of the second electron to the one-electron reduced complex of enzyme, oxygen, and substrate (O_2^-—P450—Fe^{III}—AH) must be mediated by a native electron donating ferredoxin in the well-documented case of cytochrome $P450_{CAM}$ (Gunsalus et al., 1974; Sligar et al., 1974, Lipscomb et al., 1976). The requirement of the electron transfer component to act as an effector during catalysis leads to the concept of a multicomponent cytochrome P450–dependent oxygenase system. The P450-dependent monooxygenase systems that have been characterized may be divided into two classifications based on the nature of the electron transfer components. Cytochrome $P450_{CAM}$, which hydroxylates camphor to 5-exo-hydroxycamphor (Gunsalus et al., 1974, Gunsalus and Wagner, 1978); cytochrome $P450_{scc}$, which cleaves the side chain of cholesterol to form pregnenolone (Mitani, 1979); cytochrome $P450_{11\beta}$, which hydroxylates deoxycorticosterone and other steroids at carbon 11 (Mitani, 1979); and a cytochrome P450 from kidney mitochondria that hydroxylates 25-hydroxycholecalciferol (25-hydroxy vitamin D_3) at carbon 1 (Kulkoski and Ghazarian, 1979) are all part of three-component systems containing an FAD-dependent flavoprotein, an iron–sulfur protein, and the cytochrome P450. In contrast, the liver microsomal cytochrome P450 systems are dependent upon a cytochrome P450 reductase that contains both FMN and FAD (Yasukochi and Masters, 1976; Vermilion et al., 1981). There are as yet an unknown number of multiple forms of cytochrome P450 in liver microsomes (Lu and West, 1980; Lang and Nebert, 1981; Lang et al., 1981; Guengerich et al., 1982), which catalyze a vast array of oxygenase activities, including aliphatic hydroxylation, aromatic hydroxylation and epoxidation, alkene epoxidation, oxidative dealkylation of nitrogen and oxygen, oxidative dehalogenation, and oxidation of nitrogen and sulfur (Wislocki et al., 1980).

Of even more interest than the source of electrons in the reaction scheme shown in Eq. (40) is the nature of the oxygenating species derived from P450—Fe^{II}—O_2^- complexed to the substrate. Loss of H_2O from this complex ($Fe^{II}O_2^-$ or $Fe^{III}O_2^{2-}$) yields $(FeO)^{3+}$, Fe^{II} bound to an oxene oxygen with six valence electrons. The concept of an oxenoid species of oxygen with the capability of direct insertion into carbon–carbon and carbon–hydrogen bonds has been of value in understanding P450-dependent oxygenations (Hamilton, 1974). However, evidence for a two-step mechanism involving radical intermediates has been proposed after studies of the alkylated heme obtained during the inactivation of cytochrome P450 by vinyl fluoride and propyne (Ortiz de Montellano et al., 1982; Ortiz de Montellano and Kunze, 1981), from the stereochemistry of products derived from deuterated camphor (Gelb et al., 1982), and from model studies (Groves et al., 1980; Blake and Coon, 1981; Groves, 1980). Active site acyl peroxide intermediates have also been suggested (Blake and Coon, 1981; Sligar et al., 1980). A number of reviews of cytochrome P450 oxygenation are recommended (Wislocki et al., 1980; Ullrich, 1979; Lu and West, 1980; Gunsalus et al., 1974, 1975; White and Coon, 1980; Coon and White, 1980; Gander and Mannering, 1980; Gunsalus and Sligar, 1978, Sato and Omura, 1978).

In addition to the well-characterized P450 cytochromes already mentioned, a number of interesting systems have been described in yeast, fungi, and bacteria (Sato and Omura, 1978). *Bacillus megaterium* contains a steroid 15β-hydroxylase system (A. Berg and Rafter, 1981; A. Berg et al., 1976, 1979) as well as a system that hydroxylates and epoxidizes fatty acids (Matson et al., 1977, 1980; Ruettinger and Fulco, 1981). Cytochrome P450 is involved in alkane hydroxylation in *Corynebacterium* (Cardini and Jurtshuk, 1970), *Acinetobacter* (Asperger et al., 1981), and yeast (H.-G. Muller et al., 1980); in O-dealkylation in *Norcardia* (Broadbent and Cartwright, 1974); and in aromatic hydroxylation in the fungus *Cunninghamella bainieri* (Ferris et al., 1976). Thus the sources of cytochrome P450–dependent oxygenase systems are numerous and have undoubtedly not been exhausted.

Copper-Containing and Nonheme Iron–Containing Monooxygenases. This section is devoted to copper-containing and nonheme iron–containing enzymes that catalyze the reaction shown in Eq. (36). We have chosen to discuss the copper enzyme tyrosinase (along with the internal monooxygenases) in the following dioxygenase section. In our minds, they are similar to many of the dioxygenases from the standpoint of the potential for process development in that they avoid the need for the consideration of cofactor regeneration discussed below. While reviews have covered the copper and nonheme iron monooxygenases (Ullrich and Duppel, 1975; Vanneste and Zuberbuhler, 1974), there is much recent work, particularly in the area of the methane monooxygenases, that warrants discussion.

Dopamine β-hydroxylase from mammalian adrenal medulla is a tetrameric enzyme that contains two atoms of copper per subunit (Ullrich and Duppel, 1975; Vanneste and Zuberbuhler, 1974). It is able to activate dioxygen and stereoselectively hydroxylate the methylene carbon in dopamine to generate R-(−)-norepinephrine and is also able to oxidize sulfides to sulfoxides stereoselectively (May and Phillips, 1980). Rather than utilizing NAD(P)H as the source of reducing equivalents, as shown in Eq. (36), this copper monooxygenase oxidizes ascorbate to dehydroascorbate during turnover. The methane monooxygenase system from *Methylosinus trichosporium,*

$$CH_4 + O_2 + NAD(P)H + H^+ \longrightarrow CH_3OH + H_2O + NAD(P)^+ \qquad (41)$$

has been reported to contain a copper protein (Tonge et al., 1977); however, this system is still under investigation (Higgins et al., 1981; Stirling et al., 1979).

The nonheme iron monooxygenases, like the P450 monooxygenases, are best described as monooxygenase enzyme systems, since they also require additional electron transfer proteins to provide reducing equivalents for the monooxygenase reaction (Eq. 36). With the exception of the pterin monooxygenases discussed above, the iron or iron–sulfur center appears to be sufficient to activate molecular oxygen, and the terminal electron acceptor and site of oxygenation are apparently free of additional cofactors. Methane monooxygenase is a nonheme iron monooxygenase that has received a great deal of attention recently because of its broad substrate specificity (Higgins et al., 1980; Dalton, 1980; Colby et al., 1977). While the monooxygenase functions to oxidize methane (Eq. 41) for use in the growth of methanotrophic bacteria, it is able to oxidize nongrowth substrates and synthesize primary and secondary alcohols, aldehydes, and methylketones from alkanes (Dalton, 1980; Hou et al., 1980a, b, 1981; Patel et al., 1979, 1980a, b; Stirling et al., 1979; Colby et al., 1977), phenols from aromatic compounds, and a variety of other oxygenations (Hou et al., 1979).

Several processes have been patented that are based on the oxygenative biotransformations of methane monooxygenases (Higgins, 1982; Patel and Hou, 1981a, b). These substrates do not allow growth and they cannot be metabolized to provide the reducing equivalents required for the monooxygenase reaction (Eq. 36). However, the cooxidation of energy yielding methane metabolites added to resting cells stimulates the epoxidation of propylene to propylene oxide (Hou et al., 1980a). An exception is the production of methyl ketones from alkanes, since hydroxylation to the intermediate alcohol in resting cells provides a substrate for the pyridine nucleotide–linked secondary alcohol dehydrogenase, which is also present (Patel et al., 1981); thus this process provides its own source of reducing equivalents (Hou et al., 1980a, 1981; Patel et al., 1980b). The best characterized methane monooxygenase is an iron–sulfur monooxygenase that requires an additional iron–sulfur flavoprotein to provide electrons from NADH and possibly a third component of uncertain function (Colby and Dalton, 1978; Dalton, 1980; Higgins et al., 1981).

Two other iron-dependent monooxygenase systems have been characterized that can oxygenate alkane or alkene hydrocarbons. The best-characterized oxygenase of the three is the ω-hydroxylase found in *Pseudomonas oleovorans*. First selected for its ability to grow on octane, this strain has undergone several improvements in temperature tolerance (Schwartz, 1973) and rate of oxygenation (Schwartz and McCoy, 1973). The ω-hydroxylase system has been purified and characterized as a three-component system consisting of the ω-hydroxylase, a 40,000-MW protein with a single iron atom and no acid labile sulfide; rubredoxin, an iron-containing electron transfer protein; and rubredoxin reductase, a flavoprotein that reduces rubredoxin at the expense of NADH (Griffith et al., 1978). The ω-hydroxylase system catalyzes the omega hydroxylation of 6- to 14-carbon alkanes and fatty acids (Griffith et al., 1978), the hydroxylation of branched and alicyclic hydrocarbons (Griffith et al., 1978), the 1,2-epoxidation of 6- to 12-carbon 1-alkenes (Abbot and Hou, 1973), and the stereoselective epoxidation of 1,7-octadiene to 7,8-epoxy-1-octene and 1,2,7,8-diepoxyoctane (May et al., 1976). Unlike octane and 1-octene, 1,7-octadiene will not support the growth of *Pseudomonas oleovorans;* however, it does serve as an inducer of the ω-hydroxylase system (Schwartz and McCoy, 1976). Studies of the epoxidation of deuterated 1-octene suggests that, as in the case of the P450 monooxygenases, the catalytic cycle of this nonheme iron monooxygenase operates through a two-step radical mechanism (May et al., 1977). Processes to epoxidize 1-octene in a two-phase water:octene system (de Smet et al., 1981) and to epoxidize 1,7-octadiene in a two-phase water:cyclohexane system (Schwartz and McCoy, 1977), both using whole cells, have been developed. A different three-component monooxygenase system that hydroxylates long-chain alkanes (*n*-hexadecane to *n*-hexadecanol) has been characterized in *Pseudomonas aeruginosa* and is similar to the ω-hydroxylase in *Pseudomonas oleovorans* except that it utilizes a cytochrome *c* as an electron transfer protein rather than rubredoxin (Matsuyama et al., 1981). A novel monooxygenase system in *Pseudomonas putida,* 4-methoxybenzoate O-demethylase, or putidamonooxin, is composed of an oxygen-activating protein that contains an iron–sulfur cluster and a mononuclear non-heme iron oxygen-binding site and an electron transfer protein that contains an iron–sulfur cluster and FMN (Bernhardt et al., 1975; Bernhardt and Kuthan, 1983). Finally, the liver microsomal stearyl coenzyme A desaturase may be a monooxygenase. This enzyme catalyzes the formation of olelyl-CoA from stearyl-CoA by a suggested hydroxylation–dehydration mechanism (Topham and Gaylor, 1970):

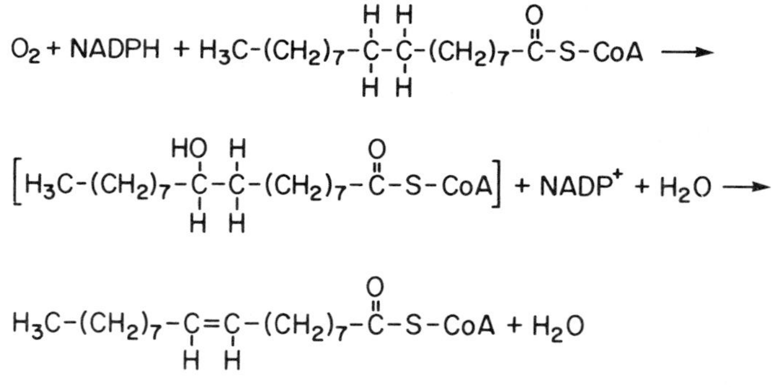

(42)

Thus oxygen would be transiently incorporated into an intermediate by this mechanism. Such a hydroxylated intermediate has not been demonstrated, however (Hamberg et al., 1974). The desaturase requires O_2 and NADPH as shown, and the desaturase system is composed of the desaturase that contains iron but no sulfide, cytochrome b_5, and a flavoprotein, cytochrome b_5 reductase (Strittmatter et al., 1974; Enoch et al., 1976; Strittmatter and Enoch, 1978). A similar cytochrome b_5–dependent system in the liver introduces the C-5 double bond into cholest-7-en-3β-ol (Reddy et al., 1977) and is analogous to the O_2- and NADPH-dependent ergosterol Δ^5-desaturase in yeast (Osumi et al., 1979).

b. Dioxygenases. By definition, the dioxygenases catalyze the incorporation of both atoms of molecular oxygen into the reaction product. Dioxygenases may or may not require an additional two electrons during turnover, depending upon whether the oxidation state of the substrate is increased by a net of four or two electrons:

$$O_2 + AH \longrightarrow AO_2H \tag{43}$$

$$O_2 + AH + NAD(P)H \longrightarrow AO_2H_2 + NAD(P)^+ \tag{44}$$

In addition, because the problem of cofactor regeneration is so important when alternative processes are considered, we have chosen to discuss in this section enzymes that incorporate only one atom of molecular oxygen into a substrate but that do not require additional reduced cofactors.

$$O_2 + AH + H^+ \longrightarrow AO + H_2O \tag{45}$$

These latter enzymes, along with the noncofactor-dependent dioxygenases, hold the greatest promise in the area of biotechnology until the issue of cofactor regeneration is successfully (and economically) resolved.

One of the reduced cofactor-requiring dioxygenases, toluene dioxygenase (Eq. 44), was mentioned in the introductory part of this section (Eq. 34). Although the dioxygenation of naphthalene is shown, this enzyme from *Pseudomonas putida* is named toluene dioxygenase after its ability to oxygenate toluene to (+)-*cis*-1(S),2(R)-dihydroxy-3-methylcyclohexa-3,5-diene. The oxygen-activating cofactor of the oxygenase is an iron–sulfur center (thus the abbreviation ISP_{tol}), and the requisite electron transfer components are a ferredoxin and an FAD-containing ferredoxin reductase (Subramanian et al., 1979, 1981). The identical kinds of proteins are found in the pyrazon dioxygenase system isolated from bacteria that degrade the herbicide pyrazon (Sauber et al., 1977). *Pseudomonas arvilla* contains an NADH-dependent dioxygenase system, benzoate 1,2-dioxygenase, which is composed of an iron–sulfur oxygenase and a single iron–sulfur flavoprotein (Yamaguchi and Fujisawa, 1981). Finally, 2-methyl-3-hydroxypyridine-5-carboxylic acid oxygenase oxidatively cleaves a hydroxypyridine ring and incorporates a molecule of oxygen into the product α-N-(acetylaminomethylene)succinic acid. This enzyme is a single flavoprotein dioxygenase that has similarities to the flavoprotein monooxygenases (Kishore and Snell, 1981a, b).

Dioxygenases that do not require reduced cofactors (Eq. 43), also utilize a wide variety of cofactors. The protocatechuate dioxygenases are three different iron-containing enzymes that cleave the aromatic ring of protocatechuate in one of three sites (between carbons 2 and 3, 3 and 4, or 4 and 5), depending upon the dioxygenase (Hayaishi et al., 1975; Wood, 1980). All three contain a nonheme iron coenzyme. The best-characterized is protocatechuate 3,4-dioxygenase, which is proposed to cleave protocatechuate via an iron-bound peroxy or endo peroxy intermediate (Bull et al., 1981; Que, 1980; Walsh, 1979; Que et al., 1977). A similar ring cleavage is affected by quercetinase, a non-blue copper protein that contains two moles of copper and cleaves the flavonol ring of quercetin (Vanneste and Zuberbuhler, 1974). Finally, prostaglandin cyclooxygenase is an example of a heme-containing dioxygenase. The cyclooxygenase contains two hemes, performs two dioxygenations of the substrate, arachidonic acid, to incorporate four atoms of oxygen into the product, and also acts as a peroxidase (Egan et al., 1979, 1980). Prostaglandin cyclooxygenase is thought to utilize a free radical mechanism for substrate oxygenation (Hemler and Lands, 1980).

From the standpoint of a simplified process, the advantage of the dioxygenase that catalyzes the reaction shown in Eq. (43) is that there is no requirement for reduced cofactors. Certain monooxygenases share this advantage and catalyze the general reaction shown in Eq. (43). A group of flavoproteins that fall into this category includes lactate monooxygenase, lysine monooxygenase, and arginine monooxygenase; these have been called internal flavoprotein monooxygenases (Massey and Hemmerich, 1975). The flavoproteins in this group all catalyze the oxidative decarboxylation of their respective substrates to yield a smaller carboxylic acid, carbon dioxide, and water. For example, lactate monooxygenase oxidizes lactate to acetate:

$$CH_3\overset{\overset{OH}{|}}{C}HCOOH + O_2 \longrightarrow CH_3COOH + CO_2 + H_2O \tag{46}$$

Mechanistically, the internal flavoprotein monooxygenases can be thought of as flavoprotein oxidases (Eqs. 20 and 21) that do not release the keto acid (or imino acid) as a product. Rather, the bound keto acid is further peroxidatively decarboxylated to yield the observed products (Massey and Hemmerich, 1975). An iron-containing oxygenase from liver tissue, α-ketoisocaproate oxygenase, apparently utilizes the reducing equivalents obtained from the oxidative decarboxylation of the keto acid α-ketoisocaproate in order to activate molecular oxygen for the further hydroxylation of the intermediate acid to form β-hydroxyisovalerate (Sabourin and Bieber, 1982):

$$(CH_3)_2CHCH_2\overset{\overset{O}{\|}}{C}COOH + O_2 \longrightarrow (CH_3)_2\overset{\overset{OH}{|}}{C}CH_2COOH + CO_2 + H_2O \tag{47}$$

Finally, the phenol *o*-monooxygenases or tyrosinases contain a binuclear copper active site and catalyze the oxidation of phenol to orthoquinone (Ullrich and Duppel, 1975):

$$\text{monophenol} + O_2 \longrightarrow \text{orthoquinone} + H_2O \tag{48}$$

It is likely that an intermediate orthodiphenol serves as a reduced cofactor for the oxygenation of a second monophenol during turnover (Walsh, 1979). Evidence from ultraviolet resonance raman spectroscopic studies of oxy-tyrosinase indicates that the active oxygen intermediate is an enzyme-bound peroxide (Eickman et al., 1978). In addition, tyrosinase can catalyze the oxidative coupling of phenols in a reaction similar to that shown for the peroxidases in Eq. (30), except that O_2 is the oxidant (Brown, 1967).

c. Potential for the Industrial Use of Oxygenases. In order to make large amounts of an oxygenase to be used in an industrial process, it must be possible to control the levels of expression of the enzyme; this implies an understanding of its genetic regulation. Many of the oxygenases that have been discussed are encoded on plasmids. Plasmids bearing the genes for cytochrome P450 CAM (Rheinwald et al., 1973), the ω-hydroxylase (Chou et al., 1974), catechol 2,3-dioxygenase (Connors and Barnsley, 1982; Austen and Dunn, 1980), and many other oxygenases have been described. Tremendous progress in the understanding of the regulation of plasmid-encoded oxygenases has been made, as for example in the case of the TOL plasmid, which encodes for oxygenases involved in toluene degradation (Worsey et al., 1978; P. A. Williams and Worsey, 1978; Franklin and Williams, 1980), and the ALK plasmid, which is involved in alkane hydroxylation and degradation (Benson et al., 1979; Fennewald and Shapiro, 1977). The ability to control genes encoding for oxygenases is shown by the ability to express toluene-degrading genes from *Pseudomonas* in *Escherichia coli* (K. Sakaguchi, 1982; Jacoby et al., 1978) and by the ability to express the genes for both cytochrome P450 CAM and the ω-hydroxylase in a single microorganism (Chou et al., 1974; Chakrabarty, 1973; Fennewald and Shapiro, 1979). A number of reviews of the diversity and applications of the degradative plasmids in *Pseudomonas* are recommended (K. Sakaguchi, 1982; Shapiro et al., 1979; P. A. Williams, 1979; Wheelis, 1975; Chakrabarty, 1976). Genetic manipulation is not restricted to prokaryotic genes, however, and several of the liver microsomal cytochrome P450-dependent monooxygenases have been cloned recently, as well as the required cytochrome P450-reductase (Fujii-Kuriyama et al., 1982; Gonzalez and Kasper, 1982; Fagan et al., 1982). Thus the prospect of bringing the tremendous range of enzymes that catalyze oxygenations under the control of industrial microbiologists is becoming a reality and must be anticipated to lead to the rapid development of processes based on this class of enzyme.

Several problems that are somewhat unique to the oxygenases must be addressed; these include uncoupling, autoinactivation, and lack of stereoselectivity. In addition, most of the monooxygenases and many of the dioxygenases require the consideration of cofactor regeneration; however, this general problem will be discussed in a separate section. Uncoupling refers to the wasteful use of reduced cofactors by monooxygenases without the productive oxygenation of the substrate. An uncoupled monooxygenase will oxidize NAD(P)H and generate H_2O_2 from O_2, thus functioning as a useless NAD(P)H oxidase. The extent of uncoupling of a monooxygenase may vary from a few percent (one or two H_2O_2 molecules formed per hundred successfully oxygenated products) up to complete uncoupling. The extent of uncoupling is often a function of the substrate that is being oxygenated, as is the case with the liver cytochrome P450 system (Nordbloom and Coon, 1977) and the flavoprotein aromatic monooxygenase, salicylate hydroxylase (Kamin et al., 1978). The problem of uncoupling, when present, will place an even greater demand on any cofactor regeneration systems that are utilized in a process.

Autoinactivation is certainly not a universal problem of the oxygenases; however, it is not surprising that some enzymes that activate molecular oxygen have been observed to undergo turnover-dependent inactivation. Two cases of autoinactivation that have been characterized involve liver microsomal cytochrome P450 (Loosemore et al., 1980) and prostaglandin cyclooxygenase (Egan et al., 1979; Hemler and Lands, 1980). Oxygenases that demonstrate autoinactivation cannot be considered for industrial-scale biotransformations, since enzymes that only function to catalyze 800–5000 turnover events before they are destroyed are too expensive to use in a biotechnological process.

Finally, the products of an enzymatic oxygenation are not always found to be 100% chirally pure. Thus epoxidation of 1,7-octadiene by the ω-hydroxylase from *Pseudomonas oleovorans* results in a product mixture of 92% (R)-7,8-epoxy-1-octene and 8% (S)-7,8-epoxy-1-octene (May et al., 1976). Similarly, sulfoxidation of ethyltolylsulfide by both the bacterial cyclohexanone monooxygenase and liver microsomal FAD monooxygenase resulted in sulfoxides of 82% (S) and 95% (R) chiral purity, respectively (Light et al., 1982). Although the chiral purity of these products is high, chemical catalysts have also been developed that are capable of generating products of high chiral purity. The vanadium-catalyzed epoxidation of acyclic allylic alcohols (Narula, 1982) and the rhodium complex–catalyzed reduction of olefinic substrates (J. Halpern, 1982) are notable examples of the increasing ability of synthetic catalysts to generate products that approach 100% chiral purity. Thus while the use of microbial systems to effect chiral oxygenations is well-established, their application cannot assure absolute stereoselectivity, and they must be compared with available chemical systems.

IV. APPLICATION OF OXYGENASES AND OXIDATIVE ENZYMES

As was shown above, there are many enzymes that catalyze oxidations that are potentially of significant economic importance. However, exploitation of their catalytic power is often difficult, and examples of process design problems are numerous. The expense of enzyme production suggests the need for repeated use; hence immobilized enzyme or cell systems must be developed. With all the reaction systems, substrates must be brought to the enzyme(s), and products and by-products must be removed. With the oxidases and oxygenases, one key substrate is molecular oxygen. Its low solubility in aqueous media presents an especially difficult substrate supply problem. Often enzyme cofactors must also be immobilized and regenerated. All of these considerations can present serious process design problems to the potential user.

It is not possible here to address completely all of the above problems. Cell and enzyme immobilization techniques have been recently reviewed by several authors (Mosbach, 1976; Abbott, 1977; Durand and Navarro, 1978; Messing, 1980; Cheetham, 1980). These publications have covered the area much better than space permits here. Problems such as biocatalyst stability, substrate delivery, and product removal are intrinsic to the challenge of immobilizing cells and enzymes and are also addressed by the reviews listed above.

A. Cofactor Immobilization and Regeneration

One area, however, that deserves special attention in this presentation is that of economically immobilizing and regenerating the enzyme cofactors that are consumed in stoichiometric quantities by many oxidative or reductive enzymatic reactions. The ones most studied have been NAD and NADP.

Chambers et al. (1978) have presented a rough but revealing analysis of the economics involved. They assume a bulk NAD cost of $100/lb and a product having a molecular weight of 180. Without regeneration, then, NAD supply contributes a cost of $370 for each pound of product. For all but a few products, such expenses are unacceptable, and cofactor regeneration and reuse are required.

As might be expected, cofactor cost will be approximately inversely proportional to the number of cofactor turnovers. For example, if the above assumptions are used, 37,000 turnovers would reduce the cofactor cost to one cent per pound of product. However, such large turnover numbers require extremely accurate regeneration. Chambers et al. (1978) showed the following table as an illustration:

Number of Regenerations	*Required Regeneration Specificity*
50	0.996
1000	0.9998
20,000	0.99999

Expressed another way: To attain an average of 20,000 regenerations, less than 10 μmoles of inactive cofactor can be allowed per mole of cofactor regenerated. Regeneration specificity and cofactor stability are thus very important economic considerations.

1. Cofactor Regeneration. Several methods of cofactor regeneration have been studied, including electrochemical, chemical, and enzymatic regeneration.

a. Electrochemical Regeneration. Electrochemical regeneration has been tested with platinum electrodes (Aiazwa et al., 1975; Coughlin et al., 1975; Coughlin and Alexander, 1975) and with carbon electrodes (Kelly and Kirwan, 1977). Turnover numbers for the former case were not presented, but in the latter, regeneration specificity was found to be only 90–95%, with maximum turnover numbers of about 60. While this may be sufficient for many chemical reactions, it does not allow the numbers of cofactor turnovers desired. As a frame of reference, turnover numbers of 10,000–20,000 have been observed with enzymatic regeneration (Chambers et al., 1978; Fink and Rodwell, 1975).

b. Chemical Regeneration. Another method of cofactor regeneration uses chemical catalysis. One example is the oxidation of NADH with molecular oxygen as the electron acceptor and phenazine methosulfate (PMS) as the electron carrier (Chambers et al., 1974; Legoy and Thomas, 1978). Many of the suggested redox dyes require photoactivation for significant NADH oxidation rates. Chambers et al. (1974) measured rates using photoactivated methylene blue, acriflavin, riboflavin, and flavin mononucleotide (FMN). Each catalyzed significant oxidation rates. Cited advantages over enzymatic regeneration were increased catalyst stability, wider range of operating conditions, ready availability, and lower cost. The disadvantage mentioned was the low molecular weights of the chemical catalysts; this leads to difficulties in immobilizing the catalysts by entrapment. In order to avoid this restriction, FMN was polymerized, and rates similar to those for the monomer were observed. However, regeneration specificity and cofactor turnover numbers were not presented for any of the cases. Also, providing the light required for many of the dyes presents serious scale-up problems. Perhaps it is these limitations that have made enzymatic cofactor regeneration the method most frequently studied.

c. Enzymatic Regeneration. Several types of enzymatic regeneration have been investigated both for the oxidation of NADH or NADPH and for the reduction of NAD^+ or $NADP^+$. Usually, dehydrogenase enzymes are used with organic molecules as the terminal electron donors or acceptors. As an alternative, however, hydrogenases can be used with molecular hydrogen as the ultimate reducing agent (Wong et al., 1981). Using H_2 provides a relatively inexpensive substrate but also adds complexity to the regeneration process. The method recommended by Wong et al. (1981) uses a hydrogenase from *Methanobacterium thermoautotrophicum* to catalyze the reduction of MV^{2+} to MV^+ (MV = methyl viologen). MV^+ is then used to reduce NAD or NADP. The latter reaction is catalyzed by the flavoenzymes lipamide dehydrogenase or ferredoxin reductase. Thus in this scheme, two enzymes and an intermediate electron carrier are required. The instability of the second enzyme adds further problems.

Enzymatic cofactor regeneration is much more straightforward when dehydrogenases are used. Several have been applied, including alcohol dehydrogenase (Wykes et al., 1972; Weibel et al., 1974; Gestrelius et al., 1975; Campbell and Chang, 1978; Legoy and Thomas, 1978; Chambers et al., 1978; Grunwald and Chang, 1979), lactate dehydrogenase (Larsson and Mosbach, 1971; P. Davies and Mosbach, 1974; Weibel et al., 1974; Wykes et al., 1975; Y. Sakaguchi and Murachi, 1980; V. Sakaguchi et al., 1981), formate dehydrogenase (Wandrey, 1979; Shaked and Whitesides, 1980; Buckmann et al., 1981; Wichmann et al., 1981), malate dehydrogenase (Grunwald and Chang, 1979; Y. Sakaguchi and Murachi, 1980), glutamate dehydrogenase (P. Davies and Mosbach, 1974; Y. Sakaguchi and Murachi, 1980; V. Sakaguchi et al., 1981), alanine dehydrogenase (P. Davies and Mosbach, 1974; Wandrey, 1979), galactose dehydrogenase (P. Davies and Mosbach, 1974), and glucose-6-phosphate dehydrogenase (Wong and Whitesides, 1981).

The alcohol dehydrogenase system has been the most popular. The enzyme is readily available, and the substrates are relatively inexpensive. The equilibrium favors the formation of ethanol and NAD^+ from acetaldehyde and NADH with an equilibrium constant of 1250 (at pH 8 and 23°C) (Backlin, 1958). This can be a major advantage or disadvantage, depending on whether oxidation or reduction is required. As an example, Chambers et al. (1978) present the case in which sorbitol is oxidized to fructose by NAD^+. The resulting NADH is then reoxidized by acetaldehyde and alcohol dehydrogenase. Although the sorbitol-to-fructose reaction is normally unfavorable (equilibrium constant = 0.11 at pH 8 and 23°C) (Chakravorty et al., 1962), when it is combined with the acetaldehyde-to-ethanol reaction, a net favorable equilibrium constant of 138 results. Thus coupling the two reactions by cofactor regeneration allows the desired conversion to proceed to near completion. Obviously, however, if NAD^+ reduction were required for regeneration, the equilibrium constant could become a disadvantage. The possibly detrimental effects of ethanol and acetaldehyde on the enzymes must also be considered.

A similar regeneration system uses lactate dehydrogenase. The equilibrium favors the formation of lactate and NAD^+ from pyruvate and NADH (equilibrium constant = 3400 at pH 8 and 25°C) (Wykes et al., 1975). Although the equilibrium constant is somewhat larger than it is for the alcohol dehydrogenase system, the substrate is much more expensive. For applications such as analytical immobilized enzyme probes (P. Davies and Mosbach, 1974; V. Sakaguchi et al., 1981), however, the equilibrium advantage may outweigh cost considerations.

One of the newest and most attractive systems for the regeneration of NADH from NAD^+ uses the formate dehydrogenase–catalyzed conversion of formate to CO_2. The substrate is inexpensive, and the product is gaseous to help drive the reaction to completion. Wandrey (1979) has suggested a formate dehydrogenase reaction to regenerate NADH for conversion of pyruvate to alanine. More recently, Wichmann et al. (1981) have studied the production of L-amino acids from α-keto acids. They also use formate dehydrogenase to regenerate the NADH required. Shaked and Whitesides (1980), however, point out that, at present, the commercial enzyme is expensive and does not accept NADP as a substrate.

The other dehydrogenases listed above have been used less frequently for cofactor regeneration. They generally suffer from higher substrate costs, but they may be useful for special applications such as immobilized enzyme probes.

2. Cofactor Immobilization. In designing a biocatalytic process using oxidizing or reducing enzymes, profitable application on a long-term basis will probably depend upon the use of continuous processing with immobilized cells or enzymes and with the necessary cofactor also retained within the reactor. While a batch process that provides the required number of cofactor turnovers might be envisioned, the enzyme and cofactor concentrations would probably be so low as to limit reaction rates severely. To allow the higher production rates brought by continuous operation, enzyme and cofactor retention within the bioreactor becomes an economic necessity.

Several methods of cofactor retention have been advanced. Because the cofactor molecular weights are relatively small, their separation from substrates and products by filtration becomes quite difficult. The general approach has thus been to increase the cofactor molecular weights. This has been effected by coupling to both insoluble and soluble supports and also by coimmobilizing the enzyme and the cofactor, usually on insoluble supports. With the insoluble supports, retention of the cofactor is then effected by the use of conventional chromatography columns; with the soluble supports, this is generally done by using ultrafiltration devices such as hollow fiber reactors. However, microencapsulation is also being studied for use with the soluble complexes.

Much of the early work focused on insoluble cofactor supports. Although biochemical spacer arms were used to improve steric availability, Larsson and Mosbach (1971) found that the bound NAD^+ showed only about 0.2% of the coenzymic function of soluble NAD^+. This low functionality may have been caused either by steric unavailability or by attachment to a position on the NAD^+ molecule necessary for coenzymic function. It has been shown that attachment to the exocyclic amino group (N-6) of the adenine moiety produces the most active immobilized cofactor (Mosbach, 1976). However, even when bound at the optimal position on the NAD^+ molecule, insoluble systems were shown not to be suitable for coenzymically active preparations (Mosbach, 1976).

Another approach that was explored was the coimmobilization of the enzyme and cofactor (Gestrelius et al., 1975; Legoy and Thomas, 1978). Gestrelius et al. immobilized both NADH and alcohol dehydrogenase to Sepharose 4B in such a manner that the coenzyme would be fixed in or near the active site of the enzyme. Cofactor regeneration is then effected by alternating the substrate between ethanol and acetaldehyde. When the NAD^+ must be reduced, the volatile acetaldehyde is removed by nitrogen sparging to drive the reaction in the desired direction. Such requirements, though, probably rule out economic applications.

Legoy and Thomas (1978) suggested the use of the electron carrier phanazine methosulfate (PMS) for regeneration of coimmobilized NAD^+ and alcohol dehydrogenase. They employ the system in an immobilized enzyme probe that uses a dissolved oxygen probe to monitor the rate of oxygen consumed as the cofactor is regenerated with PMS. Again, the complexity of the system probably limits its applicability.

The use of cofactors fixed to soluble polymers appears to be the most promising method for economic regeneration systems. Wykes et al. (1972) described the attachment succinyl–NAD^+ to polyethyleneimine (PEI). They used PEI having a molecular weight of 40,000–60,000 and retained the complex using ultrafiltration membranes. While 20–45% of the attached NAD^+ was enzymically reducible, the succinyl link to the NAD^+ was found to be labile above pH 7.

A more actively studied method is coupling the NAD^+ to soluble dextran (Weibel et al., 1974, P. Davies and Mosbach, 1974; Y. Sakaguchi and Murachi, 1980; V. Sakaguchi et al., 1981). P. Davies and Mosbach coupled NAD^+-N^6-(N-(6-aminohexyl)acetimide) to cyanogen bromide–activated dextran T40 for use in enzyme electrodes. Glutamate and pyruvate concentrations were determined with immobilized glutamate and lactate dehydrogenases coupled by the dextran-NAD^+. They also examined two enzyme reactors that use the dextran-NAD^+ for alanine production. One couples galactose and alanine dehydrogenases to produce alanine from pyruvate and ammonia. Galactose is oxidized to galactonolactone to regenerate the dextran-NADH required. The other scheme uses lactate dehydrogenase to convert lactate to pyruvate and then converts the pyruvate to alanine with alanine dehydrogenase. Here, the dextran-NADH is regenerated as the lactate is oxidized to the intermediate, pyruvate. The equilibrium is unfavorable for pyruvate formation but can be driven by high lactate concentrations and by consumption of the products by the second reaction. In this scheme, not only is the reaction pulled in the desired direction, but also there are no by-products produced by the cofactor-regenerating reaction. The pyruvate is used directly for alanine production. The dextran-NAD^+ was shown to be effective with both reaction schemes.

Weibel et al. (1974) also coupled NAD^+ to dextran T-40 through the exocyclic amino group of the adenine moiety. They found that the resultant high-molecular-weight cofactor was active with lactate and alcohol dehydrogenases but that, at pH 6, the reactions had greatly reduced k_{cat}'s and increased K_m's when compared to reactions with free NADH. The stability of the dextran-NADH, however, was similar to that of the free molecule.

More recently, Y. Sakaguchi and Murachi (1980) studied the effects of changing the spacer arm length between the N^6-substituted NAD^+ and dextran T70. The benefits varied with the type of spacer arm and also were dependent on the apoenzyme being investigated. With alkyl spacer arms, the highest coenzyme activity with rabbit muscle lactate dehydrogenase was observed with six atoms between NADH's exocyclic nitrogen and the dextran. However, when using oligoglycyl groups as spacers, the best activity was obtained with spacers of 12–15 atoms. The extent of activity for the NADH coupled by the 12-atom spacer (compared to free NADH) varied from 12% for beef liver glutamate dehydrogenase to 52% for rabbit muscle lactate dehydrogenase. This NADH derivative was also quite stable even at weakly alkaline pH.

V. Sakaguchi et al. (1981) went on to use a dextran T40-N^6-glycylglycylglycylaminoethyl-NAD^+ as the coenzyme for immobilized L-lactate dehydrogenase and L-glutamate dehydrogenase. The immobilized enzymes were packed in columns and used for continuous flow analysis for lactate, glutamate, and glutamate–oxaloacetate aminotransferase (GOT) activity in serum samples. The substrate concentrations were determined by the extent of conversion of dextran-NAD^+ to dextran-NADH as monitored by absorbance at 340 nm. The dextran-NADH was then separated from the mixture and regenerated by oxidation with phanazine methosulfate for reuse. Successful analyses and cofactor regenerations were demonstrated.

Another technique for obtaining high-molecular-weight NAD^+ has recently been published by Buckmann et al. (1981). They couple N(1)-(2-aminoethyl)-NAD^+ to carboxylated polyethylene glycol (PEG), which has a molecular weight of 10,000 or 20,000, and then rearrange to the N^6-(2-aminoethyl)-NAD^+ form. These derivations are much more stable at moderate alkaline conditions than native NAD^+ and exhibit coenzyme activities comparable to the native cofactor. This same group has used the high-molecular-weight cofactor with excellent results in a reactor for the conversion of α-ketoacids to L-amino acids (Wichmann et al., 1981). The PEG–NAD^+ is then reduced by formate dehydrogenase and recycled within an ultrafiltration membrane reactor. When used in continuous operation, nearly total substrate conversion was obtained for 28 days with approximately 3000 cofactor turnovers. Optimization should further improve the system.

Most of the work to date has used hollow fiber reactors or ultrafiltration membrane reactors for retaining the higher-molecular-weight cofactors. However, a more novel approach has also been evaluated. Campbell and Chang (1978) used microcapsules of semipermeable polymer to entrap both enzymes and high-molecular-weight cofactors. Using the polyethyleneimine-NAD^+ of Wykes et al. (1972), they demonstrated malate production with microencapsulated yeast alcohol dehydrogenase and malic dehydrogenase.

The high-molecular-weight cofactor was active and was retained within the collodion microcapsules. More recently, Grunwald and Chang (1979) have described a procedure using the recycling of dextran-NAD^+ in collodion–hemoglobin microcapsules. The hemoglobin in the microcapsules stabilizes the enzymes, and successful production of malate was again demonstrated.

In summary, considerable effort has now been applied to the development of economical systems for cofactor retention and regeneration. The successes reported suggest that practical application of complex oxidative or reductive reaction schemes employing cofactors may now be feasible for selected products.

B. Conclusions

As can be seen, there is a very wide variety of oxidative and reductive reactions that are efficiently catalyzed by enzymes. Many are potentially of commercial importance; but in order for them to be exploited, both the enzyme mechanism and the cofactors involved must first be known. While many enzymes have been investigated, there is still much work left to do. Moreover, even when the enzyme is well-understood, the practical application of these enzymes or enzyme systems will still depend upon coupling many technologies. These will include enzyme or cell immobilization techniques, an understanding of heat and mass transfer limitations in bioreactors, the ability to recycle cofactors, and effective bioreactor design, in addition to the new recombinant DNA technology and classical enzymology. While further developments in these fields are needed, the existing work provides a good basis for industrial application. The future therefore seems to hold great promise for large-scale exploitation of oxidative and reductive biocatalysis.

REFERENCES

Abbott, B. J. (1976) *Adv. Appl. Microbiol. 20,* 203–257.

Abbott, B. J. (1977) in *Annual Reports on Fermentation Processes* Vol. I (Perlman, D., and Tsao, G. T., eds.) pp. 205–233, Academic Press, New York.

Abbott, B. J., and Gledhill, W. E. (1971) *Adv. Appl. Microbiol. 14,* 249–388.

Abbott, B. J., and Hou, C. T. (1973) *Appl. Microbiol. 26,* 86–91.

Adachi, O., Matsushita, K., Shinagawa, E., and Amekama, M. (1980) *Agr. Biol. Chem. 44,* 301–308.

Adams, M. J., Buehner, M., Chandrasekhar, K., Ford, G. C., Hackert, M. L., Liljas, A., Rossmann, M. G., Smiley, I. E., Allison, W. S., Everse, J., Kaplan, N. O., and Taylor, S. S. (1973) *Proc. Nat. Acad. Sci. U.S.A. 70,* 1968–1972.

Aiazwa, M., Coughlin, R. W., and Charles, M. (1975) *Biochem. Biophys. Acta 385,* 362.

Alberti, B. N., and Klibanov, A. M. (1982) *Enzyme Microb. Technol. 4,* 47–49.

Allen, B. R., and Fink, D. J. (1980) in *Enzyme Engineering* Vol. V (Weetal, H. H., and Royer, G. P., eds.) pp. 443–445, Plenum Press, New York.

Ameyama, M., Shinagawa, E., Matsushita, K., and Adachi, O. (1981) *Agr. Biol. Chem. 45,* 851–861.
Anderson, M. S., Hall, R. A., and Griffin, M. (1980) *J. Gen. Microbiol. 120,* 89–94.
Andersson, P., and Pettersson, G. (1982) *Eur. J. Biochem. 122,* 559–568.
Anthony, C., and Zatman, L. J. (1967) *Biochemistry 104,* 960–969.
Asperger, O., Naumann, A., and Kleber, H.-P. (1981) *FEMS Microbiol. Lett. 11,* 309–312.
Austen, R. A., and Dunn, N. W. (1980) *J. Gen. Microbiol. 117,* 521–528.
Backlin, K. I. (1958) *Acta Chem. Scand. 12,* 1279.
Bailey, S. W., and Ayling, J. E. (1980) *J. Biol. Chem. 255,* 7774–7781.
Banaszak, L. J., and Bradshaw, R. A. (1975) in *The Enzymes* Vol. XI, *Oxidation-Reduction,* Part A (Boyer, P. D., ed.) pp. 369–396, Academic Press, New York.
Baratti, J., Couderc, R., Cooney, C. L., and Wang, D. I. C. (1978) *Biotech. Bioeng. 20,* 333–348.
Barker, S. A., and Shirley, J. A. (1980) in *Economic Microbiology,* Vol. V, *Microbial Enzymes and Bioconversions* (Rose, A. H., ed.) pp. 171–226, Academic Press, New York.
Beaty, N. B., and Ballou, D. P. (1981a) *J. Biol. Chem. 256,* 4611–4618.
Beaty, N. B., and Ballou, D. P. (1981b) *J. Biol. Chem. 256,* 4619–4625.
Benson, S., Oppici, M., Shapiro, J., and Fennewald, M. (1979) *J. Bacteriol.* 754–762.
Berg, A., and Rafter, J. J. (1981) *Biochem. J. 196,* 781–786.
Berg, A., Gustafsson, J.-A., Ingelman-Sundberg, M., and Carlstrom, K. (1976) *J. Biol. Chem. 251,* 2831–2838.
Berg, A., Ingelman-Sundberg, M., and Gustafsson, J.-A. (1979) *J. Biol. Chem. 254,* 5264–5271.
Berg, K. A., and Abeles, R. H. (1980) *Biochemistry 19,* 3186–3189.
Bernhardt, F.-H., and Kuthan, H. (1983) *Eur. J. Biochem. 130,* 99–103.
Bernhardt, F.-H., Pachowsky, H., and Staudinger, H. (1975) *Eur. J. Biochem. 57,* 241–256.
Blake, R. C., II, and Coon, M. J. (1981) *J. Biol. Chem. 256,* 12127–12133.
Bourdillon, C., Thomas, V., and Thomas, D. (1982) *Enzyme Microb. Technol. 4,* 175–180.
Branden, C.-I., Jornvall, H., Eklund, H., and Furugren, B. (1975) in *The Enzymes* Vol. XI (Boyer, P. D., ed.) pp. 103–190, Academic Press, New York.
Bray, R. C. (1975) in *The Enzymes* Vol. XII (Boyer, P. D., ed.) pp. 299–419, Academic Press, New York.
Bright, H. J., and Porter, J. T. (1975) in *The Enzymes* Vol. XII (Boyer, P. D., ed.) pp. 421–505, Academic Press, New York.
Britton, L. N., and Markovetz, A. J. (1977) *J. Biol. Chem. 252,* 8561–8566.
Broadbent, D. A., and Cartwright, N. J. (1974) *Microbios 9,* 119–130.
Brodelius, P., Hagerdal, B., and Mosbach, K. (1980) in *Enzyme Engineering* Vol. V (Weetal, H. H., and Royer, G. P., eds.) pp. 383–387, Plenum Press, New York.
Brown, B. R. (1967) in *Oxidative Coupling of Phenols,* (Taylor, W. I., and Battersby, A. R., eds.) pp. 167–201, Marcel Dekker, New York.
Bruice, T. C. (1980) *Acc. Chem. Res. 13,* 256–262.

Bryant, F., and Ljungdahl, L. G. (1981) *Biochem. Biophys. Res. Comm. 100,* 793–799.

Buchholz, K., and Jaworek, D. (1979) in *Enzyme Engineering* Vol. III (Pye, E. K., and Weetal, H. H., eds.) pp. 139–144, Plenum Press, New York.

Buckland, B. C., Dunnill, P., and Lilly, M. D. (1975) *Biotech. Bioeng. 17,* 815–826.

Buckmann, A. F., Kula, M. R., Wichmann, R., and Wandrey, C. (1981) *J. Appl. Biochem. 3,* 301–315.

Bull, C., Ballou, D. P., and Otsuka, S. (1981) *J. Biol. Chem. 256,* 12681–12686.

Cain, R. B. (1980) in *Hydrocarbons in Biotechnology* (Harrison, D. E. F., Higgins, I. J., and Watkinson, R., eds.) pp. 99–132, Heyden, London.

Campbell, J., and Chang, T. M. S. (1978) in *Enzyme Engineering* Vol. III (Pye, E. K., and Weetall, H. H., eds.) pp. 371–377, Plenum Press, New York.

Cardini, G., and Jurtshuk, P. (1970) *J. Biol. Chem. 245,* 2789–2796.

Carrea, G., Colombi, F., Mazzola, G., and Cremonesi, P. (1979) *Biotech. Bioeng. 21,* 39–48.

Carter, R. S., Prenosil, J. E., and Bourne, J. R. (1980) in *Enzyme Engineering* Vol. V (Weetal, H. H., and Royer, G. P., eds.) pp. 321–324, Plenum Press, New York.

Castro, C. E. (1980) *Pharmacol. Ther. 10,* 171–189.

Cedergren-Zeppezauer, E., Samana, J.-P., and Eklund, H. (1982) *Biochemistry 21,* 4895–4908.

Chakrabarty, A. M. (1973) *Proc. Nat. Acad. Sci. U.S.A. 70,* 1641–1644.

Chakrabarty, A. M. (1976) *Ann. Rev. Genet. 10,* 7–30.

Chakravorty, M., Veilga, L. A., Bacila, M., and Horecker, B. L. (1962) *J. Biol. Chem. 237,* 1014.

Chambers, R. P., Ford, J. R., Allender, J. H., Baricos, W. H., and Cohen, W. (1974) in *Enzyme Engineering* Vol. II (Pye, E. K., and Wingard, L. B., eds.) pp. 195–202, Plenum Press, New York.

Chambers, R. P., Walle, E. M., Baricos, W. H., and Choen, W. (1978) in *Enzyme Engineering* Vol. III (Pye, E. K., and Weetall, H. H., eds.) pp. 363–369, Plenum Press, New York.

Cheetham, P. S. J. (1979) *J. Appl. Biochem. 1,* 51–59.

Cheetham, P. S. J. (1980) in *Topics in Enzyme and Fermentation Biotechnology* Vol. IV (Wisaman, A., ed.) pp. 189–238, John Wiley, New York.

Cho, Y. K., and Bailey, J. E. (1977) *Biotech. Bioeng. 19,* 157–158.

Chou, G. I. N., Katz, D., and Gunsalus, I. C. (1974) *Proc. Nat. Acad. Sci. U.S.A. 71,* 2675–2678.

Chun, Y. Y., Iida, M., and Iizuka, H. (1981) *J. Gen. Appl. Microbiol. 27,* 505–509.

Colby, J., and Dalton, H. (1978) *Biochem. J. 171,* 461–468.

Colby, J., Stirling, D. I., and Dalton, H. (1977) *Biochem. J. 165,* 395–402.

Connors, M. A., and Barnsley, E. A. (1982) *J. Bacteriol. 149,* 1096–1101.

Cook, P. F., and Cleland, W. W. (1981) *Biochemistry 20,* 1790–1796.

Cook, P. F., Oppenheimer, N. J., and Cleland, W. W. (1981) *Biochemistry 20,* 1817–1825.

Coon, M. J., and White, R. E., (1980) in *Metal Ion Activation of Dioxygen* (Spiro, T. G., ed.) pp. 73–123, John Wiley, New York.

Cooney, C. L., and Hueter, J. (1974) *Biotech. Bioeng. 16,* 1045–1053.

Coughlin, R. W., and Alexander, B. F. (1975) *Biotech. Bioeng. 17,* 1379.
Coughlin, R. W., Aizawa, M., Alexander, B. F., and Charles, M. (1975) *Biotech. Bioeng. 17,* 515.
Courderc, R., and Baratti, J. (1980) *Agr. Biol. Chem. 44,* 2279–2289.
Crane, F. L., and Beinert, H. (1956) *J. Biol. Chem. 218,* 717–731.
Cremonesi, P., Carrea, G., Ferrara, L., and Antonini, E. (1975) *Biotech. Bioeng. 17,* 1101–1108.
Cromartie, T. H., and Walsh, C. (1975) *Biochemistry 14,* 3482–3490.
Cruse, W. B. T., Kennard, O., and Salisbury, S. A. (1980) *Acta Cryst. B36,* 751–754.
Dahlstrom, R. V., and Jachnig, J. H. (1973) U.S. Patent No. 3,773,621.
Dahlstrom, R. V., and Jachnig, J. H. (1974) U.S. Patent No. 3,784,445.
Dalton, H. (1980) *Adv. Appl. Microbiol. 26,* 71–87.
Davies, J., and Jones, J. B. (1979) *J. Amer. Chem. Soc., 101,* 5405–5410.
Davies, P., and Mosbach, K. (1974) *Biochem. Biophys. Acta 370,* 329–338.
De Beer, R., Duine, J. A., Frank, J., and Large, P. J. (1980) *Biochim. Biophys. Acta 622,* 370–374.
De Beer, R., Duine, J. A., Frank, J., and Westerling, J. (1983) *Eur. J. Biochem. 130,* 105–109.
de Smet, M.-J., Wynberg, H., and Witholt, B. (1981) *Appl. Env. Microbiol. 42,* 811–816.
Duarte, J. M. C., and Lilly, M. D. (1980) in *Enzyme Engineering* Vol. V (Weetal, H. H., and Royer, G. P., eds.) pp. 363–367, Plenum Press, New York.
Duine, J. A., and Frank, J. (1981a) *J. Gen. Microbiol. 122,* 201–209.
Duine, J. A., and Frank, J. (1981b) in *Microbial Growth on C1 Compounds* (Dalton, H., ed.) pp. 31–41, Heyden, London.
Duine, J. A., Frank, J., and Verweil, P. E. J. (1980) *Eur. J. Biochem. 108,* 187–192.
Duine, J. A., Frank, J., and Westerling, J. (1978) *Biochim. Biophys. Acta 524,* 277–287.
Duine, J. A., Frank, J., and Zeeland, J. K. V. (1979) *FEBS Lett. 108,* 443–446.
Dunn, M. F., Bernhard, S. A., Anderson, D., Copeland, A., Morris, R. G., and Roque, J.-P. (1979) *Biochemistry 18,* 2346–2354.
Durand, G., and Navarro, J. M. (1978) *Process Biochem. 9,* 14–23.
Durham, D. R., and Perry, J. J. (1978) *J. Bacteriol. 135,* 981–986.
Edmondson, D. E., Kenney, W. C., and Singer, T. P. (1976) *Biochemistry 15,* 2937.
Egan, R. W., Gale, P. H., and Kuehl, F. A., Jr. (1979) *J. Biol. Chem. 254,* 3295–3302.
Egan, R. W., Gale, P. H., VandenHeuvel, W. J. A., Baptista, E. M., and Kuehl, F. A., Jr. (1980) *J. Biol. Chem. 255,* 323–326.
Eickman, N. C., Solomon, E. I., Larrabee, J. A., Spiro, T. G., and Lerch, K. (1978) *J. Amer. Chem. Soc. 100,* 6529–6531.
Eirich, L. D., Vogels, G. D., and Wolfe, R. S. (1978) *Biochemistry 17,* 4583–4593.
Eklund, H., and Branden, C.-I. (1979) *J. Biol. Chem. 254,* 3458–3461.
Eklund, H., Nordstrom, B., Zeppezauer, E., Soderlund, G., Ohlsson, I., Boiwe, T., Soderberg, B.-G., Tapia, O., Branden, C.-I., and Akeson, A. (1976) *J. Mol. Biol. 102,* 27–59.
Enoch, H. G., Catala, A., and Strittmatter, P. (1976) *J. Biol. Chem. 251,* 5095–5103.

Entsch, B., Ballou, D. P., and Massey, V. (1976) *J. Biol. Chem. 251,* 2550–2563.
Entsch, B., Husain, M., Ballou, D. P., Massey, V., and Walsh, C. (1980) *J. Biol. Chem. 255,* 1420–1429.
Erbes, D. L., Burris, R. H., and Orme-Johnson, W. H. (1975) *Proc. Nat. Acad. Sci. U.S.A. 72,* 4795–4799.
Fagan, J. B., Pastewka, J. V., Park, S. S., Guengerich, F. P., and Gelboin, H. V. (1982) *Biochemistry 21,* 6574–6580.
Fennewald, M., and Shapiro, J. (1977) *J. Bacteriol. 132,* 622–627.
Fennewald, M. A., and Shapiro, J. A. (1979) *J. Bacteriol. 139,* 264–269.
Ferris, J. P., MacDonald, L. H., Patrie, M. A., and Martin, M. A. (1976) *Arch. Biochem. Biophys. 175,* 443–452.
Fink, D. J., and Rodwell, V. W. (1975) *Biotech. Bioeng. 17,* 1029.
Flashner, M. S., and Massey, V. (1974) in *Molecular Mechanisms of Oxygen Activation* (Hayaishi, O., ed.) pp. 245–284, Academic Press, New York.
Fonken, G. S., and Johnson, R. A. (1972) *Chemical Oxidations with Microorganisms* 275 pp., Marcel Dekker, New York.
Forney, F. W., Markovetz, A. J., and Kallio, R. E. (1967) *J. Bacteriol. 93,* 649–655.
Forrest, H. S., Salisbury, S. A., and Kilty, C. G. (1980) *Biochem. Biophys. Res. Comm. 97,* 248–251.
Fox, B., and Walsh, C. T. (1982) *J. Biol. Chem. 257,* 2498–2503.
Franklin, F. C. H., and Williams, P. A. (1980) *Mol. Gen. Genet. 177,* 321–328.
Fridovich, I. (1970) *J. Biol. Chem. 245,* 4053.
Fujii-Kuriyama, Y., Mizukami, Y., Kawajiri, K., Sogawa, K., and Muramatsu, M. (1982) *Proc. Nat. Acad. Sci. U.S.A. 79,* 2793–2797.
Gander, J. E., and Mannering, G. J. (1980) *Pharm. Ther. 10,* 191–221.
Gelb, M. H., Heimbrook, D. C., Malkonen, P., and Sligar, S. G. (1982) *Biochemistry 21,* 370–377.
Gestrelius, S., Mansson, M. O., and Mosbach, K. (1975) *Eur. J. Biochem. 57,* 529–535.
Gibson, D. (1971) *CRC Crit. Rev. Microbiol. 1,* 199.
Gitlitz, P. H., and Krasna, A. I. (1975) *Biochemistry 14,* 2561–2568.
Gonzalez, F. J., and Kasper, C. B. (1982) *J. Biol. Chem. 257,* 5962–5968.
Gottschall, D. W., Dietrich, R. F., Benkovic, S. J., and Shiman, R. (1982) *J. Biol. Chem. 257,* 845–849.
Gray, C. T., and Gest, H. (1965) *Science 148,* 186–192.
Griffith, G. R., Ruettinger, R. T., McKenna, E. J., and Coon, M. J. (1978) *Meth. Enzymol. 53,* 356–360.
Groves, J. T. (1980) in *Metal Ion Activation of Dioxygen* (Spiro, T. G., ed.) pp. 125–162, John Wiley, New York.
Groves, J. T., Kruper, W. J., Jr., and Haushalter, R. C. (1980) *J. Amer. Chem. Soc. 102,* 6377–6380.
Grunwald, J., and Chang, T. M. S. (1979) *J. Appl. Biochem. 1,* 104–114.
Guengerich, F. P., Dannan, G. A., Wright, S. T., Martin, M. V., and Kaminsky, L. S. (1982) *Biochemistry 21,* 6019–6030.
Gunsalus, I. C., and Sligar, S. G. (1978) *Adv. Enzymol. Relat. Areas Mol. Biol. 47,* 1–44.
Gunsalus, I. C., and Wagner, G. C. (1978) *Meth. Enzymol. 52,* 166–188.
Gunsalus, I. C., Meeks, J. R., Lipscomb, J. D., De Brunner, P., and Munck, E.

(1974) in *Molecular Mechanisms of Oxygen Activation* (Hayaishi, O., ed.) pp. 559–613, Academic Press, New York.
Gunsalus, I. C., Pederson, T. C., and Sligar, S. G. (1975) *Ann. Rev. Biochem. 44,* 377.
Halpern, J. (1982) *Science 217,* 401–407.
Halpern, M. G. (1981) *Industrial Enzymes from Microbial Sources,* pp. 3–20, Noyes Data Corp., Park Ridge, N. J.
Hamberg, M., Samuelsson, B., Bjorkhem, I., and Danielsson, H. (1974) in *Molecular Mechanisms of Oxygen Activation* (Hayaishi, O., ed.) pp. 30–86, Academic Press, New York.
Hamilton, G. A. (1974) in *Molecular Mechanisms of Oxygen Activation* (Hayaishi, O., ed.) pp. 405–452, Academic Press, New York.
Hamilton, G. A. (1976) in *Techniques of Chemistry* Vol. X, Part 2, *Applications of Biochemical Systems in Organic Chemistry* (Jones, J. B., Sih, C. J., and Perlman, D., eds.) pp. 875–899, John Wiley, New York.
Hanisch, W. H., Dunnill, P., and Lilly, M. D. (1980) *Biotech. Bioeng. 22,* 555–570.
Harris, J. I., and Waters, M. (1976) in *The Enzymes* Vol. XIII, Part C (Boyer, P. D., ed.) pp. 1–49, Academic Press, New York.
Harrison, J. E., and Schultz, J. (1976) *J. Biol. Chem. 251,* 1371–1374.
Hasegawa, S., Patel, M. N., and Snyder, R. C. (1982) *J. Agr. Food Chem. 30,* 509–511.
Hatefi, Y., and Stiggall, D. L. (1976) in *The Enzymes* Vol. XIII, Part C (Boyer, P. D., ed.) pp. 175–297, Academic Press, New York.
Hayaishi, O., Nozaki, M., and Abbott, M. T. (1975) in *The Enzymes* Vol. XII (Boyer, P. D., ed.) pp. 119–189, Academic Press, New York.
Hemler, M. E., and Lands, W. E. M. (1980) *J. Biol. Chem. 255,* 6253–6261.
Higgins, I. J. (1982) U.S. Patent No. 4,323,649; Imperial Chemical Industries Ltd., London.
Higgins, I. J., Best, D. J., and Hammond, R. C. (1980) *Nature 286,* 561–564.
Higgins, I. J., Best, D. J., Hammond, R. C., and Scott (1981) *Microbiol. Rev. 45,* 556–590.
Hill, I. D. (1967) U.S. Patent No. 3,318,781.
Holbrook, J. J., Liljas, A., Steindel, S. J., and Rossman, M. G. (1975) in *The Enzymes* Vol. XI (Boyer, P. D., ed.) pp. 191–292, Academic Press, New York.
Holland, H. L. (1981) in *The Alkaloids: Chemistry and Physiology* (Manske, R. H. F., and Rodrigo, R. G. A., eds.) p. 323, Academic Press, New York.
Hollenberg, P. F., and Hager, L. P. (1978) *Meth. Enzymol. 52,* 521–529.
Hollenberg, P. F., Rand-Meir, T., and Hager, L. P. (1974) *J. Biol. Chem. 249,* 5816–5825.
Holmgren, A. (1980) *Experientia Suppl. 36,* 149–180.
Hou, C. T., Patel, R., Laskin, A. I., and Barnabe, N. (1979) *Appl. Environ. Microbiol. 38,* 127–134.
Hou, C. T., Patel, R. N., Laskin, A. I., and Barnabe, N. (1980a) *FEMS Microbiol. Lett. 9,* 267–270.
Hou, C. T., Patel, R. N., and Laskin, A. I. (1980b) *Adv. Appl. Microbiol. 26,* 41–69.
Hou, C. T., Patel, R. N., Laskin, A. I., Marczak, I., and Barnabe, N. (1981) *Can. J. Microbiol. 27,* 107–115.

Husain, M., and Massey, V. (1979) *J. Biol. Chem. 254,* 6657–6666.
Husain, M., Schopfer, L. M., and Massey, V. (1978) *Meth. Enzymol. 53,* 543–558.
Ibrahim, M., and Schlegel, H. G. (1980a) *Biotech. Bioeng. 22,* 1877–1894.
Ibrahim, M., and Schlegel, H. G. (1980b) *Biotech. Bioeng. 22,* 1895–1906.
Ichinose, M., Gomes, B., Sanemori, H., and Yasunobu, K. T. (1982) *J. Biol. Chem. 257,* 887–888.
Iida, M., and Iizuka, H. (1981) *J. Pharmacobio-Dyn. 4,* 5–49.
Irwin, A. J., and Jones, J. B. (1976) *J. Amer. Chem. Soc. 98,* 8476–8482.
Irwin, A. J., and Jones, J. B. (1977) *J. Amer. Chem. Soc. 99,* 556–561.
Irwin, A. J., Lok, K. P., Huang, K. W.-C., and Jones, J. B. (1978) *J. Chem. Soc. Perkin I,* 1636–1642.
Iwatsuki, N., Joe, C. O., and Werbin, H. (1980) *Biochemistry 19,* 1172.
Jacobson, F. S., Daniels, L., Fox, J. A., Walsh, C. T., and Orme-Johnson, W. H. (1982) *J. Biol. Chem. 257,* 3385–3388.
Jacoby, G. A., Rogers, J. E., Jacob, A. E., and Hedges, R. W. (1978) *Nature 274,* 179–180.
Jeffery, J., Cummins, L., Carlquist, M., and Jornvall, H. (1981) *Eur. J. Biochem. 120,* 229–234.
Johnson, R. A. (1978) in *Oxidations in Organic Chemistry* Part C (Trahanovsky, W. S., ed.) pp. 131–210, Academic Press, New York.
Jones, J. B. (1980) in *Enzymic and Non-Enzymic Catalysis* (Dunnill, P., Wiseman, A., and Blakebrough, N., eds.) pp. 54–83, Ellis Harwood, Ltd., Chichester, England.
Jornvall, H., Persson, M., and Jeffery, J. (1981) *Proc. Nat. Acad. Sci. U.S.A. 78,* 4226–4230.
Kamin, H., White-Stevens, R. H., and Presswood, R. P. (1978) *Meth. Enzymol. 53,* 527–543.
Kaminski, Z. W., and Jezewska, M. M. (1982) *Biochem. J. 207,* 341–346.
Kaufman, S. (1979) *J. Biol. Chem. 254,* 5150–5154.
Kaufman, S., and Fisher, D. B. (1974) in *Molecular Mechanisms of Oxygen Activation* (Hayaishi, O., ed.) pp. 285–370, Academic Press, New York.
Kelly, R. M., and Kirwan, D. J. (1977) *Biotech. Bioeng. 19,* 1215–1218.
Kieslich, K. (1980a) in *Economic Microbiology* Vol. V, *Microbial Enzymes and Bioconversions* (Rose, A. H., ed.) pp. 370–465, Academic Press, New York.
Kieslich, K. (1980b) in *13th International TNO Conference: Biotechnology—A Hidden Past, a Shining Future* pp. 83–96, Netherlands Central Organization for Applied Science Research TNO, The Hague, Netherlands.
Kieslich, K., and Raspe, G. (1964) U.S. Patent No. 3,161,662.
Kieslich, K., and Sebek, O. K. (1979) *Ann. Rep. Ferment. Processes 3,* 275–304.
Kim, Y. M., and Hegeman, G. D. (1981a) *J. Bacteriol. 148,* 904–911.
Kim, Y. M., and Hegeman, G. D. (1981b) *J. Bacteriol. 148,* 991–994.
Kishore, G. M., and Snell, E. E. (1981a) *J. Biol. Chem. 256,* 4228–4233.
Kishore, G. M., and Snell, E. E. (1981b) *J. Biol. Chem. 256,* 4234–4240.
Klibanov, A. M., and Huber, J. (1981) *Biotech. Bioeng. 23,* 1537–1551.
Klibanov, A. M., and Morris, E. D. (1981) *Enzyme Microbiol. Technol. 3,* 119–122.
Klibanov, A. M., Alberti, B. N., Morris, E. D., and Felshin, L. M. (1980) *J. Appl. Biochem. 2,* 414–421.
Klinman, J. P. (1981) *CRC Crit. Rev. Biochem. 10,* 39–78.

Komai, H., Massey, V., and Palmer, G. (1969) *J. Biol. Chem. 244,* 1692.
Koroly, M. J., and Dempsey, M. E. (1981) *Lipids 16,* 755–758.
Krasna, A. I. (1979) *Enzyme Microbiol. Technol. 1,* 165.
Kulkoski, J. A., and Ghazarian, J. G. (1979) *Biochem. J. 177,* 673–678.
Kvassman, J., Larsson, A., and Pettersson, G. (1981) *Eur. J. Biochem. 114,* 555–563.
Lamed, R. J., Keinan, E., and Zeikus, J. G. (1981) *Enzyme Microbiol. Technol. 3,* 144–148.
Lang, M. A., and Nebert, D. W. (1981) *J. Biol. Chem. 256,* 12058–12067.
Lang, M. A., Gielen, J. E., and Nebert, D. W. (1981) *J. Biol. Chem. 256,* 12068–12075.
Larsson, P. O., and Mosbach, K. (1971) *Biotech. Bioeng. 13,* 393–398.
Lazarus, R. A., Dietrich, R. F., Wallick, D. E., and Benkovic, S. J. (1981) *Biochemistry 20,* 6834–6841.
Legoy, M. D., and Thomas, D. (1978) in *Enzyme Engineering* Vol. III (Pye, E. K., and Weetall, H. H., eds.) pp. 93–99, Plenum Press, New York.
Light, D. R., and Orme-Johnson, N. R. (1981) *J. Biol. Chem. 256,* 343–350.
Light, D. R., Waxman, D. J., and Walsh, C. (1982) *Biochemistry 21,* 2490–2498.
Lilly, M. D., Cheetham, P. S. J., and Dunnill, P. (1979) in *Enzyme Engineering* Vol. III (Pye, E. D., and Weetal, H. H., eds.) pp. 73–78, Plenum Press, New York.
Lipscomb, J. D., Sligar, S. G., Namtvedt, M. J., and Gunsalus, I. C. (1976) *J. Biol. Chem. 251,* 1116–1124.
Loosemore, M., Light, D. R., and Walsh, C. (1980) *J. Biol. Chem. 255,* 9017–9020.
Lu, A. Y. H., and West, S. B. (1980) *Pharmcol. Rev. 31,* 277–295.
Malmstrom, B. G., Andreasson, L.-E., and Reinhammar, B. (1975) in *The Enzymes* Vol. XII (Boyer, P. D., ed.) pp. 507–579, Academic Press, New York.
Marcotte, P., and Walsh, C. (1978) *Biochemistry 17,* 2864–2868.
Martin, C. K. A. (1977) in *Applied Microbiology* Vol. XXII (Perlman, D., ed.) pp. 29–58, Academic Press, New York.
Massey, V., and Hemmerich, P. (1975) in *The Enzymes* Vol. XII (Boyer, P. D., ed.) pp. 191–252, Academic Press, New York.
Massey, V., and Hemmerich, P. (1978) *Biochemistry 17,* 9–17.
Matson, R. S., Hare, R. S., and Fulco, A. J. (1977) *Biochem. Biophys. Acta 487,* 487–494.
Matson, R. S., Stein, R. A., and Fulco, A. J. (1980) *Biochem. Biophys. Res. Comm. 97,* 955–961.
Matsushita, K., Ohno, Y., Shinagawa, E., Adachi, O., and Ameyama, M. (1980) *Agr. Biol. Chem. 44,* 1505–1512.
Matsuyama, H., Hakahara, T., and Minoda, Y. (1981) *Agr. Biol. Chem. 45,* 9–14.
May, S. W. (1979) *Enzyme Microbiol. Technol. 1,* 15–22.
May, S. W., and Phillips, R. S. (1980) *J. Amer. Chem. Soc. 102,* 5983–5984.
May, S. W., Steltenkamp, M. S., Schwartz, R. D., and McCoy, C. J. (1976) *J. Amer. Chem. Soc. 98,* 7856–7858.
May, S. W., Gordon, S. L., and Steltenkamp, M. S. (1977) *J. Amer. Chem. Soc. 99,* 2017–2024.
Mayhew, S. G., Whitfield, C. D., Ghisla, S., and Schuman-Jorns, M. (1974) *Eur. J. Biochem. 44,* 579–591.

McFarland, J. T., Lee, M. Y., Reinsch, J., and Raven, W. (1982) *Biochemistry 21,* 1224–1229.

McGraw-Hill's Biotechnology Newswatch (1982) *2* (Jan. 4), p. 3.

McIntire, W., Edmondson, D. E., Singer, T. P., and Hopper, D. J. (1980) *J. Biol. Chem. 255,* 6553–6555.

McIntire, W., Edmondson, D. E., Hopper, D. J., and Singer, T. P. (1981) *Biochemistry 20,* 3068–3075.

McKenna, E. J., and Kallio (1965) *Ann. Rev. Microbiol. 19,* 183–208.

Messing, R. A. (1980) in *Reports on Fermentation Processes* Vol. IV (Tsao, G. T., ed.) pp. 105–121, Academic Press, New York.

Miall, L. M. (1980) in *Hydrocarbons in Biotechnology* (Harrison, D. E. F., Higgins, I. J., and Watkinson, R., eds.) pp. 25–34, Heyden, London.

Minato, S., Mikamj, Y., and Hayashi, K. (1974) U.S. Patent No. 3,823,070.

Mincey, T., Tayrien, G., Mildvan, A. S., and Abeles, R. H. (1980) *Proc. Nat. Acad. Sci. U.S.A. 77,* 7099–7101.

Mincey, T., Bell, J. A., Mildvan, A. S., and Abeles, R. H. (1981) *Biochemistry 20,* 7502–7509.

Mitani, F. (1979) *Mol. Cell. Biochem. 24,* 21–43.

Moore, R. J., and Wilson, J. D. (1975) *Meth. Enzymol. 36,* 466–474.

Morrison, M., and Schonbaum, G. R. (1976) *Ann. Rev. Biochem. 45,* 861–888.

Mosbach, K. (1976) in *Applications of Biochemical Systems in Organic Chemistry* Part II (Jones, J. B., ed.) pp. 969–994, John Wiley, New York.

Mosbach, K. (1982) *J. Chem. Tech. Biotechnol. 32,* 179–188.

Muller, D. (1928) *Biochemische Zeitschrift 199,* 136.

Muller, H.-G., Mauersberger, S., Schunck, W.-H., Riege, P., Honeck, H., and Huth, J. (1980) in *Biophysics and Regulation of Cytochrome P-450* (Gustafsson, J.-A., et al., eds.) Elsevier/North Holland, Biomedical Press, Amsterdam.

Nakagawa, H., Asano, A., and Sato, R. (1975) *J. Biochem. 77,* 221.

Nakajin, S., Shively, J. E., Yuan, P.-M., and Hall, P. F. (1981) *Biochemistry 20,* 4037–4042.

Narula, A. S. (1982) *Tetrahedron Lett.* (in press)

Neidleman, S. L., Cohen, A. I., and Dean, L. (1969) *Biotech. Bioeng. 11,* 1227–1232.

Neidleman, S. L., Pan, S. C., and Diassi, P. A. (1967a) U.S. Patent No. 3,316,156.

Neidleman, S. L., Pan, S. C., and Diassi, P. A. (1967b) U.S. Patent No. 3,347,927.

Neidleman, S. L., Amon, W. F., Jr., and Geigert, J. (1979) European Patent Application 0,007,176,A2.

Neidleman, S. L., Amon, W. F., Jr., and Geigert, J. (1981) U.S. Patent No. 4,284,723.

Niehaus, W. G., Frielle, T., and Kingsley, E. A. (1978) *J. Bacteriol. 134,* 177–183.

Nordbloom, G., and Coon, M. J. (1977) *Arch. Biochem. Biophys. 180,* 343–347.

Ohlsson, I., Nordstrom, B., and Branden, C.-I. (1974) *J. Mol. Biol. 89,* 339–354.

Ohta, S., Fujita, T., and Tobari, J. (1981) *J. Biochem. 90,* 205–213.

Ohta-Fukuyama, M., Miyake, Y., Emi, S., and Yamano, T. (1980) *J. Biochem. 88,* 197.

Orme-Johnson, W. H. (1973) *Ann. Rev. Biochem. 42,* 159–204.

Ortiz de Montellano, P. R., and Kunze, K. L. (1981) *Biochemistry 20,* 7266–7271.

Ortiz de Montellano, P. R., Kunze, K. L., Beilan, H. S., and Wheeler, C. (1982) *Biochemistry 21,* 1331–1339.

Osumi, T., Nishino, T., and Katsuki, H. (1979) *J. Biochem. 85,* 819–826.
Pai, E. F., and Schulz, G. E. (1983) *J. Biol. Chem. 258,* 1752–1757.
Patel, R. N., and Hou, C.-T. (1981a) U.S. Patent No. 4,268,630, Exxon Research and Engineering Co.
Patel, R. N., and Hou, C.-T. (1981b) U.S. Patent No. 4,269,940, Exxon Research and Engineering Co.
Patel, R. N., Hou, C. T., and Felix, A. (1978) *J. Bacteriol. 133,* 641–649.
Patel, R. N., Hou, C. T., Laskin, A. I., Felix, A., and Derelanko, P. (1979) *J. Bacteriol. 139,* 675–679.
Patel, R. N., Hou, C. T., Laskin, A. I., Felix, A., and Derelanko, P. (1980a) *Appl. Environ. Microbiol. 39,* 720–726.
Patel, R. N., Hou, C. T., Laskin, A. I., Felix, A., and Derelanko, P. (1980b) *Appl. Environ. Microbiol. 39,* 727–733.
Patel, R. N., Hou, C. T., Laskin, A. I., and Derelanko, P. (1981) *J. Appl. Biochem. 3,* 218–232.
Perry, J. J. (1977) *CRC Crit. Rev. Microbiol. 5,* 387–412.
Perry, J. J. (1979) *Microbiol. Rev. 43,* 59–72.
Pietruszko, R., Clark, A., Graves, J. M. H., and Ringold, H. J. (1966) *Biophys. Biochem. Res. Comm. 23,* 526.
Pietruszko, R., Crawford, K., and Lester, D. (1973) *Arch. Biochem. Biophys. 159,* 50–60.
Pirnik, M. P. (1977) *CRC Crit. Rev. Microbiol. 5,* 413–422.
Poulsen, L. L., and Ziegler, D. M. (1979) *J. Biol. Chem. 254,* 6449–6455.
Prenosil, J. E. (1978) in *Enzyme Engineering* Vol. IV (Broun, G. B., Manecke, G., and Wingard, L. B., Jr., eds.) pp. 99–100, Plenum Press, New York.
Prough, R. A., Freeman, P. C., and Hines, R. N. (1981) *J. Biol. Chem. 256,* 4178–4184.
Quayle, J. R. (1980) in *Hydrocarbons in Biotechnology* (Harrison, D. E. F., Higgins, I. J., and Watkinson, R., eds.) pp. 1–9, Heyden, London.
Que, L., Jr. (1980) *Struct. Bonding* (Berlin), *40,* 39–72.
Que, L., Lipscomb, J., Munck, E., and Wood, J. (1977) *Biochem. Biophys. Acta 485,* 60.
Reddy, V. V. R., Kupfer, D., and Caspi, E. (1977) *J. Biol. Chem. 252,* 2797–2801.
Rheinwald, J. G., Chakrabarty, A. M., and Gunsalus, I. C. (1973) *Proc. Nat. Acad. Sci. U.S.A. 70,* 885–889.
Roberts, J. E., Hoffman, B. M., Rutter, R., and Hager, L. P. (1981) *J. Biol. Chem. 256,* 2118–2121.
Ronchi, S., Minchiotti, L., Galliano, M., Curti, B., Swenson, R. P., Williams, C. H., Jr., and Massey, V. (1982) *J. Biol. Chem. 257,* 8824–8834.
Rosazza, J. P., and Smith, R. V. (1979) *Adv. Appl. Microbiol. 25,* 169–208.
Rossmann, M. G., Moras, D., and Olsen, K. W. (1974) *Nature 250,* 194–199.
Rossmann, M. G., Liljas, A., Branden, C.-I., and Banaszak, L. J. (1975) in *The Enzymes* Vol. XI (Boyer, P. D., ed.) pp. 61–102, Academic Press, New York.
Rubin, B., Van Middlesworth, J., Thomas, K., and Hager, L. (1982) *J. Biol. Chem. 257,* 7768–7769.
Ruettinger, R. T., and Fulco, A. J. (1981) *J. Biol. Chem. 256,* 5728–5734.
Ryerson, C. C., Ballou, D. P., and Walsh, C. (1982) *Biochemistry 21,* 2644–2655.
Sabourin, P. J., and Bieber, L. L. (1982) *J. Biol. Chem. 257,* 7460–7467.

Sakaguchi, K. (1982) in *Current Topics in Microbiology and Immunology* Vol. 96 (Henle, W., Hofschneider, P. H., Koprowski, H., Melchers, F., Rott, R., Schweiger, H. G., and Vogt, P. K., eds.) pp. 31–45, Springer, Berlin.

Sakaguchi, V., Sugahara, M., Endo, J., and Murachi, T. (1981) *J. Appl. Biochem. 3,* 32–41.

Sakaguchi, Y., and Murachi, T. (1980) *J. Appl. Biochem. 2,* 117–129.

Salisbury, S. A., Forrest, H. S., Cruse, W. B. T., and Kennard, O. (1979) Nature *280,* 843–844.

Samama, J.-P., Wrixon, A. D., and Biellman, J.-F. (1981) *Eur. J. Biochem. 118,* 479–486.

Sato, R., and Omura, T. (1978) *Cytochrome P-450,* Kodansha, Ltd., Tokyo, and Academic Press, New York.

Sauber, K., Frohner, C., Rosenberg, G., Eberspacher, J., and Lingens, F. (1977) *Eur. J. Biochem. 74,* 89–98.

Schlegel, H. G., and Schneider, K. (eds.) (1978) *Hydrogenases: Their Catalytic Activity, Structure and Function,* Erich Goltze, Göttingen, West Germany.

Schmidt, J., Reinsch, J., and McFarland, J. T. (1981) *J. Biol. Chem. 256,* 11667–11670.

Schwartz, R. D. (1973) *Appl. Microbiol. 25,* 574–577.

Schwartz, R. D., and Hutchinson, D. B. (1981) *Enzyme Microb. Technol. 3,* 361–363.

Schwartz, R. D., and McCoy, C. J. (1973) *Appl. Microbiol. 26,* 217–218.

Schwartz, R. D., and McCoy, C. J. (1976) *Appl. Env. Microbiol. 31,* 78–82.

Schwartz, R. D., and McCoy, C. J. (1977) *Appl. Env. Microbiol. 34,* 47–49.

Shaked, Z., and Whitesides, G. M. (1980) *J. Amer. Chem. Soc. 102,* 7105–7107.

Shapiro, J., Fennewald, M., and Benson, S. (1979) in *Genetics of Industrial Microorganisms* (Sebek, O. K., and Laskin, A. I., eds.) pp. 147–153, American Society for Microbiology, Washington, D.C.

Shiman, R., and Jefferson, L. S. (1982) *J. Biol. Chem. 257,* 839–844.

Shull, G. M., Kita, D. A., and Davisson, J. W. (1953) U.S. Patent No. 2,658,023.

Sih, C. J., and Rosazza, J. P. (1976) in *Techniques of Chemistry* Vol. X, Part 1, *Applications of Biochemical Systems in Organic Chemistry* (Jones, J. B., Sih, C. J., and Perlman, D., eds.) pp. 69–106, John Wiley, New York.

Sih, C. J., Wang, K. C., and Tai, H. H. (1968) *Biochemistry 7,* 796.

Singer, T. P., and Edmondson, D. E. (1974) *FEBS Lett. 42,* 1.

Sligar, S. G., Debrunner, P. G., Lipscomb, J. D., Namtvedt, M. J., and Gunsalus, I. C. (1974) *Proc. Nat. Acad. Sci. U.S.A. 71,* 3906–3910.

Sligar, S. G., Kennedy, K. A., and Pearson, P. C. (1980) *Proc. Nat. Acad. Sci. U.S.A. 77,* 1240–1244.

Smith, R. V., and Rosazza, J. P. (1975) *Biotech. Bioeng. 17,* 785–814.

Sofer, S. S. (1979) *Enzyme Microbiol. Technol. 1,* 3–8.

Stankovich, M. T., Schopfer, L. M., and Massey, V. (1978) *J. Biol. Chem. 253,* 4971–4979.

Steenkamp, D. J., Keaney, W. C., and Singer, T. P. (1978) *J. Biol. Chem. 253,* 2812.

Stirling, D. I., Colby, J., and Dalton, H. (1979) *Biochem. J. 177,* 361–364.

Strittmatter, P., and Enoch, H. G. (1978) *Meth. Enzymol. 52,* 188.

Strittmatter, P., Spatz, L., Corcoran, D., Rogers, M. J., Setlow, B., and Redline, R. (1974) *Proc. Nat. Acad. Sci. U.S.A. 71,* 4565–4569.

Subramanian, V., Liu, T.-N., Yeh, W.-K., and Gibson, D. T. (1979) *Biochem. Biophys. Res. Comm. 91,* 1131–1139.
Subramanian, V., Liu, T.-N., Yeh, W.-K., Narro, M., and Gibson, D. T. (1981) *J. Biol. Chem. 256,* 2723–2730.
Summers, A. O., and Sugarman, L. I. (1974) *J. Bacteriol. 119,* 242–249.
Takeda, R., Nakanishi, I., Nawa, H., Uchibayashi, M., Kusaka, T., Terumichi, J., Uchida, M., Katsumata, M., Yoshino, K., and Fujitani, H. (1962) U.S. Patent No. 3,037,915.
Tassin, J.-P., and Vandecasteele, J.-P. (1972) *Biochim. Biophys. Acta* 31–42.
Theiler, R., Cook, J. C., Hager, L. P., and Siuda, J. F. (1978) *Science 202,* 1094–1096.
Tokeida, T., Niimura, T., Takamura, F., and Yamaha, T. (1977) *J. Biochem. 81,* 851–858.
Tonge, G. M., Harrison, D. E. F., and Higgins, I. J. (1977) *Biochem. J. 161,* 333–344.
Topham, R. W., and Gaylor, J. L. (1970) *J. Biol. Chem. 245,* 2319–2327.
Tsernoglov, D., Hill, E., and Banaszak, L. J. (1972) *J. Mol. Biol. 69,* 75–87.
Ullrich, V. (1979) *Top. Curr. Chem. 83,* 67–104.
Ullrich, V., and Duppel, W. (1975) in *The Enzymes* Vol. XII (Boyer, P. D., ed.) pp. 253–297, Academic Press, New York.
Vanneste, W. H., and Zuberbuhler, A. (1974) in *Molecular Mechanisms of Oxygen Activation* (Hayaishi, D., ed.) pp. 371–404, Academic Press, New York.
Vermilion, J. C., Ballou, D. P., Massey, V., and Coon, M. J. (1981) *J. Biol. Chem. 256,* 266.
Vining, L. C. (1980) in *Economic Microbiology* Vol. V *Microbial Enzymes and Bioconversions* (Rose, A. H., ed.) pp. 523–573, Academic Press, New York.
Volc, J., Sedmera, P., and Musilek, V. (1978) *Folia Microbiol. 23,* 292–298.
Walsh, C. (1979) *Enzymatic Reaction Mechanisms,* W. H. Freeman, San Francisco.
Walsh, C. (1980) *Acc. Chem. Res. 13,* 148–155.
Wandrey, C. (1979) in *Biochemical Engineering, Annals of the New York Academy of Sciences* Vol. 326 (Vieth, W. R., Venkatasubramarian, K., and Constantinides, A., eds.) pp. 87–95, The New York Academy of Sciences, New York.
Waxman, D. J., Light, D. R., and Walsh, C. (1982) *Biochemistry 21,* 2499–2507.
Weibel, M. K., Fuller, C. W., Stadel, J. M., Buckmann, A. F. E. P., Doyle, T., and Bright, H. J. (1974) in *Enzyme Engineering* Vol. II (Pye, E. K., and Wingard, L. B., eds.) pp. 203–208, Plenum Press, New York.
Wheelis, M. L. (1975) *Ann. Rev. Microbiol. 29,* 505–524.
White, R. E., and Coon, M. J. (1980) *Ann. Rev. Biochem. 49,* 315–356.
Wichmann, R., Wandrey, C., Buckmann, A. F., and Kula, M. R. (1981) *Biotech. Bioeng. 23,* 2789–2802.
Williams, C. H. (1976) in *The Enzymes* Vol. XIII, Part C (Boyer, P. D., ed.) pp. 89–173, Academic Press, New York.
Williams, P. A. (1979) in *Genetics of Industrial Microorganisms* (Sebek, O. K., and Laskin, A. I., eds.) pp. 154–159, American Society for Microbiology, Washington, D.C.
Williams, P. A., and Worsey, M. J. (1978) in *Microbiology 1978* (Schlessinger, D., ed.) pp. 167–169, American Society for Microbiology, Washington, D.C.
Wills, C., Kratofil, P., and Martin, T. (1982) in *Genetic Engineering of Microorganisms for Chemicals* (Hollaender, A., ed.) pp. 305–329, Plenum Press, New York.

Wilson, L. D., and Harding, B. W. (1970) *Biochemistry 9,* 1615–1621.
Wislocki, P. G., Miwa, G. T., and Lu, A. Y. H. (1980) in *Enzymatic Basis of Detoxication* Vol. I (Jakoby, W. B., ed.) pp. 135–182, Academic Press, New York.
Wong, C. H., Daniels, L., Orme-Johnson, W. H., and Whitesides, G. M. (1981) *J. Amer. Chem. Soc. 103,* 6227–6228.
Wong, C. H., and Whitesides, G. M. (1981) *J. Amer. Chem. Soc. 103,* 4890–4899.
Wood, J. M. (1980) in *Metal Ion Activation of Dioxygen* (Spiro, T. G., ed.) John Wiley, New York.
Worsey, M. J., Franklin, F. C. H., and Williams, P. A. (1978) *J. Bacteriol. 134,* 757–764.
Wykes, J. R., Dunnill, P., and Lilly, M. D. (1972) *Biochim. Biophys. Acta 286,* 260–268.
Wykes, J. R., Dunnill, P., and Lilly, M. D. (1975) *Biotech. Bioeng. 17,* 51–68.
Yamada, K., Nishihara, A., Shirakawa, Y., and Nakazawa, T. (1978) Japan Patent File No. 47-57033.
Yamada, K., Hattori, T., and Shirakawa, Y. (1979) Japan Patent File No. 45-97137.
Yamaguchi, M., and Fujisawa, H. (1981) *J. Biol. Chem. 256,* 6783–6787.
Yamazaki, I. (1974) in *Molecular Mechanisms of Oxygen Activation* (Hayaishi, O., ed.) pp. 535–558, Academic Press, New York.
Yasukochi, Y., and Masters, B. S. S. (1976) *J. Biol. Chem. 251,* 5337.
Ziegler, D. M. (1980) in *Enzymatic Basis of Detoxication* Vol. I (Jakoby, W. B., ed.) pp. 201–227, Academic Press, New York.

CHAPTER
10

Chemical and Fuel Production from One-Carbon Fermentations: A Microbiological Assessment

J. Gregory Zeikus

I. INTRODUCTION

Industrial society's present dependence on petroleum has generated applied scientific interest in microbial fermentations for chemical and fuel production. Renewable resources, organic wastes, and underdeveloped fossil reserves have been proposed as the raw materials for the development of new fermentation technology. The fermentation products envisioned include various industrial chemicals, expensive biochemicals and enzymes, and methane or liquid fuels (Zeikus, 1980).

Two alternative routes for chemical and fuel production by anaerobic bacterial fermentations have been proposed (see Fig. 10.1). The direct fermentation of organic matter from various biomass components after minimal pretreatment may offer a wider array of potential end products (for example, solvents as well as acids). However, fermentation of one-carbon feedstocks generated from the pyrolysis of organic matter may have

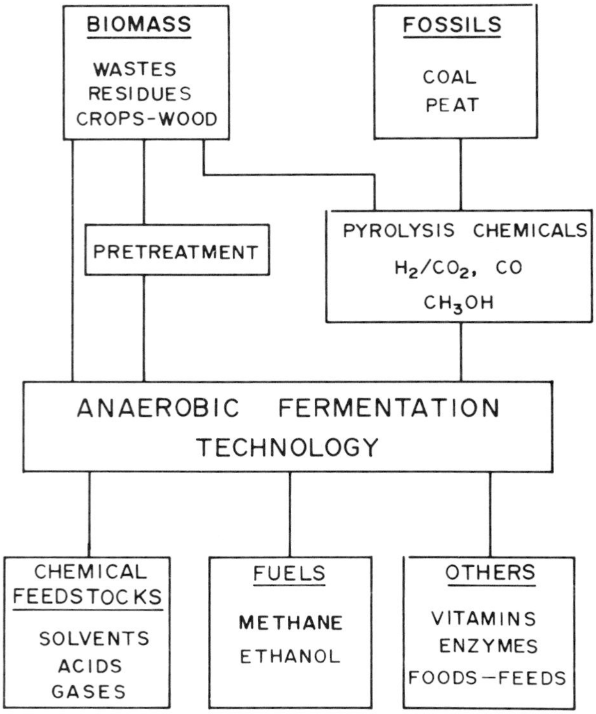

FIGURE 10.1 The relation of anaerobic one-carbon fermentations to developing technology for chemical and fuel production from renewable or underutilized resources (from Zeikus, 1980).

more advantages for industrial fermentation technology. First, one-carbon substrates represent enormously larger quantities of fermentation feedstocks than do the fermentable organic components of biomass because renewable resources (that is, lignocellulosic materials) as well as fossil fuels such as coals and peats can be pyrolyzed into synthesis gases. High-temperature gasification of solid fuels such as coal or wood produces gas mixtures with high concentrations of CO, H_2, or CO_2. Typical raw gas compositions and gasification (energy) efficiencies of oxygen- or steam-blown coal and wood gasifiers are shown in Table 10.1. Methanol is readily and inexpensively formed by thermal treating of wood or by chemical synthesis from CO/H_2 gas mixtures (Cheremisinoff, 1979). Most important, from a microbiological perspective, unicarbonotrophic bacteria that ferment one-carbon compounds as the sole carbon source for growth in the absence of O_2 are the most metabolically efficient biosynthetic life forms on earth (Zeikus, 1982). These microorganisms perform certain chemical transformation reactions with nearly maximal theoretical yield of product and with minimal energy loss from the system.

TABLE 10.1 Gas Composition and Efficiency of Coal and Wood Gasification

Gasifier	Shell-Koppers	BGC Lurgi	Winkler
Feed	Ill. #6 Coal	Ohio #9 Coal	Wood (10% moisture)
Gasification temperature	1500–1800°C	1100–1900°C	1100°C
Raw gas composition (vol. %)			
CO	65	55.5	29.2
H_2	32.1	28.3	32.6
CO_2	0.8	5.1	19.2
CH_4	—	6.3	0.5
C_2H_4	—	0.1	—
C_2H_6	—	0.4	—
N_2	0.5	4.0	11
H_2S COS	1.4	2.0	—
H_2O	—	—	18.5
Cold gas thermal efficiency (%)	83	76–85	70–80

From Datta and Zeikus (1982).

The objective of this paper is to explain the microbiological features of the gaseous and soluble one-carbon substrate fermentations and to assess their potential in chemical and fuel production. Within the total group of unicarbonotrophic microbes, two general kinds of fermentations are examined in detail. These include methane production by methanogenic bacteria (that is, the methanogens) and organic acid production by acidogenic bacteria (that is, the acidogens). Certain aspects of methanogen and acidogen metabolism and biochemistry are examined in more detail in order to assess the potential utility of these two groups in generating products other than natural gas and acetic and butyric acids. Although a wide variety of anaerobes have been recently examined for their ability to transform one-carbon compounds such as H_2/CO_2, formate, methanol, methylamine, and carbon monoxide, this review will focus on a relatively few physiologically distinct species that produce methane or organic acids via unicarbonotrophic metabolism. Unicarbonotrophic species that are most representative of the metabolic and physiological diversity that exists among acidogenic and methanogenic bacteria are listed in Table 10.2.

II. METHANOGENESIS

Methanogenic bacteria are physiologically diverse and are comprised of mesophilic and thermophilic species that grow as rods (*Methanobacterium* sp.), cocci (*Methanococcus* sp.), packets (*Methanosarcina* sp.), filaments

TABLE 10.2 Consumption of One-Carbon Substrates by Selected Anaerobic Unicarbonotrophs

	Physiological Function					
Bacterial Species	H_2/CO_2	CO	$HCOOH$	CH_3OH	CH_3NH_2	CH_4
METHANOGENS						
Methanosarcina barkeri	1	1	NR	1	1	4
Methanobacterium thermoautotrophicum	1	1	NR	NR	NR	4
Methanococcus vanneilii	1	NR	1	NR	NR	NR
Methanobacterium formicicum	1	4	1	NR	NR	4
ACIDOGENS						
Clostridium thermoautotrophicum	1	1	NR	1	NR	NR
Clostridium thermoaceticum	1	1	NR	NR	NR	NR
Acetobacterium woodii	1	1	1	1	NR	NR
Butyribacterium methylotrophicum	1	1	3	1	NR	NR

Adapted from Zeikus (1982).

Abbreviations: 1 = utilized as a carbon source and electron donor; 2 = utilized as an electron donor; 3 = utilized as a carbon source; 4 = exchange or adventitious reaction; NR = not reported.

(*Methanothrix* sp.), and spirals (*Methanospirillum* sp.). At the macromolecular level, methanogens are quite distinct from other anaerobes, contain ether-linked membrane lipids, and possess walls devoid of peptidoglycan (Balch et al., 1979).

Unicarbonotrophy unites the physiologically diverse methanogenic species. Several biochemical compounds, which appear to be widespread in methanogens, have been identified to function in expression of their unique metabolic potential. Corrinoids were first identified in *M. barkeri* as a unique cobalamin form, Factor III, which contains 5-hydroxybenzimidazole as a ligand (T. C. Stadtman, 1967). Corrinoids abounded in all methanogens examined, and *M. barkeri* contained the highest levels, but the amount varied with the carbon and electron sources used during growth (Krzycki and Zeikus, 1980). Cobalamins function in methyl transfer reactions in methanogens (T. C. Stadtman, 1967; Taylor and Wolfe, 1974; Wood et al., 1982; Kenealy and Zeikus, 1982). The presence of coenzyme M or 2-mercaptoethane sulfonic acid appears to be unique to methanogens (Balch et al., 1979). Methyl CoM is an intermediary metabolite of one-carbon metabolism in methanogens and functions in methyl transfer reactions (Balch et al., 1979; Daniels and Zeikus, 1978). The coenzyme factor F_{430} contains nickel and is a tetrapyrrole (Whitman et al., 1981; Diekert et al., 1980). F_{430} is a component of methyl CoM methyl reductase and functions in methanogenesis (Ellefson and Wolfe, 1981; Tanner, 1982). Coenzyme YFC, a pteridine, was discovered as an intermediary metabolite of one-carbon metabolism in diverse methanogens (Daniels and Zeikus, 1978). Coenzyme YFC has been identified as a unique carboxylated pteridine, carboxydihydromethanopterin (Keltjens and Vogels, 1981). Factor F_{420}, a novel deazaflavin, is a low-redox electron carrier that functions in both catabolic and anabolic reactions (Wolfe and Higgins, 1979; Tzeng et al., 1975; Zeikus et al., 1977).

Detailed studies of methanogen growth physiology have been limited primarily to *M. thermoautotrophicum,* the most prolific species, and to *M. barkeri,* the most metabolically versatile species. The only substrates known to serve as sole carbon and electron donors for growth of *M. thermoautotrophicum* at 65°C with phosphate, ammonium, and sulfide as P, N, and S sources are H_2/CO_2 and CO (Zeikus and Wolfe, 1972; Daniels et al., 1977). The maximal cell growth rate in pressure tubes on CO alone was only 1% of that on H_2/CO_2, and the strain used was not adapted to grow on CO concentrations in the headspace of greater than 30%. Growth of *M. thermoautotrophicum* strain Marburg was dependent on nickel, cobalt, and molybdenum and achieved a doubling time of 1–18 h and a cell density of 2 g cells dry weight per liter in batch culture (Schonheit et al., 1979, 1980). In a NH_4^+-limited chemostat, the growth yield of *M. thermoautotrophicum* was directly dependent on medium ammonium concentration at a doubling time of about 2 h (Kenealy et al., 1982). Thus thermophilic methanogens like *M. thermoautotrophicum* are, to date, the most prolific methanogenic, chemolithotrophic, or autotrophic microorganisms described. In general,

M. thermoautotrophicum appears as a chemolithotroph, but it is not an obligate autotroph. A significant portion of cell carbon can be derived from acetate (Zeikus et al., 1975), and cysteine can serve as an S source (Kenealy et al., 1982).

The metabolic versatility of *M. barkeri* is based on the use of a variety of one-carbon compounds, polymethylamines, and acetate either as the sole carbon and energy source during unitrophic growth or in combinations with these substrates during mixotrophic growth (Zeikus, 1982). *M. barkeri* consumes either H_2/CO_2, methanol, or methylamine as the sole carbon and electron donor for growth and can grow via simultaneous metabolism of H_2/CO_2/methanol or methanol/methylamine (Weimer and Zeikus, 1978a). The growth yield (mg cells/mmol methane formed) obtained at the end of batch cultivation was higher on methanol than H_2/CO_2. It has become apparent from studies in progress that most growth parameters (that is, μ_{max}, K_S, Y_{CH_4}, Y_S) reported in the literature (Weimer and Zeikus, 1978a, b) for growth of *Methanosarcina* strains on H_2/CO_2, methanol, or acetate may not be valid in continuous flow culture. Detailed nutritional studies aimed at growth improvement of *M. barkeri* strains indicate that cobalt, molybdenum, selenium, and nickel were required as trace elements for growth; certain strains required riboflavin; and methionine, but not cysteine, could replace Na_2S as the sulfur source (Scherer and Sahm, 1981a, b).

The overall mechanism for generating cellular energy via one-carbon transformation reactions in methanogens is complex, and at present it is a very speculative science. Two major conceptual problems have been examined: the path of carbon and electrons from energy substrate to methane and the less studied coupling of these reactions to synthesis of ATP. Barker (1956) first proposed a carbon flow pathway that unified the consumption of energy sources by methanogens. Today, Barker's scheme still serves as a useful model because the data at hand are in agreement. The thermodynamically most favored reaction, which is common to methanogenesis from all energy substrates for growth, is the reduction of a methyl-level intermediate to methane. By and large, establishing the exact biochemistry of this transformation has been a major goal of understanding methanogenesis. The exact function of coenzymes (that is, coenzyme M, vitamin B_{12}, and/or factor F_{430}) in this reaction remains to be vigorously proven by biochemists. The mechanism of gaining ATP is coupled with the conversion of one-carbon compounds to methane. It is generally assumed that a proton motive force is generated by methanogenesis and that this feature drives ATP synthesis (Zeikus, 1982). However, the reaction that is common to methanogenesis from all carbon precursors is the reduction of a methyl group to methane; enigmatically, this transformation consumes protons.

Understanding unicarbonotrophic metabolism in either methanogens or acetogens is complex because both the catabolic products and the anabolic products must be synthesized from the same precursor. For example, during growth on CO or methanol, methanogens must make CO_2 and methane as well as all the different chemical components that comprise cells. *M. barkeri* has been used as a model organism to examine the metabolic path of carbon and electrons during growth on one-carbon substrates. Both ^{14}C-methanol and $^{14}CO_2$ were fixed into similar short-term intermediary metabolites (that is, CoM derivatives, YFC, and alanine, aspartate, and glutamate) by cells grown on either methanol or H_2/CO_2 as the sole carbon and energy source (Daniels and Zeikus, 1978). During growth on methanol/CO_2, 48% of the cell carbon was derived from methanol, a value indicating that equivalent amounts of cell carbon were derived from CO_2 and an organic intermediate more reduced than CO_2 (Weimer and Zeikus, 1978a). Cell extracts lacked dye-coupled methanol dehydrogenase activity as well as activity for key cell carbon synthesis enzymes of the Calvin cycle, serine path, or hexulose path. Growth of *M. barkeri* on H_2/CO_2 was inhibited by low concentrations of iodopropane, a corrinoid antagonist (Kenealy and Zeikus, 1981). Notably, 40 μM of iodopropane inhibited growth by 80%, whereas methanogenesis was inhibited by only 30%. In the presence of acetate, iodopropane was no longer an inhibitor of growth on H_2/CO_2. This finding suggested that a cobalamin (that is, vitamin B_{12}) -dependent step was involved in the autotrophic synthesis of acetyl CoA or acetate in methanogens. Notably, about half of total cell carbon was synthesized from added ^{14}C-acetate when cells were grown on either CH_3OH or H_2/CO_2. During mixotrophic growth on H_2/CO_2/methanol/acetate, the percentages of cells derived from ^{14}C-CO_2, CH_3OH, and U-acetate were 36.0, 13.5, and 50.8, respectively. Notably, these studies clearly demonstrated that acetate addition to cells grown on H_2/CO_2/methanol greatly diminished the contribution of CO_2 and methanol to cells, but iodopropane addition did not significantly alter the distribution of label incorporated. Hence acetate appeared as a direct precursor to cell carbon, and its synthesis from one-carbon compounds is iodopropane sensitive.

M. barkeri appears to assimilate ammonia via the enzymatic activities common to other bacteria (Kenealy et al., 1982). Cell extracts contained glutamine synthetase, glutamate synthase, glutamate oxaloacetate transaminase, and glutamate pyruvate transaminase. Alanine dehydrogenase and glutamate dehydrogenase were not detectable. Hydrogen-reduced 5-deazaflavin or FMN was used as the electron donor for glutamate synthase.

The major carbon transformation reactions associated with synthesis of C_3, C_4, and C_5 units involved in the one-carbon metabolism of *M. barkeri* have been established (Weimer and Zeikus, 1979). Cell extracts of *M. barkeri* contained anabolic levels of 5-deazaflavin-linked pyruvate synthase, citrate synthase, aconitase, NADP-linked isocitrate dehydrogenase, and NAD-linked malate dehydrogenase. The function of these enzymes in cell carbon synthesis was supported by analysis of the position of label in alanine, aspartate, and glutamate when cells were grown on H_2/CO_2 in the presence of ^{14}C-acetate. The specific radioactivities of glutamate from cells grown on [1-^{14}C]- or [2-^{14}C]-acetate were approximately twice that of aspartate. The methyl and carboxyl carbons of acetate were incorporated into aspartate and glutamate to similar extents. Degradation studies revealed that acetate was not significantly incorporated into the C_1 of alanine, C_1 or C_4 of aspartate, or C_1 of glutamate. The C_5 of glutamate, however, was partially derived from the carboxyl carbon of acetate. These studies indicated that the oxidative reactions of the TCA cycle were used in *M. barkeri* to synthesize a C_5 unit (that is, glutamate) and that this metabolic path is different in *M. thermoautotrophicum* (Zeikus et al., 1977; Fuchs and Stupperich, 1978).

Cell suspensions of unicarbonotrophically grown *M. barkeri* synthesized acetate and alanine from $^{14}CO_2$ (Kenealy and Zeikus, 1982). The addition of iodopropane totally inhibited acetate synthesis but not incorporation of CO_2 into alanine, a finding indicating inhibition of a corrinoid-dependent reaction but not the CO_2 exchange function of pyruvate synthase. Cell extracts of H_2/CO_2-grown cells catalyzed the synthesis of ^{14}C-acetate from ^{14}CO ($\sim$1 nmol/min/mg protein) and isotopic exchange between CO_2 or CO and the C_1 of pyruvate. Acetate synthesis from ^{14}CO was stimulated by CH_3B_{12} but not CH_3-tetrahydrofolate or CH_3CoM. CH_3CoM and CoM were inhibitory to acetate synthesis. Cell extracts of *M. barkeri* grown on H_2/CO_2, methanol, or acetate contained high levels of phosphotransacetylase ($>$6 μmol/min/mg protein) and acetate kinase ($>$0.14 μmol/min/mg protein). Thus it was not possible to distinguish between acetate or acetyl CoA as the immediate product of two-carbon synthesis via autotrophic cell carbon fixation. In short, the evidence at hand suggests that acetate/acetyl CoA is synthesized via a C_1–C_1 condensation reaction; this condensation reaction is inhibited by iodopropane and associated with CO dehydrogenase activity, and this reaction is not required when cells are grown on H_2/CO_2 or methanol with acetate as a carbon source.

Fig. 10.2 represents a general carbon flow scheme for *M. barkeri* that illustrates the common relationships between cell carbon synthesis, methanogenesis, autotrophy, and methylotrophy (Kenealy and Zeikus, 1982). When *M. barkeri* was grown mixotrophically on $H_2/CO_2/CH_3OH$, $^{14}CH_3OH$ or $^{14}CO_2$ was equivalently incorporated into the major cellular components (that is, lipids, proteins, and nucleic acids). $^{14}CH_3OH$ was selectively incorporated into the C_3 of alanine with decreased amounts fixed in the C_1 and C_2 position, whereas $^{14}CO_2$ was selectively incorporated into the C_1 moiety with decreasing amounts assimilated into the C_2 and C_3 atoms. Notably, $^{14}CH_4$ and $^{14}CH_3$-alanine synthesized from $^{14}CH_3OH$ during growth shared a common specific activity distinct from that of CO_2 or methanol. These results suggested that common intermediates and carbon transformations linked both autotrophic (that is, CO_2) and methylotrophic (that is, CH_3OH) assimilation pathways of *M. barkeri* and that the methyl (that is, C_3) of alanine and the methyl of methane share a common precursor. Also, this represented the first evidence to indicate a biochemical unification between catabolic and anabolic one-carbon transformations in methanogens.

The biochemical mechanism that accounts for methylotrophy in *M.*

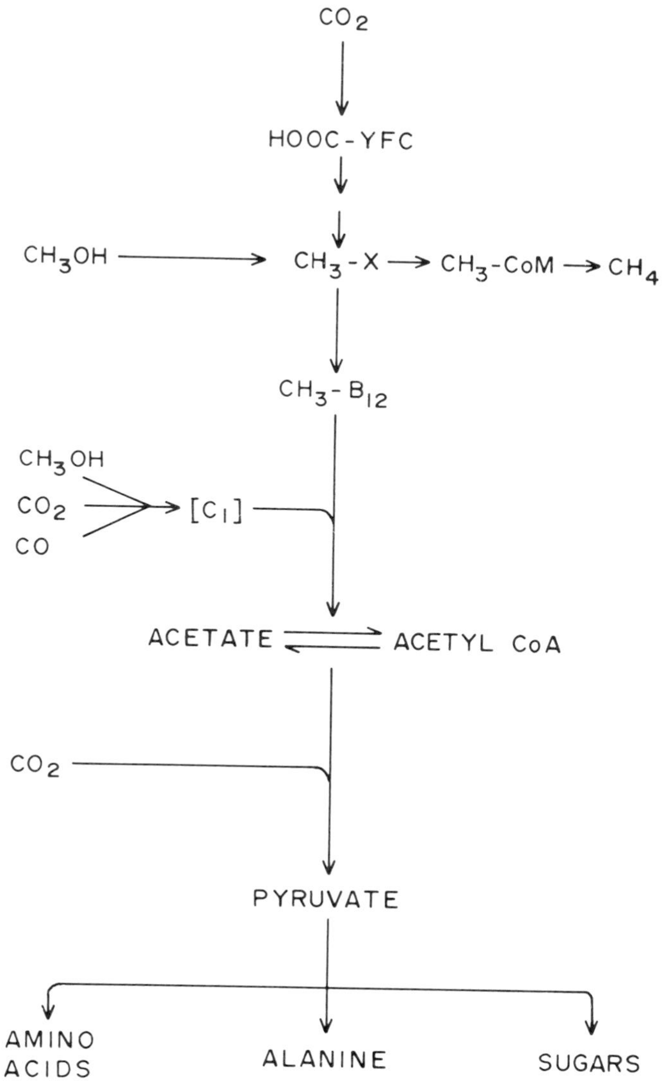

FIGURE 10.2 Unification of carbon flow pathways for methane and cell synthesis in *M. barkeri*. This model predicts common, shared one-carbon carriers for initial carbon transformations leading to methane and cell synthesis. Autotrophy and methylotrophy are accounted for by the use of common carbon carriers and a common C_1–C_1 condensation reaction. Acetyl CoA or acetate is the immediate biosynthesis product of unicarbonotrophic metabolism involving CO dehydrogenase activity. Abbreviations: B_{12}, a corrinoid; CoA, coenzyme-A; CoM, 2-mercaptoethansulfonic acid; YFC, a pteridine coenzyme (from Zeikus, 1982).

barkeri appears to be quite distinct from that described in aerobic unicarbonotrophs, which assimilate via either CO_2 fixation in the reductive pentose phosphate path, formaldehyde fixation in the hexulose phosphate path, or CO_2 and formaldehyde fixation in the serine path. *M. barkeri* assimilated cell carbon at three different one-carbon oxidation states: at the methyl level, at a C_1 unit more oxidized than methyl, and as CO_2. The key enzymatic activities in C_3 synthesis during unicarbonotrophic carbon assimilation thus appear to be methyltransferase, CO dehydrogenase, and pyruvate synthase.

III. ACIDOGENESIS

A wide variety of anaerobic bacteria produce different kinds of acid products (for example, lactic, acetic, butyric, formic, caproic, propionic, and succinic) as a consequence of fermentation. However, one diverse metabolic group of anaerobes, commonly called homoacetogens or acetogens, can grow unicarbonotrophically and, depending on the exact environmental parameters, synthesize acetic acid as the sole fermentation end product. In acetogens, the general mechanism of one-carbon transformation is characterized by the net synthesis of an acetyl CoA from either heterotrophic (for example, glucose as substrate) or unicarbonotrophic (for example, carbon monoxide as substrate) fermentations. The metabolic path for acetogenic transformation of one-carbon substrates appears at a distance to be mechanistically analogous to methanogen metabolism because a C_1–C_1 condensation reaction is used for cellular synthesis of a two-carbon unit.

As a group, acetogenic species display considerable physiological diversity (Zeikus, 1982). The DNA guanosine plus cytosine content varies from 33 to 58 mol %. This variation is as broad a range as that observed in methanogens. Their metabolic versatility is quite outstanding, and, unlike the methanogens, acetogens are not limited to one-carbon compounds and acetate as principal energy sources. Rather, acetogens proliferate unicarbonotrophically on H_2/CO_2, formate, methanol, or CO, and/or on hexose, lactate, and a diversity of other substrates; this appears as a highly species-specific property. All species form acetate as a fermentation product, but formate, butyrate, propionate, and H_2/CO_2 can be significant end products of certain species.

Acetobacterium woodii ferments H_2/CO_2, glucose, lactate, and formate (Balch et al., 1977). This species also ferments methanol or cleaves and ferments the methoxyl moieties from a variety of aromatic acids (Bache and Pfennig, 1981). Cultures can be adapted to grow on carbon monoxide as an energy source in the presence or absence of formate (R. Kerby, unpublished findings, author's lab).

Methylotrophy in acetogens was first established in *B. methylotrophicum,* which readily ferments hexose, lactate, H_2/CO_2, or methanol/CO_2/acetate. The ability of acetogens to grow via CO fermentation was just recently reported (Lynd and Zeikus, 1981). A *B. methylotrophicum* strain was adapted to grow rapidly (~12 h doubling time) on 100% CO in the culture headspace as the sole fermentation substrate. The principal end products of *B. methylotrophicum* fermentations depend on the substrate and medium pH. This species can produce either acetate or butyrate as the sole end product or can form mixtures of H_2/CO_2, butyrate, and acetate. *B. methylotrophicum* appears to be similar to *Eubacterium limosum* in general fermentation properties, including the ability to ferment methanol or CO (B. R. Genthner and Bryant, 1982; B. R. Genthner et al., 1981).

Clostridium thermoaceticum was generally considered to ferment hexoses or pentoses to acetic acid (Fontaine et al., 1942; Ljungdahl and Wood, 1969). *C. thermoaceticum* type strain Fontaine can also grow readily on H_2/CO_2 at 55°C, or it can be culturally adapted to grow on CO as an energy source, but methanol is not a fermentation substrate (Kerby and Zeikus, 1983).

Clostridium thermoautotrophicum is a thermophilic species that grows on sugars or H_2/CO_2 at temperatures greater than 60°C, and the species fermentation substrate range differs from that of *C. thermoaceticum* (Wiegel et al., 1981). Many other homoacetogenic species have been described, but detailed metabolic studies have been limited to *C. thermoaceticum, A. woodii,* and *B. methylotrophicum.*

Notably, the spore morphological properties vary considerably among different acetogenic species. *C. thermoautotrophicum* forms terminal, swollen, phase white refractile spores (Wiegel et al., 1981). The spore ultrastructure of these species is typical of other published *Clostridium* species, and it includes a phase bright cortex and multiple spore coat layers (Braun et al., 1981). However, my lab has not detected phase white refractile spores in *C. thermoaceticum* type strain Fontaine (unpublished findings, author's lab). Rather, spores appear like swollen clubs, as in *C. barkeri* (E. R. Stadtman et al., 1972), and with limited phase bright refractility, but with a definite dark external outline that delineates the spore shape. Spores of *B. methylotrophicum* are very difficult to document and are quite morphologically distinct from vegetative cells; they remain viable for long periods in spent culture fluid and are resistant to heat treatment at 75–80°C for 10 min (Zeikus et al., 1980).

One-carbon metabolism of acetogens like methanogens is associated with high intracellular levels of corrinoids (Tanner et al., 1978; Zeikus et al., 1980). The only other carbon carriers implicated in acetogen metabolism at present are pteridines, which all derive from tetrahydrofolic acid as chromophore (Tanner et al., 1978; Parker et al., 1971). By and large, the electron carriers involved in acetogen metabolism have only been characterized well in *C. thermoaceticum* and include ferredoxin (Yang et al., 1977), rubredoxin (Yang et al., 1980), cytochrome B, and menaquinone (Gottwald et al., 1975).

Detailed knowledge on the biochemistry of one-carbon transformation is limited to studies of *C. thermoaceticum* grown on hexose and not on one-carbon compounds. H. G. Wood (1952) proved that *C. thermoaceticum* accomplished a total synthesis of acetate from CO_2 by showing that some of the acetate formed by glucose in the presence of $^{13}CO_2$ was two mass units heavier than unlabeled acetate. The first clues as to the path of acetate formation from CO_2 came from the observation that ^{14}C-formate was selectively incorporated into the methyl of acetate (Lentz and Wood, 1955). Tetrahydrofolate-bound intermediates were implicated by the findings that methyl-THF was converted to acetate in cell extracts (Ghambeer et al., 1971) and that CH_3THF and 10-formyl THF became highly labeled after pulsing cells with $^{14}CO_2$ (Parker et al., 1971). The enzymes necessary to convert formate to methyl THF all required folate intermediates as carbon carriers (Andreeson et al., 1973).

The most elusive part of the acetogenic pathway is the mechanism of acetate synthesis from methyl THF, and this aspect is still not certain. ^{14}C-methyl-corrinoids acquired a very high specific activity when cell extracts were exposed to $^{14}CO_2$, and ^{14}C-acetate was formed in the presence of pyruvate from $^{14}CH_3$-corrinoids (Ljungdahl et al., 1966). Corrinoids were further implicated in acetate synthesis by the finding that transmethylation via Co-methylcobalamin was inhibited by alkylhalides such as propyliodide (Ljungdahl et al., 1965). A corrinoid containing enzyme from *C. thermoaceticum* was purified to 80% homogeneity that catalyzed the transmethylation of CH_3THF to a methyl–enzyme complex and participated in acetate synthesis from CH_3THF and pyruvate (Welty and Wood, 1978).

Acetate-transforming enzyme activity was purified into five components from crude extracts of *C. thermoaceticum,* which together catalyzed acetate synthesis from pyruvate (Drake et al., 1981). These components included F_1, a phosphotransacetylase; F_2, a methyltransferase; F_3, a CO–dehydrogenase complex; F_4, a pyruvate cleavage activity; and F_5, ferredoxin. Notably, the CO–dehydrogenase component was purified 14-fold and shown to contain nickel (Drake et al., 1980). It was recently demonstrated (Hu et al., 1982) that acetyl CoA was synthesized by components F_3 and F_2 according to the following reaction:

$$^{14}CO + CH_3THF + CoA \xrightarrow{ATP} CH_3{}^{14}CO\ CoA + THF$$

Thus the direct role for methyltransferase (F_2) and CO–dehydrogenase (F_3) in acetate synthesis was established. The F_3 enzyme complex requires further purification and elucidation; however, it appears to contain a corrinoid methyltransferase activity as indicated by the finding that propyl iodide inhibited the exchange between CO and [1-^{14}C]-acetyl CoA. As a result of these findings, Hu et al. (1982) proposed a scheme applicable for unicarbonotrophic growth of *C. thermoaceticum* on CO. The CO–dehydrogenase activity was suggested to function in the generation of [HCOOH] and reducing equivalents. The formyl intermediate is then reduced to CH_3THF and trans-methylated to CH_3 corrinoid by component F_3. Acetyl CoA results from the "transcarboxylation" of the methyl corrinoid by factor F_3. In short, acetyl CoA synthesized via CO metabolism becomes the precursor for further anabolic and catabolic reactions. However, careful analysis questions whether "transcarboxylation" is the correct term for the reaction demonstrated (that is, transcarbonylation). Ljungdahl and Wood (1982) have summarized the latest understanding on how acetate is formed during saccharide fermentations of acetogens.

Very little is known about the biochemistry of one-carbon metabolism in acetogens other than *C. thermoaceticum*. Cell extracts of *A. woodii* synthesize acetate synthesis from biocarbonate and hydrogen (Schoberth, 1977). The *C. thermoaceticum* homoacetate pathway was implicated with the synthesis of acetate by *A. woodii* because extracts contained high levels of corrinoids, formate dehydrogenase, formyl H_4 folate synthetase, methenyl H_4 folate cyclohydrolase, and methylene-H_4 folate dehydrogenase (Tanner et al., 1978).

B. methylotrophicum grows on H_2/CO_2, methanol/CO_2/acetate, glucose, or CO as sole fermentation energy sources (Lynd et al., 1982; Lynd and Zeikus, 1983). The organism also grows mixotrophically on glucose or methanol in the presence of CO, which replaces both CO_2 and acetate as the electron acceptor and inhibits both H_2 and butyrate production (Lynd et al., 1982).

CO dehydrogenase, hydrogenase, and formate dehydrogenase levels were compared in *B. methylotrophicum* grown on glucose, methanol, or CO as the electron donor. Notably, methyl viologen-linked CO dehydrogenase activity was higher during growth on methanol than CO, whereas formate dehydrogenase levels were higher on CO and lower on methanol. This finding and the absence of free formate as detectable intermediate suggested a speculative C_1 transformation scheme for CO conversion to acetate whereby the principal role for a formate dehydrogenase activity would be to couple formyl oxidation to methyl synthesis via a formyl reduction, and CO dehydrogenase would also function in transcarbonylation of a methyl intermediate (that is, direct assimilation without oxidation). The hypothesized formyl intermediate was interpreted as a single-carbon unit complexed with two electrons present in an enzyme complex or on a one-carbon carrier.

Little has actually been proven about the mechanisms of coupling ATP synthesis to one-carbon transformations in homoacetogens. Nonetheless, it would appear that ATP is derived from substrate-level phosphorylation (SLP) and electron transport–mediated phosphorylation (ETP) and perhaps by novel mechanisms (for example, transport-coupled phosphorylation). The high cell yield reported (Andreeson et al., 1973) for *C. thermoaceticum* (that is, 40–50 g/mol glucose) grown on glucose suggested a higher ATP gain per mol glucose than via other anaerobic glycolytic paths, and these authors hypothesized ETP in addition to SLP.

The stoichiometric, synthetic, and energetic relationships of *B. meth-*

ylotrophicum fermentations were determined (Lynd and Zeikus, 1983) in order to assess the percentage of substrate assimilated into cells and the apparent mechanisms for ATP synthesis (see Table 10.3). For anaerobes, acetogens are very efficient in cell synthesis reactions. When *B. methylotrophicum* was grown on methanol or glucose, more than 24% of the substrate is assimilated into cells in lieu of acid end products. Less carbon was assimilated into cells during growth on one-carbon compounds more oxidized than methanol (that is, H_2/CO_2 or CO). Most notably, during growth on one-carbon substrates, the growth yield (g cells/mol substrate) was directly related to the number of carbon-bound electrons in the substrate. Hence the highest growth yield occurred with methanol (that is, six carbon-bound electrons). During growth of *B. methylotrophicum* on methanol/acetate/CO_2 as fermentation substrates, either acetate or butyrate can be produced as the main end product by varying the electron acceptor concentrations (that is, acetate or CO_2). Acetate is the main product when acetate is limiting and CO_2 is in excess, whereas butyrate is virtually the only product when acetate is in excess and CO_2 is limiting (Lynd and Zeikus, 1983; Lynd et al., 1982).

Both SLP and ETP are suggested as mechanisms of energy conservation during growth of *B. methylotrophicum* on one-carbon substrates. The growth yield on CO (4 g/mol) was consistent with the synthesis of 0.5 mol of ATP per mol of CO (that is, 2 ATP per mol of acetate produced). This high cellular ATP yield necessitates more than SLP via acetate kinase alone. The CO-dependent reduction of a formyl intermediate to the methyl level was postulated as the driving force of ETP during growth on CO (Lynd et al., 1982) because the free energy of $CO + CH_2O + H_2O \rightarrow CH_3OH + CO_2$ (−15.4 Kcal) is sufficiently exergonic to be coupled to ATP synthesis via ETP. Neglecting cell synthesis, the stoichiometry observed for growth of *B. methylotrophicum* on methanol was 10 methanol + 2 $CO_2 \rightarrow$ 3 butyrate. If butyrate were synthesized via transcarbonylation of a methyl intermediate according to:

$$10\ CH_3OH + 2\ CO_2 \rightarrow 6\ \text{acetyl CoA} + 12\ e^-$$

$$12\ e^- + 6\ \text{acetyl CoA} \rightarrow 3\ \text{butyryl CoA}$$

$$3\ \text{butyryl CoA} + \text{ADP} + \text{P} \rightarrow 3\ \text{butyrate} + 3\ \text{ATP}$$

then an ATP gain of 0.3 ATP/methanol would be far below what is needed to explain the observed growth yield of ~8 g cells/mol CH_3OH by SLP alone, unless free formate is an intermediary metabolite of methanol oxidation. The growth yield of H_2/CO_2 (that is, ~2 g cells/mol H_2) implies greater than 1 ATP/mole acetate and ETP because, even if formate were a free intermediate, ATP would be needed to make formyltetrahydrofolate. The lower cell yield on H_2/CO_2 than CO is consistent with the thermodynamic requirement for energy consumption in the formation of a formyl intermediate because CO_2 itself cannot carbonylate a methyl intermediate without electrons.

TABLE 10.3 **Stoichiometric, Synthetic, and Energetic Relationships of *B. Methylotrophicum* Fermentations**

Energy Source	*Substrate–Product Stoichiometry*	*Carbon Recovery (%)*	*Electron Recovery (%)*	Y_s *g cells / Mol sub.*	$\Delta G^{\circ\prime}$ (Kcal/mole sub.)	ΔH (Kcal/mole sub.)	ΔG (Kcal/mole sub.)	*Cell Synthesis Efficiency (%)*	*Substrate–Cell Synthesis Conversion Ratio (%)* Carbon	*Substrate–Cell Synthesis Conversion Ratio (%)* Electrons
H_2/CO_2	$1.00H_2 + 0.517CO_2 + 0.013NH_2^+ \rightarrow 0.216CH_3COO^- + 0.002CH_3(CH_2)_2COO^- + 0.063[C_1H_{1.82}O_{0.50}N_{0.21}] + 0.231H^+$	97 ± 4	102 ± 2	1.6 ± 0.2	− 4.2	−15.0	− 3.9	63 ± 9	12	13
CO	$1.000CO + 0.027NH_4^+ \rightarrow 0.172CH_3COO^- + 0.001CH_3(CH_2)_2COO^- + 0.129[C_1H_{1.82}O_{0.50}N_{0.21}] + 0.497CO_2 + 0.200H^+$	97 ± 5	97 ± 0	3.5 ± 0.5	−10.8	−16.4	−11.0	58 ± 4	13	27
Methanol	$1.000CH_3OH + 0.183CO_2 + 0.060NH_4^+ \rightarrow 0.016CH_3COO^- + 0.234CH_3CH_2)_2COO^- + 0.284[C_1H_{1.82}O_{0.50}N_{0.21}] + 0.31OH^+$	105 ± 3	100 ± 1	7.5 ± 0.5	−10.4	−14.2	− 8.7	76 ± 3	24	20
Glucose	$1.000C_6H_{12}O_6 + 0.332NH_4^+ \rightarrow 1.593CH_3COO^- + 0.707CH_3(CH_2)_2COO^- + 0.155H_2 + 0.313CO_2 + 2.132H^+ + 1.580[C_1H_{1.82}O_{0.50}N_{0.21}]$	99 ± 8	100 ± 6	43.0 ± 0.5	−64.9	−41.2	−65.3	74 ± 17[h]	26	28

Adapted from Lynd and Zeikus (1983).

Zeikus (1982) proposed a model to explain the high efficiency of cell synthesis in homoacetogens on the basis of how carbon precursors may enter a common metabolic pathway in *B. methylotrophicum* (see Fig. 10.3). In analogy to what occurs in *M. barkeri,* both catabolism and anabolism in *B. methylotrophicum* were postulated to share common one-carbon transformations for the synthesis of acetyl CoA, the direct precursor to acetate, butyrate, and cell carbon. C_1 substrates with different oxidation states enter the path on different carriers, a condition that is required to explain the growth efficiencies on these substrates. In analogy to what occurs in *C. thermoaceticum* (Ljungdahl and Wood, 1982), synthesis of acetyl CoA would appear to occur via a C_1–C_1 condensation reaction involving an OHC-X

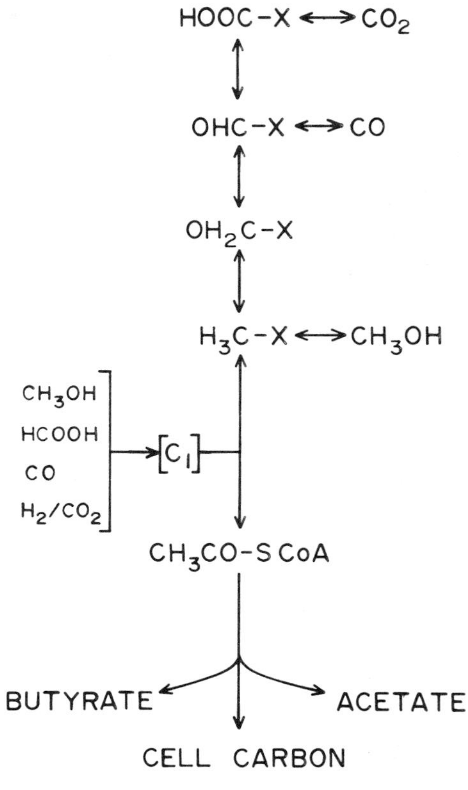

FIGURE 10.3 Unification of carbon flow pathways for catabolism and anabolism in *B. methylotrophicum*. The proposed model predicts that C_1 units are fixed on different carbon carriers or enzyme active sites (X). The key biosynthetic reaction involves acetyl-S CoA synthesis via CO dehydrogenase activity. Unicarbonotrophy is linked by a common carrier system for metabolism of different one-carbon substrates (from Zeikus, 1982).

and methyl X (that is, a transcarbonylation reaction). The cells certainly contain high levels of corrinoids and CO dehydrogenase, but detailed enzymatic studies and ^{14}C-isotope incorporation studies are required to establish that acetyl CoA synthesis in *B. methylotrophicum* is identical to that in *C. thermoaceticum*.

It is worth noting that, unlike other described methylotrophs (that is, aerobic *Methylococcus* species or anaerobic *Methanosarcina* species), *B. methylotrophicum* requires an electron acceptor to grow on methanol because the fermentation products are all more oxidized than the electron donor (that is, methanol). During mixotrophic growth on methanol, CO replaces the need for acetate or CO_2 as an electron acceptor (Lynd et al., 1982). The suggested biochemical basis for this phenomenon is also indicated by analysis of the carbon flow scheme illustrated in Fig. 10.3. During growth on CH_3OH/acetate/CO_2, methanol must be oxidized to synthesize acetyl CoA, to reduce acetate and CO_2 to butyrate, and to generate cell carbon. However, on methanol/CO, both substrates could be assimilated directly to acetyl CoA with limited methanol or CO oxidation for cell synthesis. Thermodynamically, CO oxidation is favored over methanol oxidation for the generation of reducing equivalents (that is, reduced electron carriers) at the redox state needed for anabolic pyruvate synthesis from acetyl CoA.

IV. POTENTIALS AND LIMITATIONS OF ONE-CARBON FERMENTATIONS IN BIOTECHNOLOGY

The metabolism of one-carbon compounds by aerobic microorganisms has been a very active topic in biotechnology for the past ten years. Aerobic methylotrophs have important applications in single-cell protein production from methanol (Mateles, 1979) because of their metabolic efficiency in assimilating one-carbon compounds. Based on the broad-substrate specificity of the methane monooxygenase, applications of aerobic methylotrophs have also been suggested (Patel et al., 1980) in chemical transformation reactions that require the introduction of oxygen into hydrocarbons (for example, conversion of alkanes into alcohols or methyl ketones). This latter practical application of aerobic methanotrophs requires the use of enzymes or non-growing cells.

Methanogenic and acidogenic fermentations have been suggested for anoxic, methane, or acetic acid production because of either the high yields (that is, mol product/mol substrate fermented) or the abundance and low cost of substrates (that is, organic wastes, polymeric carbohydrates, or simple pyrolysis chemicals). In addition, other features of anaerobic one-carbon fermentations are worth noting: maximal energy conservation associated with transformation of C_1 substrates, high catalytic rates for growth and metabolism, high efficiency of cell synthesis from C_1 substrates, and lack of oxygen and limited cooling requirements associated with anticipated process designs. In general terms, aerobic methylotrophic fermentations are more oxidative and result in higher amounts of CO_2 and H_2O and lesser amounts of reduced carbon end products formed per mole of C_1 substrate consumed, while anaerobic fermentations involve more direct and reductive transformation reactions that result in higher amounts of reduced carbon end products per mole of C_1 substrate consumed.

Methanogens and homoacetogens that transform one-carbon compounds have a limited amount of theoretical energy when compared to aerobes that utilize these same substrates. However, aerobes couple oxidation of H_2, CO, or methanol to reduction of O_2, whereas anaerobes link these substrate transformations to the reduction of one-carbon substrates themselves. Therefore the aerobes have a much higher degree of available theoretical chemical energy by $\Delta G^{\circ\prime}$ calculations, but anaerobes appear to be more efficient in the conservation of the biological energy available for cell carbon synthesis. Because thermodynamic calculations are independent of kinetics, the specific transformation rate for conversion of a one-carbon substrate into microbial products depends specifically on the properties of the biocatalysts involved. Thus the limited thermodynamics available to anaerobic one-carbon transformations is not necessarily associated with low reaction rates. On the contrary, the most prolific H_2-oxidizing autotrophic (that is, assimilating CO_2 as the sole carbon source) species documented is *M. thermoautotrophicum,* which displays a doubling time of about 60 min in chemostat culture, while the rapidly growing aerobic hydrogen bacterium, *Alcaligines eutrophus* (Friedrich et al., 1979) has a 192 min doubling time. However, the aerobe is constantly faced with the generation of low redox potential electrons in the presence of O_2 and the enzymatic removal of toxic active oxygen species, and it contains an active oxygenase (that is, ribulose biphosphate carboxylase) that can reoxidize reduced cell carbon formed by net CO_2 fixation. Similarly, the most prolific methylotroph is apparently not an aerobic *Methylococcus* or *Pseudomonas* species but rather *B. methylotrophicum,* which in a chemostat has a doubling time at washout of <30 min (T. Moench, unpublished findings, author's lab).

The equations shown in Fig. 10.4 indicate that more favorable thermodynamics are associated with the transformation of either H_2/CO_2, CO, or methanol by methanogens than by acetogens. However, comparison of the growth yields (for example, gram dry wt. cell/mol electron donor) of *M. barkeri* were noticeably lower than those of *B. methylotrophicum* (Weimer and Zeikus, 1978a; Lynd and Zeikus, 1983). For example, the growth yield of *B. methylotrophicum* during exponential growth was 8.2, while that of *M. barkeri* was 3.9. This suggests that basic mechanistic differences exist between energy conservation from C_1 transformation reactions in methanogens and acetogens. This indirectly implies that the efficiency and utility of conversion of one-carbon compounds to multicarbon products may be better practiced with acetogens than with methanogens.

A comparison of the thermodynamics of these transformation reactions

REACTION	$\Delta G^{\circ\prime}$ (Kcal/mol/8e⁻)
I. <u>M.</u> <u>BARKERI</u>	
A. $4CO + 2H_2O \longrightarrow CH_4 + 3CO_2$	-50.5
B. $4H_2 + HCO_3^- + H^+ \longrightarrow CH_4 + 3H_2O$	-32.7
C. $4CH_3OH \longrightarrow 3CH_4 + CO_2 + 2H_2O$	−19.1
II. <u>B.</u> <u>METHYLOTROPHICUM</u>	
A. $4CO + 4H_2O \longrightarrow CH_3COO^- + 2HCO_3^- + 3H^+$	-39.5
B. $4H_2 + 2CO_2 \longrightarrow CH_3COO^-\ H^+ + 2H_2O$	-27.6
C. $10CH_3OH + 2HCO_3^- \longrightarrow 3CH_3CH_2CH_2COO^- + H^+ + 6H_2O$	-26.0

FIGURE 10.4 Comparative energetics of C_1 fermentation reaction stoichiometrics indicative of *M. barkeri* and *B. methylotrophicum* energy metabolism (from Zeikus, 1982).

in Fig. 10.4 with cell growth yields also supports the carbon flow schemes proposed for unification of one-carbon metabolism in both *M. barkeri* and *B. methylotrophicum* (Zeikus, 1982). The thermodynamically favored order for available free energy from C_1 transformations in either methanogens or acetogens is CO > H_2/CO_2 > methanol. However, the order of growth yields obtained in *B. methylotrophicum* was methanol > CO > H_2/CO_2 (Lynd and Zeikus, 1983), and the order in *M. barkeri* (Weimer and Zeikus, 1978a) was methanol > H_2/CO_2. These facts suggest that carbon-bound electrons are conserved in methanol during cell carbon synthesis and support the concept of direct assimilation of carbon at the methyl level. Notably, the mechanism proposed for acetyl CoA synthesis in unicarbonotrophic anaerobes suggests that initial methylotrophic or autotrophic cell carbon synthesis reactions in acetogens can actually yield energy.

Metabolic products other than organic acids and natural gas, such as expensive cell products (for example, certain tricarboxylic acids, amino acids, or vitamins), are theoretically feasible from anoxic one-carbon fermentations because a common intermediary metabolite appears as a precursor to both anabolic cell carbon synthesis and catabolic reduced end product fermentation. For example, it would appear that biochemically and/or genetically altering carbon and electron flow in acetogenic methanol fermentations, the production of either butyrate, acetate, or cells could be suppressed in order to enhance production of a specific biosynthetic product (for example, citrate, glutamate, or vitamin B_{12}). In addition, the enzymes involved in redox reactions or one-carbon transformations in methylotrophic anaerobes could be potentially useful as direct catalysts for C_1 compounds, or they may serve as guides for designing better catalysts for chemical synthesis.

The first unique application of anoxic one-carbon fermentations was suggested by Wise et al. (1978). These authors proposed the utilization of methanogens to upgrade the BTU content of coal gasification mixtures, a process termed "biomethanation." Biomethanation was also associated with some other practical advantages, including the conversion of toxic CO to CH_4 and the insensitivity of methanogens to sulfide. Sulfides are commonly present in synthesis gas and often limit chemical processes that use abiological catalysts to synthesize organic chemicals from H_2 and CO. The rates of biomethanation were improved by the use of high-pressure reactors, which also lowered process capital costs by decreasing the bioreactor size. Further improvements in biomethanation technology may be gained by improving the nutritional–environmental conditions optimal for methanogenesis; by not having cells produced from synthesis gas mixtures (for example, develop a reactor with immobilized cells such that H_2/CO_2 and CO gas mixtures are only directed into CH_4 and not into cell components); and by selecting for methanogen strains that can metabolize CO and H_2/CO_2 simultaneously at high CO partial pressures. The last improvement appears to be critical for a direct continuous process linked to cell growth because methanogens, unlike acetogens, do not rapidly consume CO during growth and CO inhibits H_2/CO_2 metabolism at concentrations greater than 30% CO in the culture headspace at 1 atm total gas pressure (Daniels et al., 1977). Biomethanation is now also considered as one part of total vendable products recovery from developing anoxic biomass transformations technology (Levy et al., 1982). This total biomass conversion process is termed "biorefining," and it includes C_1 fermentations to acetic acid or methane as well as direct fermentation of biomass into other industrial chemicals (for example, ethanol, propionate).

Synthesis of products other than natural gas with methylotrophic methanogens may have some utility. For example, *M. barkeri* wild-type strains produce high amounts ($\leq$1% of total cell carbon) of vitamin B_{12} during growth on methanol (Krzycki and Zeikus, 1980). Production of vitamin B_{12} could become an economic process provided that cell yield is greatly increased and that, via strain mutation, selection and regulation of B_{12} production can be enhanced in lieu of methane formation. Notably, methanogens, unlike acetogens, do not form any known toxic, metabolic end products, and the needed approach for process development success appears to be routine to industrial microbiology.

Acetogens display more potential biocatalytic applications for anaerobic C_1 transformation reactions than do methanogens. The potential products of homoacetogenic metabolism are broader and more valuable. Organic acids are used in large volume by both food and chemical industries (Keiger, 1977). At present, the only clearly identified products are acetic and butyric acids and possibly vitamin B_{12}. Acetic acid is the major organic chemical used by the chemical industry. Butyric acid is used in much lower volumes, but its price is much higher than that of acetic acid. Studies have been initiated to find lower-cost substrates for vitamin B_{12} fermentations (Florent and Ninet, 1979; Kamikubo et al., 1978). Methane fermentations for B_{12} production with *Butyribacterium* could become more economical than *Propionibacterium* fermentations, which use more expensive substrates (for example, glycerol and lactate), provided that the B_{12} yield (that is, grams per liter) can be greatly improved.

C. thermoaceticum fermentations were first suggested as a novel process for the conversion of biomass-derived sugars into acetic acid (Wang et al., 1978). Several major biotechnological problems arise with saccharide fermentations of acetogens; these include low rates of acetic acid production of nongrowing cells; high amounts of substrate (>20%) conversion to cells during growth; low species tolerance to end products, with high proton concentration being more rate limiting; and ineffective product recovery systems at low acid concentrations (<5.0%). Recently, strains of *C. thermoaceticum* that produce acetic acid at pH values below 5.0 have been obtained (Schwartz and Keller, 1982). However, economic process technology has not been demonstrated for the recovery of dilute organic acids (<5%) from fermentation liquors. It should be noted that acetic acid waste streams of the chemical industry are often fermented to methane because technology is not available to recover the acids from dilute aqueous solutions.

The recent discoveries of homoacetic and homobutyric acid production from one-carbon fermentations promise lower substrate cost and perhaps novel process design with immobilized cells. One-carbon substrates may have more utility than saccharides because C_1 substrates are readily diffusible and do not require phosphorylation to enter the cell. Thus greater potential exists for acid production by nongrowing cells, which may give higher product/substrate ratios (that is, gram per gram) and greater tolerance to acids. Recently, a novel method for organic acid recovery in dilute solutions has been described (Datta, 1981). This method involves further chemical transformation of the spent fermentation broth with alcohols to convert organic acids into esters, which can be economically recovered from dilute aqueous solutions. The methyl and ethyl esters of volatile organic acids are potentially useful as solvents, gasoline octane improving agents, and chemical intermediates. For example, the application of this potential technique to butyric acid production from *B. methylotrophicum* methanol fermentations could yield recoverable methyl butyrate as the end product. Methyl butyrate represents a chemical feedstock for plastics manufacture, and it is more soluble in gasoline and has higher octane than ethanol. This can also be broadly viewed as a microbial technology to convert solid, underutilized organic fossil fuels and residues into high-value liquid fuels. However, most of these ideas are in early stages of development, and significant advances need to be made before any successful commercial application can evolve.

Perhaps more exciting products can be developed from the one-carbon metabolism of acetogens by regulatory or genetic manipulation of their biochemical activities. The most interesting feature to pursue is to alter normal electron and carbon flow to new products. The catabolic electron flow potential exists to biochemically engineer acetyl CoA reduction to ethanol and butyryl CoA reduction to butanol. Likewise, the metabolic potential exists, since acetyl CoA is the direct precursor to acetic acid or cell synthesis, to biochemically engineer carbon and electron flow away from acetate to the formation of ketoacids, amino acids, and vitamin B_{12}. By metabolic control, alanine has been produced as the major reduced end product of a H_2/CO_2 metabolizing acetogen (D. Gold and C. Cooney, personal communication). Unlike aerobes, acetogens have evolved highly efficient mechanisms for one-carbon transformations and also for the excretion of organic end products (for example, solvents, acids, and the like). Now it will be of interest to see if genetic and biochemical engineering can be put to practical use in applied biocatalysis of acetogens for chemical feedstocks, liquid fuels, biochemicals, or enzyme production.

In summary, future opportunities exist for application of anoxic one-carbon fermentations in the production of gaseous and liquid fuels, large-volume organic acids, chemical feedstocks, food chemicals, and fine-grade biochemicals and enzymes. However, these microbial systems need to be further understood and improved, especially in regard to the identification and enhancement of rate-limiting metabolic steps (for example, mass transfer of substrate into cells, key intracellular carbon and electron transformation reactions, and removal of acid anion and proton from cells). At present, only organic waste fermentations to methane is economical, and not anoxic organic chemicals production from C_1 transformation reactions. Hence anaerobic one-carbon fermentations are not employed in industry. Undoubtedly, more research on both the fundamentals and the applications of anoxic one-carbon fermentations by bacteria is needed before their true utility in developing biotechnological processes can be addressed.

ACKNOWLEDGMENTS

The research was supported by the College of Agricultural and Life Sciences, University of Wisconsin—Madison, and by grants from the National Science Foundation, the U.S. Department of Energy, Exxon Research Education Foundation, Shell Oil Company, the Alexander Von Humbolt Foundation, and the Institut Pasteur Foundation.

REFERENCES

Andreesen, J. R., Schaupp, A., Neurauter, C., Brown, A., and Ljungdahl, L. (1973) *J. Bacteriol. 114,* 742–751.

Bache, R., and Pfennig, N. (1981) *Arch. Microbiol. 130,* 255–261.

Balch, W. E., Schoberth, S., Tanner, R. S., and Wolfe, R. S. (1977) *Int. J. Syst. Bacteriol. 27,* 355–361.

Balch, W. E., Fox, G. E., Magrum, L. J., Woese, C. R., and Wolfe, R. S. (1979) *Microbiol. Rev. 43,* 260–296.

Barker, H. A. (1956) in *Bacterial Fermentations,* John Wiley, New York.

Braun, M., Mayer, F., and Gottschalk, G. (1981) *Arch. Microbiol. 128,* 288–293.

Cheremisinoff, N. P. (1979) in *Gasohol for Energy Production* 140 pp., Ann Arbor Science Publishers, Inc., Ann Arbor, Mich.

Daniels, L., and Zeikus, J. G., (1978) *J. Bacteriol. 136,* 75–84.

Daniels, L., Fuchs, G., Thauer, R. K., and Zeikus, J. G. (1977) *J. Bacteriol. 132,* 118–126.

Datta, R. (1981) *Biotech. Bioeng. Symp. 11,* 521–532.

Datta, R., and Zeikus, J. G. (1982) *Dev. Ind. Microbiol. 23.* (in press)

Diekert, G., Klee, B., and Thauer, R. K. (1980) *FEBS Lett. 119,* 118–120.

Drake, H. L., Hu, S. I., Wood, H. G. (1980) *J. Biol. Chem. 255,* 7174–7180.

Drake, H. L., Hu, S. I., Wood, H. G. (1981) *J. Biol. Chem. 256,* 11,137–11,144.

Ellefson, W. L., and Wolfe, R. S. (1981) in *Microbial Growth on C_1 Compounds* (Dalton, H., ed.) pp. 171–180, Heyden, London.

Florent, J., and Ninet, L. (1979) in *Microbial Technology* (Peppler, H. J., and Perlman, D., eds.) Vol. I, 2nd ed., pp. 497–517, Academic Press, New York.

Fontaine, F. E., Peterson, W. H., McCoy, E., Johnson, E., and Ritter, M. T. (1942) *J. Bacteriol. 43,* 704–715.

Friedrich, C. G., Bowien, B., and Friedrich, B. (1979) *J. Gen. Microbiol. 115,* 185–192.

Fuchs, G., and Stupperich, E. (1978) *Arch. Microbiol. 118,* 121–125.

Genthner, B. R., and Bryant, M. P. (1982) *Appl. Environ. Microbiol. 43,* 70–74.

Genthner, R. R. S., Davis, C. L., and Bryant, M. P. (1981) *Appl. Environ. Microbiol. 42,* 20–22.

Ghambeer, R. K., Wood, H. G., Schulman, M., and Ljungdahl, L. (1971) *Arch. Biochem. Biophys. 143,* 471–484.

Gottwald, M., Andreesen, J. R., LeGall, J., and Ljungdahl, L. G. (1975) *J. Bacteriol. 122,* 325–328.

Hu, S. H., Drake, H. L., and Wood, H. G. (1982) *J. Bacteriol. 144,* 440–446.

Kamikubo, T., Hayaski, M., Nishio, N., and Nagai, S. (1978) *Appl. Environ. Microbiol. 35,* 971–973.

Keiger, D. M. (1977) *Chemical & Engineering News 55,* 31–32.

Keltjens, J. T., and Vogels, G. D. (1981) in *Microbial Growth on C_1 Compounds* (Dalton, H., ed.) pp. 152–158, Heyden, London.

Kenealy, W. K., and Zeikus, J. G. (1981) *J. Bacteriol. 146,* 133–140.

Kenealy, W. R., and Zeikus, J. G. (1982) *J. Bacteriol. 151,* 932–941.

Kenealy, W. R., Thompson, T. E., Schubert, K. R., Zeikus, J. G. (1982) *J. Bacteriol. 150,* 1357–1365.

Kerby, R., and Zeikus, J. G. (1983) *Curr. Microbiol. 8,* 27–30.

Krzycki, J. D., and Zeikus, J. G. (1980) *Curr. Microbiol. 3,* 243–245.

Lentz, K., and Wood, H. G. (1955) *J. Biol. Chem. 215,* 645–654.

Levy, P. T., Sanderson, J. E., Kispert, R. G., and Wise, D. L. (1982) *Enzyme Microbiol, Technol. 3,* 207–215.

Ljungdahl, L. J., and Wood, H. G. (1969) *Ann. Rev. Microbiol. 23,* 515–537.

Ljungdahl, L. G., and Wood, H. G. (1982) in B_{12} Vol. II (Dolphin, D., ed.) pp. 165–202, John Wiley, New York.

Ljungdahl, L., Irion, E., and Wood, H. G. (1965) *Biochemistry 4,* 2771–2780.

Ljungdahl, L., Irion, E., and Wood, H. G. (1966) *Fed. Amer. Soc. Exper. Biol. 25,* 1642–1648.

Lynd, L., and Zeikus, J. G. (1981) in *Trends in the Biology of Fermentations in Fuels and Chemicals* (Hollander, A., ed.), Plenum Press, New York.

Lynd, L., and Zeikus, J. G. (1983) *J. Bacteriol. 153.*

Lynd, L., Kerby, R., and Zeikus, J. G. (1982) *J. Bacteriol. 149,* 255–263.

Mateles, R. I. (1979) in *Microbial Technology: Current State and Future Prospects* (Bull, Elwood, and Ratledge, eds.) *29th Symposium of the Society for General Microbiology* pp. 29–52, Cambridge University Press, Cambridge, England.

Parker, D. J., Wu, T. F., and Wood, H. G. (1971) *J. Bacteriol. 108,* 770–776.

Patel, R. M., Hou, C. T., Laskin, A. I., Felix, A., and Derelanko, P. (1980) *Appl. Environ. Microbiol. 34,* 720–726.

Scherer, P., and Sahm, H. (1981a) *Eur. J. Appl. Microbiol. Biotechnol. 12,* 28–35.

Scherer, P., and Sahm, H. (1981b) *Acta Biotechnol. 1,* 57–65.

Schoberth, S. (1977) *Arch. Microbiol. 114,* 143–148.

Schoenheit, P., Moll, J., and Thauer, R. K. (1979) *Arch. Microbiol. 123,* 105–107.

Schoenheit, P., Moll, J., and Thauer, R. K. (1980) *Arch. Microbiol. 127,* 59–65.

Schwartz, R. D., and Keller, F. Z. (1982) *Appl. Environ. Microbiol. 43,* 117–123.

Stadtman, E. R., Stadtman, T. K., Postan, I., and Smith, L. D. (1972) *J. Bacteriol. 110,* 758–760.

Stadtman, T. C. (1967) *Ann. Rev: Microbiol. 21L,* 121–142.

Tanner, R. S., Wolfe, R. S., and Ljungdahl, L. G. (1978) *J. Bacteriol. 134,* 668–670.

Taylor, C. D., and Wolfe, R. S. (1974) *J. Biol. Chem. 249,* 4879–4885.

Thauer, R. K. (1982) in *Anaerobic Digestion* (D. Hughes et al., eds.) pp. 37–41, Elsevier Biomedical Press, Amsterdam.

Tzeng, S. F., Bryant, M. P., and Wolfe, R. S. (1975) *J. Bacteriol. 121,* 192–196.

Wang, D., Fleishchaker, R., and Wang, G. (1978) *AIChE Symp. Ser. 182 74,* 105–110.

Weimer, P. J., and Zeikus, J. G. (1978a) *Arch. Microbiol. 119,* 468–478.

Weimer, P. J., and Zeikus, J. G. (1978b) *Arch. Microbiol. 119,* 49–57.

Weimer, P. J., and Zeikus, J. G. (1979) *J. Bacteriol. 137,* 332–339.

Welty, F. K., and Wood, H. G. (1978) *J. Biol. Chem. 253,* 5832.

Whitman, W. B., Ankswanda, E., and Wolfe, R. S. (1981) *J. Bacteriol. 149,* 852–863.

Wiegel, J., Braun, M., and Gottschalk, G. (1981) *Curr. Microbiol. 5,* 255–260.

Wise, D. L., Cooney, C. L., and Augenstein, D. C. (1978) *Biotech. Bioeng. 20,* 1153–1172.

Wolfe, R. S., and Higgins, I. J. (1979) *Int. Rev. Biochem. 21,* 267–353.

Wood, H. G. (1952) *J. Biol. Chem. 194,* 905–931.

Wood, J. M., Moura, I., Moura, J. J. G., Santos, M. H. I., Xavier, A. V., and LeGall, J. (1982) *Science 216,* 303–305.

Yang, S. S., Ljungdahl, L. G., and LeGall, J. (1977) *J. Bacteriol. 130,* 1084–1090.

Yang, S. S., Ljungdahl, L. G., Dervartanian, D. V., and Watt, G. D. (1980) *Biochim. Biophys. Acta 590,* 24–33.

Zeikus, J. G. (1980) *Ann. Rev. Microbiol. 34,* 423–464.

Zeikus, J. G. (1983) *Adv. Microb. Physiol. 24*. (in press)
Zeikus, J. G., and Wolfe, R. S. (1972) *J. Bacteriol. 109,* 707–713.
Zeikus, J. G., Weimer, P. J., Nelson, D. R., and Daniels, L. (1975) *Arch. Microbiol. 104,* 129–134.
Zeikus, J. G., Fuchs, G., Kenealy, W., and Thauer, R. K. (1977) *J. Bacteriol. 132,* 604–613.
Zeikus, J. G., Lynd, L. H., Thompson, T. E., Kryzcki, J. E., Weimer, P. J., and Hegge, P. W. (1980) *Curr. Microbiol. 3,* 381–388.

CHAPTER

11

Acetone–Butanol Fermentation

Antonio R. Moreira

I. INTRODUCTION

The production of acetone and butanol by fermentation, the well-known Weizmann process, is one of the oldest industrially practiced fermentation processes, dating as far back as 1916 (Walton and Martin, 1979). At that time, butanol was in demand for use in the synthesis of butadiene for synthetic rubber. However, during World War I, the emphasis on butanol production decreased at the same time that the demand for acetone saw a large increase, since acetone was used to dissolve cordite in the manufacture of explosives. Because this fermentation produces more butanol than acetone, much of the butanol produced then was either stored or simply disposed of while the acetone was utilized.

After World War I, the demand for acetone decreased, and the emphasis switched again to butanol production for the manufacture of butyl acetate. Butyl acetate was used mainly as a solvent for the nitrocellulose lacquers required by the automobile industry. During the early 1930's, a culture of *Clostridium saccharobutylicum* was isolated having the capability of fermenting sucrose to butanol and acetone. This development, together with a continued demand for butanol, led to the construction of new plants in

Philadelphia, Baltimore, Puerto Rico, and England utilizing molasses as the raw material.

Today, all the acetone and butanol production in the United States is based on synthetic processes that utilize petrochemical raw materials. Acetone is produced by one of two routes: the cumene hydroperoxide process or the catalytic dehydrogenation of isopropanol (Nelson and Webb, 1979). Butanol is currently produced by either the oxo process starting from propylene or the aldol process starting from acetaldehyde (Sherman, 1979). In either case, these synthetic routes have proved to be economically advantageous in comparison to the fermentation-based processes; thus all fermentative production in the United States has ceased.

In this chapter, a review is presented of the technology of the acetone–butanol fermentation as it is known today, followed by a summary of the research work with which the author has been involved in an attempt to revive this old fermentation. First, however, a brief discussion on the physical properties of butanol and acetone and the current uses and demand for these solvents is presented.

II. BACKGROUND ON THE ACETONE/BUTANOL MARKET

Acetone is the simplest and most important ketone available. It is a colorless, flammable liquid, miscible in all proportions with water and a variety of other organic solvents such as ether, methanol, and ethanol. Some of the most important physical properties of acetone are shown in Table 11.1. Acetone is used as a chemical intermediate in the manufacture of methacrylates, methyl isobutyl ketone, bisphenyl A, and methyl butynol, among others, and as a solvent for resins, paints, varnishes, lacquers, nitrocellulose, and cellulose acetate. A breakdown of the relative amounts consumed for each end-use has been compiled recently (Villet, 1981) and is shown in Table 11.2.

Butanol is a colorless liquid, miscible with most commonly used organic

TABLE 11.1 Physical Properties of Acetone and Butanol

Property	*Acetone*	*Butanol*
Formula	CH_3COCH_3	$CH_3(CH_2)_2CH_2OH$
Formula weight	58.079	74.12
Melting point at 101.3 kPa (°C)	−94.6	−90.2
Boiling point at 101.3 kPa (°C)	56.1	117.7
Specific gravity, 20°C/4°C	0.7899	0.8098
Heat of vaporization (cal/g)	119.7	141.3
Heat of fusion (cal/g)	23.42	29.93
Heat of Combustion (cal/g)	7354	8642
Vapor pressure at 20°C (kPa)	24.62	0.61
Specific heat of liquid (cal/g-K)	0.62	0.56

TABLE 11.2 Uses of Acetone

Use	*Percent of Total*
Manufacture of:	
Methacrylates	33–34
Methyl isobutyl ketone	13–14
Bisphenyl A	5–6
Methyl butynol	5
Others (methyl isobutylcarbinol, diacetone alcohol, isophorone, etc.)	9–12
Solvent for:	
Resins, paints, lacquers, varnishes	16–18
Processing use	5
Cellulose acetate	4–5
Other	1–10

solvents and with a solubility of 8 wt. % in water. A list of the most important physical properties of butanol is also shown in Table 11.1, along with the acetone data. The major end-uses for butanol are as a solvency enhancer in the formulation of nitrocellulose lacquers, as a latex in the form of butyl acrylate, and as a precursor of derivatives such as butyl acetate and dibutyl phthalate. A breakdown of the relative amounts of butanol used for different applications is shown in Table 11.3 (Villet, 1981).

Using available data (*Chemical Marketing Reporter,* 1963–1982), a time profile on the demand of both acetone and butanol over the last 20 years is shown in Fig. 11.1, while Fig. 11.2 shows the trend of the F.O.B. costs for both solvents over the same time period. It is interesting to observe that the acetone demand is about 2–3 times larger than the demand for butanol, a situation that is opposite to conventional butanol/acetone yields in a traditional fermentation, as will be shown later. Also interesting is the skyrocketing of the acetone and butanol costs in the second half of the

TABLE 11.3 Uses of *n*-Butanol

Use	*Percent of Total*
Manufacture of:	
Butyl acrylate	21
Glycol ethers	18
Dibutyl phthalate	10
Solvent for:	
General use	22
Amine resins	5
Other	24

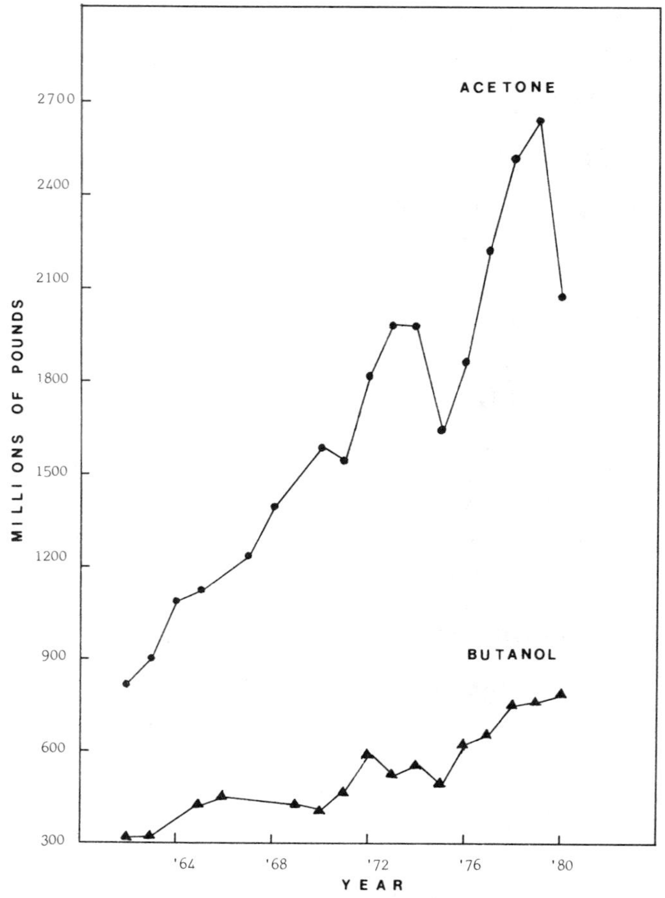

FIGURE 11.1 Annual demand for acetone and butanol.

1970's and into the 1980's as a consequence of the world oil situation. Such dramatic price increases may very well justify the revitalization of the acetone–butanol fermentation industry in a not too distant future.

III. BIOCHEMISTRY OF THE ACETONE–BUTANOL FERMENTATION

The metabolic pathway for glucose utilization by *Clostridium acetobutylicum* is shown in Fig. 11.3 and reflects the knowledge currently available from published information (Doelle, 1975; Stainier et al., 1976; Rosenfeld and Simon, 1950a, b). Glucose is initially converted to pyruvate via the Embden-

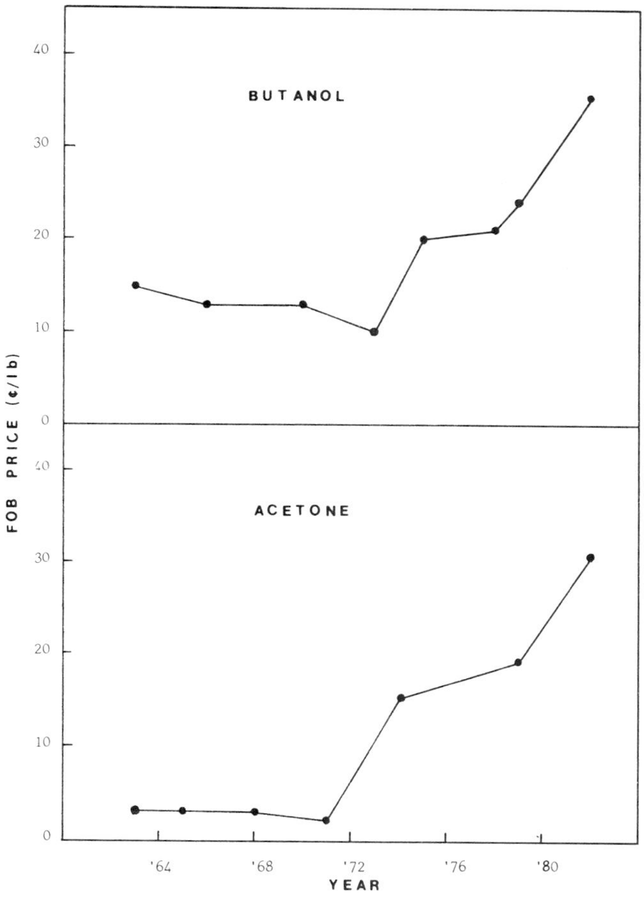

FIGURE 11.2 F.O.B. cost trends for acetone and butanol.

Meyerhof-Parnas (EMP) pathway; the pyruvate molecule is then broken down to acetyl-CoA with the release of carbon dioxide and hydrogen.

Acetyl-CoA plays a key role in the metabolism of *Clostridium acetobutylicum* by serving as a precursor to acetic acid, ethanol, and all the remaining fermentation solvents. The formation of butyric acid and neutral solvents (acetone and butanol) is thought to occur in two separate steps. Initially, two acetyl-CoA molecules combine to form acetoacetyl-CoA, thus initiating a cycle leading to the production of butyric acid. The increase in acid concentration results in a decrease in the pH of the fermentation medium from an initial value of about 6.2 to about 4.

At this point in the fermentation, a new enzyme system is activated

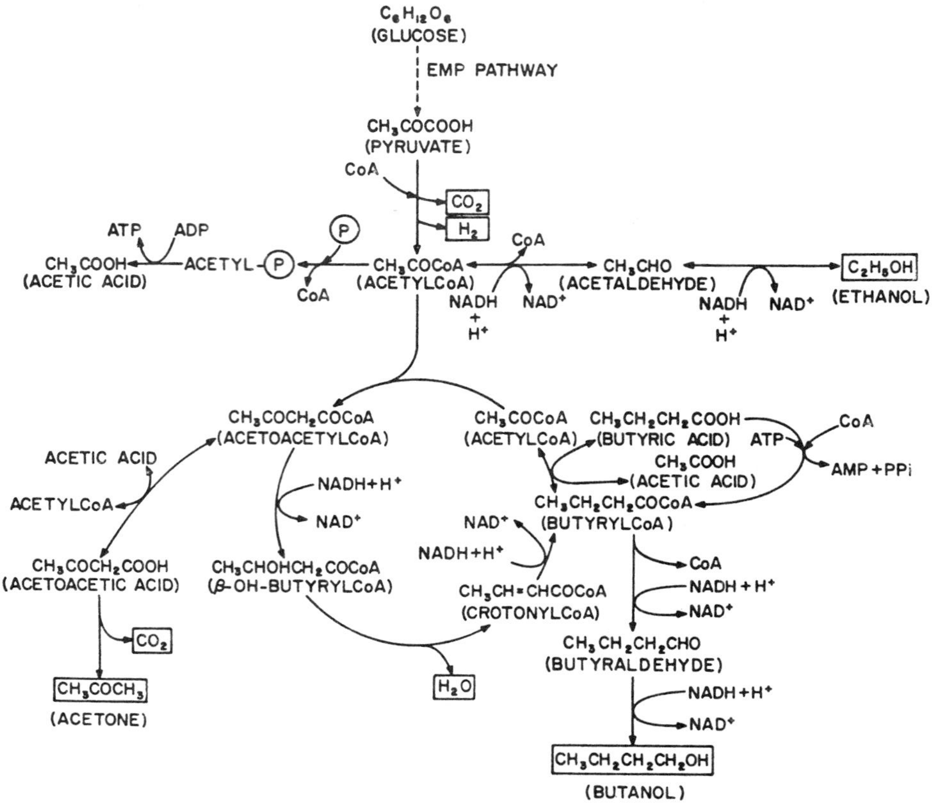

FIGURE 11.3 Metabolic pathway for glucose utilization by *C. acetobutylicum*.

leading to the production of acetone and butanol. Acetoacetyl-CoA is diverted by a transferase system to the production of acetoacetate, which can then be decarboxylated to acetone. This causes the elimination of two steps in the previous cycle that generate NAD^+. Therefore the bacteria must find an alternative way for reoxidizing the $NADH_2$. This is accomplished by reducing butyric acid to butanol through the three consecutive reactions shown in Fig. 11.3.

Isotopic tracer studies performed by Wood et al. (1945) partly confirmed that the acids are indeed precursors to butanol and acetone. However, the relationship was not as direct as would be expected. Butyric acid labeled with carbon-13 gave rise to labeled butanol, but the labeled carbon was also found in the other fermentation products. Approximately 85% of the carbon-13 was found in the butanol. The small conversion of butyric acid to acetic acid and acetone was thought to occur by a series of reversible reactions through acetoacetic acid. Wood et al. (1945) therefore concluded that butyric acid is a precursor of butanol.

Similar studies were performed by the same authors with labeled acetic acid. The carbon-13 of labeled acetic acid was recovered in butanol, acetone, acetic acid, ethanol, and carbon dioxide. Butanol contained 50% of the added carbon-13. Acetone accounted for approximately 15–19% of the added carbon-13, and an equivalent fraction was recovered in the carbon dioxide. These results indicated that acetic acid plays a complex role in the acetone–butanol fermentation and is not converted to acetone quantitatively; such complex behavior may be the major reason why acetic acid formation was rather difficult to simulate in a recent study on the mathematical modeling of the acetone–butanol fermentation (Costa, 1983).

As was mentioned previously, the shift from acid to neutral solvent formation is believed to be caused by the drop in pH of the fermentation broth. However, several studies have shown that successful solvent production can be achieved without the pH of the medium falling below 5.0 (Moreira et al., 1981; Barber et al., 1979). Pye et al. (1979b) have also determined that the maximum butanol production occurs when the pH is maintained at 5.0. Such data indicate that the mechanism regulating the production of acids and neutral solvents is much more complex than can be accounted for by pH changes alone. The levels of key enzymes associated with the shift in solvent production in *Clostridium acetobutylicum* must be closely monitored, as was recently suggested (Gottschalk and Bahl, 1981), if the biochemistry of the shift is to be understood in the future.

From the previous description, it is evident that the production of acetone and butanol by *Clostridium* bacteria follows a rather complex mechanism leading to a large number of end products—carbon dioxide, hydrogen, water, acetic acid, butyric acid, and acetoin, in addition to the desired acetone and butanol. In spite of such a large number of end products formed, it is interesting to observe that about 97% of the energy content of the glucose fermented is conserved in the chemicals produced during the fermentation (Lenz and Moreira, 1980).

IV. PROCESS ENGINEERING

A. Microorganisms and Culture Characteristics

A large number of bacteria are capable of producing small amounts of butanol; however, only certain *Clostridia* species produce butanol at high enough levels to make them commercially attractive. *Clostridium acetobutylicum* is one of the most important butanol fermenters and certainly is the microorganism mostly focused upon today for research purposes. A strain of *Clostridium acetobutylicum* is used in one of the few plants still in operation around the world today (National Chemical Products, South Africa) (Spivey, 1978). *Bergey's Manual* (Buchanan and Gibbons, 1974) describes *C. acetobutylicum* as a straight rod 0.6–0.9 by 2.4–4.7 μm. It is motile with peritrichous flagella and gram-positive with a cell wall containing DL-diaminopimelic acid. It will ferment the following carbohydrate substrates: arabinose, galactinol, fructose, galactose, glucose, glycogen, lactose, maltose, mannose, salicin, starch, sucrose, trehalose, and xylose. The optimum growth temperature is 37°C.

The *Clostridia* are anaerobic, spore-forming rods. The spores are formed when the nutrients become exhausted. Variations in the morphology of the culture during different stages of the fermentation have been observed by Spivey (1978) and Peterson and Fred (1932). These morphological variations were found to be dependent upon the type of medium and strain of microorganism used. Because the *Clostridia* are spore formers, they are easily maintained as such by using sterile soil as a carrier (Walton and Martin, 1979). A 5% (wt./vol.) corn meal medium, with 0.05% (wt./vol.) cysteine added to ensure anaerobiosis, has also been used with success in storing the cultures for over a year (Ulmer et al., 1981).

Culture degeneration has been a common problem with continued transfers of *C. acetobutylicum* (Casida, 1968; Steel, 1958; Beesch, 1952). A complete loss of solvent-forming ability may eventually occur. As a result, heat shocking is a commonly practiced technique with each culture transfer. This is accomplished by subjecting spores to boiling water for approximately 60–90 seconds followed by rapid cooling. Heat shocking rejuvenates the culture by eliminating any vegetative cells or weak spores present. Weizmann (1919) has suggested that 100–150 successive heat shockings be used to obtain a high-yielding culture.

The primary carbohydrates used for the acetone–butanol fermentation have been corn and molasses (Spivey, 1978; McCutcheon and Hickey, 1954). The fermentable sugar concentration of these substrates is set at 6.0–6.5% (wt./vol.) at the start of the fermentation. A large number of *Clostridia* strains have been isolated that can ferment molasses (Beesch, 1952; Ross, 1961). Each of such bacteria is unique in the carbohydrate source fermented, the nitrogen source utilized, and the type and distribution of solvents. All are spore-forming rods that were isolated from either soil, manure, roots of leguminous plants, cereals, cornstalks, decayed wood, or sewage. Other carbohydrate sources that may be used include wheat, rice, horse chestnuts, Jerusalem artichokes, cheese whey, cassava, and potatoes (McCutcheon and Hickey, 1954; Fouad et al., 1976; Wendland et al., 1941). Waste product–derived carbohydrates have also been successfully fermented; these include the sugars released by the acid hydrolysis of corn cobs (Langlykke et al., 1948), acid hydrolyzates of wood (Leonard et al., 1947), and waste sulfite liquor (Wiley et al., 1941).

The ratios and yield of solvents produced in the acetone–butanol fermentation depend upon several factors. These include the strain of microorganisms used, the substrate, and the fermentation temperature and pH. A theoretical calculation assuming acetone, butanol, carbon dioxide, hydrogen, and water as the only end products shows that the maximum possible yield of solvents is 0.38 g of solvents per gram of glucose fermented (Leung and Wang, 1981). In practice, yields of 30% (wt./wt.) are typically observed for a molasses fermentation with a butanol:acetone:ethanol ratio of 6:3:1. Corn fermentations usually yield 26.5% solvents with the same solvent ratio as molasses.

Maddox (1980) and Lindberg and Moreira (1982a) have obtained butanol:acetone ratios much higher than those reported above. Maddox (1980) utilized sulfuric acid whey filtrate supplemented with 0.5% (wt./vol.) yeast extract and obtained butanol levels of 1.2 wt. % with a butanol:acetone:ethanol ratio of 10:1:1. On the other hand, Lindberg and Moreira (1982a) have obtained butanol levels as high as 1.4 wt. % from plain cheese whey, with a butanol:acetone:ethanol ratio of approximately 20:1:1.

B. Batch Fermentation

All known commercial fermentation facilities installed for the production of acetone and butanol used batch technology in their operation. This fact results from the difficulty in maintaining an aseptic anaerobic fermentation operating over extended periods of time and from the tendency for *Clostridia* cultures to lose their butanol-producing capability with time.

The flowsheet for a traditional molasses acetone–butanol fermentation is shown in Fig. 11.4 and is similar to that reported for past commercial operations (Lenz and Moreira, 1980; Beesch, 1941). In this process, molasses is mixed with a recycle stream from the beer stills and then fed to cookers/sterilizers for steam sterilization at 121°C. After cooking, the sterile diluted molasses feed is sent to a battery of 16 batch fermentors (1514 m^3 of volume each); nutrients and culture inoculum (*C. saccharo-acetobutylicum*) are added; and fermentation proceeds for 48 hours. The fermentor off-gases, primarily carbon dioxide and hydrogen, are vented continuously. The fermentation broth is withdrawn batchwise and fed to the beer stills, where an approximately 40/60 (wt. %) solvent/water stream is obtained. The slops (water and stillage) from the beer stills are heat-exchanged with the feed to the stills and are either recycled for mixing with the molasses feed or sent to a battery of five multiple-effect evaporators and a spray dryer to produce a good-quality animal feed. The solvent/water stream from the beer stills is further distilled in a batch column that separates acetone (as a low boiler) from an ethanol/water azeotrope side stream and a butanol/water tail stream. The acetone and the ethanol/water azeotrope streams are sent directly to individual product storage, while the butanol/water stream is sent to two additional distillation columns and a decanter, where water is finally purged and *n*-butanol is produced as a high boiler and sent to storage.

A recent economic analysis (Lenz and Moreira, 1980) of the acetone–butanol fermentation of molasses has shown the unfavorable position of this process relative to the petrochemical-based routes even when the best

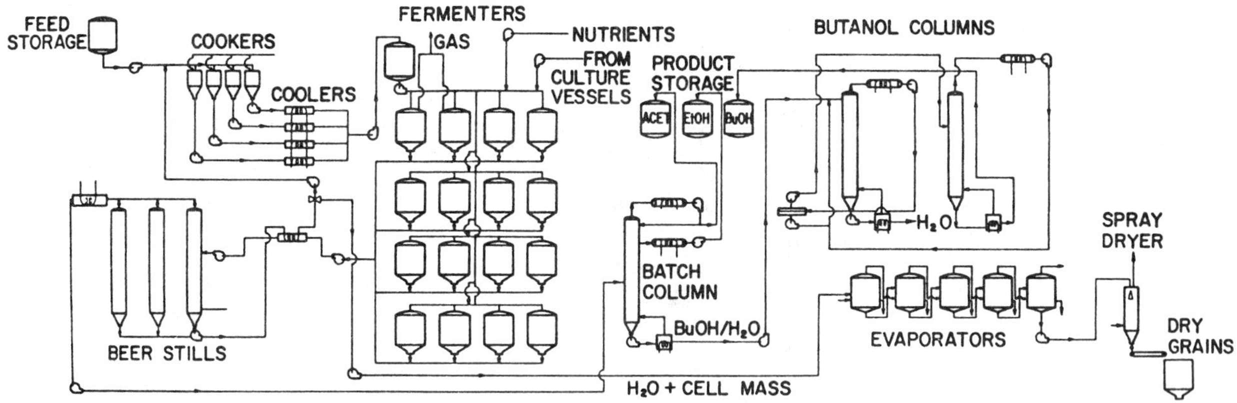

ANNUAL PRODUCTION:			
ACETONE	=	13.2×10^6 kg/yr	(29×10^6 PPY)
BUTANOL	=	26.4×10^6 kg/yr	(58×10^6 PPY)
ETHANOL	=	4.5×10^6 kg/yr	(10×10^6 PPY)
CO_2	=	72.0×10^6 kg/yr	(159×10^6 PPY)
H_2	=	1.8×10^6 kg/yr	(4.0×10^6 PPY)
DRY GRAINS	=	6.4×10^7 kg/yr	(1.4×10^8 PPY)
MOLASSES FEED	=	220,000 kg/6hr	(582×10^6 PPY @ 57% SUGAR)

ACETONE-BUTANOL FERMENTATION PROCESS FLOWSHEET

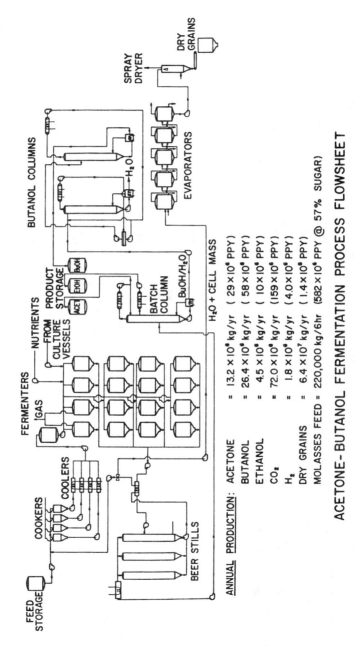

FIGURE 11.4 Flowsheet for a batch acetone–butanol fermentation.

available technology is practiced in a modernized fermentation facility. Raw materials costs accounted for 62% of the total production costs for the molasses-based production. However, this situation was dramatically improved when a waste material such as cheese whey was utilized as raw material, even when whey trucking costs were accounted for in the calculations. Theoretically, if all the by-products of a whey-based fermentation were utilized, a net return on the investment of 30% could be realized. If no credits are assigned to the gaseous by-products (CO_2 and H_2), an attractive 15% net return can still be achieved for the whey fermentation.

C. Continuous Fermentation

In spite of all the potential advantages of a continuous fermentation system over a batch process, the use of a chemostat for the acetone–butanol fermentation is still a research curiosity for the reasons previously described. In a recently published study (Andersch et al., 1982), acetone and butanol were produced in a chemostat when the pH was kept below 5.0, the optimum production occurring at a pH of 4.3. It was also found that chemostat-grown cells had lost their ability to form spores and produced low solvent concentrations when tested in a batch fermentor.

Bahl et al. (1982) used a chemostat to study the effects of pH and butyrate concentration on solvent production by *C. acetobutylicum*. Their results showed that chemostat cultures growing at pH 7 produced acids only. By lowering the pH to 4.3 and by addition of butyrate, neutral solvents were produced. Butyrate could not be replaced by acetate and achieve the same effect.

The overall volumetric productivity of neutral solvents (acetone and butanol) was recently shown to be higher for a chemostat than for a batch system (Leung and Wang, 1981). On a specific rate basis, the productivities of acetone and butanol in a chemostat reached maximum values of 0.32 g/g-h and 0.18 g/g-h, respectively, at a dilution rate of 0.22 h^{-1}. However, for a dilution rate above 0.26 h^{-1}, butyrate formation was favored. It is important to note that these studies (Leung and Wang, 1981) were performed with the culture pH maintained above 5.0, which is in disagreement with the results obtained by Andersch et al. (1982).

In another study, Conheady et al. (1981) found that butanol production in a single chemostat system could not be sustained even for short periods of time. Butyric and acetic acid production, however, continued for up to 3 weeks. The lack of heat shocking and consequent culture degeneration may very well preclude the use of continuous culture for the acetone–butanol fermentation on a commercial scale. One of the few positive reports concerning continuous acetone–butanol fermentation described a system installed in the Soviet Union in which as many as 11 fermentors were used in series and operated successfully for as long as 8 days (Hospodka, 1966).

D. Extractive Fermentation

A major drawback of the acetone–butanol fermentation is the relatively dilute concentration of solvents typically obtained in the range of 1.5–2.0 wt. % (Walton and Martin, 1979; Steel, 1958). The solvents, most notably butanol, are toxic to the bacterial cells at the levels produced during the fermentation, as will be discussed later in greater detail. Owing to such end product toxicity, the initial concentration of fermentable sugars is also kept relatively low, at 6.0–6.5 wt. %.

An interesting concept being investigated as a means of alleviating the end product toxicity is known as extractive fermentation. In this case, the fermentation broth is contacted with an organic solvent phase where butanol is preferentially soluble in relation to the aqueous phase. The organic phase must be immiscible with the water phase at all product concentrations encountered in the fermentation, nontoxic to the fermentation process, non-emulsion-forming, and easy to separate from the extracted products. With the availability of such technology, it should be possible to initiate the fermentation at a higher sugar concentration, since the toxic end products will now be distributed between two phases. Increased fermentation rates, more compact equipment, and a potentially easier recovery of the fermentation products from the extracting solvent may result in savings in both capital and operating costs.

In a rather comprehensive study (Pye and Humphrey, 1979a), many potential extractants were screened for the acetone–butanol fermentation. These included many vegetable oils, alcohols, esters, and aromatic compounds. Of all the compounds studied, dibutyl-phthalate was the best nontoxic extractant found, with a distribution coefficient for butanol of 1.8 (defined as the ratio of the butanol concentration in the organic phase to the concentration in the aqueous phase). The most favorable distribution coefficient was obtained for 2-octanol and equalled 10.4, but this secondary alcohol was toxic to the microbial cells. The longer-chain alcohols seemed to give the best distribution but were generally found to be toxic to the cell growth.

Another study, performed on ethanol–water–organic phase systems, showed some more general trends in distribution coefficients (Roddy, 1981). The level of extraction basically followed the order hydrocarbon = halocarbon < ether < ketone < amine < ester < alcohol = phosphates. There was also a noted temperature effect on the distribution coefficients, higher temperatures causing higher distribution coefficients. In some instances, increasing the temperature from 20°C to 40°C resulted in a doubling of the distribution coefficient with a particular extractant.

E. Immobilized Cell Systems

Immobilized cell reactors (ICR's) are currently receiving significant attention from the biotechnology research community. These bioreactors have the potential of achieving very high productivities in view of the very high cell densities that can be maintained and their plug flow nature, which is ideal for end product–inhibited reactions.

Very few reports are available in the literature addressing *C. acetobutylicum* immobilization (Conheady et al., 1981; Haggstrom and Molin, 1980). The results have always been quite discouraging, since immobilized cells have shown a rather weak capacity to produce neutral solvents and typically lose such a capacity in a relatively short time. In one study (Conheady et al., 1981), the immobilized cell reactor was able to sustain continuous operation for up to 300 hours, although butanol production ceased after only 170 hours and could not be restored. The maximum butanol productivity was 0.44 g/l-h. Butyrate formation was maintained over the entire operating period and reached a maximum productivity of 4.08 g/l-h. The failure of the immobilized cell reactor to maintain butanol production was attributed either to lack of ideal plug flow conditions due to cell overgrowth or to metabolic changes in the microorganisms.

The other study (Haggstrom and Molin, 1980) used spores and vegetative cells of *C. acetobutylicum* immobilized in Ca-alginate gel under nongrowth conditions. In less than 2 days after the immobilization, the butanol production rate fell to less than 25% of the original rate when using vegetative cells for the immobilization. Spore immobilization followed by viable cell growth looked like a very promising alternative, but extensive research will be required before an economically viable process can be established.

V. SUMMARY OF CURRENT RESEARCH ACTIVITIES

A. End Product Toxicity

As was stated earlier, the major drawback of the acetone–butanol fermentation is the very low concentration of the end products in the final fermented broth. The results obtained for a typical batch fermentation starting with a 6% (wt./vol.) glucose concentration are shown in Table 11.4. These data were obtained in our laboratory under controlled temperature (37°C) and pH (5.0) conditions for a totally soluble medium. During the exponential growth phase, a small increase in the concentration of butanol, acetone, and ethanol was observed, while butyric and acetic acids reached their peak concentrations. After the exponential growth phase, the concentrations of acetone, butanol, and ethanol increased rapidly to their final values. At the same time, the concentrations of butyric and acetic acids decreased to 1.3 g/l and 1.7 g/l, respectively.

In order to understand the inhibition phenomena for this fermentation, the growth rates of *C. acetobutylicum* in the presence of each fermentation product were determined through a series of challenge experiments (Costa, 1983). There appeared to be a threshold concentration that had to be reached

TABLE 11.4 Composition of Medium with Respect to Glucose, Cell Mass, Acids, and Solvents During the Course of Batch *C. Acetobutylicum* Fermentation in Normal Soluble Medium

Elapsed Time (h)	*Glucose (g/l)*	*Cell Mass (dry) (g/l)*	*Acetone (g/l)*	*Ethanol (g/l)*	*Butanol (g/l)*	*Acetic Acid (g/l)*	*Butyric Acid (g/l)*
0.0	58.4	0.0	—	—	0.20	—	—
8.5	49.3	0.4	0.08	—	0.35	0.95	0.54
11.0	46.8	1.1	0.14	—	0.41	1.97	1.26
14.0	37.8	2.5	0.07	—	0.45	2.60	3.31
17.0	28.7	3.1	1.63	0.14	2.12	4.50	4.17
21.0	8.1	3.5	1.63	0.34	6.04	4.58	2.48
24.0	3.4	3.8	2.71	0.55	7.76	3.08	1.87
33.0	0.6	3.5	3.26	—	7.51	1.58	1.55
39.0	0.4	3.1	4.07	0.62	11.18	1.74	1.33

before growth inhibition occurred. Such a concentration was found to vary with each inhibitor under study. A summary of the inhibition data collected is shown in Table 11.5.

The levels of acetone and ethanol typically seen during fermentation (4 g/l and 1 g/l, respectively) were found to be noninhibitory to cell growth. However, butyric acid, acetic acid, and butanol do reach levels during fermentation that can cause significant growth inhibition. Butanol actually reaches concentrations that cause 50% inhibition of cell growth, although it is believed that it is the combined action of all the end products that is responsible for the complete shutdown of this fermentation.

The above growth challenge experiments indicated that butanol is the major fermentation product reaching concentration levels that are significantly inhibitory to the growth of *C. acetobutylicum* cells. However, these studies did not identify the mechanism for the alcohol toxicity to the bacteria. In fact, the basic cause of alcohol toxicity in microorganisms is not known.

TABLE 11.5 Growth Inhibition Data for *C. Acetobutylicum*

End Product	*Threshold Concentration* (M)	*Concentration at Which Growth Was Inhibited by 50%* (M)	(*g/l*)
Butyric acid	0.02	0.07	6.0
Butanol	0.05	0.15	11.0
Acetic acid	0.04	0.13	8.0
Ethanol	0.25	1.10	51.0
Acetone	0.36	0.88	43.5

The bactericidal properties, as well as the hydrophobic properties, of the aliphatic alcohols have been shown to increase with the number of carbon atoms in the alcohol molecule (Duperray et al., 1976). Hydrophobicity has been correlated to a number of biological membrane properties. For example, the following concentrations of the aliphatic alcohol series cause 50% inhibition of membrane-bound adenosine triphosphatase (ATPase) on beef heart mitochondrial inner membranes: ethanol (4.9 *M*), *n*-butanol (0.4 *M*), *n*-pentanol (0.1 *M*), *n*-heptanol (0.008 *M*), and *n*-octanol (0.003 *M*). The same concentrations also caused the maximum change in the enthalpy required to orient a spin label compound in the membrane lipid bilayer (Grisham and Barnett, 1973).

Many other reports relate to subtle alterations of membrane-bound enzyme activities with concentrations of *n*-butanol less than 0.4 *M* (Grisham and Barnett, 1973; Lenaz et al., 1975, 1976; Patterson et al., 1972; Sullivan et al., 1974; Pavlasova and Harold, 1969). In many of these studies, the interaction between membrane-bound proteins and the annular lipids surrounding the proteins was found to be of major importance.

Exposure of growing cells to a medium containing aliphatic alcohols has been shown to induce changes in the lipid composition of bacterial membranes. As an example, *E. coli* cells, when adapted to ethanol, show a decrease in the ratio of saturated to unsaturated (S/U) fatty acids (Ingram, 1976), whereas alcohols longer than four carbons produced an increase in the S/U ratio. Ethanol caused a membrane-freezing effect by inhibiting saturated fatty acid synthesis in vivo (Buttke and Ingram, 1978). Butanol, on the other hand, had an effect opposite to ethanol, that is, it increased membrane fluidity (Grisham and Barnett, 1973; Patterson et al., 1972), thus causing a physiological response of cell adaptation to butanol by increasing the S/U ratio of membrane fatty acids. The membrane properties of thermophilic microorganisms would thus be desirable for butanol tolerance, since they are typically richer in saturated and branched fatty acid components (Yoshida et al., 1975; Hirata et al., 1976).

The ability of aliphatic alcohols to increase the fluidity of cell membranes is proportional to the chain length of the alcohol, probably owing to the greater solubility of longer-chain alcohols in the core of phospholipid bilayers. This is shown by the large octanol/water partition coefficient for longer chain alcohols (Moreira and Linden, 1982). Similar solubility aspects for aliphatic alcohols on membrane lipids of *S. cerevisiae* have been recently shown to be related to alcohol inhibition of the yeast glucose transport system (Leao and van Uden, 1982).

To further elucidate the mechanism for end product inhibition in *C. acetobutylicum*, a series of studies was performed in which the effect of increasing concentrations of ethanol, butanol, and hexanol on alanine uptake, 3-O-methyl-glucose uptake, and membrane-bound ATPase activity was closely monitored. Table 11.6 shows the concentrations of the various alcohols

TABLE 11.6 Concentrations of Aliphatic Alcohols that Limit Various Cellular Functions of *C. Acetobutylicum* to Half-Maximal Rates

	Concentration (M)		
Cellular Function	*EtOH*	*BuOH*	*HxOH*
Cell growth	1.10	0.15	0.015
Alanine uptake	0.91	0.13	0.006
3-O-Methyl glucose uptake	1.00	0.10	0.007
ATPase activity	0.82	0.09	0.015

that limited the cellular functions of *C. acetobutylicum* to half-maximal rates.

Two major points stand out from Table 11.6: (1) same-order-of-magnitude concentration for each individual alcohol causes similar inhibition on all the cellular functions studied; and (2) ethanol is approximately one order of magnitude less toxic than butanol, and butanol itself is one order of magnitude less toxic than hexanol. It is also evident from the data in Table 11.6 that the cell membrane seems to be the site of inhibitory interaction and that sugar transport seems to proceed via an active transport process with the ATPase enzyme components (Maloney and Wilson, 1975). The incorporation of the straight-chain alcohols into the membrane probably disrupts the necessary interactions between the membrane-bound proteins and the fluid phospholipid environment of the membrane, causing a breakdown in the sugar transport machinery and thus affecting cell viability.

As an attempt to increase the butanol tolerance of the *Clostridia* cells, an alteration in the phospholipid environment by selective incorporation of specific fatty acids into the cytoplasmic membrane of *C. acetobutylicum* was designed next. This was achieved by blocking fatty acid synthesis in biotin-deficient media (through avidin binding of yeast extract biotin) and by supplementing the fermentation medium with 10 mg/l of either oleic acid or elaidic acid. Table 11.7 shows a comparison of butanol concentrations that caused 50% inhibition of the maximal rates of cell growth, nutrient uptake, and ATP hydrolysis in preparations of *C. acetobutylicum* grown on a normal soluble medium, on a biotin-deficient oleic acid–supplemented soluble medium, and on a biotin-deficient elaidic acid–supplemented soluble medium. The data shown in Table 11.7 are extremely encouraging, since they indicate that the modified cells do indeed have an increased tolerance to *n*-butanol in terms of both cell growth and membrane-bound ATPase activity. A twofold to threefold increase in butanol tolerance seems to be a realistic achievement at the present time. Such an improvement will become much more meaningful if the modified *Clostridia* cells prove to be able to ferment *n*-butanol at an equally threefold higher concentration than normal cells.

TABLE 11.7 Comparison of *n*-Butanol Concentrations that Cause 50% Inhibition of Cellular Functions in Various *C. Acetobutylicum* Preparations

	n-Butanol Concentration (M)		
Half-Maximal Rates	*Normal*	*Oleate*	*Elaidate*
Cell growth	0.15	0.30	0.22
Alanine uptake	0.10	0.26	N.D.
3-O-Methyl glucose uptake	0.10	N.D.	N.D.
ATPase activity	0.06	0.09	0.16

This table shows a different series of experiments from those used to calculate data in Table 11.6.

B. Cheese Whey Fermentation

Over 10 billion pounds of cheese whey are wasted each year by dumping into the sewer system. Cheese whey is an opaque greenish-yellow liquid that remains in a cheese vat following the separation of the curd from the liquid when making cheese from milk. Whey contains approximately 5–7% (wt./vol.) lactose and 0.7–1% (wt./vol.) protein, along with some minerals and small amounts of fat and lactic acid. In spite of the fact that its nutrient content makes cheese whey an attractive substrate for fermentation, the cheese manufacturer has found it more convenient to pay for the high costs of disposal rather than trying to reutilize it. Cheese whey has a high biochemical oxygen demand (BOD) of 30,000–60,000 ppm (Zall, 1967), thus its high disposal surcharge.

Antipollution legislation is forcing cheese manufacturers to turn to cheese whey reutilization rather than disposal; however, the bulkiness and perishability of whey make its processing very often a marginal venture. As an example, the price of dried whey for animal feed barely covers the operating costs associated with the drying process.

Since the initial sugar level in cheese whey is about 5% (wt./vol.) and *C. acetobutylicum* is capable of fermenting lactose directly, it seems logical to consider cheese whey as an attractive raw material for the acetone–butanol fermentation. In order to assess the viability of such fermentation, bench scale studies were performed by fermenting liquid whey from a mozzarella cheese plant with *C. acetobutylicum*. These studies have shown butanol:acetone ratios as high as 12:1 to 20:1 for the cheese whey fermentation (Lindberg and Moreira, 1982a) as compared to a 3:1 butanol:acetone ratio for a glucose fermentation. The cause for such a shift in the butanol:acetone ratio has not yet been fully identified, and more recent studies (Lindberg and Moreira, 1982b) seem to indicate that it is very likely due to a combination of factors rather than to a single component present in the cheese whey. Determination of the gas phase composition for a cheese whey fermentation has also shown some differences relative to the composition typically seen during a glucose fermentation. The cheese whey fermentation showed an overall gas composition of 35% H_2/65% CO_2 (mole ratio) versus a typically 42% H_2/58% CO_2 gas composition for a glucose fermentation (Lindberg and Moreira, 1982b). The decrease in the H_2/CO_2 ratio for the cheese whey fermentation was found to be consistent with an increased butanol/butyric acid ratio for this fermentation.

C. Utilization of the Fermentor Off-Gases

The off-gas stream from a conventional acetone–butanol fermentation plant is composed mainly of carbon dioxide and hydrogen in a mole ratio, as was seen previously, of approximately 50:50. These gases are currently not utilized in any further processing step but are simply discarded to the environment. Such a gaseous stream, in addition to creating a disposal problem, represents a drop in the efficiency of the fermentation, since valuable amounts of carbon are lost as carbon dioxide.

The conversion of the acetone–butanol fermentation off-gas to methanol represents an interesting approach to this problem. In fact, such conversion was industrially practiced in the 1930's–1950's (Gabriel, 1928). Methanol is a valuable liquid fuel and chemical feedstock that is currently receiving increased attention from investigators working on the development of alternative sources of energy. Moreover, technology recently developed for the catalytic conversion of carbon dioxide and hydrogen to methanol (*Chemical Engineering*, 1980) seems extremely promising in terms of energy consumption and process economics. In a recent study (Moreira et al., 1982), it was found that a discounted cash flow rate of return (DCFRR) of 25–30% is likely to be achieved for realistic operating conditions of a methanol synthesis plant located next to an acetone–butanol fermentation facility capable of producing 25 million gallons of solvents per year. This rate of return was based on a methanol value of \$0.80/gal and assumed that the excess carbon dioxide produced in the fermentation plant would have no value. It should be stressed that the rate of return cited above refers to the *incremental* investment in a methanol plant and not to the overall economic feasibility of the fermentation facility. It seems obvious that if the fermentation facility cannot stand alone economically, it will be very difficult for by-product recovery (such as this methanol process) to make the overall system economical.

As a side result from our recent interest on the catalytic conversion of the CO_2/H_2 mixture to methanol, bench scale glucose fermentation experiments with *C. acetobutylicum* under different rates of agitation have shown that the agitation rate has a remarkable influence on the gas yield and composition from this fermentation (Moreira et al., 1982). Utilizing 1-liter fermentation flasks with 450 ml of a 5% (wt./vol.) glucose-soluble medium, the total volume of gas produced during fermentation varied from 9.4 liters at 0 RPM (revolutions per minute) to 12.9 liters for an experiment at 250 RPM. Utilizing gas phase compositional data, computations showed that a total of 3.9 liters of H_2 were released during the 0-RPM batch, while a 250-RPM experiment produced 5.5 liters of H_2. The most interesting observation was that the highest concentration of *n*-butanol occurred under no agitation, which also produced the lowest amount of H_2 in the off-gas stream. A material balance indicated, however, that less than 10% of the decreased hydrogen production can be accounted for by the increased butanol levels at lower agitation rates. This "inefficiency" of hydrogen conversion to butanol will be of significant importance in deciding on the installation of a methanol synthesis unit adjacent to an acetone–butanol fermentation plant.

VI. FUTURE OUTLOOK FOR THE ACETONE–BUTANOL FERMENTATION

Today's state-of-the-art technology on the acetone–butanol fermentation is still at the same level as it was in the 1950's. Since this fermentation was virtually abandoned once the less expensive petrochemical routes were established, no research has been carried out in this area for the last 30 years, with the exception of the very last few years, when some activity has again surfaced in a few laboratories. The lack of interest in this fermentation occurred at the same time that biochemical engineering became established as a new field and advances in biochemistry and microbiology, supported by greatly improved instrumentation, allowed for a better understanding of microbial activities.

As research focuses again on this fermentation, the potential is immense for a revitalization of the acetone–butanol fermentation industry during the late 1980's. A detailed knowledge of the *C. acetobutylicum* metabolism will be achieved, and the mechanisms that control neutral solvent formation will be understood. At the same time, the genetics of *C. acetobutylicum* will be developed. This is, at the present time, a totally unexplored area that could lead into future developments such as increased tolerance to end product concentration, thermophilic fermentation systems, and direct fermentation of recalcitrant substrates like lignocellulose. Also, further progress in the biological membranes field will allow for a better understanding of membrane composition and function in *C. acetobutylicum* and of how these are affected by the presence of the fermentation end products. This knowledge should lead to new solutions to the solvent toxicity problems.

The integration of such developments with the utilization of the least expensive raw materials should result in attractive economics for the acetone–butanol fermentation and thus provide the necessary driving force for the startup of modernized facilities that can make this fermentation again competitive with the petrochemical routes.

ACKNOWLEDGMENTS

The research results summarized in this chapter were obtained while the author was a faculty member in the Agricultural and Chemical Engineering Department at Colorado State University. Partial support for this research was provided through several contracts from the U.S. Department of Energy, the Solar Energy Research Institute, and a few private companies. The author also wishes to acknowledge many contributions to this research program received from other faculty members, postdoctoral associates, graduate students, and laboratory technicians at C.S.U. A special acknowledgment is extended to Dr. James C. Linden, who has shared with me many of the exciting as well as the occasionally depressing moments of our acetone–butanol research program at C.S.U.

REFERENCES

Andersch, W., Bahl, H., and Gottschalk, G. (1982) *Biotechnol. Lett. 4,* 29–32.

Bahl, H., Andersch, W., Brian, K., and Gottschalk, G. (1982) *Eur. J. Appl. Microbiol. Biotechnol. 14,* 17–20.

Barber, J. M., Robb, F. T., Webster, J. R., and Woods, D. R. (1979) *Appl. Environ. Microbiol. 37,* 433–437.

Beesch, S. C. (1952) *Ind. Eng. Chem. 44,* 1677–1682.

Buchanan, R. E., and Gibbons, N. E., eds. (1974) *Bergey's Manual of Determinative Bacteriology* 8th ed., pp. 556–560, The Williams and Wilkens Co., Baltimore, Md.

Buttke, T. M., and Ingram, L. O. (1978) *Biochem. J. 17,* 637–644.

Casida, L. E. (1968) *Industrial Microbiology* pp. 260–274, John Wiley, New York.

Chemical Engineering (1980) February 11, p. 49.

Chemical Marketing Reporter (1963–1982).

Conheady, J. A., Clausen, E. C., and Gaddy, J. L. (1981) "Performance of Immobilized *Clostridium acetobutylicum,*" paper presented at the AIChE Annual Meeting, New Orleans, Louisiana, November 7.

Costa, J. M. (1983) "Kinetics of the Acetone-Butanol Fermentation," M.S. thesis, Colorado State University, Dept. of Agricultural and Chemical Engineering, Fort Collins, Co.

Doelle, H. W. (1975) *Bacterial Metabolism* 2nd ed., p. 587, Academic Press, New York.

Duperray, B., Chastrette, M., Makabeh, M. C., and Pacheco, H. (1976) *Eur. J. Med. Chem-Chimica Therapeutica 11,* 433.

Fouad, M., Asou-Zeid, A. A., and Yassein, M. (1976) *Acta Biol. Acad. Sci. Hung. 27,* 107–117.

Gabriel, C. L. (1928) *Ind. Eng. Chem. 20,* 1063.

Gottschalk, G., and Bahl, H. (1981) in *Trends in the Biology of Fermentations for Fuels and Chemicals* (Hollaender, A., ed.) pp. 463–471, Plenum Press, New York.

Grisham, C. M., and Barnett, R. E. (1973) *Biochem. Biophys. Acta 311,* 417–422.

Haggstrom, L., and Molin, N. (1980) *Biotechnol. Lett. 2,* 241–246.

Hirata, H., Sone, N., Yoshida, M., and Kagawa, Y. (1976) *J. Biochem.* (Tokyo) *76,* 1157–1166.

Hospodka, J. (1966) in *Theoretical and Methodological Basis of Continuous Culture of Microorganisms* (I. Malek and Z. Fencl, eds.) pp. 611–613, Academic Press, New York.

Ingram, L. D. (1976) *J. Bacteriol. 125,* 670–678.

Langlykke, A. F., van Lanen, J. M., and Fraser, D. R. (1948) *Ind. Eng. Chem. 40,* 1716–1719.

Leao, C., and van Uden, N. (1982) *Biotechnol. Bioeng. 24,* 2601–2604.

Lenaz, G., Parenti-Castelli, G., and Sechi, A. M. (1975) *Arch. Biochim. Biophys. 167,* 72–79.

Lenaz, G., Bertoli, E., Curatola, G., Mazzanti, L., and Bigi, A. (1976) *Arch. Biochem. Biophys. 172,* 278–288.

Lenz, T. G., and Moreira, A. R. (1980) *Ind. Eng. Chem. Prod. Res. Dev. 19,* 478–483.

Leonard, R. H., Peterson, W. H., and Ritter, G. J. (1947) *Ind. Eng. Chem. 39,* 1443–1445.

Leung, J. C. Y., and Wang, D. I. C. (1981) in *Proceedings of the Second World Congress of Chemical Engineering* Vol. I, pp. 348–352, Montreal, Canada.

Lindberg, S. C., and Moreira, A. R. (1982a) "Acetone-Butanol Fermentation of Cheese Whey," paper presented at the AIChE Winter Meeting, Orlando, Florida, February 28–March 3.

Lindberg, S. C., and Moreira, A. R. (1982b) "Production of Neutral Solvents from Cheese Whey," paper presented at the American Chemical Society National Meeting, Kansas City, Missouri, September 12–17.

Maddox, I. S. (1980) *Biotechnol. Lett. 2,* 493–498.

Maloney, P. C., and Wilson, T. H. (1975) *J. Memb. Biol. 25,* 285.

McCutcheon, W. N., and Hickey, R. J. (1954) in *Industrial Fermentations* Vol. I (Underkofler, L. A., and Hickey, R. J., eds.) pp. 347–388, Chemical Publishing Co., New York.

Moreira, A. R., and Linden, J. C. (1982) "Anaerobic Production of Chemicals," paper presented at the Symposium on the Biological Basis of New Developments in Biotechnology, University of Minnesota, Minneapolis, May 25–28.

Moreira, A. R., Ulmer, D. C., and Linden, J. C. (1981) *Biotechnol. Bioeng. Symp. 11,* 567–579.

Moreira, A. R., Dale, B. E., and Doremus, M. G. (1982) *Biotechnol. Bioeng. Symp. 12,* 263–277.

Nelson, D. L., and Webb, B. P. (1979) in *Kirk-Othmer Encyclopedia of Chemical Technology* Vol. I, 3rd ed., pp. 179–190, John Wiley, New York.

Patterson, S. S., Butler, K. W., Huang, P., Belle, J., Smith, I. C. P., and Schneider, H. (1972) *Biochim. Biophys. Acta 266,* 597–602.

Pavlasova, E., and Harold, F. M. (1969) *J. Bacteriol. 98,* 198.

Peterson, W. H., and Fred, E. B. (1932) *Ind. Eng. Chem. 24,* 237–242.

Pye, E. K., and Humphrey, A. E. (1979a) "Biological Production of Liquid Fuels from Biomass," University of Pennsylvania Interim Report, p. 117, Philadelphia, February.

Pye, E. K., and Humphrey, A. E. (1979b) "Biological Production of Liquid Fuels from Biomass," University of Pennsylvania Annual Report prepared for the U.S. Department of Energy under Contract COO/4070-11, September 1, 1978–August 31, 1979.

Roddy, J. W. (1981) *Ind. Eng. Chem. Proc. Design Develop. 20,* 104–108.

Rosenfeld, B., and Simon, E. (1950a) *J. Biol. Chem. 186,* 395–404.
Rosenfeld, B., and Simon, E. (1950b) *J. Biol. Chem. 186,* 405–410.
Ross, D. (1961) *Prog. Ind. Microbiol. 3,* 73–90.
Sherman, P. D. (1979) in *Kirk-Othmer Encyclopedia of Chemical Technology* Vol. IV, 3rd ed., pp. 338–345, John Wiley, New York.
Spivey, M. J. (1978) *Proc. Biochem. 13,* 2–4, 25.
Stainier, R. Y., Adelberg, E. A., and Ingraham, J. (1976) *The Microbial World* 4th ed., Prentice-Hall, Englewood Cliffs, N. J.
Steel, R. (1958) *Biochemical Engineering: Unit Processes in Fermentation* pp. 125–148, MacMillan, New York.
Sullivan, K. H., Jain, M. K., and Koch, A. L. (1974) *Biochim. Biophys. Acta 352,* 287–297.
Ulmer, D. C., Linden, J. C., and Moreira, A. R. (1981) "The Effect of Ethanol and Butanol on Sugar Transport and Viability of *Clostridium* Species Bacteria," Annual Report to the U.S. Dept. of Energy for Subcontract XK-0-9059-1, September.
Villet, R. (1981) *Biotechnology for Producing Fuels and Chemicals from Biomass* Vol. II, *Fermentation Chemicals from Biomass* Solar Energy Research Institute Report TR-621-754, Golden, Co.
Walton, M. J., and Martin, J. L. (1979) in *Microbial Technology* (Peppler, H. J., and Perlman, D., eds.) Vol. I, pp. 187–209, Academic Press, New York.
Weizmann, C. (1919) U.S. Patent 1,315,585.
Wendland, R. T., Fulmer, E. I., and Underkofler, L. A. (1941) *Ind. Eng. Chem. 33,* 1078–1081.
Wiley, A. J., Johnson, M. J., McCoy, E., and Peterson, W. H. (1941) *Ind. Eng. Chem. 33,* 606–610.
Wood, H. G., Brown, R. W., and Werkman, C. H. (1945) *Arch. Biochem. 6,* 243–260.
Yoshida, M., Sone, N., Hirata, H., and Kagawa, Y. (1975) *J. Biol. Chem.* (Tokyo) *250,* 7910–7916.
Zall, R. R. (1967) *Milk Dealer* (December) *57,* 17.

CHAPTER
12

The Economics of Organic Chemicals from Biomass

Godfred E. Tong
Ronald P. Cannell

I. INTRODUCTION

Every organic chemical currently being produced from coal, petroleum, or natural gas can technically be produced from biomass raw materials. This statement of technology availability, combined with the economic fact that petroleum prices have escalated at least five times faster than biomass prices since 1974, triggers obvious questions. Why have biomass feedstocks such as sugar, starch, lignocellulose, plant oils, and natural rubber provided only less than 20% of the raw material needs of the U.S. organic chemical industry? What segments of the organic chemical industry are increasing or will increase economic utilization of biomass feedstocks over coal or petroleum? How significant will biomass feedstocks be to the organic chemical industry 10 years from now?

This chapter examines the answers to the above questions. The objectives of this chapter are twofold: first, to identify major areas where increased supply of biomass-derived chemicals can occur in the United States within the next 10 years, and second, to pinpoint specific groups of chemicals that will be increasingly dependent on biomass feedstock sources.

The analysis is based on the following underlying assumptions:

1. Future organic feedstocks for the U.S. chemical industry will continue

to be derived simutaneously from petroleum, coal, and biomass as they have been in the past.

2. Increased utilization of biomass feedstocks will occur when relative cost advantages can be enhanced, specifically in biomass processing and conversion plants where
   a. low capital investments for additional chemical recovery units will help improve overall product revenues;
   b. inherent compatibility exists between chemical composition of the biomass feedstock component and the targeted product, for example, carbohydrate for oxygenated chemicals and plant oils for aromatics and olefins; and
   c. further improvements in conversion efficiencies can still be made, such as in improved catalytic selectivities of plant oil cracking and improved microbial genetics to utilize pentose sugars and organic acids in by-product streams of biomass processing plants.

The economic and market data as well as technical information provided in this chapter can be most effectively used to meet the above objectives if the basic historical and economic background of the biomass processing industry is understood. The four sections providing this background are as follows.

1. *The History of Biomass Utilization*. This section describes the push–pull effects and other interactions that competing biomass industries have on feedstock supply and prices.
2. *Biomass Fractionation Processing Industries*. This section identifies the size and type of biomass refining and processing industries that existed in the United States as of 1981. It analyzes the essential characteristics of economically competitive biomass processing operations. Furthermore, it projects two potential types of lignocellulose processing involving organic chemicals production.
3. *Biomass Process Conversion Technologies*. This section identifies the thermochemical, chemical, and biochemical conversion processes that can transform biomass feedstocks into organic chemicals. The general economic characteristics of these conversion technologies are discussed.
4. *Six Entry Points for Biomass Feedstocks*. This section describes how the organic chemical industry is classified into six categories, which are the potential entry points for utilizing biomass feedstocks.

II. THE HISTORY OF BIOMASS UTILIZATION

One element common to biomass resources and petroleum resources is the availability of technology to utilize them in four major industries. Four industries compete for organic feedstocks: food and animal feed, fuel for energy, fiber and polymer, and specialty chemicals. The energy sector has traditionally controlled the price and availability of petroleum. The price of biomass, on the other hand, is controlled by the food and feed industry for agricultural feedstocks and by the timber and pulp industry in the case of forest biomass resources. The matrix of historical trends in Fig. 12.1 shows complementary use of fossil and biomass feedstocks in the four major industries.

Fig. 12.1 provides an overview of how both petroleum and biomass feedstocks have complemented one another historically in each of the four industries consuming organic feedstocks. Biomass had been the dominant feedstock for both the food and animal feed industry and the organic intermediate and specialty chemicals industries until the 1940's. By the 1960's, petroleum-derived feedstocks became an important factor in agrichemical industries such as pesticides and insecticides. Synthetic detergents took an increasing share of markets originally dominated by vegetable oil–derived soap. Food additives such as butylated hydroxyamisole antioxidants and synthetic sweeteners were introduced. In addition, petroleum-derived single-cell protein processes were developed to provide animal feed supplements. Despite the inroads made by petrochemical technology, biomass will continue to play a dominant role in both the food and animal feed and specialty chemical industries, since biomass feedstocks continue to increase in price at a slower rate than petroleum feedstocks. As indicated by the heavy lines in Fig. 12.1, the present and short-term future role of biomass in the fuel, fiber, and polymer sectors is more limited to supplementing petroleum-derived chemicals, as cited by the specific examples in Table 12.1. This is true for biomass-derived chemicals either with unique functional properties, such as natural rubber for tires, or in areas where the processing cost of petroleum-derived substitutes is too energy-intensive, as in the production of fatty acids. Both of these topics are further analyzed in later sections of this chapter.

Order of magnitude statistics of 1978 U.S. consumption of organic raw materials shown below clearly illustrate the controlling elements:

Feedstocks	*Million Tons in 1978*
A. Petroleum	
Crude oil	880
Natural gas	451
B. Agricultural	
Food and feed crops of corn and wheat	210
Agricultural straw of above crops	200
C. Forest	
Wood for pulp and paper	100
Wood for lumber	50
Forest residues	200

INDUSTRIES / FEEDSTOCKS	FOOD & FEED	FUEL	FIBER & POLYMER	SPECIALTY CHEMICALS & INTERMEDIATE CHEMICALS
FOSSIL eg. COAL, PETROLEUM	NONE → 1980	MOST 1800 → 1955	NONE → 1940	LITTLE → 1940
	SCP PROTEIN → 1980 1970	MOST 1955 → 1980	SYNTHETIC RUBBER, FIBER & FILM PLASTICS MOST 1940 → 1980	MOST 1940 → 2000 AGRICHEMICALS, DETERGENTS, SOLVENTS, PHARMACEUTICALS
BIOMASS	MOST → 1980	ALL → 1700 WOOD	NATURAL RUBBER COTTON, PULP ALL 1900 → 1980	MOST → 1940 SOAP
		LITTLE 1800 → 1980 METHANE BIOGAS, ETHANOL, METHANOL OXYGENATES	SOME 1980 → 2000 XANTHAN GUM, METHYL GLUCOSIDE	1940 → 1980 BIOLOGICALS ENZYMES ORGANIC ACIDS & SOLVENTS

* ADAPTED FROM DAVIES, D., "THE HUMANE TECHNOLOGIST", OXFORD UNIV. PRESS, p. 71 (1976).

FIGURE 12.1 Matrix of historical trends showing the complementary use of fossil and biomass feedstocks (adapted from Davies, 1976).

TABLE 12.1 List of Biomass-Derived Chemicals Blended with Petroleum Products

Product	*Use*	*Renewable Component*	*Petroleum Component*
Rubber	Tires	Natural rubber (guayule)	Isoprene
H-span	Seed-coating gels	Hydrolyzed starch	Polyacrylonitrile
Hexamethylene diamine	Nylon	Furfural	Adipic acid
Diesel fuel	Transport fuel	Coconut oil	Diesel oil
Gasohol	Transport fuel	Ethanol and oxygenates	Gasoline
Mobil 1	Synthetic lube oil	Fatty acids	Polyols
Semisynthetic penicillin	Antibiotics	Penicillin (6 APA)	Side chain chemicals
Isocyanates	Polyurethane	Methanol ethanol	Benzene, NO_2, CO

In 1978, 60–70 million tons of organic chemicals were produced in the United States. The sequence of industrial steps involved in producing final chemical products from organic feedstocks is illustrated in Fig. 12.2. This figure represents the overall structure of the chemical industry and the revenue (in the form of U.S. shipment values) generated at each step of product utilization leading to end-use markets. The third block in the figure represents the 60–70 million tons of organic chemicals produced. The production of these organic chemicals required only 6% of the 880 million tons of crude oil and 3% of the natural gas annually consumed in the United States. Agricultural feedstocks contributed less than 2 million tons, utilized for the production of organic chemicals such as industrial starch and ethanol. The balance of agricultural crops, more than 90%, were consumed by the food and feed industry. With seasonal availability and storage problems associated with harvested crops, a 5% surplus production can result in a price decline of more than 30%. Similarly, production shortages can cause significant price increases. The implication of this strong pricing dependency makes it risky to produce more than 30% of organic chemicals from agricultural resources without structural changes such as massive increases in crop production.

The forestry biomass resource situation is similar and currently contributes only a small amount to organic chemical production in the form of tall oil, lignosulfonates, and ethanol.

The first requirement for extended biomass use is the availability of a large, dependable supply of raw material. The two biomass resources that are available in quantities substantial enough to make a major impact on the organic chemical industry are agricultural straw and forest residues. However, the existence of pulp mills with an infrastructure for collection,

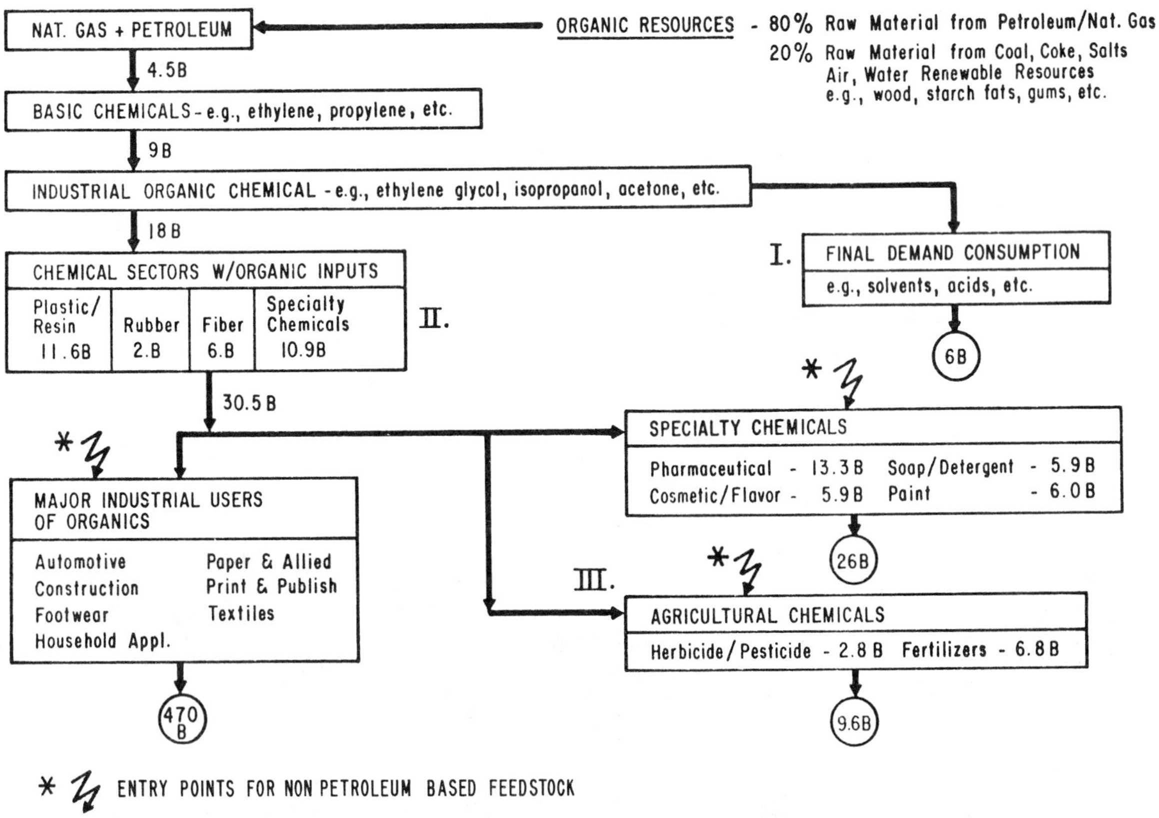

FIGURE 12.2 Flow of industrial organic chemicals from raw material to consumption (estimated 1977 value of shipments in billions of U.S. dollars).

storage, and transport, as well as the higher density of wood, favors the utilization of forest residues over agricultural straw.

A second requirement for the utilization of biomass as feedstocks for the chemical industry is competitive price. One way to compare the relative prices of various organic feedstocks is to determine the feedstock price per unit weight of carbon. Based on 1982 prices, the prices per unit of carbon for various pure organic chemical intermediates, for example, acetic acid, ethanol, methanol, and syngas, are shown in Table 12.2. These prices are fossil fuel–based. Table 12.2 shows both methanol and syngas, which can be obtained from wood gasification, to be the least expensive on a unit carbon basis. Therefore provided that these two organic chemical intermediates can be manufactured competively from biomass feedstocks, they could supply key chemical building blocks for alleviating the current dependence on ethylene and propylene. In the category of crude chemical liquid feedstocks, cane molasses, as shown in Table 12.2, has the lowest unit carbon price relative to *n*-paraffin, rice bran, and dextrose syrup. In the category of heterogeneous feedstocks, coal at 6 cents per kilogram of carbon is the least expensive carbon source, followed by wood residues at 7 cents per kilogram of carbon. This price difference, although small, reflects the higher density and higher carbon content of coal compared to wood residues.

TABLE 12.2 Comparative Value of Carbon Feedstock Sources

Type of Feedstocks	*Carbon Content (Wt. %)*	*Assumed Price (¢/kg)*	*Calculated Price per Unit Carbon (¢/kg)*
A. Pure chemicals			
1. Acetic acid	40	58	145
2. Ethanol	52	60	115
3. Methanol	37.5	24	64
4. H_2/CO syngas	40	16[1]	40
B. Crude chemicals			
1. *n*-Paraffin	85	44	52
2. Molasses	40	15[2]	38
3. White rice bran	40	18[2]	45
4. High test molasses	40	32[2]	80
5. Dextrose syrup	40	30	75
C. Heterogeneous			
1. Petroleum	85	25.0	29
2. Corn stover	28[3]	5.0[4]	18
3. Wood residues	50	3.3	7
4. Coal	75	4.4	6

Adapted from Saito (1977).

[1] Synthesis gas at $3/MSCF, Mole ratio CO to H_2 at 1.

[2] Unit price based on sugar.

[3] Hemicellulose and cellulose at 70% of dry weight.

[4] This does not include hydrolysis and purification cost for sugars.

However, both coal and wood solid residues have at least one fourth of the carbon cost of liquid petroleum feedstocks.

Summarizing, biomass in the form of forest residues is now competitive with either petroleum or coal as a feedstock source for the organic chemical industry from the standpoint of the first two requirements: quantitative feedstock availability and carbon feedstock cost. What remains to be determined is the ability of biomass resources to meet three additional requirements before they can be utilized by the organic chemical industry. These are as follows.

1. Existence of cost-competitive biomass fractionation or refining technologies.
2. Availability of efficient conversion and product recovery technologies.
3. Compatibility of biomass-derived chemical products with the existing infrastructure of the organic chemical industry.

III. BIOMASS FRACTIONATION PROCESSING

This section reviews the existing biomass processing industries and their potential to supply feedstocks to the organic chemical industry. The primary focus will be on methods to fractionate solid biomass feedstocks such as wood and agricultural crops. Biomass residue processing such as acid hydrolysis (Fiechter, 1981), enzyme hydrolysis (Fiechter, 1981), and gasification pyrolysis (Wat et al., 1981) are mentioned but are not discussed in detail, since they are developmental, and thus nonexistent, industries in the United States.

Implicit in the following discussions is the assumption that a biomass industry is not economically viable unless it processes all the fractions in the heterogeneous feedstock to the highest overall product and by-product values. This statement is particularly valid for biomass processing that involves chemical or biochemical conversion steps. This assumption is based on a historical analysis of the development of the biomass-refining industries. Two major examples of biomass processing that have economically survived are the fractionation and refining of forest biomass to pulp fiber and the wet milling of corn to food and animal feed products. Both of these fractionation and refining processes are analogous to the refining of petroleum to liquid fuels and liquid feedstocks for the petrochemical industry. This concept of complete utilization of feedstock is illustrated in Table 12.3 for agricultural and forest residues. As indicated in the table, the fractionation of each type of biomass results in principal products, co-products, and an underutilized residue. Establishing markets and generating revenue from all fractions is one of the most important factors in the development of an economical biomass industry.

Table 12.4 reviews existing biomass processing industries in the United

TABLE 12.3 Organic Chemical Potential from Biomass Fractionation Processes

Feedstock Base	*Forest Resources*		*Agricultural Resources*	
	Softwood and hardwoods		Grain crops	Sunflower, milkweed, latex-bearing crops
Fractionation process	Lignocellulose fractionation		Botanochemical refining	
	Alkali chemical pulping	Acid chemical pulping	Carbohydrate refining	Plant latex or oil solvent extraction
Principal product	Kraft pulp	Sulfite pulp	Sugar, starch, protein feed	Leaf protein[1]
Co-product organic chemical sources	Tall oil, turpenes, rosin	Sugars,[1] acetic, formic acids	Molasses for specialty chemicals via aerobic fermentation	Rubber latex, plant oils for fatty acids
Underutilized residue	Lignin	Lignosulfonates	Straw[1], bagasse	Bast fibers for paper[1]

[1] Areas requiring further technology development.

TABLE 12.4 U.S. Industry Shipment Values for Products from Selected Biomass Refining and Processing Industries

Industries	*Shipment Values (1981 million dollars)*
A. Agricultural	
1. Flour and other grain mill products	4,031
2. Wet corn milling	3,378
3. Cane sugar refining	3,699
4. Beet sugar	1,911
5. Cottonseed oil mills	976
6. Soybean oil mills	10,547
B. Forestry	
1. Market pulp	3,495
2. Paper and board	35,200

Based on U.S. Department of Commerce (1982).

States. This table shows the shipment values of the total products from the respective biomass-processing industries in 1981. This is a selected grouping of industries that are providing or can potentially provide organic feedstocks to the chemical industry. The shipment values indicate both the size of the industry and, indirectly, the amount of biomass products annually produced in the United States. The approximate range of dry weight products for the respective industries in million short tons per year are 53 for all pulps, 20 for soy products, 10 for corn wet mill products, 5.5 for cane and beet sugar products, and 2.5 for cottonseed products.

The above analysis of biomass-refining industries clearly shows that processing of forestry resources to pulp and miscellaneous allied products is the largest existing biomass-processing industry. This is true in terms of both shipment values in the range of 35 billion dollars per year and tonnage quantities of 53 million tons of pulp per year. Second, it identifies the economic importance of soybean oil mills and cottonseed oil mills. At present, these provide 10.5 billion and 0.97 billion dollars of shipment values, respectively, in the United States. The plant oil–processing industries are analogous to the botanochemical processing to be described later.

The pulping industry has been identified as a likely candidate for the biomass-refining concept and as a potential source of organic feedstock for the chemical industry. The current practice in a pulp mill is to convert approximately 50% of the wood (cellulose fraction) to pulp while the remaining 50% of the wood is dissolved in the cooking liquor. The resulting liquid, called spent liquor, contains a mixture of organic acids, five- and six-carbon sugars, methanol, furfural, lignin compounds, and inorganic compounds used in the cooking liquor. This mixture is evaporated to a combustible consistency and burned. The energy obtained from burning the organics is used in the process, and the inorganics are recovered and reused in the cooking liquor. The condensates from evaporation contain low concentrations of organic acids (primarily acetic) and are generally either discharged untreated into the environment or are treated for biochemical oxygen demand (BOD) removal. In light of the current environmental restrictions and the increased value of organic chemicals, the future economics may favor recovery of the organic chemicals and the use of an alternative fuel such as coal. (K. Baierl (pulp and paper consultant), personal communications).

Pulping technology for converting biomass to chemicals and pulp should alleviate several of the intrinsic economic problems cited in the previous section. First, the techniques for tree harvesting, chipping, handling, and storage are well-established, and the equipment is readily available. Another important factor concerns the feedstock itself. Wood does not compete with food sources and is in abundant supply. Also, the spent liquor from a pulping operation is essentially a waste product and is in a physical state that is appropriate for processing rather than existing in a crystalline, unreactive state. The capital expense for handling and hydrolyzing the wood is already provided by the pulping operation.

Fig. 12.3 and 12.4 illustrate two alternatives for lignocellulose processing. Fig. 12.3 is a process scheme adapted from Miller (1982), in which wood is fractionated by steam autohydrolysis to cellulose, lignin, and hemicellulose. The cellulose is converted to its monomer sugar, glucose, by an enzymatic or acid hydrolysis step and fermented to ethanol. The hemicellulose fraction of the wood is hydrolyzed in the process to five- and six-carbon sugars and concentrated as wood molasses, which may be used as a cattle feed. The lignin fraction of the wood is recovered in a relatively pure state and has the potential to be a feedstock for aromatic chemicals.

Fig. 12.3 is an example of a technology in the development phase to convert biomass to chemicals, whereas Fig. 12.4 illustrates how chemicals may be obtained concurrently with the existing sulfite pulping industry. This second alternative produces a pulp from the cellulose fraction of wood and ferments the hemicellulose-derived sugars to ethanol. The lignin is recovered as lignosulfonates, which have a variety of uses as binders and dispersants, or may be burned for its fuel value.

Table 12.5 is a comparison of the revenue obtained from the two lignocellulose process schemes; it highlights the increased revenue expected from chemical recovery combined with pulping. In alternative 2 with sulfite pulping, the polymeric market value of the cellulose is retained as pulp; whereas in alternative 1, the cellulose is converted into ethanol. In the conversion to ethanol, essentially 50% of the original cellulosic feedstock is lost as CO_2. This loss in alternative 1 is somewhat balanced by the assumption in alternative 2 that pure lignin will have twice the value of the lignosulfonates. However, as noted in Table 12.5, the pulp revenue is by far the greatest quantity, with a value that alone exceeds the values of all

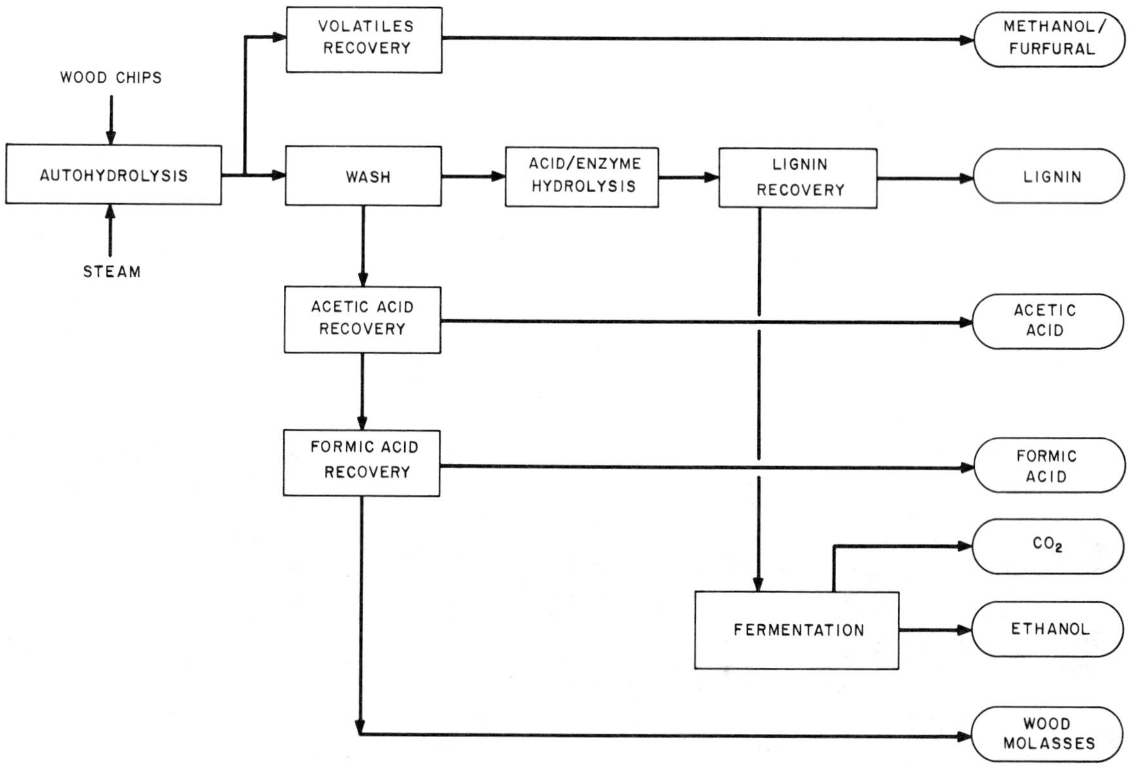

FIGURE 12.3 Alternative 1: Lignocellulose processing without pulping.

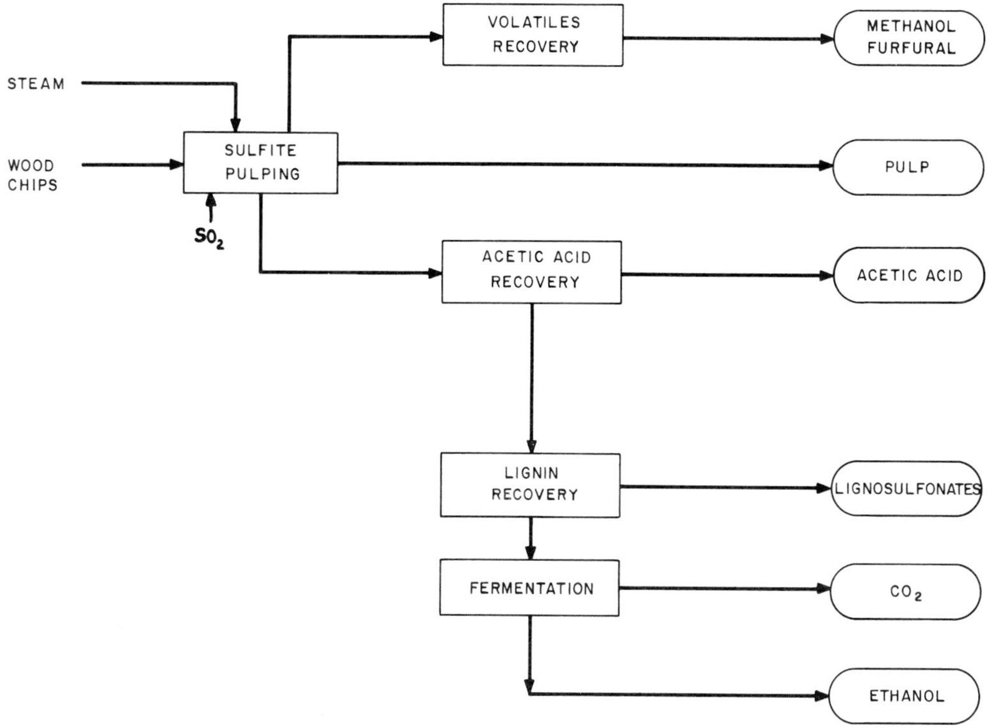

FIGURE 12.4 Alternative 2: Lignocellulose processing with pulping.

TABLE 12.5 Revenues from Lignocellulose Processing With and Without Pulping[1]

	With Sulfite Pulping		*Without Pulping*	
Product (value $/ton)	*Quantity (ton/yr)*	*Revenue (million $/yr)*	*Quantity (ton/yr)*	*Revenue (million $/yr)*
A. Polymers				
1. Pulp (510.00)	166,500	84.915	—	—
2. Lignin (400.00)	—	—	88,578	35.431
3. Lignosulfonates (200.00) (as Na salt)	104,396	20.879	—	—
B. Chemicals				
1. Acetic acid (400.00)	5,761	2.304	5,761	2.304
2. Formic acid (400.00)	100	0.040	100	0.040
3. Ethanol (545.00)	16,750[2]	9.129	38,462[3]	20.962
4. CO_2 (60.00)	16,017	0.961	36,806	2.208
5. Methanol (218.00)	1,360	0.296	1,360	0.296
6. Furfural (1,000.00)	966	0.966	966	0.966
7. Wood molasses (100.00)	—	—	63,167	6.317
C. Fuel ($9.00/MM Btu)[4]	73,936	3.726	195,600	9.858
TOTAL WOOD DERIVED PRODUCTS:	333,000	123.216	333,000	78.382

[1] Basis: 1000 ton/day dry wood feed; 333 days/yr operation; 95% recovery of acetic acid, formic acid, furfural, and lignin.
[2] Ethanol yield: 15 gal/ton wood.
[3] Ethanol yield: 35 gal/ton wood.
[4] Unutilized feedstock containing 50% moisture with fuel value of 2800 Btu/lb.

the products obtained from the process without pulping. The amounts of recoverable organic acids, methanol, and furfural are assumed to be identical in both cases and have a relatively minor impact on the revenue generated.

The quantities of chemicals and revenue listed in Table 12.5 appear to be attractive, but several economic factors must be considered to determine the overall feasibility of the lignocellulose processing schemes. For example, before considering recovery of a chemical for its chemical value, the fuel value of the chemical should be determined. If there is not a significant margin of difference, the added cost of recovery may not be economically justified.

Table 12.6 lists the chemical values and fuel values of a selection of organic chemicals potentially derived from biomass. A reference point of $9 per million Btu's is used for determining whether to burn or to recover the chemical. Chemicals such as methanol, tall oil, and soybean oil, which have a fuel value approaching $9.00 per million Btu's, would be more economically utilized as fuel rather than recovered for their chemical value. Conversely, the burning of an organic chemical such as acetic acid, valued at $43 per million Btu's, as is currently practiced in most pulp mills, results in suboptimal use of this chemical resource. The large differential of $43 and $9 per million Btu's between the chemical and fuel value of acetic acid provides an incentive for the development of a cost-effective chemical recovery technology.

In addition to a comparison of the fuel value to the chemical value, the following other economic factors will ultimately determine the most economic utilization of chemicals derived from the lignocellulose processing schemes:

TABLE 12.6 The Chemical Value Versus the Fuel Value of Selected Organic Chemicals

Organic Chemicals	*Heat of Combustion (Btu/lb)*	*Chemical Value*[1] *($/lb)*	*Calculated Fuel Value*[2] *($/million Btu)*
Methanol	9,550	0.11	11.00
Ethanol	12,765	0.26	20.00
n-Butanol	15,770	0.36	23.00
Acetic acid	6,245	0.27	43.00
Lignin	11,000	0.10 (dried)	9.00
Furfural	10,481	0.66	63.00
Tall oil (crude)	8,500	0.09	11.00
Soybean oil	130,000 Btu/gal	1.80/gal	14.00
Ethyl acetate	10,982	0.42	38.00
No. 6 fuel oil	145,600 Btu/gal	1.30/gal	$9.00/million Btu

[1] From *Chemical Marketing Reporter* (1982).

[2] Calculated values $= \dfrac{\text{chemical value}}{\text{heat of combustion}}$.

A. Marketing considerations
   1. Demand for the chemical
   2. Stability of the market
   3. Potential markets for a surplus of the chemical
B. Economies of scale
C. Availability of proven recovery technology
D. Capital cost aspects
   1. Additional cost of recovery equipment
   2. Tradeoffs between existing technology and new technology
E. Energy considerations
   1. Energy tradeoffs between chemical recovery and burning for boiler fuel value

To illustrate the influence of these factors, consider the recovery of acetic acid from spent sulfite liquor. There are 24 sulfite mills in the United States. The amount of chemicals potentially available from a large sulfite mill is shown in Table 12.5. For acetic acid, it is 5761 tons per year. As cited above, the chemical value is approximately 4.5 times the value obtained by burning acetic acid as boiler fuel, so recovery as a chemical appears to be extremely attractive at first glance. However, before considering the economics of recovering a component such as acetic acid for its chemical value, an assessment of the market condition for the chemical should be conducted. Acetic acid is a high volume chemical in the United States, with an annual consumption of 1222 million kilograms per year. Additional production by a pulp mill could be accommodated either by establishing new uses for acetic acid or by reducing production in existing petrochemical facilities. If one considered recovering a chemical such as furfural from spent liquor, the situation would be different. The furfural market is small—about one tenth of the market of acetic acid. As a result, additional production would reduce the market value and cause a surplus supply on the market.

The second most important consideration is the economy of scale. Is the quantity of chemical available for recovery large enough to justify overhead expense, the additional labor, and capital expenses? Small quantities of chemicals may not generate enough revenue to support an economical venture. In the case of recovery of acetic acid from spent sulfite liquors, the supply is approximately 30–90 kg per ton of pulp (Rydholm, 1965). The size of the pulping operation determines whether the quantity of acetic acid is adequate to justify its recovery from the spent liquor. Although existing sulfite mills are relatively small (mostly below 600 tons pulp per day capacity), the acetic acid production can be increased by a number of process options. One option is to supplement acetic acid production through wood residue gasification to synthesis gas, which can be converted to methanol, a major feedstock source for acetic acid.

Another potential barrier to organic chemical recovery is the availability of process technology. There are several alternative processes for acetic acid recovery, including solvent extraction, distillation (*Chemical & Engineering News,* 1981), and carbon adsorption (Baierl, 1979). However, only the solvent extraction process has been developed to a commercial operating scale in Europe.

Provided that technology is available, the capital and energy considerations need to be addressed next. The additional capital equipment required to recover acetic acid as a chemical must be more than compensated for by the revenue generated. Also, the additional energy required for acetic acid recovery must be evaluated on the basis of maintaining an adequate energy balance in the pulp mill.

In summary, this example is presented to illustrate guidelines that should help in assessing the economic potential of recovering chemicals from spent sulfite liquor. These guidelines are also applicable in determining the economic potential of utilizing any biomass feedstock, since similar tradeoffs and economic factors prevail in each case.

The above analysis shows the existing barriers to organic chemical recovery from sulfite pulp mills. However, as the price of fossil fuel–based organic oxygenates such as acetic acid continues to increase at rates above fuel prices, acid-based pulping processes such as sulfite pulping are expected to provide significant short-term competitive feedstock sources for acetic acid and its chemical derivatives.

Other organic chemicals currently derived from pulp mills include tall oil and resin produced annually in the amounts of 184 and 54 million kilograms, respectively. These two products together with other organic chemicals recovered as by-products of charcoal production and gum extraction from living trees had a market value of 415 million dollars in 1976.

The second potentially significant source of organic chemicals from biomass is via by-product processing of agricultural crops. Fig. 12.5 complements Table 12.3, showing the analogy between lignocellulose fractionation for wood and botanochemical refining for agricultural crops as a basis for obtaining organic chemical feedstocks. Table 12.4 shows the size of the conventional carbohydrate crop–refining and oilseed-processing industries. Both of the agricultural crop–processing industries are smaller than the pulp and paper industries. The botanochemical-refining concept is an extension of the existing agricultural crop–refining concept.

Botanochemical refining of nonedible crops is viewed as a potential source of hydrocarbon from plant latex or plant gutta and of fatty acids from plant oils. An economic analysis has been made by Buchanan et al. (1980) of USDA comparing the feasibility of plant oil and latex production from milkweed with plant gutta production from perennial grass gutta. The results of this analysis, made in 1977, favor plant oil and latex production with a rate of return (ROI) of 27% versus 15% for gutta production. In

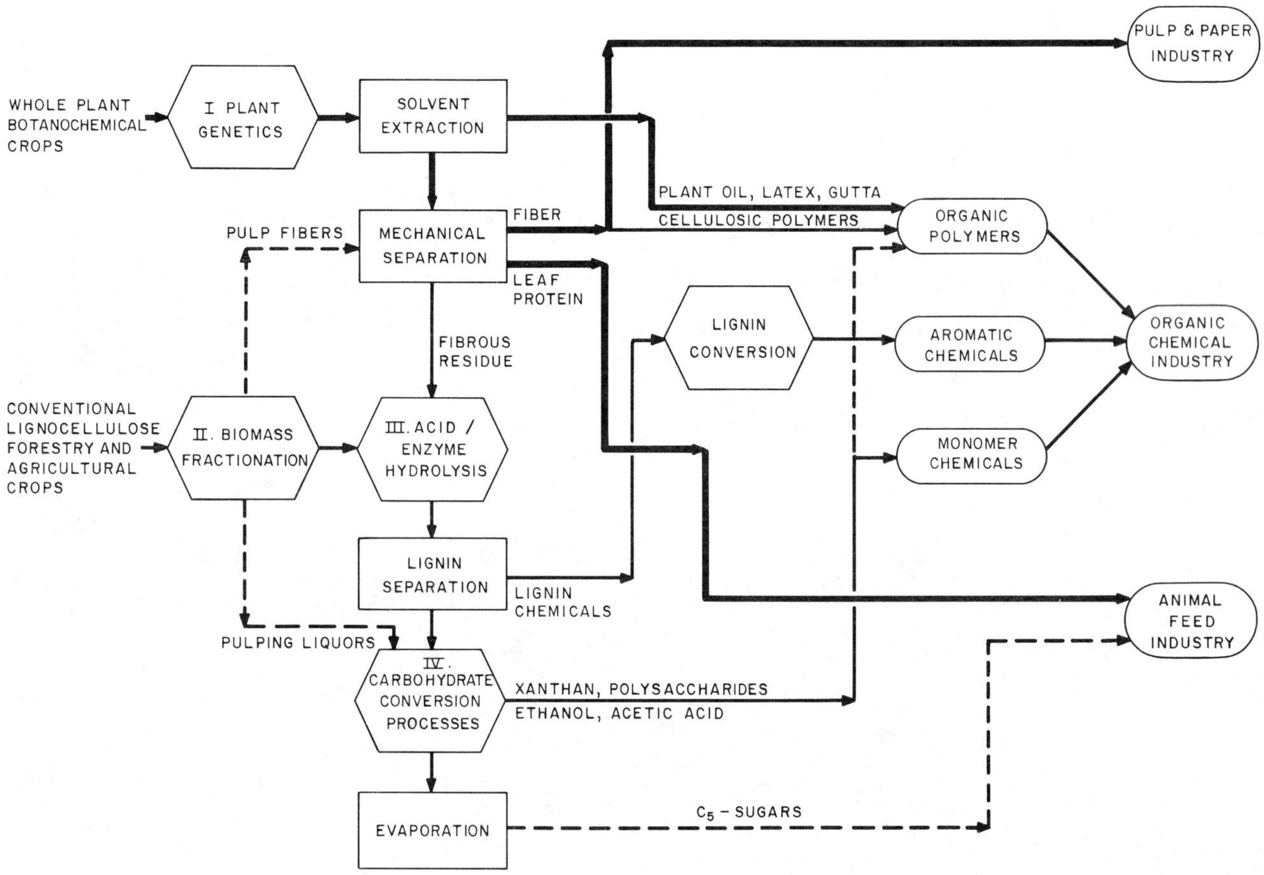

FIGURE 12.5 Lignocellulose/botanochemical refining concept.

both cases, the economics are largely dependent on the ability of botanochemical refining to provide valuable co-products. In the case of plant oil production from milkweed, the percentage distribution of product revenues was estimated to be as follows:

	Percentage of Total Revenues
Plant oil as petroleum value	21
Polyphenol as fuel value	5
Leaf meal as feed protein value	34
Bast fiber as pulp value	18
Residue as fuel value	22

While the processing technology for solvent extraction of plant oils and polyphenols will be similar to existing technology for soybean oil extraction, the major barrier to organic chemical recovery is market demonstration of the feed value of leaf meal extract and pulp value of bast fiber in milkweed. Thus the economic feasibility of utilizing unconventional plant crops for organic chemical production depends on the completion of market development work on co-products to ensure maximum utilization of all fractions of the heterogeneous feedstock.

Gasification pyrolysis of biomass residues to synthesis gas is a potential alternative to the biomass fractionation methods described here. Residue pyrolysis to synthesis gas (carbon monoxide and hydrogen) can lead to oxygenated chemicals such as methanol and its derivatives by utilizing the conversion technologies being developed for coal-derived chemicals. The economic feasibility of residue gasification to synthesis gas is highly dependent on two factors: (1) large-scale residue-processing plants of at least 1000 tons per day and (2) low biomass residue costs of no more than $30 per ton based on 1982 relative costs as compared to natural gas and coal. A detailed economic analysis for this potentially important method of processing biomass residues to chemical feedstock source can be found in Wat et al. (1981).

Enzymatic hydrolysis and acid hydrolysis are also processes for fractionating biomass, although developmental improvements are needed before the technology can be applied economically. Enzymatic hydrolysis requires work to improve the rates of reaction, the recovery of cellulose enzymes, and the yields. In addition, pretreatment steps to make the cellulase and hemicellulose accessible for enzymatic attack result in by-products and soluble lignin present in solution, which inhibit the action of the enzymes. Acid hydrolysis of biomass requires similar development work to improve yields and reduce operating cost by hydrolyzing the biomass in high solids concentration. This requires new mechanical feeding devices capable of introducing wood chips to a reactor with short residence times. Also, reactor conditions need to be optimized to minimize degradation of hemicellulose and cellulose-derived sugars.

IV. BIOMASS PROCESS CONVERSION TECHNOLOGIES

The previous section assessed biomass fractionation methods that are cost-competitive and can thus potentially provide significant fractionated organic feedstocks. These feedstocks can be obtained either as heterogeneous liquids, such as black liquors from pulp mills or molasses from sugar mills, as homogeneous solids, such as sugars or starch, or as homogeneous liquids, such as plant oils. This section reviews the characteristics of the process conversion technologies for transforming these fractionated feedstocks to organic chemicals. The characteristics, particularly energy requirements, that affect their economic competitiveness to petrochemical or coal processing are highlighted.

The process conversion technologies that are important for obtaining organic chemicals are classified according to type of feedstock:

A. Heterogeneous fractionated liquid feedstocks
   1. Anaerobic fermentation of carbohydrate feedstocks
   2. Aerobic fermentation of carbohydrate feedstocks
   3. Zeolite catalysis of plant latex or plant oils
B. Homogeneous fractionated feedstocks
   1. Dehydration
   2. Hydrogenation
   3. Oxidation
   4. Copolymerization or polymer modification
   5. Chemical catalysis
   6. Esterification
   7. Biocatalysis

Both anaerobic and aerobic fermentation are applicable to aqueous carbohydrate fractionated feedstocks. These types of process conversion technologies are cost-effective when carbohydrates such as sugars are present in relatively dilute aqueous solutions ranging from 4 wt. % to 5 wt. %—for example, sugars in pulping liquors and molasses. These processes utilize microbial catalysts that can operate in the presence of inorganic and organic impurities that generally interfere with chemical catalysts.

Anaerobic fermentation, while relatively less energy-intensive than aerobic fermentation, is limited to the production of degradative chemical products of sugars, for example, organic solvents such as ethanol, acetone, and *n*-butanol. Stoichiometric weight yields of these organic solvents from sugars are relatively low, in the range of 30–50%, and thus make these processes sugar feedstock cost–intensive. Therefore selective products such as oxygenated compounds are most suitably produced via anaerobic fermentation for reasons discussed later in the chapter.

Aerobic fermentation technology can transform sugars to a wider range of organic chemicals, including amino acids, antibiotics, and enzymes. However, it is energy-intensive, with process steam requirements of from 70–80 kilograms compared with 5–10 kilograms of steam per kilogram of product for anaerobic fermentation products. Thus conversion technology is limited to higher-valued specialty organic chemicals. More details on the aerobic fermentation characteristics are provided in the section on specialty chemicals.

The zeolite catalysis technology is relatively new and can potentially play a dominant role in the production of olefins and aromatics from plant oils and latex. Improved catalytic selectivity is required to reduce separation processing costs. Detailed discussion on the status of zeolite catalyst use is provided in the section on complements to nonoxygenated petrochemical polymers.

For homogeneous fractionated feedstocks, the range of process conversion technologies available for chemical transformation is much wider and involves more conventional chemical conversion technologies. Only the last four process conversion technologies listed above require further examination.

Polymer modification technology will play a dominant role in providing modified starches with improved functional properties. For instance, the graft polymerization of starch to acrylonitrile in the presence of ferric ammonium nitrate in nitric acid catalyst will result in the modified starch H-Span, which has water-absorbing capacities of up to 100 times its weight (see Table 12.1). The cross-linking and coprecipitation of starch-xanthate-material rubber latex will result in powered rubber (Russell, 1973). Powdered rubber with functional properties similar to powdered plastics can significantly expand the marketability of both natural rubber and starch-xanthate.

A major chemical catalyst relevant to biomass conversion to chemicals is the Cl chemical catalyst. The conversion technology has been confined to applications for enhancing the potential of coal gasification to chemicals via carbon monoxide and hydrogen as basic building blocks. These same building blocks can likewise be obtained from wood pyrolysis. Key catalytic processes utilizing carbon monoxide and hydrogen for organic chemicals synthesis will be discussed in the section on oxygenated petrochemical substitutes. The discussion will include methods of utilizing esterification technologies.

Finally, a relatively new area of biocatalysis that includes the use of immobilized enzymes or immobilized microbial cells as catalysts shows potential for effective utilization in the production of speciality chemicals. In particular, the higher specificity and reduction in costs of biocatalysts through immobilization technologies will enable production of chemicals, such as amino acids, from biomass to become more cost-effective.

V. SIX ENTRY POINTS FOR BIOMASS FEEDSTOCKS

A general understanding of the chemical components of biomass feedstocks is essential to recognition of the potential entry points into the organic chemical industry. Earlier sections identified lignocellulose feedstocks from both agricultural and forest resources as the most significant potential contributor of feedstock. The carbohydrate components of lignocellulose—cellulose and hemicellulose—are chemically the most compatible components of lignocellulose for producing three groups of organic chemicals. These are as follows: nonoxygenated petrochemicals such as ethylene, propylene, and butadiene; oxygenated petrochemicals such as organic acids, esters, and solvents; and specialty chemicals such as cellulosic polymer and products derived from aerobic fermentation. Fig. 12.6 illustrates how cellulose, hemicellulose, and starch are logical starting feedstocks for the first three segments of the organic chemical industry. The process technologies required to achieve these conversions are varied, ranging from fermentation to catalytic dehydration to pyrolysis.

The third and fourth components of lignocellulose feedstocks are lignins and oleoresins. These polymeric components are most suitable for the conversion to aromatic chemicals such as phenols and benzene. This constitutes the fourth potential entry point.

The last two potential entry points largely depend on the development of botanochemical crops, some of which already exist in commercial plantations. The plant oils, for example, palm oil, coconut oil, and soybean oil, are major sources of fatty acids—the fifth potential entry point. The plant latex crops such as natural rubber trees and guayule constitute the sixth entry point to complement nonoxygenated petrochemical polymers.

An overview of the current market volume and values of various target organic chemicals is presented in Table 12.7. These chemicals have been classified according to the six potential entry points for biomass feedstocks. Each classification will be reviewed in subsequent sections of this chapter.

VI. SUBSTITUTION OF NONOXYGENATED PETROCHEMICALS

Three of the most important examples of nonoxygenates are ethylene, propylene, and butadiene. Considerable effort has gone into evaluating the feasibility of converting biomass into these compounds, especially the conversion of biomass to ethylene. This is not surprising, since these compounds form the basic building blocks for the organic chemical industry. Ethylene is an intermediate feedstock for major chemical products such as styrene, acetaldehyde, polyvinyl chloride, polyethylene, and several others. Propylene is utilized primarily for polypropylene, acrylonitrile, and isopropylalcohol. Butadiene is used for polymers and adiponitrile (Shreve and Brink, 1977). As Section A of Table 12.7 indicates, U.S. production of each of these compounds is over one billion kilograms. Ethylene production dominates at 12.7 billion kilograms produced annually. All three compounds are currently produced from petrochemical feedstocks such as refinery cracked gas (naph-

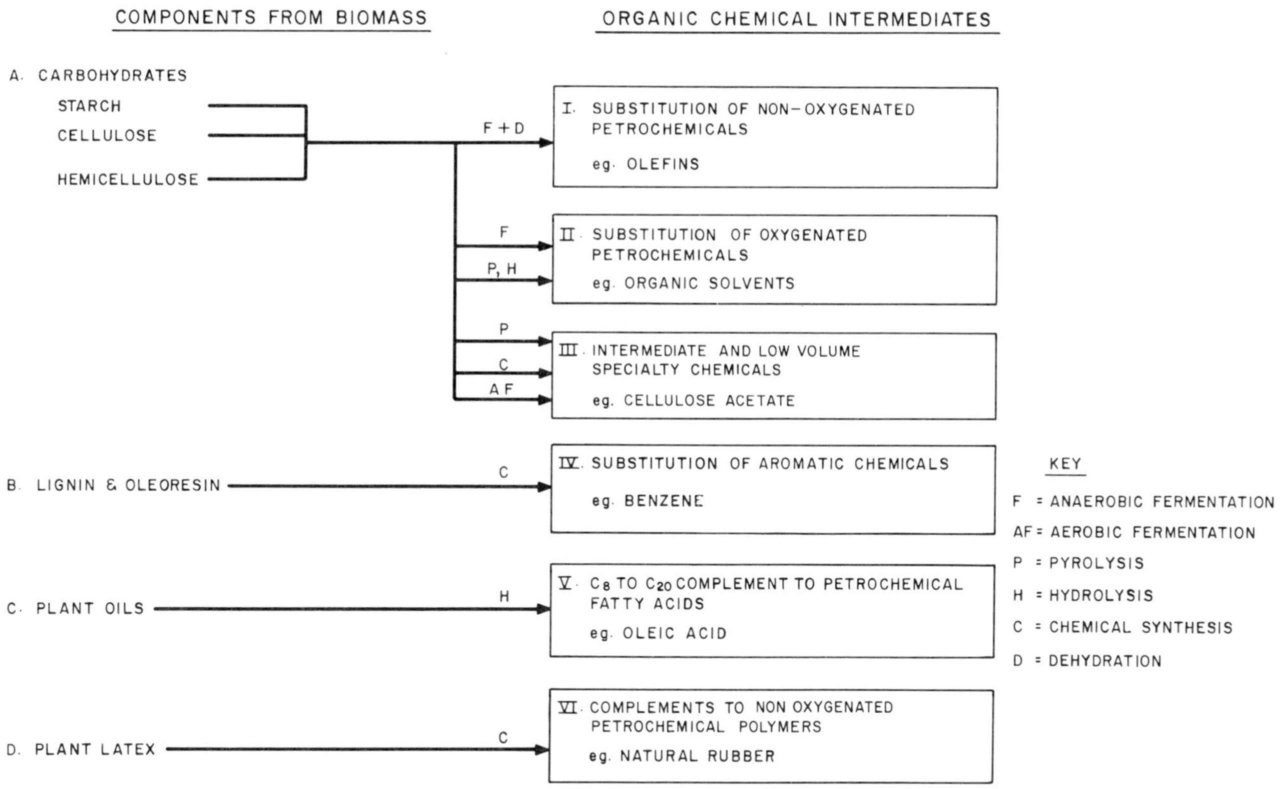

FIGURE 12.6 Alternative pathways for biomass-derived chemicals.

TABLE 12.7 Production and Price Values of Organic Chemicals That Are Potentially Derivable from Biomass

Chemicals	*U.S. Production (MM kg/yr)*	*Value ($/kg)*
A. Nonoxygenated petrochemicals		
Ethylene	12,727	0.56
Propylene	6,400	0.48
Butadiene	1,447	0.88
B. Oxygenated petrochemicals		
Methanol	3,730	0.24
Ethanol (synthetic only)	554	0.67
Isopropanol	766	0.69
n-Butanol	351	0.79
Glycerol	155	1.74
Formic acid	34	0.58
Acetic acid	1,222	0.58
Propionic acid	45	0.66
Ethyl acetate	128	0.92
Butyl acetate	127	1.03
Ethylene oxide	2,347	0.99
Propylene oxide	859	0.89
Acrylic acid	148	1.10
Adipic acid	737	1.25
Ethylene glycol	1,795	0.73
Propylene glycol	265	0.96
Acetone	989	0.70
Acetaldehyde	410	0.71
Methyl ethyl ketone	268	0.92
Furfural	145	1.45
Sorbitol (70% aqueous)	123	1.10
C. Specialty chemicals (cellulose polymers)		
Cellulose acetate	145	1.94
Rayon	300	1.70
Cellophane	41	2.90–3.50
Carboxymethyl cellulose	35	2.42
D. Aromatic chemicals		
Phenol	1,220	0.72
Benzene	4,730	0.47
E. Fatty acid feedstocks (1974 U.S. consumption)		
Soybean oil	3054	0.42
Crude tall oil	564	0.20
Cottonseed oil	575	0.44

TABLE 12.7 (Continued)

Chemicals	*U.S. Production (MM kg/yr)*	*Value ($/kg)*
Coconut oil	224	0.50
Linseed oil	103	0.62
Corn oil	227	0.55
Peanut oil	58	0.66
Castor oil	57	0.95
Tallow, inedible, and grease	1,156	0.37
F. Nonoxygenated polymers		
Natural rubber (1972 U.S. Consumption)	640	0.97

Data obtained from *Chemical Marketing Reporter* (1982), Technical Insights, Inc. (1980), Goheen and Hoyt (1981), *Predicast Base Book* (1981), Shreve and Brink (1977), and Boyles (1982a).

tha) or from ethane and propane. The latter is still the most economical route, although with continued cost increases of petrochemicals and reduced liquid natural gas availability, the economics are expected to shift in favor of alternative feedstocks. Biomass could readily supply this alternative feedstock, and the technology is available to produce the nonoxygenated intermediates. However, due to its oxygenated form and competition from other feedstocks, biomass is not expected to provide a short-term alternative for the United States. The reasons for this are low yields and feedstock costs.

The three nonoxygenates, ethylene, propylene, and butadiene, can be derived from biomass in two steps. The first step consists of anaerobic fermentation of carbohydrate (derived from biomass) to either ethanol, isopropanol, or 2,3 butanediol. The second step involves the dehydration of the alcohols to the respective products, ethylene, propylene, and butadiene. We will limit our choice of product to ethylene, since the fermentation of carbohydrates to ethanol is practiced commercially, and yields are relatively high (ranging from 88% to 95% of the theoretical conversion).

The economics of producing ethylene from carbohydrates are dominated by the cost and nature of the feedstock (Brant, 1979). Glucose, the major carbohydrate used in the fermentation, has an atomic carbon-to-oxygen ratio of 1:1 (see Table 12.8) or an oxygen content of 53 wt. % oxygen. Therefore if the process were 100% efficient, converting all the carbon and hydrogen to ethylene, the maximum yield from carbohydrate would be 47%, owing to the lost oxygen. The conversion steps involved reduce this figure considerably. The anaerobic fermentation step has a maximum sugar-to-ethanol weight yield of 51%, and in practice, yields of 0.45–0.48 lb of ethanol from one pound of carbohydrate are considered high. The balance of the feedstock is converted primarily to CO_2 as dictated by the stoichiometry of the reaction. Since the theoretical weight yield of ethanol to ethylene is

TABLE 12.8 Hydrogen-to-Carbon and Oxygen-to-Carbon Ratios of Organic Feedstocks and Products

Type	*H/C Atomic Ratio*	*O/C Atomic Ratio*
A. Feedstocks		
1. Crude oil	1.6	—
2. Tar sands		
a. Bitumen	1.5	0.01
b. Synthetic crude	1.9	0.00
3. Coal	0.8	0.03–0.22
4. Methane	4.0	0.00
5. Glucose	2.0	1.00
6. Xylose	2.0	1.00
7. Lignin	1.1	0.37
B. Chemical Products		
1. Methanol	4.0	1.0
2. Ethanol	3.0	0.5
3. Acetic acid	2.0	1.0
4. Ethylene glycol	3.0	1.0

61%, the maximum yield of ethylene from sugar is 29%. In addition to the problem of low weight yields, the least expensive carbohydrate feedstock, molasses, on a unit carbon basis (see Table 12.2) costs 1.3 times more than petroleum feedstock. It is apparent that the cost of carbohydrate feedstocks relative to petroleum feedstocks will have to be significantly reduced before ethylene can be competitively produced from ethanol. A similar analysis holds true for the production of propylene, butadiene, and other nonoxygenated compounds.

Using a linear numerical model of the chemical industry, Palsson et al. (1981) concluded that the fermentation products ethanol, isopropanol, *n*-butanol, and 2,3 butanediol would have to be available at 20–40% of their current chemical values. In this study, a computer was programmed to select the most economical technology given various economic factors and the interaction between products. Ethanol as a feedstock for ethylene was predicted to have no significant impact on the market until its price is reduced to one fifth of its current price. This is mainly due to items discussed above and partly due to the current ethylene production system, which provides valuable propylene and butadiene by-product credits (Palsson et al., 1981). A more likely scenario is the production of ethylene from alternative petrochemical sources. As reserves of crude oil and gas are depleted, tar sands and shale oil are expected to provide feedstocks for petrochemical refining. The synthetic oil that can be extracted from these two feedstocks is similar in composition to heavy crude oil and would be suitable for use in conventional refineries. Concurrent with this development, coal gasification technology and Fisher-Tropsch processes could provide alternative routes to ethylene.

In summary, it is doubtful that carbohydrate biomass feedstocks will be economically utilized for the production of nonoxygenated petrochemicals. In addition to the unfavorable feedstock economics, it is difficult to collect enough biomass in one central location to take advantage of economy of scale. This is a necessity, since the hydrocarbon-processing facilities produce on the order of one hundred million pounds per year.

Thus unless there is a critical shortage of petrochemical feedstocks resulting from a cutoff in supply or from war, it is unlikely that carbohydrate biomass feedstocks will be utilized for the production of nonoxygenated petrochemicals. However, as discussed in Section XI on complements to nonoxygenated petrochemical polymers, downstream hydrocarbon polymers can be produced directly from biomass in the form of plant latex. In addition, downstream oxygenated petrochemicals currently based on ethylene can be produced from biomass by the process technologies to be described in the next section.

VII. SUBSTITUTIONS OF OXYGENATED PETROCHEMICALS

The previous section illustrated the high feedstock cost sensitivity of carbohydrate conversion to nonoxygenated chemicals. This section focuses on two groups of oxygenated chemicals: solvents and acids and their corresponding derivatives. These oxygenates typically have atomic oxygen-to-carbon ratios ranging from 0.5 to 1.0 such as those listed in Table 12.8 for methanol, ethanol, and acetic acid. These oxygen-to-carbon ratios closely match those of the biomass feedstocks—lignin, glucose, and xylose, which have ratios ranging from 0.37 to 1.0. Therefore the high cost sensitivity of nonoxygenated organics to carbohydrate costs due to a mismatch in chemical composition will be less of a detrimental factor limiting the economic feasibility of producing oxygenates from biomass.

The relative importance of oxygenated chemicals can be seen from examining the annual U.S. production and market values of those shown in Section B of Table 12.7. For the purpose of determining the feasibility of producing oxygenates from biomass, only ethanol, methanol, and acetic acid, out of the 21 oxygenates listed there, will be examined in detail here. These three oxygenates have been selected for the following reasons:

1. They are key intermediates for the production of other oxygenated derivatives, for example, organic esters from alcohols and acids and ethylene glycol from methanol.
2. They are end products in themselves for chemicals, solvents, or liquid fuels applications.
3. These chemicals will benefit the most from the successful development of new process technologies, for example, genetic engineering for ethanol, Cl catalysis for methanol, and esterification for acetic acid.

4. They are already important chemicals with annual market volumes above 0.5 billion kilograms and have relatively high market values (except for methanol) above 50 cents per kilogram.

In summary, this group of oxygenates has favorable characteristics that strengthen the feasibility of their being produced economically from biomass feedstocks.

Ethanol production based on fermentation processes using feedstocks from various by-product carbohydrate processing (for example, corn wet milling, sulfite pulp milling, cheese whey) as well as co-product production from corn dry milling plants has already begun to dominate the chemical market for industrial ethanol. This shift from petroleum to biomass feedstock is primarily due to the recent escalation of ethylene prices. At 57 cents per kilogram, ethylene, the feedstock cost component for ethylene-derived ethanol, has reached the equivalent carbohydrate cost for corn (at $2.60 per bushel)-derived ethanol. Second, with government subsidies for grain-based ethanol plants, the net capital cost has been favorably reduced. These two factors have enabled grain-based ethanol to be produced cost-competitively with ethylene-derived ethanol.

Oxygenated derivatives of ethanol such as acetaldehyde and ethyl acetate (see Fig. 12.7) have the potential for cost-effective production, since they involve processes in which high weight yields of derivatives from ethanol are obtainable.

The two competing commercial processes for acetaldehyde are ethylene oxidation and ethanol oxidation. For ethylene oxidation, 0.67 kg of ethylene per kilogram of product is required versus 1.1 kg of ethanol per kilogram of product for ethanol oxidation. Although this inherent conversion yield factor favors the conventional ethylene-based process, this process has a higher-pressure (100 kpa) reaction condition. A comparative analysis of these two processes (Brant, 1979) has concluded that when ethylene-based acetaldehyde production cost reaches 73 cents per kilogram, the ethanol-based acetaldehyde can be cost-competitive if ethanol is obtainable at or below a cost of 57 cents per kilogram.

Acetic acid from biomass was discussed in the section on biomass fractionation. The potential recovery of acetic acid present in the form of sodium acetate in black liquors of kraft pulp mills is in the range of 2 billion kilograms per year. An obstacle to acetic acid recovery from black liquors is the development of a cost-effective solvent extraction process. Secondary potential sources of acetic acid from biomass include the direct fermentation of sugars to acetic acid by microorganisms (for example *Cl. Thermoaceticum*), recovery from evaporator condensates, and methanol carbonylation via synthesis gas from wood gasification. The first two sources involve acetic acid recovery from dilute aqueous solution in concentrations of 1–5 wt. %. The development of non-energy-intensive acetic acid recovery from dilute streams is essential for competitive economics. One potential process is

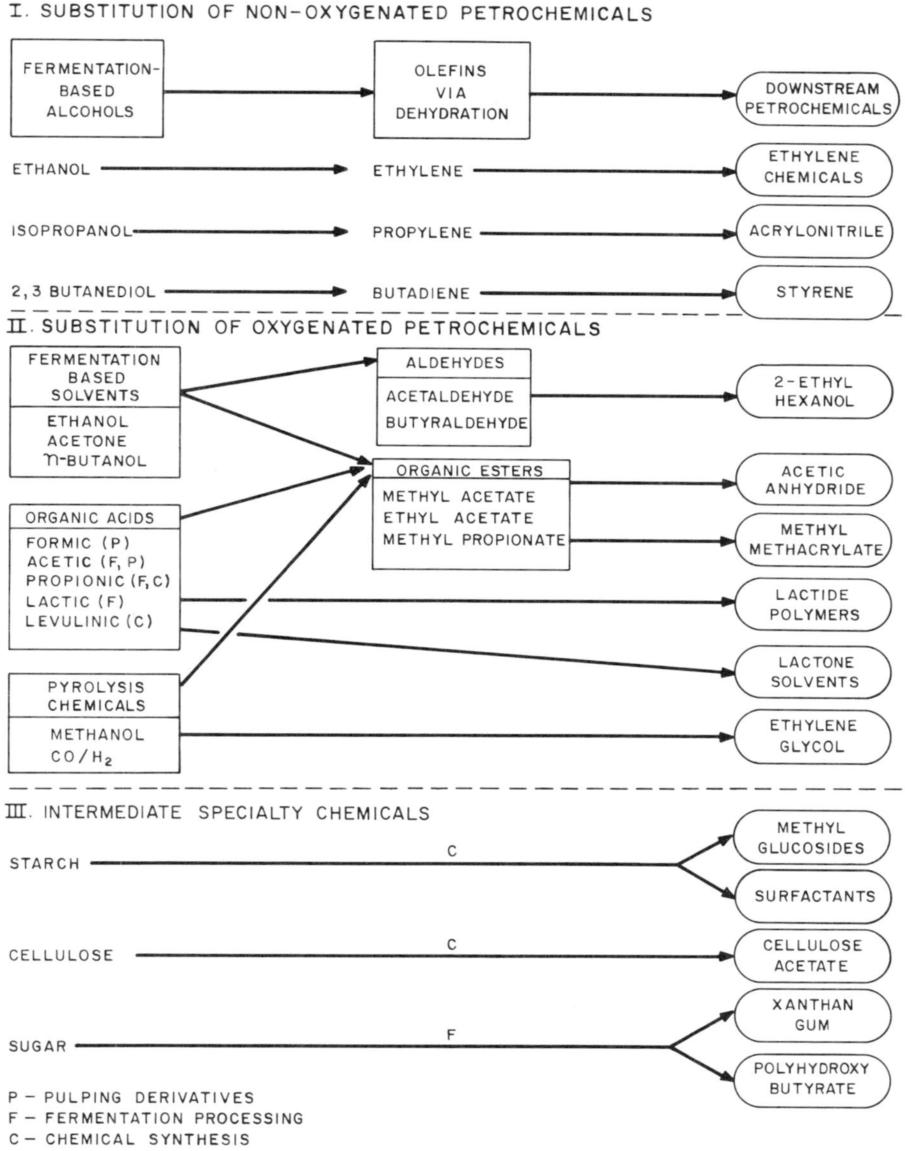

FIGURE 12.7 Carbohydrate utilization pathways.

the recovery of acetic acid in the form of organic esters, for example, methyl acetate or ethyl acetate by adsorption–regeneration (Baierl, 1977) using activated carbon. The main advantages of recovery in the form of methyl acetate instead of acetic acid are the significantly lower heat of vaporization of the ester compared to the acid and the higher market value of methyl acetate. Since 35% of current acetic acid end-use is either as acetic anhydride or as acetic esters, the recovery of acetic acid in the form of organic esters matches well the market applications of this chemical.

Biomass gasification to synthesis gas (carbon monoxide and hydrogen) is a potential source of methanol. Both gasification and subsequent gas reforming to methanol are commercially proven processes. The production of methanol from forest or agricultural residues can be economically competitive with existing natural gas–based processes if large-scale methanol plants (at least 50 million gallons per year) are considered and low-cost biomass residues (below $30 per ton) can be located. Before commercial feasibility, biomass-based methanol plants require further escalation in natural gas prices and improvements in residue collection logistics and harvesting hardware systems. Biomass-based methanol is favored by innovative catalytic processes for methanol derivatives (Sherwin, 1981). Acetic anhydride and vinyl acetate are two methanol derivatives obtained through the catalytic reaction of methyl acetate and synthesis gas. Assuming that an economically competitive recovery process for obtaining biomass-derived acetic acid as methyl acetate can be developed, both acetic anhydride and vinyl acetate can be major oxygenates obtainable from biomass residues.

VIII. SPECIALTY CHEMICALS

The third pathway for biomass-derived chemicals, illustrated in Fig. 12.6, is the conversion of carbohydrates to specialty chemicals. Included in this category are products derived by aerobic fermentation such as enzymes, amino acids, nucleotides, organic acids, and cellulosic polymers derived from dissolving pulp. Numerous other specialty chemicals exist that may be derived from biomass; but for the purpose of identifying economic characteristics, this discussion is limited to the products mentioned.

The future for biomass-derived specialty chemicals is expected to depend heavily upon aerobic fermentation processes. It is therefore important to review the economic factors that determine the feasibility of such processes. The following five major characteristics govern the economics of aerobic fermentation processes:

1. Unique functional properties of the fermentation product
2. Weight conversion of carbon feedstocks to products
3. Capital costs for grass roots fermentation and recovery facility
4. Energy requirements per unit of product

5. Fermentation operating cost, including raw materials, labor cost, and utilities

The high capital and production costs of fermentation processes make it essential that the fermented products have values of at least $1.00/lb. Table 12.9, listing various aerobic fermentation products and other specialty chemicals, illustrates this point. In order to command such high prices, the chemicals must have unique functionality properties that are difficult to reproduce by petrochemical processes.

Perhaps the most obvious and often overriding factor in evaluating the

TABLE 12.9 Market Information for Low-Volume Specialty Chemicals

	Annual Market Volume (1000 lb)	*Cost ($/lb)*	*Value (millions of dollars)*
Amino acids			
Glutamate	600,000	1.80	1,080.0
Methionine[1]	210,000	1.40	294.0
Lysine	129,000	2.10	258.0
Threonine	300	35.18	16.2
Tryptophan (animal feed)	225	43.18	9.7
Nucleotides			
5 IMP	4,000	12.00	48.0
5 GMP	2,000	12.00	24.0
Enzymes			
Urokinase	—	—	89.5
Rennin[1]	24	696.00	40.0
Glucose Isomerase	100	400.00	40.0
Amyl Glucosidase	600	20.00	12.0
Alfa Amylase	600	19.33	11.6
Papain[1]	200	59.00	11.8
Vitamins			
Vitamin C	90,000	4.50	405.0
Vitamin B-12	22	6,991.60	153.8
Vitamin E[1]	3,641	29.00	105.6
Pesticides			
Microbial	N/A	N/A	25.0
Aromatic	N/A	N/A	75.0
Biopolymers			
Xanthan gum	30,800	8.70	268.0
Dextran[2]	132	15.38–43.44	2.0–5.7

Data from Office of Technology Assessment (1981).
[1] Products derived chemically.
[2] From Westmoreland (1981).

economics of a fermentation process is the weight conversion of the carbon feedstock to the product. This criterion is commonly referred to as the yield, and it will determine the quantity of raw material required. Once the yield and value of the feedstock and product are established, the first economic evaluation of the process can be made. If the revenue of products and by-products does not exceed the cost of the feedstock by at least 50–100%, continuing further is likely to be a wasted effort.

In the case of aerobic fermentation for the production of specialty chemicals, the yields may vary from 10% to over 95% on the basis of weight of sugar converted to product. In general, for antibiotics, low carbon yields of 10–20% are common from sugar feedstocks, but the low yield is counterbalanced by the higher value of the product. Commodity compounds such as citric acid, glutamic acid, and gluconic acid are of less value but are produced in yields ranging from 60% to 90%. This weight conversion of sugar is critical. In a large plant, a 10% difference may result in an annual change in revenue of over 1 million dollars. For example, consider the production of xanthan gum at 10,000 tons per year. If produced at a yield of 50%, this would require 10,000/0.5 = 20,000 tons per year of sugar. If the compound could be produced at a 60% yield, 10,000/0.6 = 16,666 tons per year of sugar would be required. This difference at $0.25/lb of sugar results in a net savings of over $1.6 million per year. Although oversimplified, this calculation emphasizes the importance of development efforts to increase yields in aerobic fermentation processes.

In addition to feedstock cost, the fermentation process requires nutrients (nitrogen, phosphorus, trace elements) and additives for pH and foam control. Generally, the total cost of raw materials consists of 40–60% feedstock, 10–40% nutrients, and 10–30% additive chemicals. As an example, the cost breakdown of raw materials for xanthan gum production is approximately 45% carbohydrate, 40% nutrient, and 15% additives.

The second important criterion for an economic assessment is the capital cost requirements for a fermentation and recovery facility. This is reported by Reisman and Bartholomew to range from $20.00 to $50.00 per liter of installed fermenter capacity (Peppler and Perlman, 1979). The capital is roughly independent of the product produced. The low range of $20.00 applies to large-scale plants of 5000 ton/year capacity or greater. For specialty chemicals production, most aerobic fermentations are operated batchwise to minimize the risk of contamination and mutation of the desired strain of organism. As a result, the scale-up of capital costs usually requires an exponent of 0.75 compared to a scale-up factor of 0.6 in continuous process plants (Peppler and Perlman, 1979).

As a third consideration, energy requirements for an aerobic fermentation process and the recovery of products are relatively high per unit of product produced. This is apparent from the table below, which lists the steam requirements for some common fermentation and synthetic petroleum-derived products.

Aerobic Fermentation Products (Tong, 1979)	*Process Steam Required (lb Steam/lb Product)*
Xanthan gum	88
Penicillin G	73
Anaerobic Fermentation Products	
Ethanol	5
Acetic acid (Tong, 1979)	9
Petroleum-Derived Solvents	
Ethylene	10
Synthetic ethanol (Chilton, 1960)	3

The high steam consumption in aerobic fermentation processes is due to process requirements for steam sterilization, aeration, and mechanical refrigeration, and for energy needed in recovery operations such as solvent distillation, centrifugation, evaporation, and drying. Energy requirements are strongly dependent on the concentration of the product obtained in the broth, the rate at which the product is formed, and the desired physical properties of the products. Often, the energy requirements may be reduced by integrating the fermentation process with the recovery process. For example cells can be retained in the reactor by immobilization, thereby eliminating the need for recovering them from the fermented broth. Another example would be fermenting at higher temperatures to reduce cooling load and energy for preheating prior to distillation and evaporation steps. Genetic engineering techniques also promise to reduce energy requirements by allowing the cells to tolerate higher concentrations of end products and higher fermenting temperatures and by increasing the rates of production.

The final economic consideration is the operating costs, including labor, utilities, and raw materials. With the exception of labor, these costs have been discussed previously. However, taken collectively, these three items have important implications. In general, the operating costs are primarily a function of the yield and the fermentation cycle time for fermentation. As noted by Reisman and Bartholomew (Peppler and Perlman, 1979), the cost of labor, utilities, and materials would not vary greatly if put on a per unit volume, per unit fermentation cycle time basis. This assumes that the feedstocks in each case are of equivalent value. A significant difference in the overall production cost is therefore likely to result from recovery and purification of the product after fermentation. The cost of recovery is a strong function of the final concentration of the product in the fermenter and the desired physical properties of the product, that is, moisture content, purity, crystallinity, and so on.

An example of two products that require similar processing steps for recovery are lysine and glutamic acid. Any differences in production cost are therefore expected to be a direct function of the differences in productivity of the two compounds. This is reflected in the bulk selling price of the two compounds. Glutamic acid is on the market at $1.80 per pound, while lysine sells for approximately $5.50 per pound (*Chemical Marketing Reporter*, 1982). As expected, glutamic acid has a productivity of about 1.5 g/l hr, which is 3 times the productivity of lysine at 0.5 g/l hr.

In summary, it is important to analyze all four economic factors discussed above for a complete economic picture for specialty chemicals produced by aerobic fermentation. However, since all four factors are interrelated, any single factor will often provide the incentive to discount the process entirely or to investigate further the economic feasibility.

Another series of specialty compounds that may be produced from biomass are cellulosic polymers. These include cellulose acetate, carboxymethyl cellulose (CMC), rayon, and cellophane. Market information on these compounds can be found in Section C of Table 12.7. So far, the discussion has focused on the use of carbohydrates derived from starch or cellulose. However, rather than being broken down, the cellulose may be utilized in its polymeric form for the production of the above-mentioned chemicals. For instance, cellulose acetate is made from cellulose and acetic acid and may be formed into a fiber or used as a thermoplastic. CMC has many uses in detergents, textiles, and foods. Cellophane is used as a transparent plastic-like film. Rayon has desirable fibrous properties. All four compounds are produced from dissolving pulp that has undergone extensive bleaching treatment for the removal of essentially all impurities, leaving behind exclusively cellulose. The market for these products has been declining owing to competition with petroleum-based polymers. In addition, making dissolving pulp is a high-cost and energy-intensive process. For example, steam requirements to manufacture the cellulosic polymer, rayon are approximately 70 lb of steam per pound of product (Chilton, 1960), while polyethylene can be produced using only 12 lb of steam per pound of product. The situation may be different, however, in developing countries where petroleum sources are limited and trees and other forms of biomass are plentiful. Also, as petroleum products continue to escalate in price, the cellulose polymers are expected to regain their popularity.

IX. SUBSTITUTION OF AROMATIC CHEMICALS

Earlier in this chapter, lignin biomass was identified as a potential feedstock for the production of aromatic compounds. All plants consist essentially of 15–30% lignin. Thus it is apparent that lignin sources are plentiful. However, the lignin is tightly bound within the lignocellulose structure of plants, and considerable energy and processing are required to liberate it. This energy is already expended in the pulping process. For every ton of pulp produced, approximately 0.5 ton of lignin is dissolved into the cooking solvent. In 1981 alone, approximately 43 million tons of pulp were produced, yielding 24 million tons of lignin by-product (Allan, 1982).

Very little of the lignin produced is commercially used as chemical feedstock. Less than 450,000 tons were sold in various forms for uses in binders, dispersants, drilling muds, and other specialty applications (Goheen and Hoyt, 1981). Most of the lignin generated in pulp mills is concentrated and burned as boiler fuel, which provides upwards of 50% of the mill's steam requirements. As mentioned earlier, in the section on biomass refining and lignocellulose processing, lignin could be chemically processed to aromatic chemicals as an alternative to burning.

The technical feasibility of this option has been evaluated at Hydrocarbon Research, Inc. (HRI) and Crown Zellerbach Corporation (C.Z.). A process developed at HRI, called the Lignol Process, involves a two-step reaction. The first step is lignin hydrocracking; the second is hydrodealkylation to the higher phenols. Overall yields of products are 20.2 wt. % phenol, 14.4% benzene, 13.1% fuel oil, and 29.1% fuel gas. Hydrocracking uses an iron-on-alumina catalyst under 69 atm of hydrogen and 440°C. This step produces a mixture of monophenols in a 37.5 wt. % yield consisting of phenol, cresols, xylenols, ethylphenols, and propylphenols. The hydrodealkylation step converts the cresols and alkyl aromatics to phenol and benzene. The hydrogen and fuel requirements may be supplied internally from the fuel oil, which reduces the net fuel oil production from 13.1% to 10.9% (Huibers and Jones, 1980).

The lignol process as described above was developed in the mid-1960's; however, HRI has recently updated a cost analysis of the process. For a facility capable of processing 160,000 tons per year of lignin, the capital investment in 1980 dollars was estimated at $44.8 million. HRI assumed phenol and benzene to be valued at 79.2¢/kg and 49.7¢/kg, respectively. Providing lignin could be purchased for 12.1¢/kg, a 25% return on investment was calculated (Parkhurst et al., 1980).

Technically and economically, given the assumption on product values and feedstock cost, the process appears to be feasible. There are, however, certain obstacles in a practical sense that could render this concept infeasible. First of all, it would require the cooperation of a kraft mill to supply the alkali lignin feedstock. This means that the kraft mill, which is highly integrated and dependent upon the lignin for boiler fuel, would have to be modified to accommodate the lignin recovery equipment and that an additional boiler fuel supply from coal or other sources would have to be provided to make up for the lost lignin. It is also questionable whether kraft lignin would be purchased for less than the assumed 7¢/lb because of the additional cost for lignin recovery. Two precipitation stages would be required with the addition of flocculating chemicals. Furthermore, lignin would have to be dried to a low moisture content suitable for the lignol process. In addition, enough lignin would have to be collected in one area to take advantage of the economy of scale.

Overall, the process deserves further consideration, provided that a kraft mill is willing to alter its balance of energy supply and provide lignin below 10¢/lb. This could be possible with certain improvements in the process and in certain circumstances where lignin removal can increase the capacity of the mill.

One such improvement would be to alter the catalyst selectivity to favor phenol, since phenol is valued at 13.5¢/lb higher than benzene. As listed in Section D of Table 12.7, the U.S. production of phenol is approximately 1.34 million tons per year. The lignol process could provide this by utilizing 6.63 million tons of lignin at the current yields of 20.2 wt. %. If selectivity could be improved to obtain a 37.5% yield of pure phenol, the annual U.S. demand could be met with 3.6 million tons of lignin, or about 15% of the lignin produced in U.S. pulp mills. An alternative to increased yields of phenol that would also enhance the economics would be the establishment of a specialty market for the intermediate alkyl phenol products (cresol, propyphenol, ethyphenol) produced in the hydrocracking step. Finally, it may be advantageous for a kraft mill to recover lignin in order to reduce the load on the recovery furnace. This furnace is a very capital-intensive piece of equipment, and its capacity is often the limiting factor in the overall capacity of the mill.

Lignin is not the only biomass feedstock available for conversion to aromatic compounds. Hydrocarbon-like plant materials such as natural rubber, copaiba oil, various plant esters, and plant extracts may be catalytically processed to a range of aromatic compounds. Mobil Research and Development Corporation has been conducting experimental work on converting various plant materials to aromatics with the use of a zeolite catalyst. A variety of triglycerides and esters from corn oil, castor oil, and jojoba oil were completely converted to hydrocarbons, water, and carbon oxides. Yields of aromatic compounds ranged from 34% to 60% of the original feedstock. Likewise, plant extracts from E. Lathyrus and G. Squarrosa were converted by using a fluid bed type reactor over a zeolite catalyst to obtain yields of aromatics as high as 40%. In addition to this, copaiba oil has been converted to aromatics in yields of 55–66% (Haag et al., 1980).

These options for the production of aromatics are technically possible, but the high cost of feedstock renders the economics unfavorable in comparison with existing petroleum feedstock costs. One factor that could affect this is the development of the botanochemical-refining concept described in Fig. 12.5. Under this concept, the biomass crop is first refined to produce the plant oil, fiber, and leaf protein fractions by solvent extraction and mechanical processing. The fiber is used for pulp, the plant oil for its chemical aromatic value, and the plant protein for its animal feed value. This type of processing is similar to the existing soybean-processing industry, in which soybean is refined to soybean oil and soybean meal to maximize the overall product value and minimize the unit processing costs.

In conclusion, the short-term potential of biomass derived aromatic chemicals is through the chemical processing of lignin, a by-product of the existing pulp-milling industry. The second alternative biomass source is a long-term potential requiring improvements in plant genetics or oil crops and the development of a botanochemical-refining process.

X. FATTY ACID CHEMICALS

Table 12.10 lists the annual U.S. consumption and current end-uses for the three major products potentially derived from animal fats and plant oils. All three products, fatty acids, fatty alcohols, and glycerol, are produced today from both petrochemical raw materials and natural feedstocks. Approximately 35% of the 1300 million pounds of fatty acids annually produced in the United States are produced from tall oil (a by-product of the kraft pulping process) and tallow from animal fat. The remainder is synthesized from petrochemicals. Fatty alcohols are also produced from both petrochemical feedstocks and natural oils and fats. Glycerin is made in equal amounts from petrochemicals and natural fats and oils (Shreve and Brink, 1977).

As Table 12.10 indicates, all three products have wide application in many industries. Of significance is the synthetic lubrication market for fatty acids. The U.S. lubricant demand is on the order of 9 billion pounds, with synthetics (nonpetroleum-derived lubricants) accounting for close to 2%. This market is expanding at a rate of 10% per year and could potentially be supplied by fatty acids.

Although petrochemical processes involving *n*-paraffin or alpha-olefin feedstocks have been developed to provide an alternative supply for fatty acids and fatty alcohols, the trends of production economics favor the use of plant oil feedstocks. Table 12.11 illustrates this by comparing the production economics of fatty acids from synthetic versus plant oil feedstocks. The economics for producing this group of oxygenated chemicals favors the use of plant oils for the following two reasons.

1. The carbon source raw material dominates the economics, since it is at least 60% of the total production cost in both cases. This, coupled with

TABLE 12.10 Chemical Uses for Fats and Oils

	1974 Annual U.S. Consumption (MM lb)	*Current Uses*
Fatty acid	1300	Surfactants, plasticizers, greases/synthetic lubricants
Fatty alcohols (C10 or greater)	400	Surfactants, plasticizers
Glycerol	300	Drugs, cosmetics, toothpaste, food, beverages, alkylresins

Adapted from Shreve and Brink (1977).

TABLE 12.11 Comparative Economics for Fatty Acid (C8-C14) Production Synthetic Process Vs. Plant Oil Hydrolysis

	Synthetic Route (¢/lb of product)	*Natural Route (¢/lb of product)*
	1978	
Raw materials	24	37
Balance of production costs	14	3
Total	38	40
	1982	
Raw materials	42	30
Balance of production costs	14	3
Total	56	33

Synthetic route is by alpha olefin carbonylation; natural route is by coconut oil hydrolysis. Balance of production costs includes utilities, operating costs, and overhead expenses. By-product credit is not considered. Synthetic feedstock: alpha olefins; natural feedstock: coconut oil.

the fact that petroleum feedstock prices have increased eightfold in the past 8 years while plant oils have less than doubled, makes plant oils an attractive feedstock (*Chemical Week,* 1982a).

2. The operating, utility, and overhead expenses of converting petrochemicals to fatty acids are at least 3 times higher than the hydrolysis cost for producing fatty acids from plant oils.

In addition to the primary reasons, secondary factors favoring plant oils as feedstock sources for fatty acids and their derivatives are as follows.

1. There is a substantial alternative supply of sources of different plant oils that can substitute for the tall oil and tallow (see Section E of Table 12.7) (Shreve and Brink, 1977).
2. The processing of plant oils for fatty acids and fatty alcohols generates glycerol as a valuable by-product. Every 10 lb of fatty acid produced generates 1 lb of glycerin (*Chemical Week,* 1982a). Glycerin, as a by-product is recovered at costs below the cost of producing the synthetic product. As a result, this by-product credit further reduces the feedstock cost of plant oil by 15%.
3. Recent developments in agricultural genetics and enzyme technology are available to provide plant oils with the appropriate carbon chain lengths for special industrial uses. In addition to this, new technology is in the development stages for extraction of soluble polyphenols, whole plant oils, and isoprene polymers, which could be utilized for chemical feedstocks (Buchanan et al., 1980).

XI. COMPLEMENTS TO NONOXYGENATED PETROCHEMICAL POLYMERS

Certain types of botanochemical crops produce hydrocarbons called plant latex, which may offer an attractive complement to nonoxygenated petrochemical polymers. Plant latex is actually an emulsion of hydrocarbons in water. The material contains very little oxygen; hence high yields of nonoxygenated polymer may be obtained from the starting feedstock.

Commercially, the most important form of plant latex is obtained from the natural rubber tree, *Hevea brasiliensis,* and is the major source of natural rubber for the United States. In Malaysia, where the trees are cultivated, improved agronomic practices have led to yields of 2000 lb/acre-year (Calvin, 1980). The *Hevea* tree produces a high molecular weight (10^4–10^6) isoprenoid hydrocarbon $(C_5H_8)_n$ (Haag et al., 1980). This compound is oxygenated to a very small extent, resulting in high yields; hence it is a suitable alternative for synthetically produced rubber. Natural rubber sources have accounted for 20–30% of rubber consumption in the United States in 1970–1974 (Shreve and Brink, 1977). This percentage is expected to increase as petrochemical feedstocks become less competitive with natural sources of rubber; it would be further enhanced if a domestic source of natural rubber could be successfully marketed in the United States.

One such source is the guayule plant. This crop is similar to the rubber tree (*Hevea*), with natural rubber constituting 8–26% of the plant's dry weight. Guayule grows naturally throughout Texas and Mexico and could be cultivated throughout Southern California. Before guayule can become an economically important crop, research efforts in marketing, processing, and agronomic improvements must be conducted (Battelle Columbus Laboratories, 1981).

In addition to producing high-molecular-weight rubber compounds, other plants are capable of producing hydrocarbons of lower molecular weight. This plant oil may be cracked to blends of hydrocarbons similar to the composition of crude oil. One particular species under study, *Euphorbia lathyris,* produces oil in the yield of 6–10 barrels per acre. Yields are expected to improve significantly with genetic selection and plant breeding. When catalytically cracked, the oil has the following composition (Calvin, 1980):

Ethylene	10%
Propylene	10%
Toluene	20%
Xylene	15%
C_5-C_{20}	21%
Coke	5%
C_1-4 alkanes	10%
Fuel	10%

The oil is produced as 8–12% of the plant's dry weight. This plant has the potential for complementing supplies of hydrocarbon, provided that the crop is successfully marketed and strains are developed that are resistant to common pathogens.

A major problem with all botanochemical crops is in having an adequate and stable supply. Another obstacle to commercialization is the time and cost of developing new harvesting techniques.

XII. SUMMARY AND CONCLUSION

As is evident from a review of this chapter, biomass feedstocks currently contribute to the production of organic compounds and are expected to further displace petrochemical feedstocks as economic conditions dictate in the future. Three major industries are expected to be involved: the pulping industry, the oilseed-milling industry, and an industry handling forest residues. The economic driving forces as well as the obstacles associated with each developing biomass industry are summarized below:

1. As the largest biomass processor, the pulping industry can potentially supply the most significant short-term chemical feedstocks based on organics such as sugars, acids, and lignins present in the spent liquors. In the example shown in Table 12.5, the product revenue distribution of a sulfite pulp mill can be 26% for chemicals and 69% for pulp products. The major driving forces for this trend are processing modifications in pulp mills that can lead to increased pulp throughput capacity and reduced environmental treatment costs resulting from organic chemicals recovery. The obstacles to increased pulping-derived chemicals are lack of proven energy-efficient chemical recovery processing and efficient anaerobic fermentation processes for hexose and pentose sugars present in the spent liquors.
2. The second major biomass processor, with an annual shipment value of $11.5 billion (see Table 12.4) is the oilseed-milling industry. The second trend is the development of the botanochemical-processing industry based on unconventional oilseed- or latex-bearing plants analogous to the existing oilseed-milling industry. The potential product revenue distribution of a botanochemical plant would be 44% animal feed protein, 27% plant oils or latex, and 29% for fiber and phenols. The major driving force for this trend is the discovery of unique functional properties of plant oil derivatives such as fatty acids for high-value product applications, for example, plasticizers or surfactants. The major obstacles are agronomic problems of domesticating unconventional crops and development of acceptable animal feed protein co-products from these crops.

3. The third trend is the development of forest residue gasification to synthesis gas. The product revenue will be 100% dependent on the chemical value of methanol produced from synthesis gas. The major driving forces for this are the development of methanol liquid fuel applications and the existence of government subsidies for biomass-derived methanol. The major obstacle is the lack of infrastructure for forest residue collection.

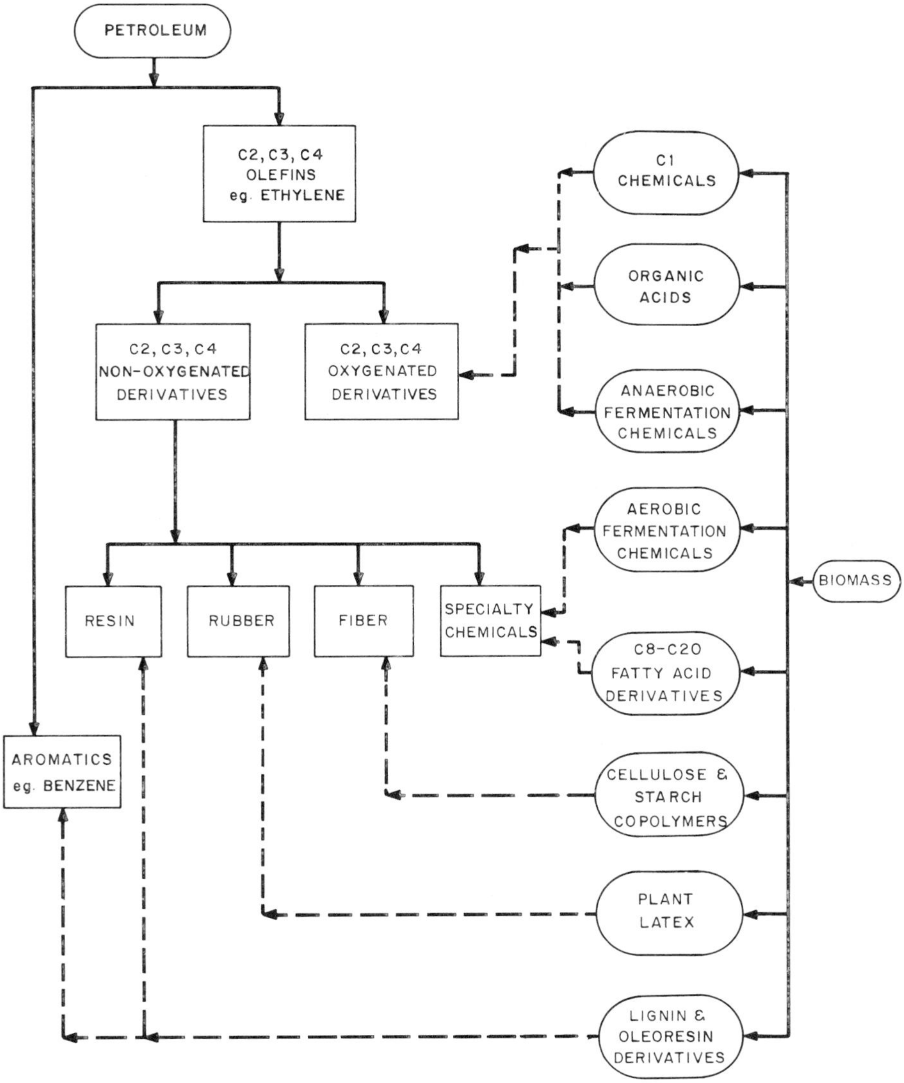

FIGURE 12.8 Most probable economic pathways for biomass-derived organic chemicals.

The three developmental trends are dependent on three separate groups of biomass sources—wood, unconventional crops, and forest residues. The organic chemicals derivable from each of the above categories are nearly independent of each other, with the exception of aromatics.

Pulping-derived chemicals are limited to oxygenated organics such as ethanol and acetic acids and possibly aromatics from lignin conversion. Gasification-derived chemicals are limited to methanol and its derivatives, for example, acetic acid, formaldehyde, and ethylene glycol. Botanochemical-derived chemicals can provide the broadest base groups of organics, including olefins, aromatics, and fatty acids.

This analysis has showed the current sources of biomass-derived chemicals to be small and largely dependent on feedstock sources based on processing of by-products of agricultural crops and pulp mills, which has led to chemicals such as specialty fermentation chemicals and tall oil derivatives. While the economics of these existing biomass-derived chemicals will continue to be favorable, significant organic chemicals from biomass can occur in the United States only when the processing of co-products such as pulp-derived chemicals and botanochemical-derived chemicals is fully developed. The economic feasibility of both these co-product processing industries is dependent not only on increasing petroleum costs but also on technology and new product application developments. Thus this shift from by-product processing to co-product processing will be gradual because of these developmental time requirements. It will, however, lead to a broader complementary role of biomass-derived chemicals in the organic chemical industry as shown in Fig. 12.8.

REFERENCES

*Adams, R. P. (1982) IGT Energy from Biomass and Wastes Symposium, January, Orlando, Fla.

Allan, D. R. (1982) *Pulp and Paper* (February), 135–137.

*Alverson, K. H., Parker, H. E., and Preibisch, R. A. (1971) *Encyclopedia of Polymer Science and Technology* Vol. XIV, John Wiley and Sons, New York, pp. 757–767.

Baierl, K., U.S. Patent 4,155,849.

*Baptist, J. N., U.S. Patent 3,036,959.

Battelle Columbus Laboratories (1981) *Biomass Technologies: An International Investigation Report* Vol. IV: *United States,* September 24, Columbus, Ohio.

Boyles, D. (1982a) SCI Symposium on Opportunities in Biotechnology for the Chemical Industry, April, London.

*Boyles, D. (1982b) SCI Symposium on Opportunities in Biotechnology for the Chemical Industry. April, London.

Brandt, D. (1979) *The Future of Carbohydrate Based Chemicals,* Stone and Webster Engineering Corp. inhouse report.

* These sources not referenced in the text are listed for suggested reading.

*Bratt, L. C. (1979) *Pulp and Paper* (June), 102–109.

Buchanan, R. A., Otey, F. H., and Hamerstrand, G. E. (1980) *Ind. Eng. Chem. Prod. Res. Dev. 19,* 478–483.

*Calvin, M. (1978) *Chemical & Engineering News* (March 20), 30–36.

Calvin, M. (1980) 1980 Scientific Conference, Basic Chemicals from Carbohydrates, the Corn Refiners Association Inc., June 25–27.

*Chauvel, A., Leprince, P., Barthel, Y., Raimbault, C., and Arlie, J. (1981) *Manual of Economic Analysis of Chemical Processes,* McGraw–Hill, New York.

Chemical & Engineering News (1981) June 8, 1981.

Chemical Marketing Reporter, January 4, 1982, pp. 42–51.

Chemical Week (1982a) (April), 40–44.

**Chemical Week* (1982b) (February 24), 47–52.

**Chemical Week* (1982c) (February 24), 30.

Chilton, C. H. (1960) *Cost Engineering in the Process Industries,* McGraw–Hill, New York.

*Coover, H. W., and Hart, R. C. (1982) *Chem. Eng. Progr.* (April), 72–75.

*Dale, B. E. (1980) 180th National Meeting of the American Chemical Society, August 24–29, San Francisco.

Davies, D. (1976) *The Humane Technologist,* Oxford University Press, Oxford, England.

*Davies, D. (1979) *Chemtech* (November), 660–663.

*D'Ianni, J. D. (1980) *Chemtech* (January), 23–25.

**European Chemical News* (1981) (April 13), 19.

**European Chemical News* (1982) (May 3), 15.

**European Chemical News* (1981) Biotechnology to account for only 5% of Organics Sector by 1990 (special report). (October 26).

*Faust, U. (1982) Symposium on Opportunities in Biotechnology for the Chemical Industry, April, London.

Fiechter, A. (1981) *Advances in Biochemical Engineering* Vol. XX, *Bioenergy* pp. 201–203, Springer, New York.

*Forester, G. L., Peters, J. F., Schroer, B. J., and Ziemke, M. C. (1982) IGT Energy from Biomass and Wastes Symposium, January, Orlando, Fla.

*Garcia, L. K. (1982) *Research and Development, Mexico* (February), 32–33.

Goheen, D. W., and Hoyt, C. H. (1981) in *Kirk & Othmer Encyclopedia of Chemical Technology* Vol. XIV, 3rd ed., John Wiley and Sons, New York, pp. 294–311.

*Gold, D. S., Goldberg, I., and Cooney, C. L. (1980) Symposium on Alternative Feedstocks for Petrochemicals, American Chemical Society, Division of Petroleum Chemistry, August 24–29, San Francisco.

*Goldstein, I. S. (1980a) in *Corn Annual* pp. 21–23, Corn Refiners Association, Inc. Washington, D. C.

*Goldstein, I. S. (1980b) Symposium on Alternative Feedstocks for Petrochemicals, American Chemical Society, Division of Petroleum Chemistry, August 24–29, San Francisco.

Haag, W. O., Rodewald, P. G., and Weisz, P. B. (1980) Symposium on Alternative Feedstocks for Petrochemicals, American Chemical Society, Division of Petroleum Chemistry, August 24–29, San Francisco.

*Hines, D. A. (1980) *Enzyme Microb. Technol. 2* (October), 327–329.

*Howells, E. R. (1982) SCI Symposium on Opportunities in Biotechnology for the Chemical Industry, April, London.

*Huber, Joan E., ed. (1980) *Kline's Guide to the Pulp and Paper Industry* 4th ed., Charles M. Kline & Co., Fairfield, N.J.

Huibers, D. J., and Jones, M. W. (1980) *Can. J. Chem. Eng. 58* (December), 718–722.

*Katzen, R., Frederickson, R., and Brush, B. F. (1980) *Chem. Eng. Progr.* (February), 62–67.

*Kindel, S. (1981) *Technology* (November/December), 62–74.

*Knifton, J. F. (1981) *Hydrocarbon Processing* (December), 113–117.

*Kohn, P. M. (1979) *Chemical Engineering* (January 29), 49–52.

*Kovaly, E. A. (1982) *Chemtech* (August), 486–489.

*Lenz, T. G., and Moreira, A. R. (1980) *Ind. Eng. Chem. Prod. Res. Dev. 19,* 478–483.

*Lipinsky, E. S. (1981) *Science 212* (June), 1465–1471.

*Lipinsky, E., Birkett, H., Polack, J., Atchison, J., Kresovich, S., McClure, T., Lauhon, W. (1978) *Sugar Crops as a Source of Fuels* Vol. II, TID-29400/2, U.S. Department of Energy, Washington, D.C.

*Meegan, Mary K., ed. (1977) *Kline's Guide to the Chemical Industry* 3rd ed., Charles M. Kline & Co., Fairfield, N.J.

Miller, D. R. (1982) IGT Energy from Biomass and Wastes Symposium, January, Orlando, Fla.

*Morgan, R. P., and Schultz, E. B. (1981) *Chemical & Engineering News* (September 7), 69–77.

Office of Technical Assessment (1981) *Impacts of Applied Genetics Microorganisms, Plants, and Animals,* Report No. OTA-HR-132.

Palsson, B. O., Fathi-Afshar, S., Rudd, D. F., and Lightfoot, E. N. (1981) *Science 213* (July 31), 513–517.

Parkhurst, H. J., Huibers, D. T., and Jones, M. W. (1980) Symposium on Alternate Feedstocks for Petrochemicals, American Chemical Society, Division of Petroleum Chemistry, August 24–29, San Francisco.

Predicast Base Book (1981) Market Volumes for Organic Chemicals, Predicast, Inc., Cleveland, Ohio.

Reisman and Bartholomew (1979) in Peppler, H. J., and Perlman, D., *Microbial Technology* Vol. II, Academic Press, New York, pp. 463–496.

Russell, C. R. (1973) in *58th Annual Meeting of the American Association of Cereal Chemists* p. 280, November 4–8, St. Louis, Missouri.

Rydholm, S. A. (1965) *Pulping Processes,* Wiley Interscience, New York, p. 95.

Saito, T. (1977) *Proc. Biochem.* (March), 17–19.

Sherwin, M. B. (1981) *Hydrocarbon Processing* (March), 79–84.

Shreve, R. N., and Brink, J. A. (1977) *Chemical Process Industries* 4th ed., McGraw-Hill, New York.

*Swartz, R. (1978) Meeting of the American Chemical Society, September 11, Miami.

*Tate, D. C. (1969) *Kirk & Othmer Encyclopedia of Chemical Technology,* Vol. XIX, 2nd ed., John Wiley and Sons, New York.

Technical Insights, Inc. (1980) *Biomass Process Handbook,* Fort Lee, N.J.

Tong, G. E. (1979) Industrial Chemicals from Fermentation. *Enzyme Microb. Technol. 1* (July), 173–179.

U.S. Department of Commerce (1982) *Industrial Outlook Report,* U.S. Government Printing Office, Washington, D.C.

Wat, E., et al. (1981) in *Proceedings of 13th Biomass Thermochemical Conversion Meeting,* Battelle Pacific Northwest Report, October.

Westmoreland, K. W. (1981) "Selecting Dextran Sucrase Catalyst to Facilitate Product Separation," Master's thesis, Virginia Polytechnic Institute and State University, Blacksburg.

INDEX